Insect Outbreaks

Edited by

PEDRO BARBOSA

Department of Entomology
University of Maryland
College Park, Maryland

JACK C. SCHULTZ

Department of Entomology
Pesticide Research Laboratory
Pennsylvania State University
University Park, Pennsylvania

ACADEMIC PRESS, INC.
Harcourt Brace Jovanovich, Publishers
San Diego New York Berkeley Boston
London Sydney Tokyo Toronto

Cover illustration courtesy of Dr. John Davidson

ACADEMIC PRESS, INC.
1250 Sixth Avenue, San Diego, California 92101

United Kingdom Edition published by
ACADEMIC PRESS INC. (LONDON) LTD.
24–28 Oval Road, London NW1 7DX

Library of Congress Cataloging in Publication Data

Insect outbreaks.

Includes index.
1. Insect populations. 2. Insects—Ecology.
3. Insect pests. I. Barbosa, Pedro, Date .
II. Schultz, Jack C.
QL496.15.I57 1987 595.7'05248 87-1405
ISBN 0–12–078148–4 (alk. paper)

PRINTED IN THE UNITED STATES OF AMERICA

87 88 89 90 9 8 7 6 5 4 3 2 1

Contents

15 The Role of Drought Stress in Provoking Outbreaks of Phytophagous Insects

WILLIAM J. MATTSON AND ROBERT A. HAACK

Part IV EVOLUTIONARY CONSEQUENCES

16 Insect Population Dynamics and Induction of Plant Resistance: The Testing of Hypotheses

ERKKI HAUKIOJA AND SEPPO NEUVONEN

17 Geographic Invasion and Abundance as Facilitated by Differential Host Plant Utilization Abilities

J. MARK SCRIBER AND JOHN H. HAINZE

18 Phenotypic Plasticity and Herbivore Outbreaks

PEDRO BARBOSA AND WERNER BALTENSWEILER

19 Genetic Change and Insect Outbreaks

CHARLES MITTER AND JOHN C. SCHNEIDER

20 Evolutionary Processes and Insect Outbreaks

NILS CHR. STENSETH

Contributors

Numbers in parentheses indicate the pages on which the authors' contributions begin.

Werner Baltensweiler (469), Institut für Phytomedizin, ETH-Zentrum/CLS, CH-8092 Zurich, Switzerland

Pedro Barbosa (469), Department of Entomology, University of Maryland, College Park, Maryland 20742

Alan A. Berryman (3), Department of Entomology, Washington State University, Pullman, Washington 99164

Brent Brodbeck (347), Department of Biological Science, Florida State University, Tallahassee, Florida 32306

A. F. G. Dixon (313), School of Biological Sciences, University of East Anglia, Norwich NR4 7TJ, England

Paul W. Ewald (269), Department of Biology, Amherst College, Amherst, Massachusetts 01002, and Department of Pure and Applied Biology, Imperial College, London SW7 2BB, England

Stanley H. Faeth (135), Department of Zoology, Arizona State University, Tempe, Arizona 85287

David N. Ferro (195), Department of Entomology, University of Massachusetts, Amherst, Massachusetts 01003

Robert A. Haack (365), U.S. Department of Agriculture, Forest Service, North Central Forest Experiment Station, East Lansing, Michigan 48823

John H. Hainze[1] (433), Department of Entomology, University of Wisconsin, Madison, Wisconsin 53706

Erkki Haukioja (411), Laboratory of Ecological Zoology, Department of Biology, University of Turku, SF-20500 Turku 50, Finland

[1] Present address: S.C. Johnson and Son, Inc., Racine, Wisconsin 53402.

Peter J. Martinat[2] (241), Department of Entomology, University of Maryland, College Park, Maryland 20740

Richard R. Mason (31), U. S. Department of Agriculture, Forest Service, Pacific Northwest Research Station, Forestry and Range Sciences Laboratory, LaGrande, Oregon 97850

William J. Mattson (365), U.S. Department of Agriculture, Forest Service, North Central Forest Experiment Station, East Lansing, Michigan 48823

Charles Mitter (505), Department of Entomology, University of Maryland, College Park, Maryland 20742

Judith H. Myers (173), The Ecology Group, Departments of Zoology and Plant Science, University of British Columbia, Vancouver 8, British Columbia, Canada V6T 2A9

Seppo Neuvonen (411), Laboratory of Ecological Zoology, Department of Biology, University of Turku, SF-20500 Turku 50, Finland

Philip J. Nothnagle (59), Windsor, Vermont 05089

Denis F. Owen (81), Department of Biology, Oxford Polytechnic, Headington, Oxford OX3 0BP, England

Stuart L. Pimm (99), Graduate Program in Ecology and Department of Zoology, University of Tennessee, Knoxville, Tennessee 37996

Peter W. Price (287), Department of Biological Sciences, Northern Arizona University, Flagstaff, Arizona 86001, and Museum of Northern Arizona, Flagstaff, Arizona 86001

Andrew Redfearn (99), Graduate Program in Ecology, University of Tennessee, Knoxville, Tennessee 37996

Stephen J. Risch (217), Division of Biological Control, University of California, Berkeley, California 94720

John C. Schneider (505), Department of Entomology, Mississippi State University, Mississippi State, Mississippi 39762

Jack C. Schultz (59), Department of Entomology, Pesticide Research Laboratory, Pennsylvania State University, University Park, Pennsylvania 16802

J. Mark Scriber[3] (433), Department of Entomology, University of Wisconsin, Madison, Wisconsin 53706

[2] Present address: Chemical and Agricultural Product Division, Abbott Laboratories, North Chicago, Illinois 60064.

[3] Present address: Department of Entomology, Michigan State University, East Lansing, Michigan 48824.

Nils Chr. Stenseth (533), Division of Zoology, Department of Biology, University of Oslo, Blindern, N-0316 Oslo 3, Norway

Donald Strong (347), Department of Biological Science, Florida State University, Tallahassee, Florida 32306

Paul W. Wellings[4] (313), CSIRO Division of Entomology Research Station, Warrawee, NSW 2074, Australia

Richard G. Wiegert (81), Department of Zoology, University of Georgia, Athens, Georgia 30602

[4] Present address: CSIRO Division of Entomology, Canberra, ACT 2601, Australia.

Preface

The abundance of insects can change dramatically from generation to generation; these generational changes may occur within a growing season or over a period of years. Such extraordinary density changes or "outbreaks" may be abrupt and ostensibly random, or population peaks may occur in a more or less cyclic fashion.

Outbreak species have often devastated economically important and aesthetic resources and thus have received a great deal of attention. Although the factors thought to trigger changes in abundance have been detailed in various studies over past decades, relatively little synthesis or retrospective analysis has been attempted. It is critical that new perspectives be developed from a synthesis of data and speculation to reshape or create new theory and working hypotheses for the future. The most basic of questions still remain to be addressed: Why do numbers change abruptly from generation to generation? What is an outbreak? Why are numbers stable in one population and highly variable in another population of the same species? Are there underlying similarities among all outbreak populations or among all outbreak species?

Recently a number of important research contributions have been made in entomology and ecology which provide new and important insights into the phenomenon of outbreaks. We believe that new theories and findings can broaden our scope beyond the restrictive views of the past which have become dogma rather than testable hypotheses.

The goal of this book is to update and advance current thinking on the phenomenon of insect outbreaks. We have not attempted to provide comprehensive coverage of the subject. Instead, the contributors have attempted to review relevant literature in order to generate a synthesis providing new concepts and important alternatives for future research. More important, they have been urged to present new ideas or syntheses that might stimulate advances in thinking and experimentation. We hope that the observations, speculation, and dissent evident in this volume will stimulate interest in the outbreak phenomenon and attract new researchers, new approaches, and new ideas to this fascinating problem.

We are grateful for the patient help and support of the editorial staff of Academic Press, and especially for the enthusiastic, patient, and critical help of Thelma Brodzina in preparing the manuscript.

Pedro Barbosa
Jack C. Schultz

Part I

INTRODUCTION TO OUTBREAKS

Chapter 1

The Theory and Classification of Outbreaks

ALAN A. BERRYMAN

Department of Entomology
Washington State University
Pullman, Washington 99164

I. INTRODUCTION

From the ecological point of view an outbreak can be defined as an explosive increase in the abundance of a particular species that occurs over a relatively short period of time. From this perspective, the most serious outbreak on the planet earth is that of the species *Homo sapiens*. From the more narrow perspective of *Homo*, however, an outbreak is an increase in the population of an organism that has a deleterious influence on human survival and well-being; such an organism is called a pest.

Outbreaks of pestiferous organisms have plagued humans from time immemorial. Locusts and mice have periodically destroyed their crops; bacteria and protozoans have decimated their populations; mosquitoes and blackflies have provided relentless annoyance. It is not surprising, therefore, that outbreak phenomena have occupied the attention of innumerable ecologists, pathologists, entomologists, epidemiologists, and mathematicians. Only recently, however, has a rigorous general theory of population outbreaks begun to emerge. This theory is outlined in the present chapter.

One important product of a rigorous and general theory should be an ability to recognize different types of outbreak phenomena. In other words, we should be able to classify outbreaks into recognizable groupings. Although a number of attempts have been made to do this in the past, few classification schemes have been rigorous or general. In this chapter I present a classification scheme based on a general theory of outbreaks that, I believe, has both of these important qualities.

II. HYPOTHESES FOR OUTBREAK CAUSATION

A number of hypotheses have been formulated in the past in an attempt to explain the causes of pest outbreaks. These hypotheses have usually resulted from research on particular kinds of organisms (rodents, insect herbivores, viruses, etc.) or on particular kinds of ecological processes (predator–prey or herbivore–plant interactions, genetic adaptation, etc.). The main weakness of these hypotheses is that they attempt to explain all outbreak phenomena (ultimate causation) from experience with particular kinds of outbreaks (proximate causation). If a general theory is to be developed, however, it should explain proximal hypotheses within a general framework. Therefore, the major hypotheses of proximate causation are summarized below:

(H1) Outbreaks are caused by dramatic changes in the physical environment. Included in this group would be those explained by the sun-spot theory (Elton, 1924), the theory of climatic release (Greenbank, 1956), and the theory of environment (Andrewartha and Birch, 1984).

(H2) Outbreaks are caused by changes in intrinsic genetic (Chitty, 1971) or physiological (Wellington, 1960; Christian and Davis, 1971) properties of individual organisms in the population.

(H3) Outbreaks result from trophic interactions between plants and herbivores or prey and predators. This hypothesis arises from the mathematical analysis of trophic interactions that produce large-amplitude population cycles under certain conditions (Lotka, 1925; Volterra, 1926; Nicholson and Bailey, 1935).

(H4) Herbivore outbreaks are due to qualitative or quantitative changes in host plants, which are usually caused by environmental stresses (White, 1978; Mattson and Addy, 1975).

(H5) Outbreaks are the result of particular life history strategies being more common among pest species, for example, *r* strategists or opportunistic species (Southwood and Comins, 1976; Rhoades, 1985).

(H6) Outbreaks result when pest populations escape from the regulating influence of their natural enemies (Holling, 1965; Morris, 1963; Takahashi, 1964; Isaev and Khlebopros, 1977).

(H7) Outbreaks occur when populations cooperatively overwhelm the defensive systems of their hosts (Thalenhorst, 1958; Berryman, 1982a,b).

III. GENERAL THEORY OF OUTBREAKS

Whenever we talk about an outbreak, we are thinking about a specific pest organism. Thus, the appropriate model to consider is one for the dynamics of a single species. We can formulate such a model as follows. Let N be the density of the population, G its genetic composition, and F the favorability of its environment; that is, F represents all the physical and biotic components of the environment that affect the reproduction, survival, and dispersal of the species. We can then write the general single-species population model

$$r = dN/N\,dt = f(N, G, F) \tag{1}$$

where r, the specific growth rate of the population, is expressed as a function of population density, genetics, and environmental favorability. If we assume for the present that genetic and environmental factors remain constant over time, we can examine the simple relationship between the specific growth rate and population density, that is, the density-dependent growth characteristics. Under these assumptions we can expand Eq. (1) by Taylor's theorem to yield

$$r = a_0 + a_1N + a_2N^2 + a_3N^3 + \cdots + a_iN^i \tag{2}$$

where a_0 is the specific growth rate of the population at the limit $N \rightarrow 0$, and $a_1 \cdots a_i$ are coefficients describing the interactions among members of the population. In order to have a biologically reasonable population model, however, we need to place certain constraints on Eq. (2):

(C1) $dN/dt = 0$ when $N = 0$, or populations do not grow when no organisms are present.

(C2) $dN/dt = 0$ when $N >> 0$, or populations cannot grow *ad infinitum*, and therefore the specific growth rate must approach zero at some relatively large population size, say K. In this case dN/dt and r become zero when $N = K$.

The simplest form of Eq. (2) that satisfies these constraints is

$$r = a_0 + a_1 N \tag{3}$$

provided that $a_0 > 0$ and $a_1 < 0$. Knowing that an equilibrium population density K occurs when $r = 0$ (C2), we can solve Eq. (3) for K,

$$\begin{aligned} 0 &= a_0 + a_1 K \\ K &= -a_0/a_1 \end{aligned} \tag{4}$$

where K is, of course, a positive number because $a_1 < 0$.

Equation (3) is equivalent to the familiar Verhulst "logistic" (Verhulst, 1838) and is represented diagrammatically in Fig. 1a. We see that the specific growth rate of the population is positive when $N < K$ and negative when $N > K$. Thus, if the starting population density N_0 is below or above K, the population will grow or decline toward the equilibrium density K (Fig. 1b).

We can ask the question: Given that a population obeys the elemental logistic equation, how can we generate an outbreak? The only way we can do this is by increasing the equilibrium density K (e.g., from K_1 to K_2 at time t_1; Fig. 1c). The outbreak can obviously be terminated by reversing this procedure. Now since $K = -a_0/a_1$, the equilibrium population density can be raised by increasing the value of a_0, the maximum specific rate of increase of the population, or by decreasing the value of a_1, the negative interaction among members of the population. Because these two parameters are likely to be affected by environmental and genetic effects, it seems reasonable to make the assumptions

$$a_0 = g(G, F) \tag{5a}$$
$$a_1 = h(G, F) \tag{5b}$$

In other words, both the maximum rate of increase of the population in the absence of density-dependent effects a_0 and the negative density-dependent interaction coefficient a_1 are assumed to be functions of population genetics and environmental favorability. Under these assumptions, K can increase under the following conditions:

1. The environment becomes more favorable for the reproduction and/or survival of the species. This explanation is in line with hypothesis (H1).
2. An increase occurs in the frequency of genes for high reproductive rates, reduced resource utilization (lower intraspecific competition), better defense or escape from natural enemies, resistance to harsh physical conditions, and so on. This explanation is in line with hypothesis (H2).

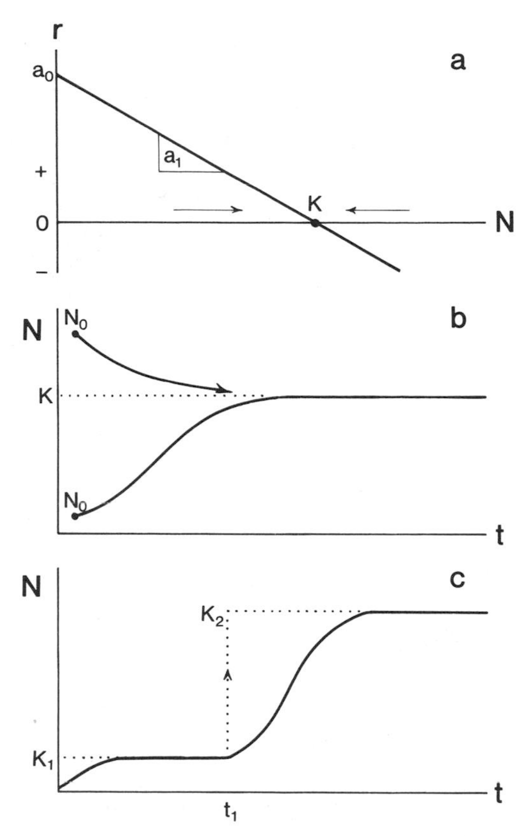

Fig. 1. (a) In the "logistic" equation the specific growth rate of the population r declines linearly with population density N, producing a single equilibrium point K; (b) populations grow or decline toward the equilibrium point K; (c) an outbreak occurs if K suddenly increases from K_1 to K_2 at time t_1.

In general, then, when the dynamics of a population are governed by the elemental logistic equation, outbreaks can be caused only by large and rapid alterations in the environment or the genetic composition of the population.

A. Time Delays

Hutchinson (1948) was apparently the first to explore the effects of time lags on the performance of the elemental logistic equation. Time delays are introduced into the equation if the species has discrete generations (many insects) or breeding seasons (many vertebrates) or if there are delays in the response of density-dependent factors. For example, the regeneration of food supplies, numerical responses of natural enemies,

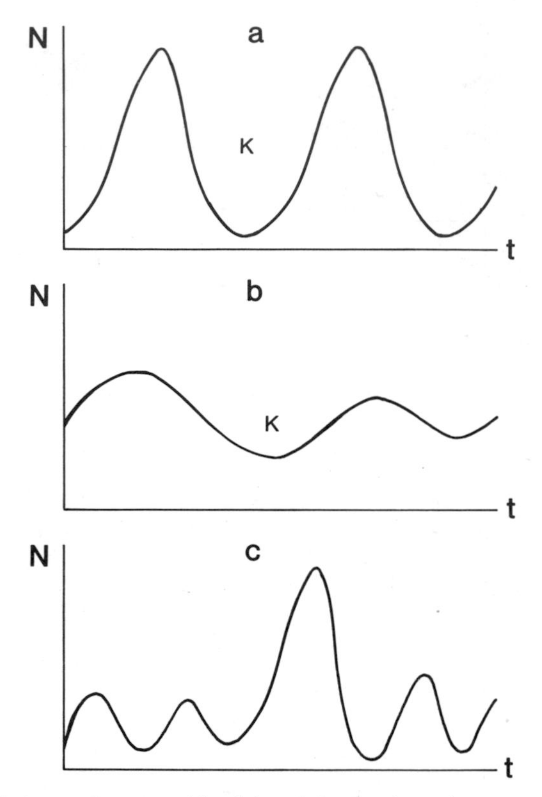

Fig. 2. Population cycles caused by delayed density-dependent negative feedback may have a high (a), low (b), or variable (c) amplitude depending on environmental or genetic conditions.

and relaxation of induced host defenses all create time delays in the density-dependent feedback process (Berryman, 1978a,b, 1981, 1986a). We can introduce the concept of time delays into Eq. (3) by

$$r_t = a_0 + a_1 N_{t-\tau} \tag{6}$$

where $N_{t-\tau}$ is the density of the population τ time periods in the past.

Hutchinson (1948) and others (e.g., Wangersky and Cunningham, 1957; May, 1974a; May *et al.*, 1974; Berryman, 1978b, 1981, 1986a) have shown that logistic models with time lags often produce cyclic population dynamics (Fig. 2) and that the amplitude and period of the cycle are related to the product

$$\tau K a_1 \quad \text{or} \quad \tau(-a_0) \tag{7}$$

If this quantity is sufficiently small (<-2.7), the dynamics can become chaotic, in which case the equation generates aperiodic outbreaks (May, 1974b). Many ecologists, however, consider chaotic dynamics to be the result of biologically unreasonable parameter values (see Hassell *et al.*, 1976, and Berryman, 1978b).

The analysis of the time-delayed logistic equation demonstrates that regular cyclic outbreaks (and possible irregular chaotic outbreaks) can occur in species with discrete generations or breeding cycles or when density-dependent factors respond with a time lag. For example, violent population cycles of the larch budmoth, *Zeiraphera diniana* Gn., seem to be caused by delays in the response and relaxation of defensive reactions in larch foliage (Benz, 1974; Fischlin and Baltensweiler, 1979). This explanation is in line with hypotheses (H3) and (H4).

It is also apparent that genetic and environmental factors have a decisive effect on the amplitude and period of regular cyclic outbreaks [Eq. (7)]. For instance, outbreaks will be more severe and less frequent when a_0 is large (favorable environment or where genotypes with high reproductive rates are common) or when a_1 is large (environments in which competition is severe or enemies are very effective or where genotypes susceptible to competition or predation are frequent). Alternatively, we may find severe outbreaks in environments that create longer delays in the regulatory process (large τ), as appears to be in the case in birch stands infested by the autumnal moth, *Epirrita autumnata* Bkh. (Haukioja, 1980).

B. Positive Density Dependence

Up to this point we have considered only the simplest case of our general population model [Eq. (2)]. Since there is no reason to expect that all natural populations obey the elemental "logistic" equation, we now expand the model (Lotka, 1925; Berryman, 1983) to

$$r = a_0 + a_1N + a_2N^2 + a_3N^3 \qquad (8)$$

This equation also meets constraints (C1) and (C2) if $a_0 > 0$ and $a_3 < 0$. Under certain conditions, however, the equation has some very interesting properties. For example, when $a_1 < 0$, $a_2 > 0$, and $a_1 > a_2 > a_3$, a system emerges that may have three equilibrium points (Fig. 3a); the first equilibrium L is potentially stable, the second T is unstable, and the third K is again potentially stable. The dynamics of populations obeying this model are quite complex, especially if time delays are present in the density-dependent responses in the vicinity of L and K. For example, when time lags are relatively short, a population starting at a sparse

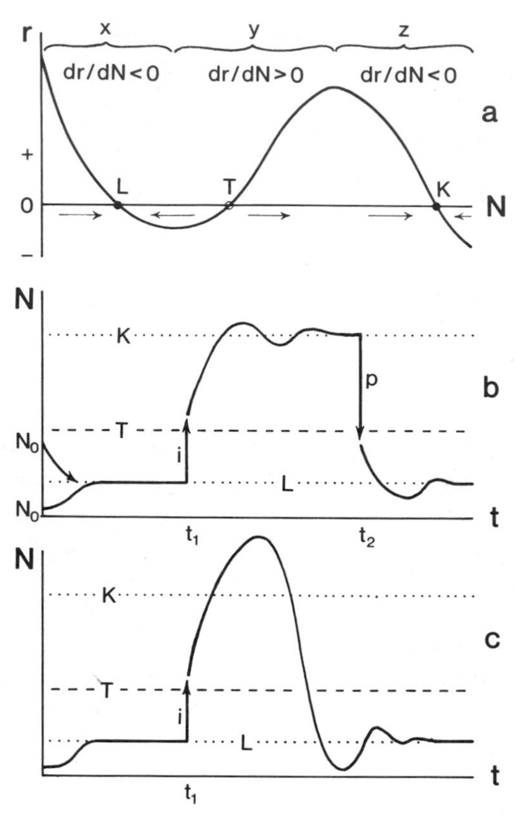

Fig. 3. (a) In this extension of the "logistic" equation, the specific growth rate r increases with population density in the density range y, and this creates an unstable equilibrium point or threshold at T. This threshold separates a low-density equilibrium L from the outbreak equilibrium K. (b) A population starting at a density $N_0 < T$ grows to the low-density equilibrium L, but if displaced above T by an immigration i at time t_1, it grows to the outbreak equilibrium K. This process is reversed if the population is reduced below T by a pesticide application p at time t_2. (c) When long time lags are present in the density-dependent factors acting on outbreak populations, the population will cycle around K and may automatically return to the low-density equilibrium L.

density ($N_0 < T$) will grow or decline asymptotically to the low-density stable equilibrium L (Fig. 3b). However, if this population is disturbed so that $N > T$, say by an immigration of insects i at time t_1 (Fig. 3b), it will grow to the outbreak density K. Conversely, if this dense population is reduced, say by an insecticide application p at time t_2 (Fig. 3b), it will automatically decline to its low-density equilibrium L. What we have in effect is a multiple-equilibrium model (Morris 1963; Takahashi, 1964; Southwood and Comins, 1976; Berryman, 1978a; Clark *et al.*, 1979). Outbreaks with this self-sustaining pattern seem to be characteristic of some insects that defoliate deciduous trees; in these cases the foliage is replaced annually and trees usually survive repeated heavy attack [e.g., the gypsy moth, *Lymantria dispar* (L.); Campbell, 1967].

In other systems, where the food supply of the insect is severely affected during outbreaks, long time delays may be introduced into the negative feedback around the upper equilibrium K (e.g., bark beetles and certain defoliators that kill their host trees so that the food supply is replaced only after many insect generations; Berryman, 1982c; Clark *et al.*, 1979). In these cases the outbreak will follow a cyclical trajectory following population displacement from the lower equilibrium (Fig. 3c).

Other scenarios are possible if long time delays are present at one or both equilibria, including continuous cyclical outbreaks or permanent outbreaks. For the present, however, I leave these possibilities to the imagination of the reader.

A critical part of our expanded model is the presence of an unstable equilibrium point T that separates low-density behavior from outbreak dynamics. For this reason T is sometimes referred to as the epidemic or outbreak *threshold* (Berryman, 1981, 1982a). The important question, of course, is: What biological conditions create this outbreak threshold?

In Fig. 3a, the specific growth curve $r(N)$ is divided into three sections x, y, and z. In the first and third sections, the curve slopes downward (the second derivative $dr/dN < 0$), indicating that negative density-dependent factors are dominating, for example, intraspecific competition, density-induced responses of natural enemies or food organisms. In the middle section, however, the slope of the curve is positive ($dr/dN > 0$), which indicates that the addition of individuals to the population increases the reproduction and/or survival of their cohorts. In other words, over this range of population density, the members of the population are *cooperating* rather than competing with one another (Berryman, 1981, 1984). For example, insects may cooperate with one another in overwhelming the defenses of their hosts (bark beetles; Raffa and Berryman, 1983) or conditioning their food supply (aphids; Dixon and Wratten, 1971). Alternatively, insects may cooperate in defense against predators (sawflies; Tostowaryk, 1972). Cooperative effects are not necessarily the outcome of adaptive behavioral traits, however, but may also result from certain interaction characteristics. Thus, insects may escape from the regulating influence of predators and parasitoids when their numbers are high because the functional responses of natural enemies become saturated (Holling, 1965; Isaev and Khlebopros, 1977). I have used the expressions "inadvertent cooperation" and "the advantage of numbers" to describe this phenomenon (Berryman, 1981).

The coincident-equilibria model incorporates the remainder of the proximate hypotheses for outbreak causation, that is, (H6), escape from natural enemies; (H7), overcomming host resistance; and because cooperative adaptations are often associated with opportunistic species, (H5).

It is perhaps obvious that the occurrence of three coincident equilibrium points in Fig. 3a depends on the values of the a parameters. These parameters are defined as a_0, the maximum specific growth rate of the population; a_1, the specific negative density-dependent effect that dominates at sparse population densities; a_2, the positive density-dependent (cooperative) effect that dominates at intermediate densities; and a_3, the negative density-dependent effect that dominates at very high population densities. All these parameters are determined by the genetic composition of the population and are also affected by environmental conditions. For instance, if the environment becomes more favorable (large a_0), the dip in the curve in Fig. 3a may be raised above the equilibrium line ($r = 0$ line). When this happens, an outbreak will be induced because $r > 0$ for all $N < K$ and only one high-density equilibrium point remains (Fig. 4a). Alternatively, if the parameter a_2 is small (small cooperative effect), we may find a system with only one low-density equilibrium point because $r < 0$ for all $N > L$ (Fig. 4b). Thus, outbreaks may be initiated and terminated by alterations in the favorability of the environment or the genetic makeup of the population.

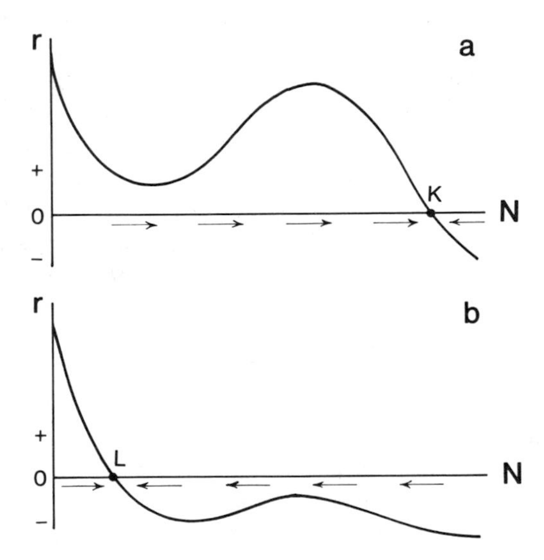

Fig. 4. Environmental or genetic changes may alter the parameter values of the model in Fig. 3, causing a permanent outbreak (a) or a stable sparse population (b).

In more general terms, outbreaks can be set in motion by the following changes:

1. Increasing environmental favorability (more or better food, fewer enemies, etc.) → larger a_0, smaller a_1 and/or a_3.

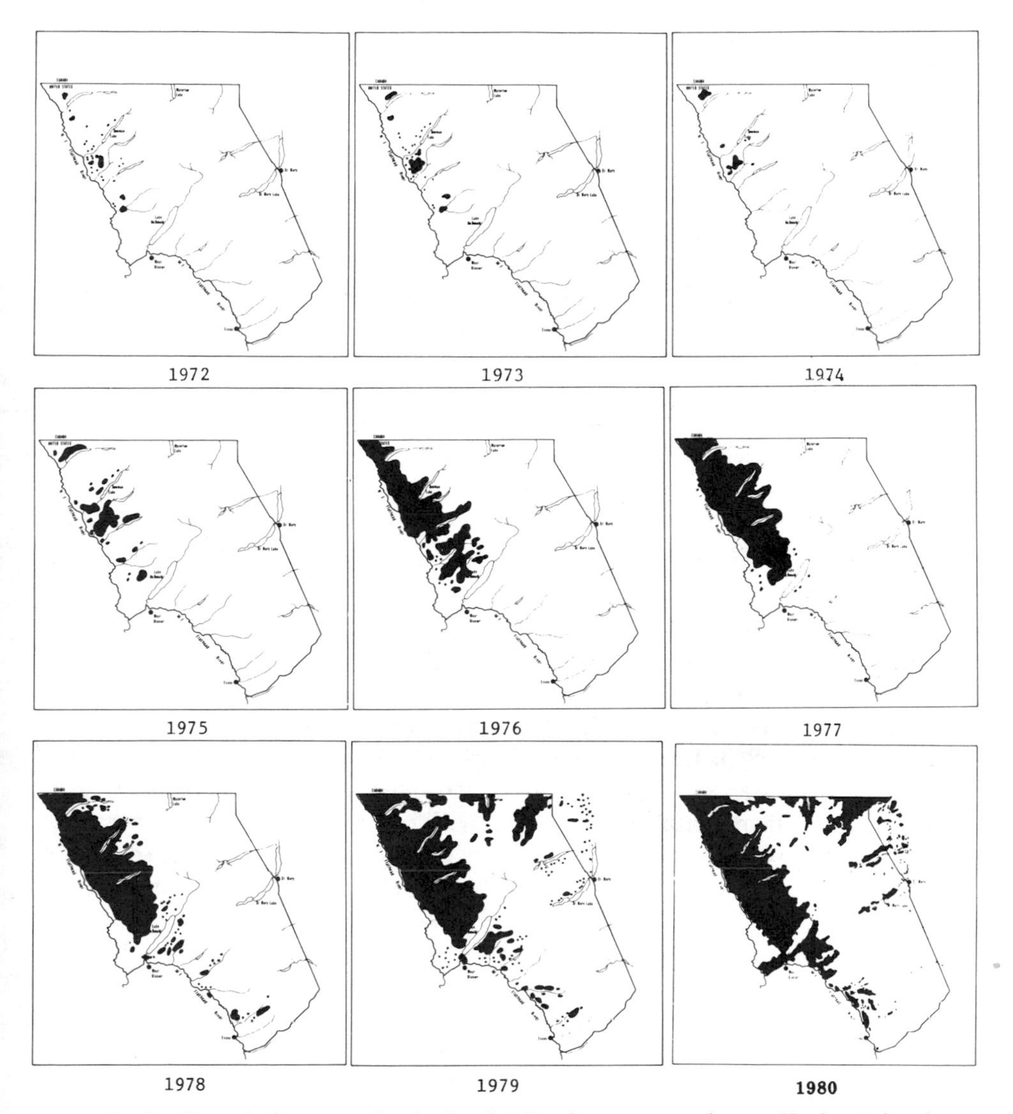

Fig. 5. Spread of a mountain pine beetle, *Dendroctonus ponderosae* Hopk., outbreak through the lodgepole pine, *Pinus contorta* Dougl. var. *latifolia* Engl., forests of Glacier National Park. From McGregor *et al.* (1983).

2. More fecund genotypes → larger a_0.
3. Genotypes less susceptible to natural enemies → smaller a_1.
4. Genotypes with better cooperative interactions or adaptations → larger a_2.

It is also important to note that outbreaks can be initiated by influxes (immigration) of individuals into a given area. For example, if a population is at the low-density equilibrium L (Fig. 3), it can be increased above the outbreak threshold T by an immigration from surrounding regions (i in Figs. 3b,c). In this way an outbreak can expand from a local epicenter, eventually encompassing a vast area. Such outbreaks will proceed in a wavelength fashion through space (Fig. 5).

Finally, we note that the initial outbreak (at the epicenter) can be set in motion by short-term (temporary) environmental or genetic disturbance. It is necessary only for the disturbance to raise the density of the population above the outbreak threshold T (Figs. 3b,c). After this the outbreak can proceed automatically even if the environment (or genetics) returns to the preoutbreak status.

IV. OUTBREAK CLASSIFICATION

If we accept the preceding theoretical arguments, we can identify three important features that determine the space–time dynamics of outbreaks:

1. Some outbreaks are self-perpetuating in that once initiated they tend to continue in both time and space (Fig. 5). These outbreaks are driven by positive feedback processes that operate at relatively high population densities and that give rise to bimodal r functions (Figs. 3 and 4). Because of their expansive nature, such outbreaks are often termed "eruptive" (Berryman, 1986b; Berryman and Stark, 1985).
2. Other outbreaks are not self-driven but are entirely dependent on external environmental or internal genetic conditions. These outbreaks arise and subside as their driving forces change in time and space, do not spread autonomously from their points of origin, and have unimodal r functions with little evidence of cooperative effects at relatively high densities (Figs. 1 and 2). Because these outbreaks merely track environmental gradients in time and space, I have termed them "gradient" outbreaks (Berryman, 1986b; Berryman and Stark, 1985).
3. Irrespective of whether the outbreak is of the gradient or eruptive type, its temporal behavior, at any one locality, is determined largely by time lags in negative feedback processes regulating population growth, larger delays giving rise to greater oscillations around equilibrium (Figs. 2a and 3c). Time lags are usually due to time-dependent responses of natural enemy or host populations. We further note that population cycles

TABLE 1

General Scheme for Classifying Pest Outbreaks

1. Outbreaks do not spread from local epicenters to cover large areas	*Gradient*
a. Short delays in the response of density-dependent regulating factors	
i. Outbreaks restricted to particular kinds of environments (site or space dependent)	Sustained gradient
ii. Outbreaks follow changes in environmental conditions (time dependent)	Pulse gradient
b. Long delays in the response of density-dependent regulating factors	
i. Outbreaks restricted to particular environments (site or space dependent)	Cyclical gradient
ii. Outbreaks follow changes in environmental conditions (time dependent)	Pulse gradient
2. Outbreaks spread out from local epicenters to cover large areas	*Eruptive*
a. Short delay in the density-dependent response at the low-density equilibrium	
i. Short delay in the density-dependent response at the high-density equilibrium	Sustained eruption
ii. Long delay in the density-dependent response at the high-density equilibrium	Pulse eruption
b. Long delay in the density-dependent response at the low-density equilibrium	
i. Short delay in the density-dependent response at the high-density equilibrium	Permanent eruption
ii. Long delay in the density-dependent response at the high-density equilibrium	Cyclical eruption

can be maintained, amplified, or suppressed by alterations in the physical environment and, possibly, by changes in gene frequencies (Fig. 2; Berryman 1978b, 1981, 1986a).

These theoretical conclusions lead logically to the proposition that all pest outbreaks can be classified according to certain behavioral features, particularly their tendency to spread from epicenters (the gradient–eruptive dichotomy) and their tendency to cycle around equilibrium (the short–long negative feedback delay dichotomy). Furthermore, environmental variations play an important role in triggering all types of outbreaks but the eruptive kinds, being self-perpetuating, are often insensitive to subsequent environmental variations. The consideration of these dichotomies and sensitivities leads to the recognition of seven classes of outbreaks (Table 1), as illustrated below with particular reference to phytophagous forest insects:

TABLE 2

Some Forest Insects That Exhibit Gradient Outbreaks

Insect	Key regulatory variable(s)	Environmental factor(s) causing gradients	Outbreak class[a]	Key reference
Rhyacionia buoliana Schiff.	Food: susceptible (poorly defended) pine shoots	Soil, weather, exposure	SG	Heikkenen (1981)
Eucosma sonomana Kft.	Food: susceptible (poorly defended) pine shoots	Soil moisture, weather	SG	Stoszek (1973)
Hylobius abietis L.	Food: tree stumps	Logging	PG	Bejer-Peterson *et al.* (1962)
Scolytus ventralis LeC.	Food: dead or dying *Abies* sp.	Weather, defoliation, pathogens, logging	PG	Berryman (1973)
Trypodendron lineatum (Oliv.)	Food: fresh conifer logs	Logging, windthrow	PG, SG[b]	Prebble and Graham (1957)
Conophthorus resinosae Hopk.	Food: cone crop size	Weather, stand age, flowering periodicity	PG, SG[b]	Mattson (1980)

Zeiraphera diniana Gn.	Food: quality of larch foliage (delayed defensive responses to defoliation)	Elevation	CG	Baltensweiler *et al.* (1977)
Orgyia pseudotsugata McD.	Food: foliage quantity Enemies: virus, parasitoids	Soil moisture, site exposure	CG, PG[c]	Mason and Luck (1978)
Epirrita autumnata Bkh.	Food: quality of birch foliage (delayed responses to defoliation)	Latitude, elevation	CG	Haukioja (1980)
Acleris variana (Fern)	Enemies: parasitoids (delayed responses to prey density)	Weather, forest maturity	CG, PG[c]	Morris (1959)

[a] SG, Sustained gradient; PG, pulse gradient; CG, cyclical gradient.
[b] These insects may exhibit sustained gradients when their food supply is maintained at a consistent high level by human activities, e.g., log storage areas or commercial seed orchards.
[c] Cycle amplitude probably altered by weather, particularly winter and spring precipitation and/or temperature.

1. *Sustained gradient outbreaks:* Persistent high-density pest populations often associated with stressed or unhealthy hosts, for example, plants growing on suboptimal sites. These pests have little impact on the survival of their hosts and are not strongly affected by density-dependent parasitism and/or predation. Examples are particularly evident among shoot- and fruit-infesting forest insects (Table 2).

2. *Pulse gradient outbreaks:* Irregular short-lived pest outbreaks associated with changes in the abundance or quality of food (or other resources) brought about by external environmental disturbances (warm, dry weather, gales, infestations of other insects or pathogens, etc.). Many cone and seed insects, "nonagressive" bark beetles, and some cyclical forest defoliators exhibit pulse gradient outbreaks (Table 2). In addition, outbreaks of this type are characteristic of many insect pests of annual agricultural crops. In these cases the environment experiences drastic temporal alterations, being very unfavorable at certain times (after the crop is harvested) and very favorable at others (monocultures of susceptible host plants).

3. *Cyclical gradient outbreaks:* Outbreaks of short duration, usually 2–3 generations, that occur at regular intervals, usually every 8–11 generations, and that are often associated with certain site conditions (e.g., soil type, elevation, slope, latitude). Numerical responses of natural enemies or delayed defensive responses of the hosts are usually the major factors involved in population regulation. Most forest insects exhibiting cyclical gradients seem to be defoliators that do not cause excessive mortality among their host populations during outbreaks (Table 2). Host mortality is often prevented by virus epizootics or dramatic increases in other natural-enemy populations, which prevents repetitive host defoliation.

4. *Sustained eruptive outbreaks:* Outbreaks that spread from local epicenters to cover large areas and that persist at outbreak levels in any one place for several to many years (Fig. 3b). Outbreaks of these pests rarely cause extensive mortality among their hosts, except after many years of attack. Natural enemies are often important at sparse densities, or the pest may have strong cooperative behavior. Most forest insects exhibiting sustained eruptions seem to be defoliators that do not cause extensive host mortality (Table 3).

5. *Pulse eruptive outbreaks:* Outbreaks that spread from epicenters but go through a pulselike cycle at any one locality (Fig. 3c). These outbreaks often cause extensive mortality among the hosts. Natural enemies or cooperative behavior are usually important in population regulation. Examples can be found in forest defoliators that cause extensive host mortality, "aggressive" bark beetles, and insects that transmit plant pathogens (Table 3).

TABLE 3

Examples of Forest Insects That Exhibit Eruptive Outbreaks

Insect	Regulatory factors acting at		Main cooperative effect	Outbreak class[a]	Key reference
	Low density	High density			
Choristoneura occidentalis Freem.	Bird or ant predation	Food depletion parasitism	Escape from predation	SE	Campbell *et al.* (1983)
Lymantria dispar (L.)	Vertebrate predation	Food depletion, virus disease	Escape from predation	SE	Campbell (1975, 1979)
Didymuria violescens Leach	Bird predation	Food depletion	Escape from predation	SE	Readshaw (1965)
Neodiprion swainei Midd.	Vertebrate predation	Food depletion, parasitism	Escape from predation	SE	McLeod (1979)
Dendroctonus ponderosae Hopk.	Host resistance	Food depletion	Overwhelming host resistance	PE	Raffa and Berryman (1983)
Ips typographus L.	Host resistance	Food depletion	Overwhelming host resistance	PE	Thalenhorst (1958)
Choristoneura fumiferana (Clems.)	Bird predation	Food depletion	Escape from predation	PE	Clark *et al.* (1979)
Cardiaspina albitextura Taylor	Bird predation, parasitoids	Food depletion	Escape from predation	PE, CE	Clark (1964)
Monochamus alternaturs Hope	Host resistance	Food depletion	Spreading pine-wood nematode	PE	Kobayashi *et al.* (1984)
Sirex noctilio F.	Host resistance, parasitism	Food depletion, nematode infection, parasitism	Spreading fungus pathogen	PE	Taylor (1983)

[a] SE, Sustained eruption; PE, pulse eruption; CE, cyclical eruption.

6. *Permanent eruptive outbreaks:* Outbreaks that spread from local epicenters but remain at outbreak levels thereafter. This behavior seems possible only if the pest has no impact on the reproduction and survival of its host. I know of no examples of this outbreak type.

7. *Cyclical eruptive outbreak:* Short-lived outbreaks (2–3 generations) that occur at regular intervals (8–11 generations) and spread out from epicenters. Outbreaks do not have a severe impact on the survival of their hosts and are often terminated by population explosions of natural enemies particularly viruses. Clark (1964) suggests that the psyllid *Cardiaspina albitextura* may occasionally exhibit cyclical eruptions (Table 3).

V. IDENTIFYING OUTBREAK CLASSES

Having erected a classification scheme, it should now be possible to identify the type of outbreak exhibited by specific insect pests. Both qualitative and quantitative data are useful in carrying out this identification.

A. Qualitative Data

Qualitative information on the frequency and spatial dynamics of outbreaks, the biology and behavior of the insect pest (particularly cooperative behavior), the interactions between the pest and its host (defensive responses and host mortality), and the interactions between the pest and its natural enemies (functional and numerical responses) can be used to deduce the potential behavior of a particular population (Table 4). For example, gypsy moth outbreaks often spread from their origin, low densities seem to be maintained by vertebrate predators, dense populations are often reduced by viral pathogens and food depletion, and host mortality is apparent only after repeated heavy defoliation (Campbell, 1967, 1975, 1979). This population system, with relatively short delays at both high and low densities (Table 4), should exhibit sustained eruptions (Table 1). As another example, consider the red pine cone beetle population studied by Mattson (1980). In this case the insect is resource-limited, the number of cones (food) being replaced yearly, but the size of the cone crop is unpredictable. Since there is no evidence that outbreaks spread, we expect the cone beetle population to exhibit pulse gradients in response to cone crop variations (Tables 1 and 4). Finally, the larch budmoth population in Switzerland seems to be regulated by delayed host defensive responses and to have little impact on the survival of its host plant (Fischlin and Baltensweiler, 1979). Therefore, it should exhibit cyclical gradient or cyclical eruptive outbreaks depending on whether or not the outbreaks

TABLE 4

Relative Density-Dependent Effect and Reaction Time (Time Lag) of Factors Acting at High and Low Pest Population Densities

Density-dependent factor	Low-density equilibrium		High-density equilibrium	
	Density dependence	Time lag (no. of generations)	Density dependence	Time lag (No. of generations)
Insectivorous vertebrates	Negative	<1	Positive	<1
Prey-specific arthropods	Negative	<1[a]	Negative[b]	>1
Nonspecific arthropods	Negative	<1	Negative[b]	>1
Pathogens	None[c]	<1	Negative[b]	<<1
Plant defensive reactions				
Immediate reactions	Negative	<1	Negative[d]	<1
Delayed reaction/relaxation	Negative	>1	Negative	>1
Food depletion				
Food replaced rapidly	None	<1	Negative	<1
Food replaced after several or many generations	None	>1	Negative	>>1

[a] Less than one generation if two or more natural-enemy generations are produced per prey generation or if the natural enemy has a sigmoid functional response.

[b] Unless parasitoids or predators do not respond numerically to high prey densities because of other limiting factors such as territoriality and alternative food.

[c] Unless pathogens are transmitted vertically and at low host densities.

[d] Immediate defensive reactions can have a positive feedback effect if high pest densities saturate the defensive capacity (e.g., some bark beetles).

spread out, a debatable point. Isaev *et al.* (1984) consider the budmoth to be an eruptive insect because there is a delay in the cycle as one progresses from the French to Austrian Alps, suggesting an eastward spread. However, Fischlin and Baltensweiler (1979) constructed a model that simulated budmoth cycles independent of immigration and emigration. This controversy does not reflect a lapse in the theory but rather the shortage of data. In fact, many of the cases in Tables 2 and 3 are only tentatively classified because of incomplete information. The theory, however, can be used to develop hypotheses that must be tested before a correct classification can be formulated.

B. Quantitative Data

Time-series data that record the density of the population in question over a relatively long period can be used to construct "phase portraits" in

r–N space (Isaev and Khlebopros, 1977; Isaev *et al.*, 1984). This is done by plotting the specific growth rate of the population for each generation (r_t) on population density N_t. The value of r_t for organisms with discrete generation can be approximated by

$$r_t = \ln(N_{t+1}/N_t) \tag{9}$$

where t is the generation number. The phase portrait often provides clues to the type of outbreak being observed.

For example, Fig. 6a shows the phase portrait for the red pine cone beetle population studied by Mattson (1980). We see that the specific growth rate of the population (r_t) fluctuates sharply around equilibrium ($r_t = 0$), indicating that the population is responding very quickly (within one generation) to changes in its environment. This leads to the hypothesis that the bettle population is reacting to temporal variations in a resource, such as food. To test this hypothesis we plot r_t against the number of cones available per beetle and observe a fairly convincing relationship (Fig. 6b; $r^2 = .63$). Because there is no evidence that cone beetle outbreaks spread out from local epicenters or that they are maintained at low densities by natural enemies or host resistance, we conclude that the population exhibits pulse gradient outbreaks (Tables 1 and 2) and perhaps

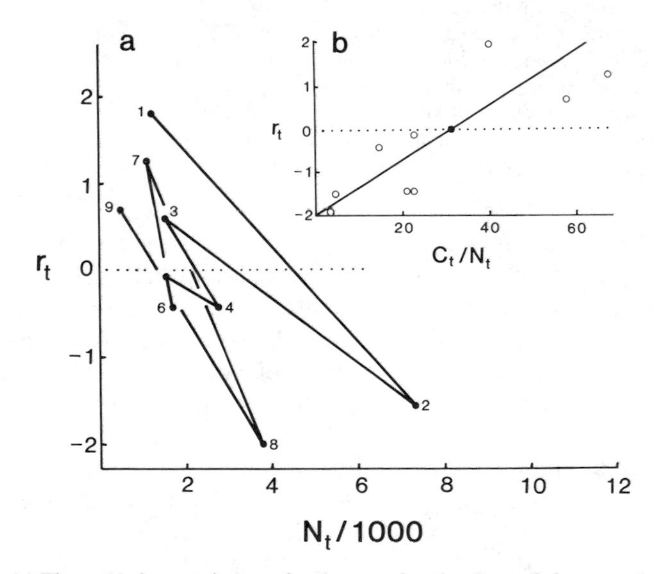

Fig. 6. (a) The r–N phase trajectory for the cone beetle, *Conophthorus resinosae*, over a period of nine generations; (b) relationship between the number of cones available per beetle and the specific growth rate of the population. At equilibrium the specific growth rate is zero. Data from Mattson (1980).

sustained outbreaks in areas where cone crops are consistently high (e.g., in commercial seed orchards).

As another example, consider the phase portrait for the black-headed budworm studied by Morris (1959) (Fig. 7a). The wide orbit of this trajectory shows no tendency to stabilize at a low density. This indicates that the outbreaks are cyclic and are caused by delayed feedback with the host plant or natural enemies. The length of the delay sometimes can be determined by plotting r_t on population density from previous generations (i.e., plot r_t on N_{t-1}, N_{t-2}, . . . , $N_{t-\tau}$), the appropriate lag being that which most clearly compresses the data to a single line (Royama, 1977; Berryman, 1981). In the case of the black-headed budworm, a time lag of one generation provides the best compression (r_t on N_{t-1}, Fig. 7b). Such delay would be expected if the population were regulated by the numerical responses of natural enemies with generation spans equal to that of their hosts. In fact, we see that the specific growth rate of the budworm is

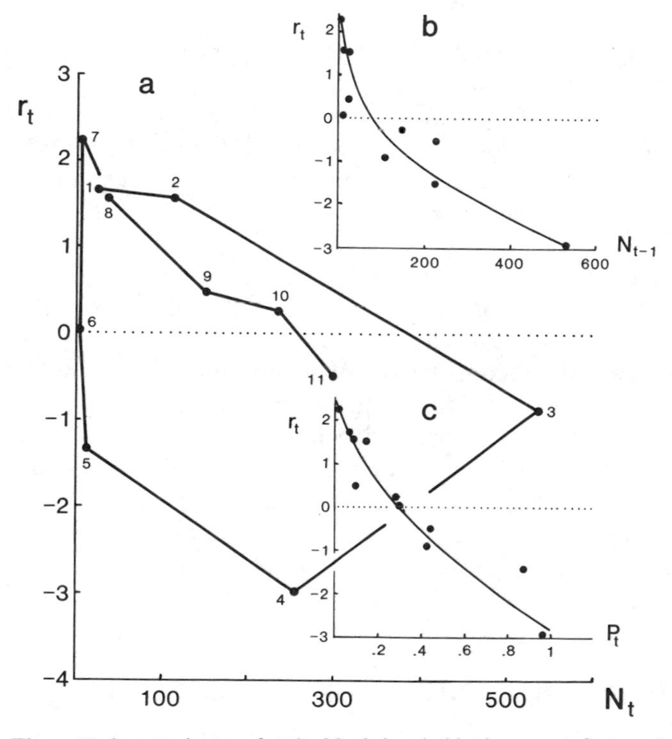

Fig. 7. The *r–N* phase trajectory for the black-headed budworm, *Acleris variana*, over a period of 11 years when r_t is plotted first on N_t (a) and then on N_{t-1} (b), and finally the relationship between r_t and the probability of a larva being parasitized P_t (c). Data from Morris (1959).

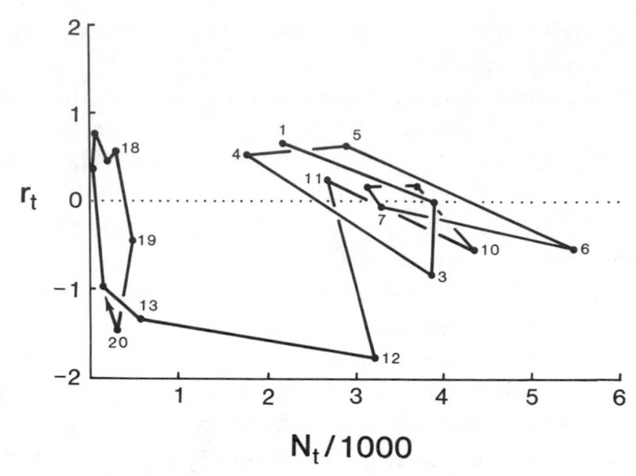

Fig. 8. The *r–N* phase trajectory for the gypsy moth, *Lymantria dispar,* over a 20-year period showing two equilibrium points and sustained outbreaks. Data from Campbell (1967).

strongly associated with the probability of parasitization by insect parasitoids (Fig. 7c; r^2 = .86).

Similar analysis of the larch budmoth phase portrait indicates a two-generation delay (Royama, 1977). Delays of this length seem to be more likely in herbivore–plant interactions because host responses often require more than one insect generation to be initiated and relax (Table 4). This interpretation is in line with that of Benz (1974) and Fischlin and Baltensweiler (1979). Note, however, that long time lags may be artifacts of the recursive equation, and conclusions should be treated cautiously and backed by biological reasoning (Royama, 1977).

As a final example, consider the phase portrait of the gypsy moth (Fig. 8), which shows clear evidence of two equilibrium points. Because this population remained at outbreak densities for many years, I conclude that the gypsy moth exhibited a sustained eruptive outbreak (Table 1). However, the gypsy moth phase trajectory does show a tendency to cycle around its equilibrium points. If the amplitude of these cycles could be altered, say by weather variations, then cyclical or even permanent eruptions would be possible in this population system (Table 1).

VI. COMPARISON OF OUTBREAK CLASSIFICATIONS

It has long been recognized that some animal populations remain at relatively constant densities, others cycle with rather regular periodicity, and still others exhibit aperiodic fluctuations (e.g., see Fig. 5 in Leopold, 1947). Only recently, however, have attempts been made to classify insect outbreaks. Perhaps the simplest classification was that developed by Shepherd (1977) for forest insect defoliators. He identified two major types of outbreaks, spreading and nonspreading, which are equivalent to my eruptive–gradient dichotomy. He then subdivided each of these into outbreaks associated with certain edaphic or topographic features, zone-restricted outbreaks associated with biogeoclimatic regions, and host-restricted outbreaks that are limited only by the distribution of the host plant.

McNamee *et al.* (1981) developed another scheme for classifying defoliator outbreaks based on the equilibrium manifolds for pest, host foliage, parasitoid–pathogen complex, and forest. From this they identified three major classes of outbreak behavior: (1) short and frequent outbreaks, (2) long-lasting outbreaks at frequent to moderate intervals, and (3) long-lasting outbreaks occurring at irregular intervals. All three outbreak types are, as far as I can see, of the eruptive type, being similar to my cyclical and sustained eruptions.

The final classification system I discuss is that of Isaev *et al.* (1984), who identified four outbreak classes on the basis of a theoretical analysis similar to that developed in this chapter [Eq. (1)]. Not surprisingly, their outbreak classes are almost identical to mine in concept if not in terminology. The terminology problems are, in my view, a result of difficulty of translating from the original Russian manuscripts. The equivalence of the terms is as follows: prodromal outbreak = gradient outbreak; the (eruptive) outbreak proper = pulse or sustained eruption; the permanent (eruptive) outbreak = cyclical eruption; the reverse (eruptive) outbreak = permanent eruption. The only missing element in the Isaev *et al.* (1984) classification is the subdivision of gradient outbreaks into the sustained, pulse, and cyclical types. Otherwise the classification scheme presented in this chapter is conceptually identical to that of Isaev and Khlebopros (1977) and is quite similar in many ways to Shepherd's (1977). However, Shepherd's scheme and that of McNamee *et al.* (1981) are meant to apply only to insect defoliators and thus lack the breadth necessary for a general classification scheme.

VII. CONCLUSIONS

A consideration of theoretical population models has enabled us to develop a general system for classifying pest outbreaks. In its simplest sense, this classification is based on the expansive capacity of the population (presence of outbreak thresholds) and time lags present in the density-dependent feedback loops (Table 1). The behavior of the outbreak is a result of the interaction structure of the system (parasitoid–prey, predator–prey, herbivore–plant, etc.; Table 4), as well as on the adaptive characteristics of all the species involved (cooperative behavior, defensive adaptations, host-finding behavior, reproductive potentials, etc.). The classification system has been derived from fundamental ecological principles and should therefore be general enough to apply to any organism that exhibits outbreak behavior, that is, insects feeding on forest trees, agricultural crops, and on other animals, as well as disease-causing pathogens and even *Homo sapiens*. (We are definitely eruptive, but which subclass are we? One would hope the sustained or permanent type.)

A conceptually sound general classification system is of considerable practical value. With the proposed system we have only to deal with seven types of problems, rather than to treat each problem as unique. Hence, similar methods can often be used to prevent or control insect outbreaks within a particular class. In addition, a theoretical understanding of the outbreak types enables the manager to design preventive or therapeutic treatments for that particular class of outbreak (Berryman, 1986b). For example, suppression with insecticides may be a useful tactic against sustained or pulse eruptions, provided that the insecticides do not disrupt the low-density regulating factors. They are of little use, however, against the other classes unless treatment is applied continuously. On the other hand, cultural practices aimed at improving host resistance are useful strategies for lowering the probability of all types of outbreaks. Finally, the general theory of outbreaks and its resulting classification scheme also provide a basic conceptual framework for designing methods for assessing the risk of insect outbreaks from environmental variables (Berryman, 1982a, 1984, 1986a,b; Berryman and Stark, 1985).

ACKNOWLEDGMENTS

The theoretical investigations presented in this chapter as well as the resulting classification scheme emerged from many interactions and discussions with my friends and colleagues but most notably with Professor Alexander Isaev, director of the V. N. Sukachev Institute of the Siberian Branch of the USSR Academy of Sciences. Our discussion and correspondence began at the IUFRO World Congress in Oslo in 1976 and has been a driving force in my thinking over the years. This chapter is dedicated to the spirit of scientific comradeship that persists in a world divided by ideological and political tensions.

REFERENCES

Andrewartha, H. G., and Birch, L. C. (1984). "The Ecological Web." Univ. of Chicago Press, Chicago, Illinois.

Baltensweiler, W., Benz, G., Bovey, P., and Delucchi, V. (1977). Dynamics of larch bud moth populations. *Annu. Rev. Entomol.* **22,** 79–100.

Bejer-Peterson, B., Juutinen, P., Kangas, E., Bakke, A., Butovitsch, V., Eidmann, H., Heqvist, K. J., and Lekander, B. (1962). Studies on *Hylobius abietis* L. I. Development and life cycle in the Nordic countries. *Acta Entomol. Fenn.* **17,** 1–106.

Benz, G. (1974). Negative Ruckkoppelung durch Raum-und Nahrungskonkurrenz sowie zylkisch Veranderung der Nahrungsgrundlage als Regelprinzip in der Populationsdynamik des Grauen Larchwicklers, *Zeiraphera diniana* Guenee). *Z. Angew. Entomol.* **76,** 196–228.

Berryman, A. A. (1973). Population dynamics of the fir engraver, *Scolytus ventralis* (Coleoptera: Scolytidae). I. Analysis of population behavior and survival from 1964 to 1971. *Can. Entomol.* **105,** 1465–1488.

Berryman, A. A. (1978a). Towards a theory of insect epidemiology. *Res. Popul. Ecol.* **19,** 181–196.

Berryman, A. A. (1978b). Population cycles of the Douglas-fir tussock moth (Lepidoptera: Lymantriidae): The time-delay hypothesis. *Can. Entomol.* **110,** 513–518.

Berryman, A. A. (1981). "Population Systems. A General Introduction." Plenum, New York.

Berryman, A. A. (1982a). Biological control, thresholds, and pest outbreaks. *Environ. Entomol.* **11,** 544–549.

Berryman, A. A. (1982b). Mountain pine beetle outbreaks in Rocky Mountain lodgepole pine forests. *J. For.* **80,** 410–413.

Berryman, A. A. (1982c). Population dynamics of bark beetles. *In* "Bark Beetles of North American Conifers. A System for the Study of Evolutionary Ecology" (J. B. Mitton and K. B. Sturgeon, eds.), pp. 264–314. Univ. of Texas Press, Austin.

Berryman, A. A. (1983). Defining the resilience thresholds of ecosystems. *In* "Analysis of Ecosystems: State-of-the-Art in Ecological Modelling" (W. K. Lauenroth, G. C. Skogerboc, and M. Flug, eds.), pp. 570–60. Elsevier, Amsterdam.

Berryman, A. A. (1984). Threshold theory and its applications to pest population management. *In* "Pest and Pathogen Control: Strategy, Tactics and Policy Models" (G. R. Conway, ed.), pp. 40–57. Wiley-IIASA, Int. Inst. Appl. Syst. Anal., Wiley (Interscience), Chichester.

Berryman, A. A. (1986a). On the dynamics of blackheaded budworm populations. *Can. Entomol.* **118,** 775–779.

Berryman, A. A. (1986b). "Forest Insects: Principles and Practice of Population Management." Plenum Press, New York.

Berryman, A. A., and Stark, R. W. (1985). Assessing the risk of forest insect outbreaks. *Z. Angew. Entomol.* **99,** 199–208.

Berryman, A. A., Stenseth, N. C., and Wollkind, D. J. (1984). Metastability of forest ecosystems infested by bark beetles. *Res. Popul. Ecol.* **26,** 13–29.

Berryman, A. A., Stenseth, N. C., and Isaev, A. S. (1987). Natural regulation of herbivorous forest insect populations. *Oecologia (Berlin)* **71,** 174–184.

Campbell, R. W. (1967). The analysis of numerical change in gypsy moth populations. *For. Sci. Monogr.* **15,** 1–33.

Campbell, R. W. (1975). The gypsy moth and its natural enemies. *Agric. Inf. Bull. (U.S., Dep. Agric.)* **381,** 1–27.

Campbell, R. W. (1979). Gypsy moth: Forest influence. *Agric. Inf. Bull.* (*U.S., Dep. Agric.*) **423,** 1–44.
Campbell, R. W., Beckwith, R. C., and Torgerson, T. R. (1983). Numerical behavior of some western spruce budworm (Lepidoptera: Tortricidae) populations in Washington and Idaho. *Environ. Entomol.* **12,** 1360–1366.
Chitty, D. (1971). The natural selection of self-regulatory behavior in animal populations. *In* "Natural Regulation of Animal Populations" (I. A. McLaren, ed.), pp. 136–170. Atherton Press, New York.
Christian, J. J., and Davis, D. E. (1971). Endocrines, behavior and population. *In* "Natural Regulation of Animal Populations" (I. A. McLaren, ed.), pp. 69–98. Atherton Press, New York.
Clark, L. R. (1964). The population dynamics of *Cardiaspina albitextura* (Psyllidae). *Aust. J. Zool.* **12,** 362–380.
Clark, W. C., Jones, D. D., and Holling, C. S. (1979). Lessons for ecological policy design: A case study of ecosystem management. *Ecol. Modell.* **7,** 1–53.
Dixon, A. F. G., and Wratten, S. D. (1971). Laboratory studies on aggregation, size and fecundity in the black bean aphid, *Aphis fabae* Scop. *Bull. Entomol. Res.* **61,** 97–111.
Elton, C. (1924). "Voles, Mice and Lemmings." Cramer, Weinheim, U.K.
Fischlin, A., and Baltensweiler, W. (1979). Systems analysis of the larch bud moth system. Part I: The larch–larch bud moth relationship. *Bull Soc. Entomol. Suisse* **52,** 273–289.
Greenbank, D. O. (1956). The role of climate and dispersal in the initiation of outbreaks of the spruce budworm in New Brunswick. I. The role of climate. *Can. J. Zool.* **34,** 453–476.
Hassell, M. P., Lawton, J. H., and May, R. M. (1976). Patterns of dynamical behaviour in single-species populations. *J. Anim. Ecol.* **45,** 471–486.
Haukioja, E. (1980). On the role of plant defenses in the fluctuation of herbivore populations. *Oikos* **35,** 202–213.
Heikkenen, H. J. (1981). The influence of red pine site quality on damage by the European pine shoot moth. *Gen. Tech. Rep. WO—U.S., For. Serv.* [*Wash. Off.*] **GTR-WO-27,** 35–44.
Holling, C. S. (1965). The functional response of predators to prey density and its role in mimicry and population regulation. *Mem. Entomol. Soc. Can.* **45,** 3–60.
Hutchinson, G. E. (1948). Circular causal systems in ecology. *Ann. N.Y. Acad. Sci.* **50,** 221–246.
Isaev, A. S., and Khlebopros, R. H. (1977). Inertial and non-inertial factors regulating forest insect population density. *In* "Pest Management" (G. A. Norton and C. S. Holling, eds.), pp. 317–339. Int. Inst. Appl. Syst. Anal., Laxenburg, Austria.
Isaev, A. S., Khlebopros, R. G., Nedorezov, L. V., Kondakov, Y. P., and Kiselev, V. V. (1984). "Population Dynamics of Forest Insects" (in Russian). Nauka Publ. House, Sib. Div., Novosibirsk.
Kobayashi, F., Yamane, A., and Ikeda, T. (1984). The Japanese pine sawyer beetle as the vector of pine wilt disease. *Annu. Rev. Entomol.* **29,** 115–135.
Leopold, A. (1947). "Game Management." Scribner's New York.
Lotka, A. J. (1925). "Elements of Physical Biology." Williams & Wilkins, Baltimore, Maryland.
McGregor, M. D., Oakes, R. D., and Meyer, H. E. (1983). Status of mountain pine beetle, Northern Region, 1982. *USDA For. Serv., North. Reg. State Priv. For. Rep.* **83–16.**
McLeod, J. M. (1979). Discontinuous stability in a sawfly life system and its relevance to pest management strategies. *Gen. Tech. Rep. WO—U.S., For. Serv.* [*Wash. Off.*] **GTR-WO-8,** 68–81.

McNamee, P. J., McLeod, J. M., and Holling, C. S. (1981). The structure and behavior of defoliating insect/forest systems. *Res. Popul. Ecol.* **23,** 280–298.

Mason, R. R., and Luck, R. F. (1978). Population growth and regulation. *USDA For. Serv. Tech. Bull.* **1585,** 41–47.

Mattson, W. J. (1980). Cone resources and the ecology of the red pine cone beetle, *Conophthorus resinosae* (Coleoptera: Scolytidae). *Ann. Entomol. Soc. Am.* **73,** 390–396.

Mattson, W. J., and Addy, N. D. (1975). Phytophagous insects as regulators of forest primary production. *Science* **190,** 515–522.

May, R. M. (1974a). "Stability and Complexity in Model Ecosystems." Princeton Univ. Press, Princeton, New Jersey.

May, R. M. (1974b). Biological populations with nonoverlapping generations: Stable points, stable cycles, and chaos. *Science* **186,** 645–647.

May, R. M., Conway, G. R., Hassell, M. P., and Southwood, T. R. E. (1974). Time-delays, density-dependence and single-species oscillations. *J. Anim. Ecol.* **43,** 747–770.

Morris, R. F. (1959). Single-factor analysis in population dynamics. *Ecology* **40,** 580–588.

Morris, R. F., ed. (1963). The dynamics of epidemic spruce budworm populations. *Mem. Entomol. Soc. Can.* **31,** 1–332.

Nicholson, A. J., and Bailey, V. A. (1935). The balance of animal populations. *Proc. Zool. Soc. London,* Pt. 1, pp. 551–598.

Prebble, M. L., and Graham, K. (1957). Studies of attack by ambrosia beetles in softwood logs on Vancouver Island, British Columbia. *For. Sci.* **3,** 90–112.

Raffa, K. F., and Berryman, A. A. (1983). The role of host plant resistance in the colonization behavior and ecology of bark beetles (Coleoptera: Scolytidae). *Ecol. Monogr.* **53,** 27–49.

Readshaw, J. L. (1965). A theory of phasmatid outbreak release. *Aust. J. Zool.* **13,** 475–490.

Rhoades, D. F. (1985). Offensive–defensive interactions between herbivores and plants: Their relevance in herbivore population dynamics and ecological theory. *Am. Nat.* **125,** 205–238.

Royama, T. (1977). Population persistence and density-dependence. *Ecol. Monogr.* **47,** 1–35.

Shepherd, R. F. (1977). A classification of western Canadian defoliating forest insects by outbreak spread characteristics and habitat restriction. *Minn., Agric. Exp. Stn., Tech. Bull.* **310,** 80–88.

Southwood, T. R. E., and Comins, H. N. (1976). A synoptic population model. *J. Anim. Ecol.* **65,** 949–965.

Stoszek, K. J. (1973). Damage to ponderosa pine plantations by the western pine-shoot borer. *J. For.* **71,** 701–705.

Takahashi, F. (1964). Reproduction curve with two equilibrium points: A consideration of the fluctuation of insect population. *Res. Popul. Ecol.* **6,** 28–36.

Taylor, K. L. (1983). The *Sirex* woodwasp: Ecology and control of an introduced forest insect. *In* "The Ecology of Pests: Some Australian Case Histories" (R. L. Kitching and R. E. Jones, eds.), pp. 231–248. CSIRO, Melbourne.

Thalenhorst, W. (1958). Grunduge der Populations dynamik des grossen Fitchenborkenkafers *Ips typographus* L. *Schriftenr. Forstl. Fak. Univ. Goettingen* **21.**

Tostowaryk, W. (1972). The effect of prey defense on the functional response of *Podisus modestus* (Hemiptera: Pentatomidae) to densities of the sawflies *Neodiprion swainei* and *N. pratti banksianae* (Hymenoptera: Neodiprionidae). *Can Entomol.* **104,** 61–69.

Verhulst, P. F. (1838). Notice sur la loi que la population suit dans son accroissement. *Corresp. Math. Phys. (Paris)* **10,** 113–121.

Volterra, V. (1926). Variazoni e fluttuazioni del numero d'individui in specie animali conviventi. *Atti R. Accad. Naz., Lincei, Mem. Cl. Sci. Fis., Mat. Nat.* **2,** 31–113.

Wangersky, P. J., and Cunningham, W. J. (1957). Time lags in population models. *Cold Spring Harbor Symp. Quant. Biol.* **22,** 329–338.

Wellington, W. G. (1960). Qualitative changes in natural populations during changes in abundance. *Can. J. Zool.* **38,** 289–314.

White, T. C. R. (1978). The importance of a relative shortage of food in animal ecology. *Oecologia* **33,** 71–86.

Chapter 2

Nonoutbreak Species of Forest Lepidoptera

RICHARD R. MASON

U.S. Department of Agriculture, Forest Service
Pacific Northwest Research Station
Forestry and Range Sciences Laboratory
La Grande, Oregon 97850

I. INTRODUCTION

The topic of this chapter differs from the general theme of other chapters in that it focuses on species of phytophagous insects that seldom increase to outbreak densities. Because of the obvious breadth of the subject, I have limited the review to defoliating Lepidoptera, emphasizing the less conspicuous species in coniferous forests of the northwestern United States. These species, although not extensively studied, are common associates of some of the most destructive outbreak species of forest defoliators in the world. The chapter concentrates on the abundance and pattern of species in this group and, I hope, calls attention to their possible stabilizing role in the forest community.

An impressive variety of caterpillars are found on the foliage of trees, yet surprisingly few species ever reach outbreak numbers (see Chapter 6, this volume). Ecologists have recognized for years that any random sample of a natural community usually yields many species with a few individuals and only a few species with many individuals (Preston, 1948; Williams, 1960). The corollary is that the number of species that increase to high population densities is small compared with the vast majority of species that persist over time at low and stable densities.

The order Lepidoptera totals about 120,000 species worldwide (Gilbert, 1979), with an estimated 15,000 or more occurring in North America (Munroe, 1979). The Canadian Forest Insect Survey has recorded a minimum of 989 lepidopteran species from forest trees; only a fraction of these species ever occur in large enough populations to be of economic significance (McGugan, 1958; Munroe, 1979; Prentice, 1962, 1963, 1966). In the western United States, Furniss and Carolin (1977) listed only 15 species or species groups of Lepidoptera that are considered major pests of forest trees because of their occasional high densities. Despite the notoriety of several destructive species, most forest Lepidoptera seldom reach outbreak proportions. Nonoutbreak species, nonetheless, are integral parts of a complex community that has a direct influence on numbers of the more eruptive species (Graham, 1956; Munroe, 1979).

Nature of Nonoutbreak Species

The term "nonoutbreak" in this chapter is used in a narrow sense to describe only species that have never or only rarely caused noticeable tree defoliation. This limited definition is necessitated by historical precedence in the use of the word "outbreak" to represent conditions in which trees have been conspicuously defoliated. The number of some defoliators may occasionally far exceed their usual equilibrium density, but if significant defoliation does not occur they are still considered nonoutbreak species.

By definition the numbers in populations of nonoutbreak species fluctuate at lower magnitudes than do those of their outbreak counterparts. Southwood (1962) has called this group the "denizens of permanent habitats" because of the relative stability of the populations. On the *r–K* continuum the traits of such species are probably closer to those of *K*-selected individuals (MacArthur, 1960; MacArthur and Wilson, 1967). That is, compared with *r*-selected species, *K* strategists generally have a lower fecundity, a higher survival rate of immatures, and more specialized food habits (Southwood, 1977). Because their populations are usually nearer equilibrium density, *K* strategists are appropriately called "equilibrium species" as opposed to "opportunistic species," which are more capable of quickly colonizing new habitats (MacArthur, 1960). Outbreak

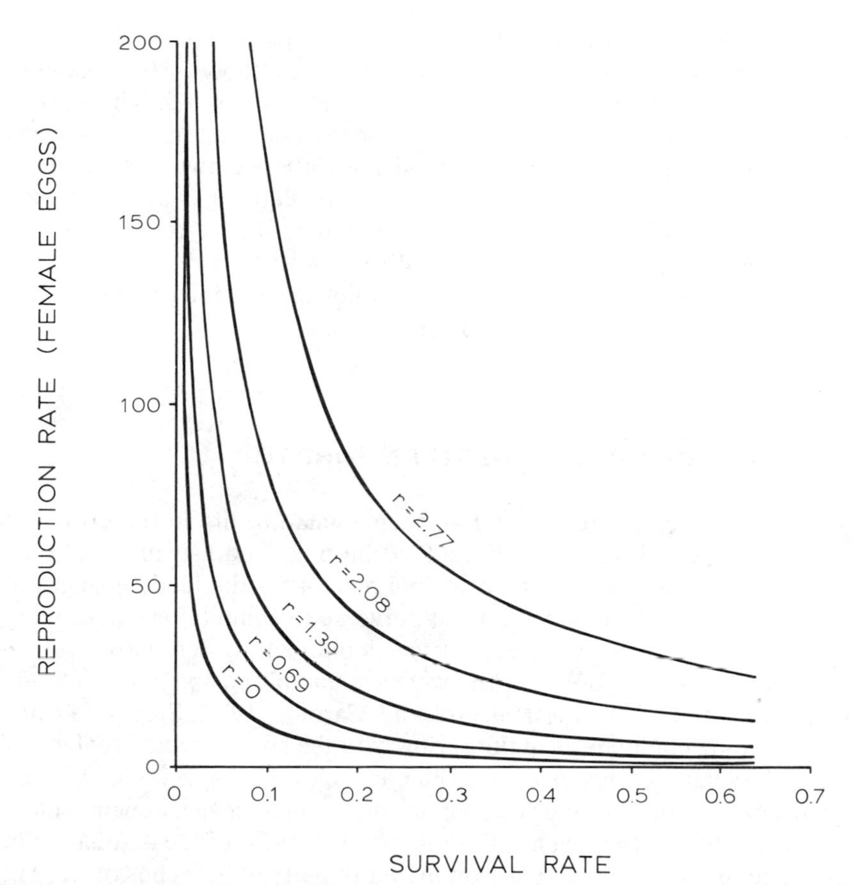

Fig. 1. Intrinsic rate of natural increase r for different rates of generation survival and reproduction in univoltine populations of defoliating Lepidoptera.

species of defoliators are classic opportunists, whereas nonoutbreak species better fit the equilibrium species mold (Conway, 1976).

Nonoutbreak species can also be characterized by their relatively low rates of population change. Working with models for single-species populations, May (1974, 1975a) demonstrated that different rates of natural increase r led to various types of dynamic behavior in simulated populations. In general, he found that populations over time reached a low stable equilibrium when r was relatively low and exhibited cyclic or sporadic outbreaklike behavior when it was high.

For univoltine species the rate of increase of a natural population is equivalent to

$$r = \ln(RS)$$

where R is the number of female eggs produced by a female adult and S the proportion of these progeny that survive to adulthood. The trade-offs between reproduction and survival that are needed to maintain a given rate of increase are obvious (Fig. 1). Like many outbreak insects, species with a high fecundity can achieve rapid population growth much easier than species with a low fecundity whose populations cannot increase unless survival is high. Low reproduction rates may be responsible for limiting the populations of many defoliators to a stable equilibrium mode. These topics are reviewed further in the following section for the typical coexisting defoliator species in western forests.

II. DEFOLIATOR GUILD IN WESTERN FORESTS

Most forest Lepidoptera spend all their immature life in the crown of host trees, where developing larvae feed on buds and leaf or needle tissues. Foliated branches are the principal habitat for the feeding stages of larvae, and they comprise the logical universe to sample for determining larval abundance. The vast majority of lepidopteran defoliators in the temperate coniferous forests of the west are univoltine and overwinter as either eggs or small larvae (Furniss and Carolin, 1977; Stevens *et al.*, 1984). Feeding commences in the spring with the swelling and breaking of buds and issuance of new needles. The life cycles of all species are closely synchronized with foliage development, and many species depend solely on the availability of young needles for survival. Most of the defoliators in a forest community can be found on the outer parts of branches of trees in early summer. The most common species often occur on every branch, whereas the rare species can be detected only after examining many branches.

Although many species of defoliating Lepidoptera have been described, few studies have quantified their interrelations in a natural community. Most species rarely become abundant enough to have an economic impact and have not been studied in any detail. Classic exceptions are the western spruce budworm, *Choristoneura occidentalis* Freeman, and the Douglas fir tussock moth, *Orgya pseudotsugata* (McDunnough), whose outbreaks periodically damage mixed conifer forests of Douglas fir, *Pseudotsuga menziesii* (Mirb.) Franco, and true fir, *Abies* spp., particularly grand fir, *Abies grandis* (Dougl.) Lindl., and white fir, *Abies concolor* (Gord. and Glend.) Lindl. ex Hildebr., in western North America. Both species feed on the foliage of new shoots in early summer and often inhabit the same trees. Because of their economic importance, outbreaks have frequently been treated by broadcast spraying of insecticides. The effectiveness of control projects is usually evaluated by extensive field sampling of the pest population in treated areas. In a few instances non-outbreak Lepidoptera have also been identified and counted with the target insect (Carolin, 1980; Carolin and Coulter, 1971; Herman and Bulger, 1979; Markin, 1982; Williams and Walton, 1968). The recent emphasis on integrated pest management has likewise stimulated an independent interest in the nonoutbreak species component of budworm and tussock moth systems (Dahlsten *et al.*, 1978; Volker, 1978). However, few quantitative data are available for coexisting Lepidoptera in natural fir forests.

A. Species of Lepidoptera Coexisting on Fir

The variety of lepidopteran species reported to feed on fir suggests a potentially rich community of defoliators in mixed fir forests. In a handbook of lepidopteran larvae associated with the western spruce budworm, Stevens *et al.* (1984) conservatively list 32 species and species groups that are likely to occur in opening buds or on new shoots of Douglas fir and true firs in the West (also see Carolin and Stevens, 1981). Table 1 is a compilation of all determined species of Lepidoptera that have been recorded by several authors in systematic samplings or controlled studies. Individuals known to feed on fir but that were not observed in these studies or were not identified to the species level are omitted. The table lists 33 species found over parts of five northwestern states. All are foliage herbivores during the early summer, and many appear to be ecological homologs (Debach, 1966; Johnson and Denton, 1975; Klomp, 1968; Markin, 1982). Larvae of some of the olethreutids and tortricids also mine buds at the same time in the spring before foliage issuance. Only five on the list are recognized as commonly or occasionally increasing to outbreak numbers. Outbreaks among the remaining 28 species are either rare or unknown. In other words, 85% of the recorded defoliators on Douglas fir and true fir are nonoutbreak. Very rare individuals undoubtedly also exist but were not recorded (Karban and Ricklefs, 1983; Preston, 1948), and if the list were expanded to include these and undetermined species the proportion of nonoutbreak species might be far more than 90%.

According to the records summarized in Table 1, the common outbreak species are more widely distributed geographically than the nonoutbreak species. The only exception is *Choristoneura retiniana* (Walsingham), whose range is restricted to northeastern California, southern Oregon, and parts of Nevada (Furniss and Carolin, 1977; Powell, 1980; Volney *et al.*, 1983). The fact that the nonoutbreak species appear to have a more

TABLE 1

Partial List of Defoliating Lepidoptera Collected as Larvae from Foliage Samples of Douglas Fir, Grand Fir, and White Fir in Systematic Sampling Studies

Family, species	Geographic distribution					Outbreak history				Feeding characteristics			Average no. eggs laid[a]	References[b]
	N. California	Oregon	Washington	Idaho	W. Montana	None	Rare	Occasional	Common	Buds	Foliage	Cones		
Gelechiidae														
Coleotechnites sp. near *piceaella* (Kearfott)		o	o			x					*			1
Geometridae														
Caripeta divisata (Wlk)				o	o	x					*			2
Enypia sp. near *griseata* Grossbeck		o	o			x				*	*			1,3
Enypia packardata Taylor	o					x				*	*			4
Eupithecia annulata (Hulst)	o	o	o	o	o	x				*	*			1,2,3,4
Gabriola dyari Taylor				o		x					*			1,2
Lambdina fiscellaria lugubrosa (Hults)		o	o	o	o		x[c]				*			1,2,5
Melanolophia imitata (Walker)	o		o	o	o			x			*		80	1,2,4,6
Nematocampa filamentaria Gn.				o	o		x				*			2,7
Nepytia freemani Munroe				o	o		x				*		67	2,8
Nepytia phantasmaria (Strecker)		o					x				*		69	1,9
Pero behrensarius Pack.				o	o	x					*			2
Semiothisa signaria dispuncta (Walker)				o	o	x					*			2
Synaxis pallulata Hulst.	o					x					*			4
Lymantriidae														
Orgyia pseudotsugata (McDunnough)	o	o	o	o	o				x		*		200	1,2,4,7,10

Noctuidae														
Achytonix epipaschia (Grote)		○	○			x				*	*			1,11
Anomogyna mustelina Smith			○			x				*	*			1,3
Syngrapha angulidens (Smith)					○	x					*			5
Syngrapha celsa sierrae (Otto)	○					x				*	*			3,4
Egira (= *Xylomyges*) *simplex* (Walker)		○	○				x			*	*			1,3
Olethreutidae														
Griselda radicana (Heinrich)		○	○	○	○	x				*	*			1,5,7,11
Zeiraphera hesperiana Mutuura and Freeman		○	○	○	○		x							1,5,7,11[d]
Zeiraphera improbana (Walker)			○			x[e]				*	*			1,3
Pyralidae														
Dioryctria reniculelloides Mutuura and Munroe		○	○	○	○	x				*	*	*		1,5,7
Tortricidae														
Acleris gloverana (Walsingham)	○	○	○	○	○				x	*	*		88	1,2,4,7,11,12
Argyrotaenia dorsalana (Dyar)		○	○			x				*	*			1,11
Argyrotaenia provana (Kearfott)	○	○				x				*	*			1,3,4
Clepsis persicana (Fitch)		○				x					*			1
Choristoneura lambertiana (Busck)	○					x					*			4
Choristoneura occidentalis Freeman		○	○	○	○				x	*	*		150	1,2,5,7,11,13
Choristoneura retiniana (Walsingham) = *viridis* Freeman	○	○						x		*	*			1,4
Choristoneura rosaceana (Harris)		○				x					*			1
Lasiocampidae														
Tolype distincta French					○	x					*			5

[a] Published data on egg production are lacking for most species.
[b] Key to references: 1, Carolin (1980); 2, Volker (1978); 3, Stevens *et al.* (1984); 4, Dahlsten *et al.* (1978); 5, Williams and Walton (1968); 6, Silver (1962); 7, Markin (1982); 8, Klein and Minnoch (1971); 9, Wickman and Hunt (1969); 10, Wickman and Beckwith (1978); 11, Carolin and Coulter (1971); 12, Miller (1966); 13, Fellin and Dewey (1982).
[c] Outbreaks common on western hemlock.
[d] Reported as *Z. griseana*.
[e] Outbreaks occasional on western larch.

limited distribution could be because they are generally rarer and less likely to be found in a sample. A narrow distribution could also imply a higher degree of specialization among nonoutbreak species, which restricts their population to a more specific set of environmental conditions (Fox and Morrow, 1981). Such specialized characteristics comprise some of the usual criteria describing equilibrium species (Conway, 1976). The Douglas fir tussock moth was reported in all the samplings, although its populations were at relatively low densities. The cosmopolitan nature of the tussock moth even at low numbers suggests that it may have a wide tolerance for different environments, which is an expected characteristic of opportunistic species (McNaughton and Wolf, 1970; Pianka, 1970).

Information on fecundity is scarce for most of the species recorded in Table 1. The outbreak species are the only group that has been studied in any detail, so estimates of egg production are available for three of those species. Although the comparison with so few data is weak, average egg production of the outbreak species exceeded that of the nonoutbreak species by about 2 : 1. If this ratio held for all species, it would support the hypothesis of a lower reproduction rate in equilibrium species. At equivalent survival rates, such lower rates of reproduction would result in a smaller rate of increase *r*, which could be partly responsible for the relative stability of nonoutbreak populations (Fig. 1). There are problems, however, in drawing conclusions from such comparisons, because egg production is not absolute for a species but is modulated by the density and habitat conditions of each population (Labeyrie, 1978; Watt, 1960). Nonetheless, fecundity is ultimately adjusted by the average death rate of a species, and its value should be a useful indicator of the harshness of a particular environment (Price, 1974).

B. Species Associated with Budworm Outbreaks

Few quantitative data on the coexistence of lepidopteran defoliators in specific forest communities have been published. This is unfortunate because the abundance of the economically more important species could be affected by the interactions resulting from feeding on the same foliage resource and sharing the same natural enemies. Some groups, particularly phenotypically similar species, may even tend to be regulated as a whole rather than as individual species (Hebert *et al.*, 1974; Markin, 1982). The only published studies on coexistence (Carolin and Coulter, 1971; Markin, 1982; Williams and Walton, 1968) have been conducted in connection with control projects or insecticide tests, and they focused primarily on the western spruce budworm; however, other Lepidoptera were also recorded as a secondary interest. In each of the studies foliage

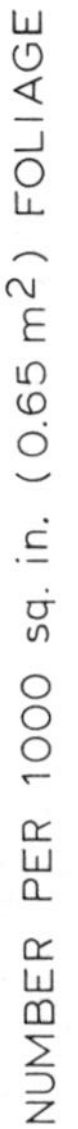

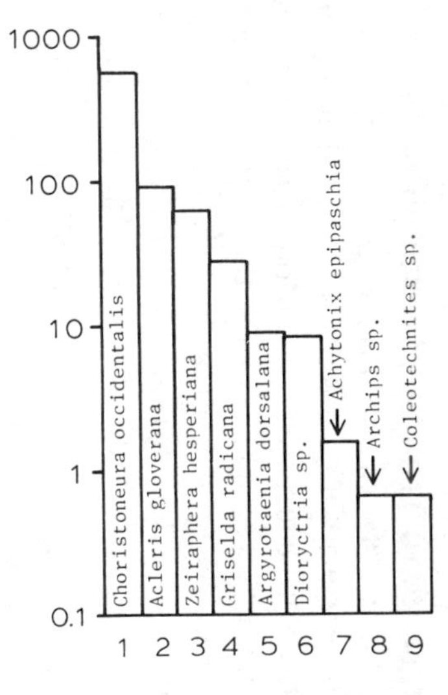

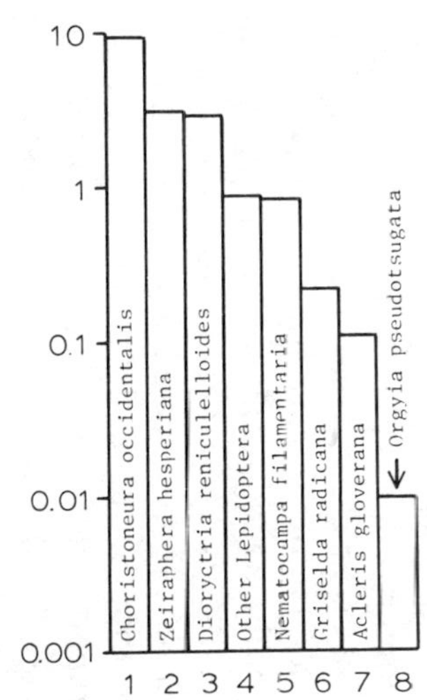

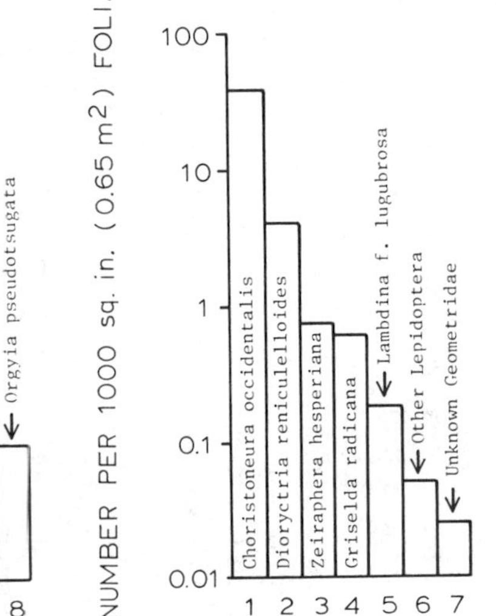

Fig. 2. Relative abundances of lepidopteran species in three outbreaks of the western spruce budworm. Graphs are plotted from data of: A, Carolin and Coulter (1971); B, Markin (1982); C, Williams and Walton (1968).

samples were systematically clipped from tree crowns on a series of plots, and all lepidopteran larvae were identified and counted. The population density of each species was then expressed in terms of a standard unit of foliage.

A compilation of the relative abundance of Lepidoptera on all plots is shown for each of these studies in Fig. 2. Carolin and Coulter (1971) recorded nine species of Lepidoptera in their samples in a 1958 budworm outbreak in eastern Oregon. Markin (1982) identified a minimum of seven species from an outbreak in southeastern Idaho, but more were probably present, as reflected in the category "other Lepidoptera." Williams and Walton (1968) listed at least six species on their plots in Montana, but they also have an "other" category for additional rare species. These findings indicate that as many as 10 species of Lepidoptera, most of which are nonoutbreak, may coexist on fir foliage in budworm outbreaks. Budworm larvae dominated, but other species accounted for 13 to 46% of the total number of Lepidoptera in the samples. It is obvious that when there is a large proportion of associated species they may contribute substantially to the observed defoliation in budworm outbreaks (Markin, 1982).

At least three of the nonoutbreak species, *Zeiraphera hesperiana* (Mutuura and Freeman), *Griselda radicana* (Heinrich), and *Dioryctria reniculelloides* Matuura and Munroe, were common to all three studies, suggesting their wide distribution in the intermountain area. The same species have also been reported as common associates of budworm in the Southwest (Stevens *et al.*, 1983). Both *Z. hesperiana* and *D. reniculelloides* also occasionally feed with the western budworm on the same new needle growth, and outbreaks have been treated with insecticides where the density of *D. reniculelloides* was higher than that of the budworm (Johnson and Denton, 1975). The black-headed budworm, *Acleris gloverana* (Walsingham), an occasional outbreak species in coastal forests, was common on the eastern Oregon plots and was also present in the Idaho sample but was not recorded in the Montana study. This species also sometimes feeds in large numbers on the same foliage as the western budworm, particularly on subalpine fir, *Abies lasiocarpa* (Hook.) Nutt. (Carolin, 1980). The other potential outbreak speices, the Douglas fir tussock moth, was recovered only on the Idaho plots and in very low numbers (Fig. 2B).

III. ORGANIZATION OF DEFOLIATOR COMMUNITIES

In each of the studies described in Section II,B, a similar number of defoliator species apparently coexisted in the stands at one time. An obvious question is, Why this number when a much larger array of species are capable of feeding on fir? Were other species excluded by the outbreak densities of budworm, or is there an intrinsic organization to the herbivore community that limits the number of coexisting species? To explore these questions further, I will use a data set that my colleagues and I collected from nonoutbreak communities in 1978.

We sampled the larvae of Lepidoptera in nine forest stands shortly after buds had burst and while new shoots were elongating. Larval abundance of most univoltine defoliators is highest at this time of early summer and then diminishes dramatically as mortality accumulates through the season. In each stand three lower crown branch tips (45 cm long) on each of 400 randomly selected Douglas fir and true fir trees were beaten over a dropcloth. All lepidopteran larvae that dropped onto the cloth were collected in alcohol and later sorted. Because of the difficulty of recognizing immatures, only the common species were positively identified by their binomial name. Similar to the procedures used in other community studies (Futuyma and Gould, 1979), the procedure here was to sort less known forms to only family and probable species on the basis of morphological differences. Although none of the populations sampled was presently at outbreak density, each of the stands had a history of budworm or tussock moth outbreaks. Specifics for each stand are given in Table 2.

A total of 1079 lepidopterans were collected from the nine stands for an average density of 0.93 larva per square meter of branch area. The number of species ranged from 5 to 12 for eight of the stands, but 26 species were collected from the unusually rich community at Mare's Egg Spring. The common defoliator families (Table 1), Tortricidae, Geometridae, Noctuidae, and Lymantriidae, were well represented. Two-thirds of the time the most abundant lepidopteran on a site was an outbreak-prone species such as the Douglas fir tussock moth, *O. pseudotsugata,* or the black-headed budworm, *A. gloverana*. The remainder were principally nonoutbreak species.

A. Relative Abundance of Lepidopteran Larvae

The stability of populations depends to some degree (mostly unknown) on the abundance of the uncommon species, and, therefore, one can best describe communities by considering their full distribution of species' relative abundance (May, 1975b; Price, 1971). MacArthur (1957, 1960) proposed a distribution for predicting the relative abundance of species whose resources occur randomly along a continuum but where there is no overlap in their utilization of the resources. The expected abundance of a

TABLE 2

Sampled Stands and Abundance of Leipdoptera in Nine Nonoutbreak Populations, 1978[a]

Stand	National forest (state)	Elevation (m)	Host tree species[b]	Lepidoptera sampled			Most abundant Lepidoptera	
				Total	No. species	Density[c]	Species	Percent dominance[d]
Iron Mt.–Plummer Ridge	Eldorado (CA)	1707–1737	Ac	420	12	3.26	*Orgyia pseudotsugata*	92
Mare's Egg Spring	Winema (OR)	1275	Ac, Pm	415	26	3.22	*Acleris gloverana*	73
Chiloquin Ridge	Winema (OR)	1690	Ac, Pm	105	8	0.81	*Acleris gloverana*	65
Corral Creek	Modoc (CA)	1740	Ac, Pm	38	8	0.29	*Acleris gloverana*	39
Stowe Reservoir	Modoc (CA)	1860	Ac, Pm	22	7	0.17	*Orgyia pseudotsugata*	32
King Mountain	Malheur (OR)	1770	Ag, Pm	26	5	0.20	Unknown Geometridae	58
High Ridge	Umatilla (OR)	1380	Ag, Pm	11	5	0.09	Unknown Geometridae	55
Wood Butte	Wallowa–Whitman (OR)	1340	Ag, Pm	27	12	0.22	Unknown Geometridae	22
Sled Springs	Wallowa–Whitman (OR)	1480	Ag, Pm	15	7	0.12	*Choristoneura occidentalis* and unknown Geometridae	27[e]

[a] See Figs. 3 and 4 for a comparison of relative abundances of species.
[b] Ac, *Abies concolor*; Ag, *Abies grandis*; Pm, *Pseudotsuga menziesii*.
[c] Number of larvae per square meter of branch area.
[d] Most abundant species as a percentage of the total number of sampled Lepidoptera.
[e] Represents only one of the two dominant species.

species in a sample calculated in multiples n_j of the average number of individuals per species is given by

$$n_j = \frac{N}{S} \sum_{i=1}^{j} \frac{1}{S - i + 1}$$

where for a sample of N individuals, S is the total number of species and i the order of a species in the sequence from the rarest ($i = 1$), through the species in question ($i = j$), to the most common ($i = S$) (Hutchinson, 1978; Whittaker, 1975). When abundance is plotted over the logarithm of species rank, the relation is almost linear. This distribution, commonly called the "broken-stick" model, has been successfully fitted to samples of a variety of natural organisms (Hairston, 1959; King, 1964; Turner, 1961). MacArthur's (1960) original premise was that the contiguous, nonoverlapping niches generated by the model would be a good description of the expected relative abundance of equilibrium species. He also theorized that opportunistic species could not be expected to conform to this distribution but are more likely to fit a log-normal distribution. Some have questioned the true biological meaning of the model's distribution, and MacArthur himself eventually dismissed the concept (Cohen, 1968; Hairston, 1969; MacArthur, 1966; Webb, 1974). Nevertheless, May (1975c) concludes that relatively simple communities are often statistically characterized by the broken-stick distribution, and conformity indicates that a major resource is being divided among the member species, although not necessarily according to the biological assumptions of MacArthur. Despite the ambiguity, there is agreement that such quantitative descriptions of species relationships are useful in the prediction and analysis of communities (DeVita, 1979; Hutchinson, 1978; May, 1975c; Pielou, 1969; Whittaker, 1975).

The broken-stick model was initially purported to describe equilibrium species, so it is desirable to see how it compares with the distribution of species observed in the stands described in Table 2. Figure 3 shows the ways in which species abundances were graduated in the six most equitable stands, i.e., where dominance by a single species was minimum. The expected abundances according to the broken-stick model were also calculated and are plotted for comparison. The small χ^2 values indicate that the sequences of observed abundances agree reasonably well with the model predictions expected for equilibrium populations. Although a few of the species were potential outbreak insects, they were presently at low densities and were ranked in the same line with other species.

When a potential outbreak species strongly dominated the other Lepidoptera, the broken-stick series no longer fit the observed abundances.

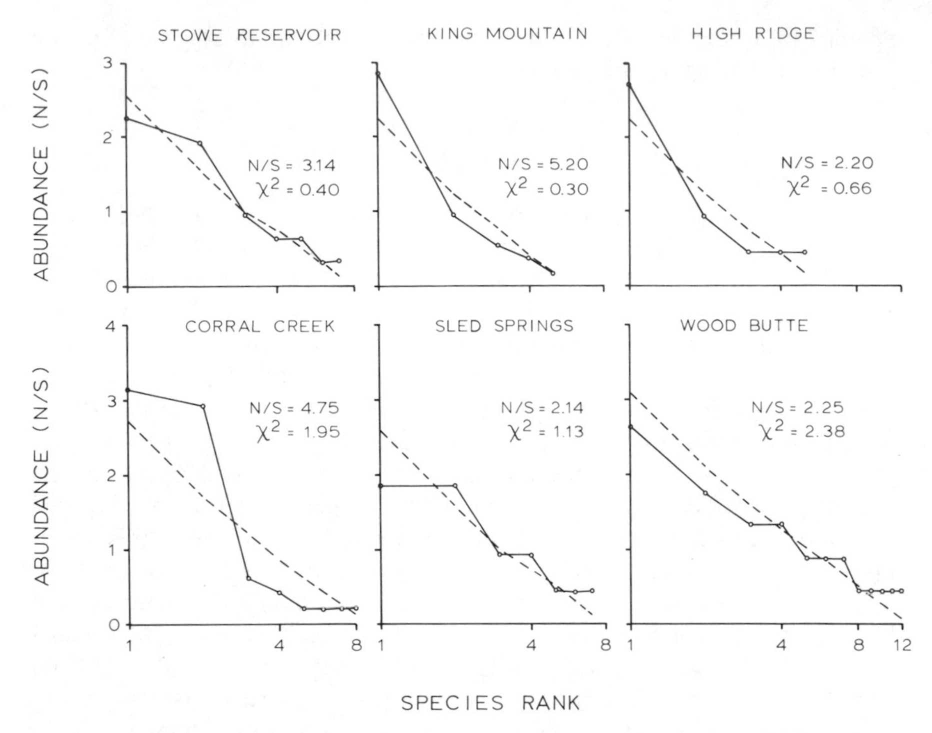

Fig. 3. Observed relative abundance of lepidopteran species on fir (solid lines) compared with abundance calculated from the broken-stick distribution (dashed lines) for six forest stands with equitable species. See Table 2 for a description of sample stands.

Examples are the first three stands in Table 2, where each dominant species departs from the linear pattern of the equilibrium community (Fig. 4A). None of the populations of these species could be labeled an outbreak by economic standards, yet each dominant species clearly differed from the abundance patterns of the other species. The poor fit of the model in Fig. 4A suggests that environmental resources were not randomly divided among all species but were disproportionately used by the outbreak species. However, when the dominant species was removed from each analysis, the broken-stick distribution gave an excellent fit to the remaining species (Fig. 4B). These results bear out MacArthur's (1960) hypothesis that the dominant species in a community are mainly opportunistic and can be discounted in studies because their number is determined mostly by the vagaries of weather and other environmental factors that have recently favored a population increase. May (1975c) also supports this view by saying that "dominant

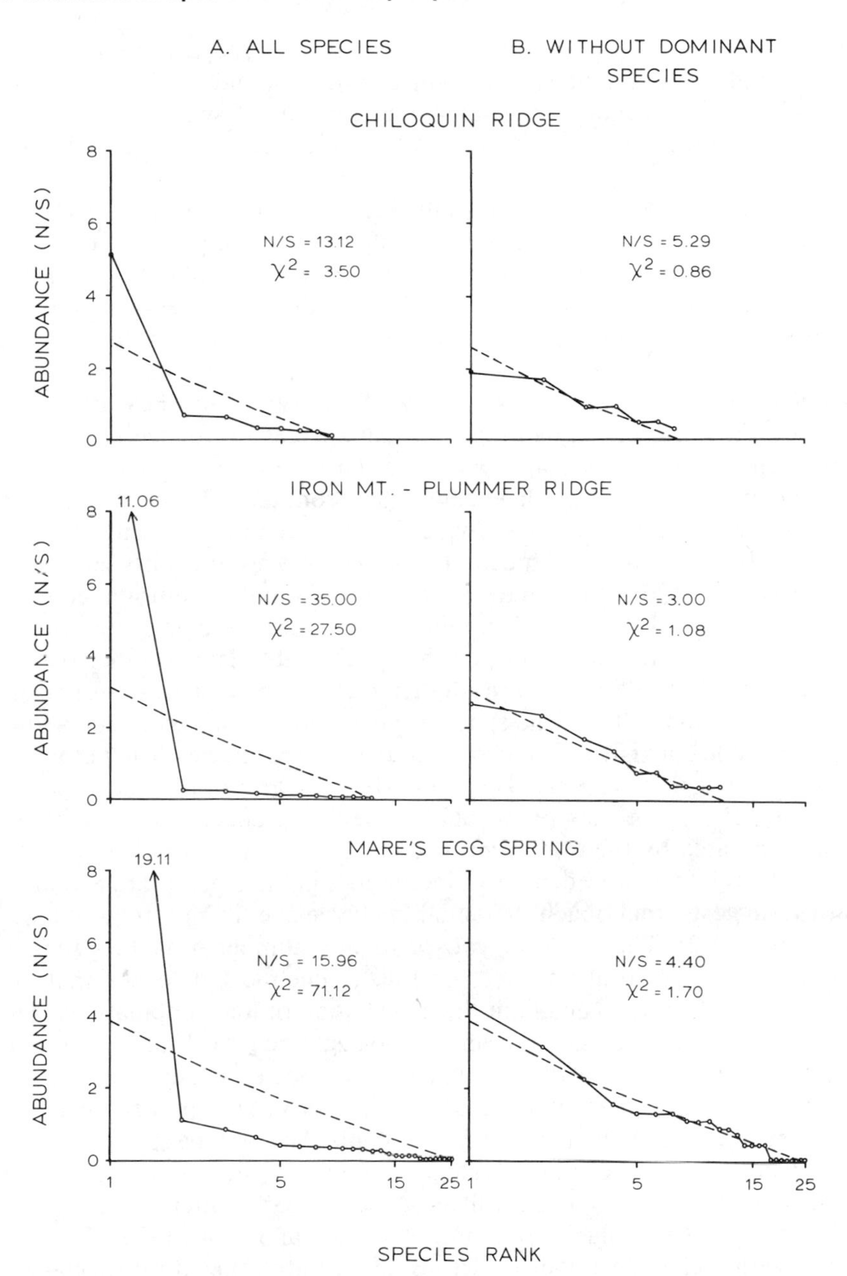

Fig. 4. Observed relative abundance of lepidopteran species on fir (solid lines) compared with abundance calculated from the broken-stick distribution (dashed lines) for three forest stands with dominant species. Comparisons are for: A, all species; B, dominant species excluded. See Table 2 for a description of sample stands.

species are simply those that have recently enjoyed a large *r*, and at different times different species will be most abundant." In contrast, the relative abundance of the remaining permanent species may reveal much about the basic structure of the whole community (MacArthur, 1960).

What can be learned from the analysis presented here? First, relatively few defoliator species may simultaneously inhabit many forest communities. Although the number varied, there was an average of only 10 coexisting species for each of the nine sampled stands. This average compares favorably with the number of species recorded from similar stands in the earlier published studies (Fig. 2). [In the only other comparable field studies of forest defoliators that I know of, McLeod and Blais (1961) also found an average of only 10 lepidopteran species on foliage samples of spruce in Quebec and Klomp (1968) reported 11 or fewer moths coexisting on pine in Holland.] The richest community contained 26 species of Lepidoptera, which was more than twice the number in the community with the next largest number of species (Table 2). Second, the fairly good fit of the broken-stick distribution to the observed data of equilibrium communities indicates that some environmental resource is being apportioned among the nonoutbreak species, although it is not clear whether the apportionment is random or the result of some biological process. A reasonable explanation is that niche space has been divided in a way that allows for the stable coexistence of a group of defoliating species, each of which is specialized in a way that keeps interference at a minimum (Fox and Morrow, 1981). An example would be coexisting species that are physically separated by their different feeding sites (Stevens *et al.*, 1984). Finally, the relatively low density of individuals in the lepidopteran communities suggests that much potential niche space is not fully utilized (Lawton, 1982). This could be caused by any number of factors such as harsh climate, natural enemies, or foliage chemical defenses that adversely affect survival and limit the distribution of local populations. The result could be the creation of an "ecological vacuum," making surplus space available for the more opportunistic species (Southwood *et al.*, 1974). When environmental conditions are favorable for their increase, these species are then more likely to colonize the vacant space and gain dominance. Typical of the *r* strategist, dominant species probably have broader niches than their subordinates, which enables them to take better advantage of the available space (McNaughton and Wolf, 1970). Communities with niche space that is already more fully utilized by specialized species may be better able to resist such habitat domination by a single species (Debach, 1966; Hardin, 1960).

B. Elusive Role of Competition

Any discussion of community organization must eventually address the roles of interspecific and intraspecific competition. The role that competitive interactions play in structuring communities has been the subject of heated controversy for years (Connell, 1983; Roughgarden, 1983; Simberloff, 1983; Schoener, 1982). Hairston *et al.* (1960) first proposed that competition is not usually an important factor in the regulation of herbivores, which more often seem to be predator-limited. In a convincing review Lawton and Strong (1981) concluded that competition has only a minimal effect on community patterns of foliage-feeding insects. There is a growing consensus that natural enemies and foliage chemistry, acting alone or together, influence the population dynamics of insect defoliators much more than do competitive interactions (Haukioja, 1980; Lawton and McNeill, 1979; Rhoades, 1979; Schultz and Baldwin, 1982, 1983; Strong, 1983; White, 1978). The community data reported in the previous section also seem to support this hypothesis insofar as the density of Lepidoptera appeared to be much lower than the carrying capacity of the environment.

Competition for food is not expected to be a factor when the sum of all individuals averages fewer than one larva per square meter of foliage (Table 2). The rarest of the nonoutbreak species must persist at densities that are even far below this average. Consequently, unless much foliage is resistant to feeding or is nutritionally inadequate, there is little chance that competition for food is evoked in nonoutbreak populations. Nor would there be significant competition for space or other resources at the low larval densities that are usually observed. Competition may occur in subtle ways, however, such as adult defense of territory, which could result in self-regulation through exclusion or emigration of another species (Baker, 1983; Dethier and MacArthur, 1964; Ito, 1980). Direct evidence of competition, or "blood on the ground" so to speak, may be rare (May, 1978).

The picture changes dramatically under outbreak conditions when all contemporary herbivores are obviously affected by tree defoliation caused by the outbreak species (Stevens *et al.*, 1984). Not only is there serious competition for food, but other environmental resources must also be shared with the dominant species. Schoener (1982) refers to such a time of scarcity as a "crunch," a period during which temporary selection is for characteristics other than interspecific competition but that are quickly obscured once the crunch ends. If true, this means that outbreaks may have had only minimal effect on phenotypes and that subordinate defoliators have evolved mainly in noncompetitive environments.

A few published data show that the population trends of some forest Lepidoptera are synchronous over time, especially in the Geometridae in British Columbia (Blitz and Ross, 1958; Harris *et al.*, 1982) and the Tortricidae in Oregon and Washington (Carolin, 1980). The similarity of oscillations of the western spruce budworm and its associated defoliators led Carolin (1980) to hypothesize that they must be controlled by the same factors. Carolin's data not only showed a synchrony in trend, but also suggested that the associated Lepidoptera may have been more persistent as a group than the budworm. When field data were first collected, budworm populations had already increased to a high density. They were then followed by an increase in the populations of other species. Although fewer in number, the associated species actually had a higher rate of increase once the outbreak was underway and after a 1957 peak declined at a slower rate than the budworm populations (Table 3). The nonoutbreak component at all study areas increased from 11 to 29% of the total Lepidoptera over the 4-year period from 1955 to 1958. The synchronous increases in numbers indicated that the environment was initially favorable for all species, but the faster decline of budworm populations after 1957 suggests that under identical crowded conditions the budworm was the poorer competitor. This is the behavior that might be expected in *r*- versus *K*-selected populations and seems to be an example of the relative competitive abilities of the two evolutionary strategies in a natural outbreak situation.

TABLE 3

Comparison of Observed Rates of Natural Increase *r* in Outbreak Populations of the Western Spruce Budworm (WSB) and Associated Defoliators (AD) in Northeastern Oregon[a]

	1955–1956		1956–1957		1957–1958		Mean	
Study area	WSB	AD	WSB	AD	WSB	AD	WSB	AD
Baker	−0.21	0.55	0.97	0.85	−0.66	−0.72	0.03	0.23
Chesnimnus	0.07	1.74	−0.30	1.30	—	—	−0.12	1.52
Dale	0.69	0.19	0.25	1.19	−1.91	−0.86	0.32	0.17
Dixie	—	—	—	—	−0.96	0.69	−0.96	0.69
Joseph	−0.32	1.20	0.66	1.26	−0.66	0.09	−0.11	0.85
Mean	0.06	0.92	0.40	1.15	−1.05	−0.20	—	—

[a] Rate of increase $r = \ln(N_t/N_{t-1})$, where N_t is population density in year t. Rates are calculated from Carolin's (1980) data.

Despite the fact that competition is an important process at high densities, evidence suggests that forest defoliators live most of the time in an environment relatively free of competition. When the factors known to influence competition such as density, food, and space are considered, it appears unlikely that competitive interactions occur frequently enough to have a strong influence on community organization (Lawton, 1982). The rare species especially may have developed specialized behaviors and feeding habits that virtually insure them against any interspecific competition. Such elimination of niche overlap through evolution could be the ultimate achievement in competitive ability (Fox and Morrow, 1981).

C. Primacy of Natural Enemies

If interspecific competition is not a determining factor, what controls the structure of lepidopteran communities in mixed conifer forests? Lawton and McNeill (1979) concluded that, in general, the trophic levels above and below a herbivore are more important than interactions at its own trophic level; that is, they believe that the vertical effects of predation, including parasitization, by natural enemies and the changing chemistry of food plants have a stronger role in determining community structure than does the horizontal effect of competition (also see Connell, 1975; Lawton and Strong, 1981; Strong, 1983; Chapter 12, this volume). This is similar to the earlier view of Hairston *et al.* (1960), except that those authors did not originally stress the potential role of host plant quality. The vagaries of weather also influence herbivore density by affecting insect survival in a variety of ways (Andrewartha and Birch, 1954; Schwerdtfeger, 1958). Natural enemies combined with weather and food quality are of the greatest importance in the organization of communities of phytophagous Lepidoptera (Price *et al.,* 1980; Strong *et al.,* 1984).

Populations of nonoutbreak species of forest Lepidoptera have seldom been studied in detail, so there are few data quantifying their mortality factors. Ample data, however, are available on the population dynamics of the Douglas fir tussock moth and western spruce budworm that identify natural enemies, especially arthropod and avian predators, as major causes of their mortality (Dahlsten *et al.,* 1977; Campbell *et al.,* 1983; Mason, 1977; Mason *et al.,* 1983). It has been shown in numerous controlled studies that the survival of these species is unnaturally high when immature stages are protected from native enemies by exclusion cages or barriers (Campbell and Torgersen, 1982; Torgersen and Campbell, 1982; Mason, 1981; Mason and Torgersen, 1983; Mason *et al.,* 1983). Although the tussock moth is a common outbreak species, some communities have been studied where it permanently coexists at low stable numbers with the equilibrium species. I studied the dynamics of one such population near Fort Klamath, Oregon, for eight generations while densities fluctuated, but always within a narrow range. Insect parasites and bird predators were positively identified as enemies of all developmental stages, and they annually accounted for 36% of the mortality per generation. An average of 62% of the mortality each year was from larvae that simply disappeared from the foliage between sample periods. In separate experiments it was deduced that a large portion of this disappearance was from predation by insects, spiders, and birds (Mason and Torgersen, 1983; Torgersen *et al.,* 1984). In total, natural enemies each year accounted for a great share of the per generation mortality and appeared to be responsible for the regulation of tussock moth populations at low densities. Because of their polyphagous habits, many of the same predators also prey on other available Lepidoptera and may be similarly responsible for constraining their populations. Insect parasites are more specific in their selection of hosts than are predators, but most forest Lepidoptera are also attacked by a wide assortment of hymenopteran and dipteran parasites that exert various degrees of population control (Barbosa, 1977; Carolin and Coulter, 1959, 1971; Dean and Ricklefs, 1980; Torgersen, 1981).

A preponderance of evidence supports the view that natural enemies are a principal force in keeping populations of forest Lepidoptera at low densities. It would be a mistake, however, to expect such natural control to be some simple function of predator–parasite density. Relations between herbivores and their natural enemies are often complex and clouded with high variances and nonlinearities (Strong, 1983). Increased competition at the parasite or predator trophic level may even have a destabilizing effect that results in less control of the prey populations (Dean and Ricklefs, 1980; Strong *et al.,* 1984; Zwolfer, 1963; Watt, 1965). Commonness or rarity of a species is the result of a variety of interacting factors (many of which are discussed in other chapters in this volume) that affect survival and reproduction. It is the combination of these sometimes incredibly complex interactions that ultimately produce the assemblages of species observed in lepidopteran communities.

IV. CONCLUDING REMARKS

From a practical standpoint the important question is how nonoutbreak species in the community affect populations of the major outbreak species. Studies in applied forest entomology have too often dealt with outbreak defoliators in a kind of vacuum without any consideration of other species at the same trophic level. The evolution of herbivore communities has been largely ignored in studies of coniferous forests, yet it may be critical for learning how to manage the problem species. The evidence reviewed in this chapter suggests that for each outbreak species there are at least nine lepidopteran defoliators that have never reached outbreak numbers in mixed fir forests. Little is known about the life history of most of these species, but owing to their equilibrium status in the community they appear to share many of the traits of typical *K* strategists. Although there is still much speculation on the subject, these equilibrium species probably contribute most to the stability of potential outbreak species and to the community at large by occupying niche space and providing alternate hosts for natural enemies.

Conventional wisdom holds and some empirical evidence supports the view that outbreak species of forest defoliators are most stable in communities that are relatively rich in other Lepidoptera (Munroe, 1979; Voute, 1946; 1964; Watt, 1965). To a great extent lepidopteran richness probably reflects overall species diversity, so that the real issue is how stability is related to diversity, a complex question that has been debated for decades without the emergence of a simple answer (Elton, 1958; Graham, 1939, 1956; May, 1973; Pimentel, 1961).

Many interior fir forests apparently support herbivore communities that have evolved to a relatively few coexisting species of defoliating Lepidoptera. These species are maintained most of the time at numbers far below the density at which larval food becomes inadequate. One explanation for their sparseness is that the site-carrying capacity for equilibrium species may actually be much lower than is first implied by the apparent abundance of foliage. Equilibrium species may be as much limited by their own specialized life history and feeding tolerances as they are constrained by natural enemies and weather. All host foliage is not equally available for food because of differences in chemical quality (Janzen, 1979). As a result many fir forests must have a large amount of unused niche space that can be occupied only by the species with broader feeding habits. When conditions favor high generation survival such as might occur when predation is temporarily relaxed, the generalist or opportunist species with their potential for a high *r* become the best candidates for colonizing the vacant space. Communities that are already relatively rich with permanent species have less unused space and probably offer more resistance to a buildup of the outbreak species. Perry and Pitman (1983), for example, postulate that the ecosystem of coastal Douglas fir var. *menziesii* maintains populations of the western spruce budworm at low densities because of their species-rich communities, despite a lack of chemical defense in the foliage. This supports the general hypothesis that biological interactions including predation and competition for niche space are likely to be most intense under relatively mild conditions (Connell, 1975) such as those found in the mesic environment of coastal forests. In the harsher climate of interior forests, where the number of interacting species is lower, the physical factors of weather and foliage chemistry may have a stronger role in complementing the control by natural enemies.

Perhaps the most important contribution of nonoutbreak Lepidoptera is providing an alternative food base for insect parasites and vertebrate and invertebrate predators (Elton, 1958; Munroe, 1979; Volker, 1978). The relative permanence of equilibrium species on trees and other host plants in the habitat undoubtedly helps maintain a consistent and diverse pool of natural enemies that is critical to checking population increases of outbreak species. The paucity of equilibrium Lepidoptera and associated predatory fauna in harsh environments like upper slopes and dry ridgetops in the West might explain the particular instability of populations of outbreak species on these sites.

Many of the concepts discussed in this chapter cannot easily be subjected to rigorous testing, but there are gaps in our knowledge that can be improved. For example, in preparing this chapter I found that there is a surprising lack of basic biological information on the nonoutbreak defoliators, and virtually nothing is known about their population dynamics. Whether these species are highly organized or just random assemblages of individuals, they are inevitably interrelated in the guild of lepidopterous defoliators. If we are ever to answer the difficult questions about population and community stability, much more attention will have to be given in the future to studying the uncommon along with the common species.

ACKNOWLEDGMENTS

I thank my colleagues H. G. Paul, T. R. Torgersen, and B. E. Wickman for their valuable assistance in collecting the 1978 field data and V. M. Carolin, D. A. Perry, and B. E. Wickman for their reviews of an earlier draft.

REFERENCES

Andrewartha, H. G., and Birch, L. C. (1954). "The Distribution and Abundance of Animals." Univ. of Chicago Press, Chicago, Illinois.

Baker, R. R. (1983). Insect territoriality. *Annu. Rev. Entomol.* **28,** 65–89.

Barbosa, P. (1977). *r* and *K* strategies in some larval and pupal parasitoids of the gypsy moth. *Oecologia* **29,** 311–327.

Blitz, W. E., and Ross, D. A. (1958). Population trends of some common loopers (Geometridae) on Douglas-fir, 1949–1956, in the Okanagan–Shuswap area. *For. Biol. Div. Ottawa, Bi-Mon. Prog. Rep.* **14,** 2–3.

Campbell, R. W., and Torgersen, T. R. (1982). Some effects of predaceous ants on western spruce budworm pupae in north central Washington. *Environ. Entomol.* **11,** 111–114.

Campbell, R. W., Torgersen, T. R., and Srivastava, N. (1983). A suggested role for predaceous birds and ants in the population dynamics of the western spruce budworm. *For. Sci.* **29,** 779–790.

Carolin, V. M. (1980). Larval densities and trends of insect species associated with spruce budworms in buds and shoots in Oregon and Washington. *U.S., For. Serv., Res. Pap. PNW* **273,** 1–18.

Carolin, V. M., and Coulter, W. K. (1959). The occurrence of insect parasites of *Choristoneura fumiferana* (Clem.) in Oregon. *J. Econ. Entomol.* **52,** 550–555.

Carolin, V. M., and Coulter, W. K. (1971). Trends of western spruce budworm and associated insects in Pacific Northwest forests sprayed with DDT. *J. Econ. Entomol.* **64,** 291–297.

Carolin, V. M., and Stevens, R. E. (1981). Key to large Lepidoptera larvae on new foliage of Douglas-fir and true firs. *U.S., For. Serv., Rocky Mt. For. Range Exp. Stn.* Res. Note **RM-401,** 1–4.

Cohen, J. E. (1968). Alternate derivation of a species-abundance relation. *Am. Nat.* **102,** 165–172.

Connell, J. H. (1975). Some mechanisms producing structure in natural communities: A model and evidence from field experiments. *In* "Ecology and Evolution of Communities" (M. L. Cody and J. M. Diamond, eds.), pp. 460–490. Harvard Univ. Press, Cambridge, Massachusetts.

Connell, J. H. (1983). On the prevalence and relative importance of interspecific competition: Evidence from field experiments. *Am. Nat.* **122,** 661–696.

Conway, G. (1976). Man versus pests. *In* "Theoretical Ecology: Principles and Applications" (R. M. May, ed.), pp. 257–307. Saunders, Philadelphia, Pennsylvania.

Dahlsten, D. L., Luck, R. F., Schlinger, E. I., Wenz, J. M., and Copper, W. A. (1977). Parasitoids and predators of the Douglas-fir tussock moth, *Orgyia pseudotsugata* (Lepidoptera: Lymantriidae), in low to moderate populations in central California. *Can. Entomol.* **109,** 727–746.

Dahlsten, D. L., Schlinger, E. I., Luck, R. F., and Williams, C. B. (1978). "Investigation of Endemic *Orgyia pseudotsugata* Populations with Emphasis on the Parasitoids, Predators, and Associated Pest Complex on White Fir, *Abies concolor,* in California," Final Prog. Rep. USDA/DFTM R&D Program. University of California, Berkeley.

Dean, J., and Ricklefs, R. E. (1980). Population size, variability, and aggregation among forest Lepidoptera in southern Ontario. *Can. J. Zool.* **58,** 394–399.

Debach, P. (1966). The competitive displacement and coexistence principles. *Annu. Rev. Entomol.* **11,** 183–212.

Dethier, V. G., and MacArthur, R. H. (1964). A field's capacity to support a butterfly population. *Nature (London)* **201,** 728–729.

DeVita, J. (1979). Niche separation and the broken-stick model. *Am. Nat.* **114,** 171–178.

Elton, C. (1958). "The Ecology of Invasions by Animals and Plants." Methuen, London.

Fellin, D. G., and Dewey, J. E. (1982). Western spruce budworm. *USDA For. Serv. Insect Dis. Leafl.* **53,** 1–10 (rev.).

Fox, L. R., and Morrow, P. A. (1981). Specialization: Species property or local phenomenon. *Science* **211,** 887–893.

Furniss, R. L., and Carolin, V. M. (1977). Western forest insects. *Misc. Publ.—U.S., Dep. Agric.* **1339,** 1–654.

Futuyma, D. J., and Gould, F. (1979). Associations of plants and insects in a deciduous forest. *Ecol. Monogr.* **49,** 33–50.

Gilbert, L. E. (1979). Development of theory in the analysis of insect–plant interactions. *In* "Analysis of Ecological Systems" (D. J. Horn and R. D. Mitchell, eds.), pp. 117–154. Ohio State Univ. Press, Columbus.

Graham, S. A. (1939). Forest insect populations. *Ecol. Monogr.* **9,** 301–310.

Graham, S. A. (1956). Forest insects and the law of natural compensations. *Can. Entomol.* **88,** 45–55.

Hairston, N. G. (1959). Species abundance and community organization. *Ecology* **40,** 404–416.

Hairston, N. G. (1969). On the relative abundance of species. *Ecology* **50,** 1091–1094.

Hairston, N. G., Smith, F. E., and Slobodkin, L. B. (1960). Community structure, population control, and competition. *Am. Nat.* **94,** 421–425.

Hardin, G. (1960). The competitive exclusion principle. *Science* **131,** 1292–1297.

Harris, J. W. E., Dawson, A. F., and Brown, R. G. (1982). The western hemlock looper in British Columbia, 1911–1980. *Can. For. Serv., Pac. For. Res. Cent.* [*Rep.*] *BC-X* **BC-X-234,** 1–18.

Haukioja, E. (1980). On the role of plant defenses in the fluctuation of herbivore populations. *Oikos* **35,** 202–213.

Hebert, P. D. N., Ward, P. S., and Harmsen, R. (1974). Diffuse competition in Lepidoptera. *Nature (London)* **252,** 389–391.

Herman, S. G., and Bulger, J. B. (1979). Effects of a forest application of DDT on nontarget organisms. *Wild. Monogr.* **69,** 1–62.

Hutchinson, G. E. (1978). "An Introduction to Population Ecology." Yale Univ. Press, New Haven, Connecticut.

Ito, Y. (1980). "Comparative Ecology" (J. Kikkawa, ed. and transl.) Cambridge Univ. Press, London and New York.

Janzen, D. H. (1979). New horizons in the biology of plant defenses. *In* "Herbivores: Their Interaction with Secondary Plant Metabolites" (G. A. Rosenthal and D. H. Janzen, eds.), pp. 331–350. Academic Press, New York.

Johnson, P. C., and Denton, R. E. (1975). Outbreaks of the western spruce budworm in the American Northern Rocky Mountain Area from 1922 through 1971. *Gen. Tech. Rep. INT—U.S., For. Serv.* [*Wash. Off.*] **INT-20,** 1–144.

Karban, R., and Ricklefs, R. E. (1983). Host characteristics, sampling intensity, and species richness of Lepidoptera larvae on broad-leaved trees in southern Ontario. *Ecology* **64,** 636–641.

King, C. E. (1964). Relative abundance of species and MacArthur's model. *Ecology* **45,** 716–727.

Klein, W. H., and Minnoch, M. W. (1971). On the occurrence and biology of *Nepytia freemani* (Lepidoptera: Geometridae) in Utah. *Can. Entomol.* **103,** 119–124.

Klomp, H. (1968). A seventeen-year study of the abundance of the pine looper, *Bupalus piniarius* L. (Lepidoptera: Geometridae). *Symp. Entomol. Soc. London* **4,** 98–109.

Labeyrie, V. (1978). The significance of the environment in the control of insect fecundity. *Annu. Rev. Entomol.* **23,** 69–89.

Lawton, J. H. (1982). Vacant niches and unsaturated communities: A comparison of bracken herbivores at sites on two continents. *J. Anim. Ecol.* **51,** 573–595.

Lawton, J. H., and McNeill, S. (1979). Between the devil and the deep blue sea: On the problem of being a herbivore. *In* "Population Dynamics" (R. M. Anderson, B. D. Turner, and L. R. Taylor, eds.), pp. 223–244. Blackwell, Oxford.

Lawton, J. H., and Strong, D. R., Jr. (1981). Community patterns and competition in folivorous insects. *Am. Nat.* **118,** 317–338.

MacArthur, R. H. (1957). On the relative abundance of bird species. *Proc. Natl. Acad. Sci. U.S.A.* **45,** 293–295.

MacArthur, R. H. (1960). On the relative abundance of species. *Am. Nat.* **94,** 25–36.

MacArthur, R. H. (1966). Note on Mrs. Pielou's comments. *Ecology* **47,** 1074.

MacArthur, R. H., and Wilson, E. O. (1967). "The Theory of Island Biogeography." Princeton Univ. Press, Princeton, New Jersey.

McGugan, B. M. (1958). Forest Lepidoptera of Canada recorded by the Forest Insect Survey. Papilionidae, Pieridae, Nymphalidae, Lycaenidae, Hesperiidae, Sphingidae, Saturnidae, Citheroniidae, Nolidae, Arctiidae. Vol. 1. *Publ.—Can. Dep. Agric.* **1034,** 1–76.

McLeod, J. M., and Blais, J. R. (1961). Defoliating insects on field spruce in Quebec. *For. Entomol. Pathol. Branch, Ottawa, Bi-Mon. Prog. Rep.* **17,** 2.

McNaughton, S. J., and Wolf, L. L. (1970). Dominance and the niche in ecological systems. *Science* **167,** 131–139.

Markin, G. P. (1982). Abundance and life cycles of Lepidoptera associated with an outbreak of the western spruce budworm, *Choristoneura occidentalis* (Lepidoptera: Tortricidae), in southeastern Idaho. *J. Kans. Entomol. Soc.* **55,** 365–372.

Mason, R. R. (1977). Advances in understanding population dynamics of the Douglas-fir tussock moth. *Bull. Entomol. Soc. Am.* **23,** 168–171.

Mason, R. R. (1981). Host foliage in the susceptibility of forest sites in central California to outbreaks of the Douglas-fir tussock moth, *Orgyia pseudotsugata,* (Lepidoptera: Lymantriidae). *Can. Entomol.* **113,** 325–332.

Mason, R. R., and Torgersen, T. R. (1983). Mortality of larvae in stocked cohorts of the Douglas fir tussock moth, *Orgyia pseudotsugata* (Lepidoptera: Lymantriidae). *Can. Entomol.* **115,** 1119–1127.

Mason, R. R., Torgersen, T. R., Wickman, B. E., and Paul, H. G. (1983). Natural regulation of a Douglas-fir tussock (Lepidoptera: Lymantriidae) population in the Sierra Nevada. *Environ. Entomol.* **12,** 587–594.

May, R. M. (1973). "Stability and Complexity in Model Ecosystems." Princeton Univ. Press, Princeton, New Jersey.

May, R. M. (1974). Biological populations with nonoverlapping generations: Stable points, stable cycles, and chaos. *Science* **186,** 645–647.

May, R. M. (1975a). Biological populations obeying difference equations: Stable points, stable cycles, and chaos. *J. Theor. Biol.* **51,** 511–524.

May, R. M. (1975b). Successional patterns and indices of diversity. *Nature (London)* **258,** 285–286.

May, R. M. (1975c). Patterns of species abundance and diversity. *In* "Ecology and Evolution of Communities" (M. L. Cody and J. M. Diamond, eds.), pp. 81–120. Harvard Univ. Press, Cambridge, Massachusetts.

May, R. M. (1978). The evolution of ecological systems. *Sci. Am.* **239,** 160–175.

Miller, C. A. (1966). The black-headed budworm in eastern Canada. *Can. Entomol.* **98,** 592–613.

Munroe, E. (1979). Lepidoptera. *In* "Canada and Its Insect Fauna" (H. V. Danks, ed.), Entomol. Soc. Can. Mem. No. 108, pp. 427–481.

Perry, D. A., and Pitman, G. B. (1983). Genetic and environmental influences in host resistance to herbivory: Douglas-fir and the western spruce budworm. *Z. Angew. Entomol.* **96,** 217–228.

Pianka, E. R. (1970). On *r*- and *K*-selection. *Am. Nat.* **104,** 592–597.

Pielou, E. C. (1969). "An Introduction to Mathematical Ecology." Wiley (Interscience), New York.

Pimentel, D. (1961). Species diversity and insect population outbreaks. *Ann. Entomol. Soc. Am.* **54,** 76–86.

Powell, J. A. (1980). Nomenclature of neartic conifer-feeding *Choristoneura* (Lepidoptera: Torticidae): Historical review and present status. *USDA For. Serv. Gen. Tech. Rep. PNW* **PNW-100,** 1–18.

Prentice, R. M. (1962). Forest Lepidoptera of Canada recorded by the Forest Insect Survey. Nycteolidae, Noctuidae, Notodontidae, Liparidae. Vol. 2. *Can. Dep. For. Bull.* **128.**

Prentice, R. M. (1963). Forest Lepidoptera of Canada recorded by the Forest Insect Survey. Lasiocampidae, Thyatiridae, Drepanidae, Geometridae. Vol. 3. *Can., For. Branch, Dep. Publ.* **1013.**

Prentice, R. M. (1966). Forest Lepidoptera of Canada recorded by the Forest Insect Survey. Microlepidoptera. Vol. 4. *Can., For. Branch, Dep. Publ.* **1142,** 77–840.

Preston, F. W. (1948). The commoness and rarity of species. *Ecology* **29,** 254–283.

Price, P. W. (1971). Toward a holistic approach to insect population studies. *Ann. Entomol. Soc. Am.* **64,** 1399–1406.

Price, P. W. (1974). Strategies for egg production. *Evolution* **28,** 76–84.

Price, P. W., Bouton, C. E., Gross, P., McPheron, B. A., Thompson, J. N., and Weis, A. E. (1980). Interactions among three trophic levels: Influence of plants on interactions between insect herbivores and natural enemies. *Annu. Rev. Ecol. Syst.* **11,** 41–65.

Rhoades, D. F. (1979). Evolution of plant chemical defense against herbivores. *In* "Herbivores: Their Interaction with Secondary Plant Metabolites" (G. A. Rosenthal and D. H. Janzen, eds.), pp. 3–54. Academic Press, New York.

Roughgarden, J. (1983). Competition and theory in community ecology. *Am. Nat.* **122,** 583–601.

Schoener, T. W. (1982). The controversy over interspecific competition. *Am. Sci.* **70,** 586–595.

Schultz, J. C., and Baldwin, I. T. (1982). Oak leaf quality declines in response to defoliation by gypsy moth larvae. *Science* **217,** 149–151.

Schultz, J. C., and Baldwin, I. T. (1983). Changes in tree quality in response to defoliation. *USDA For. Serv. Gen. Tech. Rep. NE* **NE-85,** 83–86.

Schwerdtfeger, F. (1958). Is the density of animal populations regulated by mechanisms or by chance? *Proc. Int. Congr. Entomol. 10th, 1956,* Vol. 4, pp. 115–122.

Silver, G. T. (1962). The green-striped forest looper on Vancouver Island. *Proc. Entomol. Soc.* **59,** 29–32.

Simberloff, D. (1983). Competition theory, hypothesis testing, and other community ecological buzzwords. *Am. Nat.* **122,** 626–635.

Southwood, T. R. E. (1962). Migration of terrestrial arthropods in relation to habitat. *Biol. Rev. Cambridge Philos. Soc.* **37,** 171–214.

Southwood, T. R. E. (1977). Habitat, the templet for ecological strategies? *J. Anim. Ecol.* **46,** 337–365.

Southwood, T. R. E., May, R. M., Hassell, M. P., and Conway, G. R. (1974). Ecological strategies and population parameters. *Am. Nat.* **108,** 791–804.

Stevens, R. E., Carolin, V. M., and Stein, C. (1983). Lepidoptera associated with western spruce budworm in the Southwestern United States. *J. Lepid. Soc.* **37,** 129–139.

Stevens, R. E., Carolin, V. M., and Markin, G. P. (1984). Lepidoptera associated with western spruce budworm. *U.S. Dep. Agric., Agric. Handb.* **622,** 1–63.

Strong, D. R., Jr. (1983). Natural variability and the manifold mechanisms of ecological communities. *Am. Nat.* **122,** 636–660.

Strong, D. R., Jr., Lawton, J. H., and Southwood, R. (1984). "Insects on Plants: Community Patterns and Mechanisms." Harvard Univ. Press, Cambridge, Massachusetts.

Torgersen, T. R. (1981). Parasite records for the Douglas-fir tussock moth. *USDA For. Serv. Gen. Tech. Rep. PNW* **PNW-123,** 1–38.

Torgersen, T. R., and Campbell, R. W. (1982). Some effects of avian predators on the western spruce budworm in northcentral Washington. *Environ. Entomol.* **11,** 429–431.

Torgersen, T. R., Thomas, J. W., Mason, R. R., and Van Horn, D. (1984). Avian predators of Douglas-fir tussock moth, *Orgyia pseudotsugata* (McD), in southwestern Oregon. *Environ. Entomol.* **13,** 1018–1022.

Turner, F. B. (1961). The relative abundance of snake species. *Ecology,* **42,** 600–602.

Volker, K. C. (1978). Ecology of parasites and predators of the Douglas-fir tussock moth in the Pacific Northwest. Ph.D. Thesis, University of Idaho, Moscow.

Volney, W. J. A., Liebhold, A. M., and Waters, W. E. (1983). Effects of temperature, sex, and genetic background on coloration of *Choristoneura* spp. (Lepidoptera: Tortricidae) populations in south-central Oregon. *Can. Entomol.* **115,** 1583–1596.

Voute, A. D. (1946). Regulation of the density of the insect-populations in virgin-forests and cultivated woods. *Arch. Neerl. Zool.* **7,** 435–470.

Voute, A. D. (1964). Harmonious control of forest insects. *Int. Rev. For. Res.* **1,** 325–383.

Watt, K. E. F. (1960). The effect of population density on fecundity in insects. *Can. Entomol.* **92,** 674–695.

Watt, K. E. F. (1965). Community stability and the strategy of biological control. *Can. Entomol.* **97,** 887–895.

Webb, D. J. (1974). The statistics of relative abundance and diversity. *J. Theor. Biol.* **43,** 277–291.

White, T. C. R. (1978). The importance of a relative shortage of food in animal ecology. *Oecologia* **33,** 71–86.

Whittaker, R. H. (1975). "Communities and Ecosystems." Macmillan, New York.

Wickman, B. E., and Beckwith, R. C. (1978). Life history and habits. *U.S. Dep. Agric., Tech. Bull.* **1585,** 30–36.

Wickman, B. E., and Hunt, R. H. (1969). Biology of the phantom hemlock looper on Douglas-fir in California. *J. Econ. Entomol.* **62,** 1046–1050.

Williams, C. B. (1960). The range and pattern of insect abundance. *Am. Nat.* **94,** 137–151.

Williams, C. B., Jr., and Walton, G. S. (1968). Effects of Naled and Zectran on the budworm *Choristoneura occidentalis* and associated insects in Montana. *J. Econ. Entomol.* **61,** 784–787.

Zwolfer, H. (1963). The structure of the parasite complexes of some Lepidoptera. *Z. Angew. Entomol.* **51,** 346–357.

Chapter 3

What Is a Forest Pest?

PHILIP J. NOTHNAGLE

Windsor, Vermont 05089

JACK C. SCHULTZ

Department of Entomology
Pesticide Research Laboratory
Pennsylvania State University
University Park, Pennsylvania 16802

I. IS BEING A PEST A "TACTIC"?

In most communities, some species are generally common while others are usually rare. Although fluctuations in density must occur in all populations from generation to generation, some species exhibit changes in density that are extreme compared with those of related species, even in the same habitat.

Most North American tree-feeding, defoliating Lepidoptera species exhibit narrow density fluctuations (Watt, 1965). However, a few of these species exhibit strongly bimodal population trends; in any one year or place, they may be either so rare as to be undetectable or so abundant as to defoliate their hosts completely (Campbell and Sloan, 1978). Fluctuations of five to six orders of magnitude over 3 to 5 years can be observed, although relatively high densities are sometimes maintained over a period of several years (Campbell, 1981). Because of the aesthetic or economic damage done by these high-density populations, such forest defoliators are usually thought of as pests. Because of their characteristically rapid increase and decline, they are referred to as "irruptive" or "cyclic" pests, depending on the temporal regularity of their irruptive peaks, or "outbreaks."

Several authors have suggested that the ability to increase rapidly enough to achieve "outbreak" status represents an evolutionary "tactic" or "strategy" involving a consistent suite of biological, ecological and/or life history traits (see Chapter 20, this volume). For example, Southwood *et al.* (1974) suggested that insect pests might exhibit the "*r*-selected" characteristics proposed by MacArthur and Wilson (1967). These "*r* strategists" would have high reproductive capacities, short life spans, relatively low competitive abilities, and weakly developed defensive characteristics. According to this view, forest pest species might be expected to have higher fecundities than would benign relatives; they might also develop relatively rapidly, feed on high-quality food, and be unlikely to employ chemical antipredator devices (see Chapter 2).

Schneider (1980) suggested that several forest defoliators possess traits that make variable food resources (tree leaves) more "apparent," or easier to locate and exploit predictably. In this view, a suite of traits possessed by certain Lepidoptera species are seen as adaptations for exploiting a highly variable, unpredictable resource. Presumably, pest species are occasionally more effective than other species at exploiting these resources, and at these times outbreaks occur. A wide variety of alternative viewpoints, ranging from totally stochastic to nearly deterministic, can be described (see Chapters 1, 2, and 20).

Unfortunately, current knowledge of the biology and natural history of forest Lepidoptera is too incomplete for us to assess the validity of these assertions. Although much is known about some irruptive pest species, the majority of the nonirruptive species remain virtually uncharacterized in any meaningful way. Hence, comparing the biological traits of pest and nonpest species is very difficult. Nonetheless, if the ability to undergo population outbreaks derives from having certain biological traits, we should see convergence or parallelism among outbreak species. The occurrence of these common traits in several independent lineages (convergence) would be the most powerful evidence that it is these traits that permit a species to become a "pest." However, even if the potentially important traits were phylogenetically constrained (Gould and Lewontin, 1979; Stearns, 1980) and pest species occurred only within a restricted lineage, we could still infer their importance (Stearns, 1980). Some taxonomic groups may exhibit outbreaks while others do not, but we can still ask what is different about the pest lineages as a means of gathering clues about underlying causal mechanisms.

This chapter presents an attempt to identify life history and other ecologically relevant traits held in common by defoliating Macrolepidoptera that exhibit irruptive population dynamics in North America. Our intent is to identify any consistent patterns that link these pest taxa and to infer their possible functional significance in causing or permitting outbreaks. Comparisons with nonoutbreak taxa are speculative at this time, but some interesting contrasts do exist, and we discuss these. Our goal is to answer a basic question about defoliating caterpillars: What is a pest?

II. EXAMINING THE FOREST PEST LITERATURE

The USDA reports annually on forest insect and disease conditions [U.S. Department of Agriculture (USDA), 1962–1981]. These data are gathered from a large number of observers in several agencies and represent a broad view of insect attack on forested systems that is relatively unbiased by the economic significance of particular forest types.

We defined an "outbreak pest" species as one that was identified in these records as having caused noticeable defoliation during at least two of the years surveyed. A literature search for biological attributes (Section III) of these species was carried out and was supplemented by inquiries directed to researchers currently working on species on our pest list (see Appendix A); many basic observations have not been published, even for common, serious pests. The paucity of published information about many Microlepidoptera forced us to restrict our analyses to Macrolepidoptera.

Our data set (Table 1) includes information about life histories, whether or not the pest is native to North America, feeding behavior, diet, apparent antipredator defenses of larvae, adult fecundity, egg deposition pattern, and degree to which adult females disperse by flight. Host plants were designated as primarily coniferous or deciduous, and we were able to estimate diet breadth as the number of plant species comprising 95% of a species' diet (of observations) by adding the Canadian Forest Insect Survey data (Prentice, 1962, 1963) to our other sources. Various sources tended to be quite consistent with one another. A weighted estimate of host plant species shade tolerance was obtained from a tolerance table given by Baker (1950). This measure was taken as an indicator of host plant successional status and other potentially significant ecological traits.

For statistical treatment, discontinuous variables (e.g., overwintering stage) were assigned integer values, and multivariate methods were applied as described by Holmes *et al.* (1979a), following the general procedures given in Cooley and Lohnes (1971). We attempted to assign integer values to discontinuous variables in some sort of biologically or physi-

TABLE 1

Biological Traits Collated from the Literature for Each "Outbreak Pest" Species and Trait State Scoring for MVA

1. Native to North America? (1, yes; 2, no)
2. Number of eggs per female; average fecundity (continuous)
3. Mean egg group size; number of eggs per group (continuous)
4. Overwintering stage (1, egg; 2, larva; 3, pupa; 4, adult; 5, mixed)
5. Feeding type (1, solitary; 2, gregarious; 3, tent making)
6. Generations per year (continuous)
7. Female dispersal ability (1, apterous; 2, poor flier; 3, flies; 4, strong flier)
8. Diet breadth; number of host plant species in 95% of feeding observations (continuous)
9. Phenological age of preferred tissue (1, early; 2, midseason; 3, early + mid; 4, late)
10. Shade tolerance of dominant hosts (1, very intolerant; 2, intolerant; 3, intermediate; 4, tolerant)
11. Coniferous or deciduous hosts (1, coniferous; 2, deciduous; 3, both)
12. Larval antipredator adaptations (1, matches background; 2, specific substrate mimic; 3, conspicuous defense—colors, hairs, spines, etc.)

cally meaningful way. For example, life history stages were given numbers that increase in life stage order (i.e., egg, 1; larva, 2; etc.). Continuous variables (e.g., fecundity) were entered as raw values, unless preliminary analyses indicated depature from parametric criteria (e.g., skewness); in such cases data were transformed until parametric criteria could be met.

III. WHAT THE LITERATURE TELLS US ABOUT FOREST PESTS

Forty-one macrolepidopteran species met our "pest" criteria and were mentioned in the USDA reports during two or more years (Table 2; Appendix B). Of these, four defoliated primarily range shrubs, even though they sometimes defoliated trees at very high densities; these were excluded from further consideration. Of the remaining 37 species, adequate data could be acquired for only 29. However, five of the eight species for which data were inadequate have a close congener among the 29 with complete data, and the remaining three are lymantriids having life histories very similar to lymantriids occurring in the complete data set (Ferguson, 1978). Hence, we can infer some similarities among all taxa. The 37 species exhibiting outbreaks represent well under 2% of tree-feeding Macrolepidoptera species in America north of Mexico (J. E. Rawlins, unpublished data).

Biological characteristics of the forest pest species are listed in Table 2. The outbreak species are primarily native and univoltine: Only two lymantriids, *Lymantria dispar* and *Leucoma salicis,* are introduced, and 24 species are univoltine. Three species are multivoltine over parts of their ranges; *Coloradia pandora* (Saturniidae) has a 2-year life cycle; and *Heterocampa manteo* (Notodontidae) has been reported to exhibit a variable diapause and a two-year life cycle on occasion. The univoltine species appear to have only one generation over considerable climatic ranges (e.g., elm spanworm, forest tent caterpillar).

Average total fecundity of females varies by more than an order of magnitude among the pest species (Table 2); the range is from 58 to 880 eggs per female. Average fecundity is significantly correlated ($r = .58, p < .05$, Pearson product–moment correlation) with average egg group size, so species with high fecundities tend to deposit eggs in groups. Egg masses are common in these forest pest species; 18 of the species lay eggs in groups comprising 15% or more of their total fecundity, and 9 species lay all their eggs in one mass.

Fourteen species overwinter as eggs, and 11 as pupae. Three species overwinter as larvae, and the biennial *C. pandora* overwinters first as a larva, then as a pupa. None of the forest pest species overwinters as an adult, even though this is a common habit among tree feeders (Schweitzer, 1977; Forbes, 1948, 1954, 1960).

Larval characteristics can be summarized from Table 2. When gregarious and tent-making forms are considered together, there are about equal numbers of species that feed in groups or as solitary individuals. Larval species that generally match their background and those that are defended by hairs or spines or are brightly colored are equally likely to be solitary or gregarious. Those that exhibit patterns, shapes, or behaviors suggesting specific protective resemblance of host plant substrates are largely solitary.

Larval diet breadths vary from monophagy to broad polyphagy (Table 2). However, whereas 2 species are restricted to a single host plant species and 3 have 95% or more of their feeding observations on 15 or more species, the remaining 24 species are moderately polyphagous (2–14 plant species comprise 95% of records). Sixteen species feed primarily on newly flushed leaves of their hosts, and only one species clearly prefers late-season leaves (Table 3). Fifteen species feed on trees that are shade intolerant or very shade intolerant; This total increases to 22 if species feeding on trees intermediate in shade tolerance are included (Table 3). There is an evident tendency for these pest species to feed on younger leaf tissues and shade-intolerant tree species.

There is also an evident association between life history patterns and leaf age preference (Table 4). Species that overwinter as eggs feed almost

TABLE 2

Macrolepidoptera "Outbreak Pests"[a] Used in Analysis and Their Biological Attributes[b]

Date of first outbreak	Family, Species	Trait state												In CFIS[d]
		1	2	3	4	5	6	7	8	9	10	11	12	
	Pieridae													
1882	*Neophasia menapia* (Feld and Feld)	1	49	16	1	2	1	3	3	2	2	1	1	Yes
	Saturniidae													
1850s	*Dryocampa senatoria* (J. E. Smith)	1	250	250	3	2	1	3	4	2	3	2	3	Yes
1893	*Coloradia pandora* Blake	1	153	20	5	2	.5	4	3	3	2	1	3	No
	Lasiocampidae													
—	*Malacosoma americanum* (Fab.)	1	200	200	1	3	1	3	8	1	1	2	3	Yes
1929	*M. constrictum* Stretch	1	250	250	1	3	1	3	9	1	1	2	3	Yes
ca. 1646	*M. disstria* Hubner	1	170	170	1	1	1	4	10	1	1	2	3	Yes
—	*M. californicum* (Dyer)	1	172	172	1	3	1	3	12	1	1	2	3	Yes
	Lymantriidae													
1900	*Orgyia pseudotsugata* (McD.)	1	149	149	1	1	1	1	1	1	2	1	3	Yes
1870s	*Lymantria dispar* (L.)	2	750	750	1	1	1	2	60	1	3	3	3	No
1920	*Leucoma salicis* (L.)	2	650	650	2	1	1	3	6	3	1	2	3	Yes
	Notodontidae													
—	*Datana integerrima* G. & R.	1	880	120	3	2	2[c]	2	3	2	2	2	3	Yes
1900	*Heterocampa guttivitta* (Wlkr.)	1	500	1	3	1	1	2	14	2	3	2	2	Yes
—	*H. manteo* Doubleday	1	500	300	3	2	1[c]	3	7	2	4	2	3	Yes
1884	*Symmerista canicosta* Franc.	1	319	66	3	2	1	3	4	2	3	2	2	Yes

	Dioptidae													
—	*Phryganidia californica* Pack.	1	225	60	2	1	2	2	8	4	3	2	1	No
	Arctiidae													
1887	*Halisidota argentata* Pack.	1	300	100	2	2	1	4	6	3	3	1	3	Yes
ca. 1820	*Hyphantria cunea* (Drury)	1	600	600	3	3	2.5[c]	3	16	2	2	2	3	Yes
	Geometridae													
—	*Alsophila pometaria* (Harris)	1	112	112	1	1	1	1	14	1	3	2	1	Yes
ca. 1830	*Ennomos subsignarius* (Hbn.)	1	250	70	1	2	1	3	14	1	4	2	1	Yes
ca. 1900	*Hydria prunivorata* (Ferguson)	1	450	154	3	3	1	3	3	2	2	2	1	Yes
1912	*Lambdina fiscellaria* (Guenee)	1	58	1	1	1	1	2	8	1	4	3	2	Yes
—	*Erannis tiliaria* Harris	1	300	6	1	1	1	1	18	1	3	2	2	Yes
1940s	*Nepytia freemani* Monroe	1	67	3	1	1	1	3	7	1	4	1	2	Yes
—	*N. phantasmaria* (Strecker)	1	69	1	1	1	1	3	4	1	4	1	2	Yes
ca. 1900	*Operophtera bruceata* (Hulst)	1	120	1	1	1	1	1	5	1	2	2	2	Yes
	Semiothisa sexmaculata Pack.	1	140	1	3	1	1	3	2	1	1	1	2	Yes
ca. 1800	*Paleacrita vernata* (Peck)	1	250	10	3	1	1	1	10	1	3	2	2	Yes
1969	*Phaeoura mexicanaria* (Grote)	1	345	160	3	1	1	3	1	2	2	1	2	No
1941	*Rheumaptera hastata* (L.)	1	70	1	3	2	1	3	8	1	2	2	2	No

[a] See text for criteria.
[b] Trait numbers and states as in Table 1; see Appendix A for references.
[c] Variable number of generations per year.
[d] Canadian Forest Insect Survey.

TABLE 3

Number of Pest Species in Host Tree Shade Tolerance–Tissue Age Classes

		Tissue age			
Shade tolerance	Newly flushed	Mid-season	Both	Late season	Row total
Very intolerant	6	0	1	0	7
Intolerant	2	5	1	0	8
Intermediate	4	3	0	0	7
Tolerant	2	1	1	1	5
Very tolerant	2	0	0	0	2
Column total	16	9	3	1	29

exclusively on newly flushed leaf tissue, whereas those overwintering as pupae prefer midseason leaves; one species preferring late-season leaves overwinters as a larva and is likely to complete development on young leaves (J. C. Schultz, unpublished observations). The species feeding on early leaves are often those in which the female is apterous (Table 5). Species whose females are reported as "flier" or "strong flier" are about evenly divided between early-season and midseason foliage preferences or exhibit mixed preferences. Of the 10 species in which the female is a poor flier or does not fly, 7 prefer new foliage. Although it is difficult to compile the necessary information for nonpest species, it would appear that flightlessness and/or weak flight is more common (more than one-third of the species) in this pest group than among tree-feeding North American Macrolepidoptera at large (Forbes, 1948, 1954, 1960; Schneider, 1980).

TABLE 4

Number of Pest Species in Tissue Age–Overwintering Stage Classes

	Overwintering stage					
Tissue age	Eggs	Larvae	Pupae	Adults	Mixed	Row total
Newly flushed	13	0	3	0	0	16
Midseason	1	0	8	0	0	9
Mixed (early + mid)	0	2	0	0	1	3
Late	0	1	0	0	0	1
Column total	14	3	11	0	1	29

TABLE 5

Number of Pest Species in Foliage Age–Female Dispersal Classes

Foliage age	Female dispersal characteristics				
	Apterous	"Poor fliers"	"Fliers"	"Strong fliers"	Row total
New	5	2	8	1	16
Midseason	0	2	7	0	9
Mixed	0	0	1	2	3
Late	0	1	0	0	1
Column total	5	5	16	3	29

A multivariate factor analysis organizes the more significant correlations among the traits measured within this group of species (Table 6). Eggs per female, eggs per group, number of host species in 95% of diet, and tendency to feed on deciduous trees all load most heavily on factor 1, indicating statistical association among them. Females producing many

TABLE 6

Rotated Factor Pattern from Factor Analysis of 29 Pest Species Using 11 Biological Traits[a]

Trait	Factor				
	1	2	3	4	5
Eggs per female	.68	—	—	—	—
Eggs per group $\bar{x}$	.86	—	—	—	—
Overwintering stage	—	.80	—	—	—
Feeding types	—	—	.85	—	—
Generations per year	—	—	—	.80	—
Female dispersal	—	—	.76	—	—
Host species	.86	—	—	—	—
Tissue age	—	.83	—	—	—
Host shade tolerance	—	—	—	—	.87
Conifer/deciduous	.67	—	—	—	—
Larval defenses	—	—	—	—	.66
Percentage of trace	26.4	22.3	14.0	8.8	7.7
Cumulative percentage	26.4	48.7	62.8	71.7	85.6

[a] Trait state values were converted to z scores to normalize variances. Factor pattern is based on correlation matrix. Only maximum loadings are presented for each trait; 85.6% total variance in correlation matrix is accounted for.

eggs tend to lay them in batches, their larvae tend to have broad host ranges, and they feed mainly on deciduous trees.

Heaviest loadings on factor 2 indicate an association between overwintering stage and age of preferred tissue, as already described. Feeding types (gregarious, tent making, solitary) and the female's flight ability are associated by factor 3; flightless or weak-flying females produce gregarious larvae. Voltinism seems relatively independent of other traits in this analysis (factor 4). Factor 5 indicates that most larvae exhibiting specific antipredator adaptations such as hairiness, spines, or bright coloration feed on early-season to midseason leaves.

IV. WHAT IS A FOREST PEST?

A. Three General Patterns

Our picture of species exhibiting irruptive population dynamics is not consistent with an *r*-selection–*K*-selection model. This model would lead us to expect *r*-selected outbreak species to (1) have greater fecundities, (2) have larger body sizes, (3) be multivoltine, (4) exhibit broader host preferences (greater diet breadth), and (5) invest more in defense against natural enemies than do presumably *K*-selected nonoutbreak species (MacArthur and Wilson, 1967; Pianka, 1970; Southwood *et al.*, 1974; Chapters 2 and 20, this volume). Quantitative comparison is handicapped by a lack of data for nonpest species. However, outbreak species exhibit fecundities that span almost the entire range known for Lepidoptera. For example, there are five pest species with fecundities under 75 eggs per female, and four species with more than 600 per female. This characteristic does not segregate *r*-selected pest species from *K*-selected nonpests.

We do not have quantitative information on the body sizes of forest pest Lepidoptera in North America. There are two saturniid species in our list; members of this family tend to be large. The two pest saturniids are among the smaller moths in their family, contrary to *r*–*K* selection predictions. Since body size and fecundity tend to be positively correlated in Lepidoptera (Haukioja and Neuvonen, 1985), it is not surprising that our pest species—which do not, as a group, exhibit particularly great fecundities—do not have unusually large bodies.

At the beginning of the chapter, we included Microlepidoptera in our survey of pests, before excluding them for lack of data. The number of microlepidopteran species meeting our forest pest criteria was not significantly different from the number of Macrolepidoptera, despite what is almost certainly a longer North American species list for the former

(Hodges *et al.*, 1983). Similarly, the expectation that "opportunistic" *r*-selected pests should have short generation times and tend to be multivoltine runs contrary to this lack of dominance by the typically multivoltine Microlepidoptera and a paucity of multivoltine species in our macrolepidopteran pest list.

The theory of *r–K* selection predicts broad diets for "opportunistic" species (Southwood *et al.*, 1974). There is a tendency for the species in our pest list to be polyphagous, although there are several monophagous species as well. The observation (e.g., Prentice, 1962, 1963; Chapter 2, this volume) that the host ranges and diet breadths of many nonirruptive lepidopteran species are identical to or broader than those of our pests suggest that diet breadth alone is not a good predictor of the ability to exhibit irruptive population dynamics.

The positive correlation between fecundity and defensive characteristics of larvae and eggs seen in our analysis also runs counter to *r–K* predictions. It is difficult to know the cost of producing spines, and so on, but the positive association we find between fecundity and the occurrence of presumed defenses makes it apparent that any costs are not reflected in fecundity. Moreover, since many egg masses are protected by spumaline or other secretions, and fecundity is correlated with producing eggs in masses, high-fecundity species also appear to invest in egg protection. Pest species do not gain fecundity at the expense of these attributes. Perhaps there is some gain in survivorship that balances these costs, or perhaps the "principle of resource allocation" (R. H. MacArthur and Levins, unpublished, in Cody, 1966) is merely a postulate (Rollo, 1986).

Overall, then, we find little support in our data set for the description of forest pest species as "opportunisitic" or "*r*-selected." The pest species surveyed here are either indistinguishable from other species in terms of important *r*- or *K*-selected traits or possess traits that conflict with theoretical predictions. Lepidoptera species we defined as forest pests do not seem distinctively *r*-selected in the ways predicted by the literature.

Instead, there are three suites of cocorrelated characteristics that occur among these species, mostly within but also between phylogenetic lineages. This suggests that there may be a limited number of ways in which to be a forest pest.

One important pattern appears to be related to host plant exploitation and the ability to deal with variation in environmental—especially food—quality. The most obvious piece of this picture involves coordination between life history stages and host plant phenology. The lymantriid and lasiocampid species share a suite of traits that may coordinate larval activity with peak food quality. They overwinter as eggs in a large protected mass, the eggs hatch early in the spring, and the larvae are gregarious. As suggested by Schweitzer (1977) and Schneider (1980), overwintering as an egg provides superior synchrony between egg hatch and budbreak. For species known to be sensitive to subsequent decline in leaf quality (e.g., gypsy moth, fall cankerworm), this timing can be critical to larval survivorship and growth.

Exploiting early-season foliage exposes larvae to the risk of hatching before food is available and to cold temperatures (Feeny, 1970; Wint, 1983). The ability to feed temporarily on less suitable foliage (i.e., broad diet; Wint, 1983), the ability to disperse as larvae (Wint, 1983; Lance and Barbosa, 1981), and behavioral thermoregulation or thermal insensitivity (Wellington, 1964; Knapp and Casey, 1986) are all shared by species in this group and apparently mitigate these problems. Although predation pressure in some forests may not be greatest early in the season (as migrants are arriving, before nesting has occurred; Holmes and Sturges, 1975; Holmes *et al.,* 1979b), the conspicuous, gregarious larval feeding habits of these species may have favored the evolution of the antipredator traits seen in this group. Predation risk should be especially high for species feeding into midseason (e.g., on continuously flushing trees) when nestlings' demand for food causes predation pressure to peak. Of course, all of these spring-feeding species achieve maximum size and hence potential value to predators at about the time of bird nesting and fledging.

The early-season group also prefers shade-intolerant, early-successional tree species have a long growing season and (frequently) continuously flushing growth habits (Table 3; Marks, 1975). Such tree species have early-flush or young leaves available for a comparatively long period (Marks, 1975), widening the "window" of availability to insects preferring them (Niemela, 1983). Insects sensitive to declining leaf quality would, in effect, have more resources available to them on these trees, and their life history traits would synchronize larval feeding and high food quality. This chance of missing a food-quality "window" would be reduced on shade-intolerant trees.

A second mixed group of families in our list (notodontids, arctiids, one dioptid) could be viewed as relatively insensitive to declining food quality. These species overwinter as larvae or pupae, lay eggs in masses, and feed on midseason to late-season leaves of middle- to late-successional tree species. The connection between overwintering form and feeding on older leaf material is clear in this group; they are not able to match host budbreak closely. All of the multivoltine species in our list fall into this group, reinforcing the impression that their larvae can tolerate seasonal variation in host plant quality. Some lepidopteran larvae that feed on late-season tree leaves exhibit food utilization efficiencies similar to those of species exploiting early leaves (Schroeder and Malmer, 1980; Lawson *et al.,* 1984).

The third major grouping in our pest sample, all geometrids, is biologically heterogeneous, exhibiting various combinations of characteristics seen in the other groups. Our geometrids feed primarily on early-season tissues, but on both early- and late- successional tree species. They are uniformly univoltine and undefended by spines, hairs, and so on. This combination of traits suggests that they may be sensitive to food quality and predation, restricting activity to early spring when leaf quality may be high and predation (at least from birds) low. Some species lay eggs in masses, some singly; several have very low fecundities (<70 eggs per female). Those species overwintering as eggs and feeding on spring tissues could be grouped with lymantriids and lasiocampids, but they do not have those families' high fecundities and prefer late-succesional, not early-successional, tree species.

The fact that clustering according to life history and host plant characteristics produces groups that are also organized along taxonomic lines suggests that there are phylogenetic constraints associated with being a forest pest. Within these constraints there is still some variation; few species in each family are pests, for example. Within some lineages, especially the geometrids, different suites can arise, each of which results in pest status (by our definition). We have identified several quite different suites of traits, some restricted to a lineage, others not, that are correlated with the capacity for irruptive population dynamics. These observations raise the question, Are there any useful generalizations about the tactics or syndromes involved in being an irruptive forest pest, or is each pest unique?

B. Risk Management by Outbreak Species

Den Boer related the numerical stability of a population to the number of factors that influence its survival and reproduction. Starting with a verbal model based on the premise that an industry producing only one article has a higher risk of bankruptcy than does a more diversified one (den Boer, 1968), Reddingius and den Boer (1970) later developed the idea more rigorously through computer simulation. Their analyses indicated that both the range of densities and variance in reproductive rates of populations decline as the number of randomly varying factors that could influence them increases. The approach of Reddingius and den Boer has been called the "spreading of risk" model, because individuals in populations influenced by multiple factors experience risk (e.g., of death) from multiple sources. In such populations, individuals are susceptible to risk from many sources, and variation in one of these factors can balance variation in others; the result would be little net effect of any single factor on population dynamics and relatively low, stable populations.

We propose that the fundamental difference between species or populations exhibiting irruptive dynamic behavior and those with more stable dynamics lies in the ways they manage risk. Because irruptive and stable populations often occupy the same habitats and microhabitats (Chapter 2) and are exposed to the same environmental variation, they must be differentially susceptible to risk. More specifically, the forest Lepidoptera species we have examined here appear to exhibit life history traits that concentrate risk in one or a few environmental factors to which they are exposed and susceptible. We believe that pest species are "risk concentrators" and nonpest species are "risk spreaders."

In our data set, the traits shared by the lasiocampids and lymantriids seem to minimize the risks of encountering or being influenced by poor food quality (matching the relatively broad early phenological window of early successional tree species), overwintering mortality (overwintering in dessication- or frost-resistant forms), and predation and parasitism on larvae, adults, and eggs (larval defenses, very brief adult period, eggs in protected masses). The remaining "risky" environmental variation that may influence population dynamics could be unpredictable fluctuation in host phenology and weather. These species match spring phenology better than most and may more often "hit" optimal food quality, but they must sometimes "miss." Because they are not adapted to late-season foliage traits, such a miss should have a strong impact on performance.

The impact of a phenological mismatch could be reduced growth and/or reproduction (e.g., Feeny, 1970; Hough and Pimentel, 1978; Wint, 1983), but there is little evidence that shifts in food quality alone could yield the dramatic (several orders of magnitude) changes in population density that define an outbreak. Such changes almost certainly have to come about via altered survivorship (and/or migration); we doubt that mortality caused by leaf quality alone would yield such dramatic results. For example, we calculate that increasing gypsy moth fecundity from the lower, naturally occurring level to the highest observed in nature or the laboratory (Campbell, 1981) between two seasons could not possibly yield the population density increase seen as outbreaks build. It is clear that survivorship must also change dramatically to yield these population dynamics.

In the case of spring-feeding pests, it seems much more likely that mortality risks (e.g., from weather) that necessarily attend hitting an early phenological "window" would be the driving variable in population dynamics: The important effect is being "between the devil and the deep blue sea" (Lawton and McNeil, 1979), not food quality per se. Temporal and spatial variation in food quality is probably important in augmenting these risks (Schultz, 1983a).

Alternatively, placing eggs in dense masses and feeding gregariously as larvae may subject these species to the increased risk of pathogen contagion that could make their population dynamics pathogen driven (May, 1985). Faeth (Chapter 6) points out that pathogen-driven models can generate irruptive dynamics because of their oversimplicity; they ignore too many other influences. We postulate that some subset of species in nature may have "oversimplified" the situation by overcoming (by adaptation?) most of these other influences, reducing the possible "regulatory" factors to pathogens alone.

Our mixed species assemblage of later-season pests appears to have resolved the problem of poor food quality and exhibits effective antipredator traits or life histories that reduce contact with vertebrate predators (birds). Eggs and larvae are not grouped in all these species, so pathogen contagion and impact may not often be important for some but could remain so for others. It seems unlikely that midseason to late-season variation in host plant phenology is influential for these species. However, these species spend long periods as adults (searching first for mates, then for oviposition sites) and overwintering as fungus-, cold-, and dessication-susceptible pupae or adults. We suspect that overwintering mortality plays a major role in driving population fluctuations of these species.

The geometrid pest species in our sample appear sensitive to host-quality variation and tend to aim at the early-season host leaves. They overwinter in several forms, mainly as eggs or young larvae; this improves the spring phenological match. These pest species feed before the major predator population buildup; their relatives that feed later are not pests. Risks to which they remain subject vary among the species and include overwintering conditions (for those that overwinter as larvae) and unpredictable spring phenology.

Small-scale variation in spring host quality (e.g., among trees) may be especially important to those species with flightless females and that overwinter as eggs. In the lymantriid–lasiocampid group, females choose hosts for oviposition the season before their offspring feed on those hosts, so their larvae must seek better food if their hatch site turns out to be unsuitable. Although there may be strong selection on these species for a close physiological match to host phenology (e.g., *Alsophila pometaria;* Schneider 1980; Mitter and Futuyma, 1977), missing budbreak in the host must have a devastating impact on the success of populations of species in which only the larva can disperse.

All three of the above groups of pest species subject themselves to considerable risk by putting a substantial portion of their eggs into one mass. Whereas Stamp (1980) found that 5–10% of butterfly species lay eggs in masses and Hebert (1983) identified 7% of moth species feeding on

Canadian trees that produce egg clusters of 10 or more, 68% of the species in our pest list do so. Any environmental variable influencing egg or larval success will do so in a concentrated way when offspring are massed. Massing eggs or clustering larvae may reduce the impact of parasitoids or predators, may reduce the impact of microhabitat variation, and may improve foraging efficiency among variable leaves (Schultz, 1983b). However, together with the loss of oviposition site selection by the female, massing offspring on individual hosts could make variable host quality (phenologically generated or otherwise) very influential. We do not find data to support the alternative assertion that loss of female flight is associated with increased fecundity, and the species in our list are not pests because of high fecundity.

Our survey suggests that forest pest species are those whose population dynamics can be influenced by only one or a few factors, as opposed to the many that probably influence most herbivorous insects. This may be fortuitous, or it may come about via adaptation to various mortality sources. We note that the literature suggests that many of the present-day forest pest species have become so only recently, since anthropogenic forest habitat modification has become widespread (Table 2). Perhaps forest habitats have become less variable or more suitable for some insect species. One persuasive example is the loss of the American chestnut and its replacement in most habitats by several oak species that are highly valued by the gypsy moth. Early studies demonstrated that the preference of the gypsy moth larva for chestnut leaves was very low (Mosher, 1915) compared with that of oaks and other species. It is possible that forest composition has shifted to one more favorable for this species since the chestnut blight epidemic (see Campbell, 1979); certainly the extent and intensity of gypsy moth outbreaks have increased since that time.

Whatever the cause, we believe that pest species are risk concentrators, not risk spreaders, that *K*-factor (Varley *et al*,. 1973) studies should reveal few important mortality sources for these species, and that one of the most influential environmental factors is likely to be variation in host plant quality for young larvae. Pest species have apparently been able to reduce many risks but are at the mercy of the few to which they remain susceptible; it is variation in these few factors that drives the irruptive population dynamics typical of pest species.

ACKNOWLEDGMENTS

We are especially grateful to those experts who were willing to be prodded for information and observations about their favorite pests. Discussions with R. T. Holmes, P. Barbosa,

M. Montgomery, and M. C. Rossiter contributed to idea development. This project was supported by NSF Grants DEB-8022174 (R. T. Holmes, J. C. Schultz) and BSR-840028 (JCS), U.S. Forest Service Coop. Agreement 23-974, and Pootatuck, Inc. (PJN). This is a contribution of the Hubbard Brook Ecosystem Study and the Pennsylvania State University Gypsy Moth Research Laboratory and is authorized as paper no. 7563 in the journal series of the Pennsylvania Agricultural Experimental Station.

REFERENCES

Baker, F. S. (1950). "Principles of Silviculture." McGraw-Hill, New York.

Campbell, R. W. (1979). Gypsy Moth: Forest influence. *Agric. Inf. Bull. (U.S., Dep. Agric.)* **423.**

Campbell, R. W. (1981). Evidence for high fecundity among certain North American gypsy moth populations. *Environ. Entomol.* **10,** 663–667.

Campbell, R. W., and Sloan, R. J. (1978). Numerical bimodality among North American gypsy moth populations. *Environ. Entomol.* **7,** 641–646.

Cody, M. L. (1966). A general theory of clutch size. *Evolution* **20,** 174–184.

Cooley, W. W., and Lohnes, P. R. (1971). "Multivariate Data Analysis." Wiley, New York.

den Boer, P. J. (1968). Spreading of risk and stabilization of animal numbers. *Acta Biotheor.* **18,** 165–194.

Feeny, P. (1970). Seasonal changes in oak leaf tannins and nutrients as a cause of spring feeding by wintering moth caterpillars. *Ecology* **51,** 565–581.

Ferguson, D. C. (1978). "The Moths of American North of Mexico Fasc. 22.2 Noctuoidea." E. W. Classey and the Wedge Entomol. Found., Oxford.

Forbes, W. T. M. (1948). Lepidoptera of New York and neighboring states. II. Geometridae, Sphingidae, Notodontidae, Lymantriidae. *Mem.—N.Y., Agric. Ex. Stn. (Ithaca)* **274.**

Forbes, W. T. M. (1954). Lepidoptera of New York and neighboring states. III. Noctuidae. *Mem.—N.Y., Agric. Exp. Stn. (Ithaca)* **329.**

Forbes, W. T. M. (1960). Lepidoptera of New York and neighboring states. IV. Agaristidae through Nymphalidae. *Mem.—Agric. Exp. Stn. (Ithaca)* **371.**

Gould, S. J., and Lewontin, R. C. (1979). The spandrels of San Marco and the Panglossian paradigm: A critique of the adaptationist programme. *Proc. R. Soc. London Ser. B,* **205,** 581–598.

Haukioja, E., and Neuvonen, S. (1985). The relationship between size and reproductive potential in male and female *Epirrita autumnata* (Lep., Geometridae). *Ecol. Entomol.* **10,** 417-1-427-4.

Hebert, P. D. N. (1983). Egg dispersal patterns and adult feeding behavior in the Lepidoptera. *Can. Entomol.* **115,** 1477–1481.

Hodges, R. W., Dominick, T., Davis, D. R., Ferguson, D. C., Franclemont, J. G., Munroe, E. G., and Powell, J. A. (1983). "Check List of the Lepidoptera of America North of Mexico." E. W. Classey and the Wedge Entomol. Found., London.

Holmes, R. T., and Sturges, F. W. (1975). Bird community dynamics and energetics in a northern hardwoods ecosystem. *J. Anim. Ecol.* **44,** 175–200.

Holmes, R. T., Bonney, R. E., Jr., and Pacala, S. W. (1979a). Guild structure of the Hubbard Brook bird community: A multivariate approach. *Ecology* **60,** 512–520.

Holmes, R. T., Schultz, J. C., and Nothnagle, P. (1979b). Bird predation on forest insects: An exclosure experiment. *Science* **206,** 462–463.

Hough, J. H., and Pimentel, D. (1978). Influence of host foliage on development, survival, and fecundity of the gypsy moth. *Environ. Entomol.* **7,** 97–101.

Knapp, R., and Casey, T. M. (1986). Thermal ecology, behavior and growth of gypsy moth and eastern tent caterpillars. *Ecology* **67,** 398–608.

Lance, D., and Barbosa, P. (1981). Host tree influences on the dispersal of first instar gypsy moth, *Lymantria dispar* (L.). *Ecol. Entomol.* **6,** 411–416.

Lawson, D. L., Merritt, R. W., Martin, M. M., Martin, J. S., and Kukos, J. J. (1984). The nutritional ecology of larvae of *Alsophila pometaria* and *Anisota senatoria* feeding on early- and late-season oak foliage. *Entomol. Exp. Appl.* **35,** 105–114.

Lawton, J. H., and McNeill, S. (1979). Between the devil and the deep blue sea: On the problem of being an herbivore. *In* "Population Dynamics" (R. M. Anderson, B. D. Turner, and L. R. Taylor, eds.), pp. 223–244. Blackwell, Oxford.

MacArthur, R. H., and Wilson, E. O. (1967). "The Theory of Island Biogeography." Princeton Univ. Press, Princeton, New Jersey.

Marks, P. L. (1975). On the relation between extension growth and successional status of deciduous trees of the northeastern United States. *Bull. Torrey Bot. Club* **15,** 837–845.

May, R. M. (1985). Regulation of populations with non-overlapping generations by microparasites: A purely chaotic system. *Am. Nat.* **125,** 573–584.

Mitter, C., and Futuyma, D. (1977). Parthenogenesis in the fall cankerworm *Alsophila pometaria* (Lepidoptera: Geometridae). *Entomol. Exp. Appl.* **21,** 192–198.

Mosher, F. H. (1915). Food plants of the gypsy moth in America. *U.S., Dep. Agric., Bull.* **250.**

Niemela, P. (1983). Seasonal patterns in the incidence of specialism: Macrolepidopteran larvae on Finnish deciduous trees. *Ann. Zool. Fenn.* **20,** 199–202.

Pianka, E. R. (1970). On *r*- and *k*-selection. *Am. Nat.* **104,** 592–597.

Prentice, R. M. (1962). Forest Lepidoptera in Canada. Vol. 2. *Can. Dep. For. Bull.* **128.**

Prentice, R. M. (1963). Forest Lepidoptera in Canada. Vol. 3. *Can., For. Branch, Dep. Publ.* **1013.**

Reddingius, J., and den Boer, P. J. (1970). Simulation experiments illustrating stabilization of animal numbers by spreading of risk. *Oecologia* **5,** 240–284.

Rollo, C. D. (1986). A test of the principle of allocation using two sympatric species of cockroaches. *Ecology* **67,** 616–628.

Schneider, J. C. (1980). The role of parthenogenesis and female aptery in microgeographic, ecological adaptation in the fall cankerworm, *Alsophila pometaria* Harris (Lepidoptera: Geometridae). *Ecology* **61,** 1082–1090.

Schroeder, L., and Malmer, M. (1980). Dry matter, energy and nitrogen conversion by Lepidoptera and Hymenoptera larvae fed leaves of black cherry. *Oecologia* **45,** 63–71.

Schultz, J. C. (1983a). Impact of variable plant defensive chemistry on susceptibility of insects to natural enemies. *ACS Symp. Ser.* **208,** 37–54.

Schultz, J. C. (1983b). Habitat selection and foraging tactics of caterpillars in heterogeneous trees. *In* "Variable Plants and Herbivores in Natural and Managed Systems" (R. F. Denno and M. S. McClure, eds.), pp. 61–90. Academic Press, New York.

Schweitzer, D. F. (1977). Life history strategies of the Lithophanini (Lepidoptera: Noctuidae, Cuculliinae), the winter moths. Dissertation, University of Massachusetts, Amherst.

Southwood, T. R. E., May, R. M., Hassell, M. P., and Conway, G. R. (1974). Ecological strategies and population parameters. *Am. Nat.* **108,** 791–804.

Stamp, N. E. (1980). Egg deposition patterns in butterflies: Why do some species cluster their eggs rather than deposit them singly? *Am. Nat.* **115,** 367–380.

Stearns, S. C. (1980). A new view of life-history evolution. *Oikos* **35,** 266–281.

U.S. Department of Agriculture (USDA) (1962–1981). "Forest Insect Conditions in the United States (1962-1981)." USDA, Washington, D. C.

Varley, G. C., Gradwell, G. R., and Hassell, M. P. (1973). "Insect Population Ecology." Blackwell, Oxford.

Watt, K. E. F. (1965). Community stability and the strategy of biological control. *Can. Entomol.* **97,** 887–895.

Wellington, W. G. (1964). Qualitative changes in populations in unstable environments. *Can. Entomol.* **96,** 436–451.

Wint, W. (1983). The role of alternative host-plant species in the life of a polyphagous moth, *Operophtera brumata* (Lepidoptera: Geometridae). *J. Anim. Ecol.* **52,** 439–450.

APPENDIX A: REFERENCES USED TO CONSTRUCT TRAIT MATRIX

Neophasia menapia (Felder and Felder)

Cole, W. E. (1971). Pine Butterfly. *U.S. For. Serv., For. Pest Leafl.* **66.**

Stretch, R. H. (1882). Notes on *Pieris menapia. Papilio* **2,** 103–110.

Anisota senatoria (J. E. Smith)

Felt, E. P. (1905). Insects affecting park and woodland trees. *N.Y. State Mus. Mem.* **8.**

Hitchcock, S. W. (1961). Egg parasitism and larval habits of the orange-striped oakworm. *J. Econ. Entomol.* **54,** 502–503.

Coloradia pandora Blake

Carolin, V. M., and Knopf, J. A. E. (1968). The Pandora moth. *U.S., For. Serv., For. Pest Leaf.* **114.**

Wygant, N. D. (1941). An infestation of the Pandora moth, *Coloradia pandora* Blake, in lodgepole pine in Colorado. *J. Econ. Entomol.* **34,** 697–702.

Malacosoma americanum Fab.

Stehr, F. W., and Cook, E. F. (1968). A revision of the genus Malacosoma in North America (Lepidoptera: Lasiocarpidae): Systematics, biology, immatures, and parasites. *Bull.—U.S. Natl. Mus.* **276.**

Malacosoma constrictum Stretch

Seltzer, M. J. (1968). The Great Basin tent caterpillar in New Mexico: Life history, parasites, disease, and defoliation. *U.S., For. Serv., Res. Pap. RM* **RM-39.**

Malacosoma disstria Hbn.

Sippell, W. L. (1962). Outbreaks of forest tent caterpillar, *Malacosoma disstria* Hbn, a periodic defoliator of broad-leaved trees in Ontario. *Can. Entomol.* **94,** 408–416.

Witter, J. A., Mattson, W. J., and Kulman, H. M. (1975). Numerical analysis of a forest tent caterpillar (Lepidoptera: Lasiocampidae) outbreak in northern Minnesota. *Can Entomol.* **107,** 837–854.

Malacosoma californicum (Dyar)

Myers, J. M. (1981). Interactions between western tent caterpillars and wild rose: A test of some general plant–herbivore hypotheses. *J. Anim. Ecol.* **50,** 11–25.

Seltzer, M. J. (1971). Western tent caterpillar. *U.S., For. Serv., For. Pest Leafl.* **119.**

Stehr, F. W., and Cook, E. F. (1968). A revision of the genus *Malacosoma* in North America (Lepidoptera: Lasiocampidae): Systematics, biology, immatures and parasites. *Bull.—U.S. Natl. Mus.* **276.**

Orgyia pseudotsugata (McD.)
Eaton, C. B., and Struble, G. R. (1957). The Douglas-fir tussock moth in California. *Pan-Pac. Entomol.* **33,** 105–108.
Mason, R. R. (1976). Life tables for a declining population of the Douglas-fir tussock moth in northeastern Oregon. *Ann. Entomol. Soc. Am.* **69,** 948–958.
Wickman, B. E., Mason, R. R., and Thompson, C. G. (1973). Major outbreaks of the Douglas-fir tussock moth in Oregon and California. *USDA For. Serv. Gen. Tech. Rep. PNW* **PNW-5.**
Lymantria dispar (L.)
Doskotch, R. W., O'Dell, T. M., and Goodwin, P. A. (1977). Feeding response of the gypsy moth larva, *Lymantria dispar,* to extracts of plant leaves. *Environ. Entomol.* **6,** 563–566.
Leucoma salicis L.
Wagner, T. L., and Leonard, D. E. (1979). Aspects of mating, oviposition, and flight in the satin moth, *Leucoma salicis* (Lepidoptera: Lymantriidae). *Can. Entomol.* **111,** 833–840.
Datana integerrima Grote and Robinson
Ferris, M. E., and Appleby, J. E. (1979). The walnut caterpillar, *Datana integerrima* G & R. *USDA For. Serv. Gen. Tech. Rep.* **NC-52.**
Ferris, M. E., Appleby, J. E., and Weber, B. C. (1982). Walnut caterpillar. *USDA For. Serv., For. Insect Dis. Leafl.* **41.**
Heterocampa guttivitta Walker
Allen, D. C. (1972). Insect parasites of the saddle prominent, *Heterocampa guttivitta* in the northeastern United States. *Can. Entomol.* **104,** 1609–1622.
Heterocampa manteo Doubleday
Surgeoner, G. A., and Wallner, W. E. (1978). Evidence of prolonged diapause in prepupae of the variable oak leaf caterpillar, *Heterocampa manteo. Environ. Entomol.* **7,** 186–188.
Wilson, L. F., and Surgeoner, G. A. (1979). Variable oak leaf caterpillar. *USDA For. Serv., For. Insect Dis. Leafl.* **67.**
Symmerista canicosta Franclemont
Anderson, J. F., and Kaya, H. K. (1978). Field and laboratory biology of *Symmerista canicosta. Ann. Entomol. Soc. Am.* **71,** 137–142.
Phryganidia californica Packard
Puttrick, G. M. (1986). Utilization of evergreen and deciduous oaks by the Californian oak moth *Phryganidia californica. Oecologia* **68,** 589–594.
Puttick, G. M., and Wickman, B., personal communication.
Wickman, B. E. (1971). California oakworm. *U.S., For. Serv., For. Pest Leafl.* **72.**
Hyphantria cunea Drury
Morris, R. F. (1967). Influence of parental food quality on the survival of *Hyphantria cunea. Can. Entomol.* **99,** 24–33.
Oliver, A. D. (1964). A behavioral study of two races of the fall webworm, *Hyphantria cunea* (Lepidoptera: Arctiidae) in Louisiana. *Ann. Entomol. Soc. Am.* **57,** 192–194.
Rossiter, M. C., unpublished data.
Halisidota argentata Pack
Silver, G. T. (1958). Studies on the silver-spotted tiger moth, *Halisidota argentata* Pack. (Lepidoptera: Arctiidae), in British Columbia. *Can. Entomol.* **90,** 65–80.
Alsophila pometaria Harris
Futuyma, D., personal communication.
Moore, G. E., and Drooz, A. T. (1974). Rearing the fall cankerworm on a natural diet. *U.S., For. Serv., Res. Note SE* **SE-197.**

Schneider, J. C. (1980). The role of parthenogenesis and female aptery in microgeographic, ecological adaptation in the fall webworm, *Alsophila pometaria* Harris (Lepidoptera: Geometridae). *Ecology* **61,** 1082–1090.

Ennomos subsignarius (Hbn.)

Drooz, A. T. (1980). A review of the biology of the elm spanworm (Lepidoptera: Geometridae). *Great Lakes Entomol.* **13,** 49–53.

Hydria prunivorata (Ferguson)

Schultz, D. E., and Allen, D. C. (1975). Biology and descriptions of the cherry scallop-shell moth, *Hydria prunivorata* (Lepidoptera: Geometridae) in New York. *Can. Entomol.* **107,** 99–106.

Lambdina fiscellaria (Guenee)

Carroll, W. J. (1956). History of the hemlock looper, *Lambdina fiscellaria fiscellaria* (Guenee), (Lepidoptera: Geometridae) in Newfoundland, and notes on its biology. *Can. Entomol.* **88,** 587–599.

Erannis tiliaria (Harris)

McGuffin, W. C. (1977). Guide to the Geometridae of Canada (Lepidoptera). II. Subfamily Ennominae 2. *Mem. Entomol. Soc. Can.* **101.**

McGuffin, W. C., personal communication.

Martineau, R., and Monnier, C. (1963). The basswood looper in 1962 in Quebec. *Can. Dep. For. Bimon. Prog. Rep.* **19,** 3.

Nepytia freemani Monroe

Klein, W. H., and Minnoch, M. (1971). On the occurrence and biology of *Nepytia freemani* (Lepidoptera: Geometridae) in Utah. *Can Entomol.* **103,** 119–124.

Nepytia phantasmaria (Strecker)

Wickman, B. E., and Hunt, R. H. (1969). Biology of the phantom hemlock looper on Douglas-fir in California. *J. Econ. Entomol.* **62,** 1046–1050.

Operophtera bruceata (Hulst)

Brown, C. E. (1962). The life history and dispersal of the bruce spanworm, *Operophtera bruceata* (Hulst), (Lepidoptera: Geometridae). *Can. Entomol.* **94,** 1103–1107.

Semiothisa sexmaculata Packard

McGuffin, W. C. (1972). Guide to the Geometridae of Canada. (Lepidoptera). II. Subfamily Ennominae 1. *Mem. Entomol. Soc. Can.* **86.**

McGuffin, W. C., personal communication.

Paleacrita vernata (Peck)

McGuffin, W. C. (1977). Guide to the Geometridae of Canada. (Lepidoptera). II. Subfamily Ennominae 2. *Mem. Entomol. Soc. Can.* **101.**

McGuffin, W. C., personal communication.

Phaeoura mexicanaria (Grote)

Dewey, J. E. (1972). A pine looper, *Phaeoura mexicanaria* (Grote) in southeastern Montana with notes on its biology. *Ann. Entomol. Soc. Am.* **65,** 306–309.

Rheumaptera hastata (L.)

Werner, R. A. (1979). Influence of host foliage on development, survival, fecundity, and oviposition of the spear-marked black moth, *Rheumaptera hastata* (Lepidoptera: Geometridae). *Can. Entomol.* **111,** 317–322.

APPENDIX B: SPECIES FOUND IN U.S. FOREST SERVICE DAMAGE REPORTS BUT DROPPED FROM ANALYSIS

Species	Reason	Notes
Nymphalidae		
Nymphalis californica (Bud)	Eats *Ceanothus* spp. (range plants)	—
Saturniidae		
Hemileuca nevadensis Stretch	Lack of data	Eggs in cluster
Lasiocampidae		
Malacosoma incurvum (Hyman Edwards)	Lack of data	Eggs in cluster
Lymantriidae		
Dasychira plagiata (Walker)	Lack of data	—
Dasychira grisefacter Dyar	Lack of data	—
Orgyia vetusta (Bud)	Range plants	—
Nygmia phaeorrhoea (Don.) (introduced spp.)	Lack of data	Egg mass
Arctiidae		
Halisidota ingens Edws.	No data	—
Geometridae		
Anacamptodes clivinaria (Guenee)	Feeds mainly on *Cerocampus* and *Ceanothus*	Eggs in clusters of 350
Lambdina athasaria (Guenee)	No data	—
L. punctata (Hulst.)	Range plants in Utah	Eggs single
Nepytia canosaria (Walker)	No data	—

Chapter 4

Leaf Eating as Mutualism

DENIS F. OWEN

Department of Biology
Oxford Polytechnic
Headington, Oxford OX3 0BP, England

RICHARD G. WIEGERT

Department of Zoology
University of Georgia
Athens, Georgia 30602

I. INTRODUCTION

Many multicellular plants can be grazed, even severely grazed, without being killed. The death of an individual as a result of its leaves being chewed is unusual, and when it does occur the plant is often growing in unnatural circumstances or is one whose genetic makeup has been deliberately or accidentally altered by selection or hybridization; in other words, it is a crop or plantation tree.

Nearly all outbreaks of defoliating caterpillars occur on trees growing in plantations or orchards or in monocultures of field crops. Exceptions are rare, and it is certainly possible for a working field ecologist to live a lifetime without seeing a single destructive outbreak of insects in a wholly natural situation. In terrestrial ecosystems, a remarkably small proportion of living plant material is eaten by consumers. The figure is around 2% in many forests, rising to 20 to 60% in Australian *Eucalyptus* forests (Fox and Macauley, 1977) and in "natural" grassland. (We place "natural" in quotation marks because truly natural grassland is hard to find and has rarely been investigated to determine the effects of consumers.) In the past quarter of a century various theories have been put forward to explain this low level of primary consumption.

One hypothesis, originally advocated by Hairston *et al.* (1960), is that, at a trophic level, terrestrial primary consumers are limited by predators and parasitoids and not by available food. This means that these consumers rarely have a significant effect on the food supply. Outbreaks of a particular species may occur when something has "gone wrong," in particular when population regulation by predators and parasitoids has failed for some reason.

A second explanation is that plants protect themselves with a battery of physical and chemical defenses to an extent that substantial defoliation is impossible. Many leaves are edible only when they are young; the older, tougher, and more toxic leaves are virtually unusable, and so the apparent abundance of greenery is deceptive. This view acknowledges that plant feeding is physiologically difficult and that plant-feeding insects, in particular, are severely restricted in what they can utilize.

The third explanation, related to the second, is that individual plants, particularly long-lived perennials such as trees, are mosaics of resistance to consumer attack. This state of affairs is produced in the plants either by somatic mutation or by environmentally induced within-individual variation. According to this view, developed especially by Whitham (1981, 1983) and Whitham and Slobodchikoff (1981), long-lived perennials can cope with the high evolutionary potential possessed by plant-eating insects with rapid generation times.

The second of these three possibilities has received the most attention. Much current thinking about plant–insect relationships advocates the idea of an arms race, the plants nearly always defending themselves with secondary chemical compounds, thousands of which are known. The literature is replete with terms like "antiherbivore defenses," "antiherbivore adaptations," and "chemical defenses." However, despite much detailed knowledge of secondary-compound chemistry, there is little direct evidence that, in the interaction between eaters and eaten, the plants are always on the defensive. In most cases this is simply deduced from the inference that any herbivore must reduce plant fitness and that the plant must, in consequence, evolve defensive mechanisms.

II. MUTUALISM

Our first publication suggesting that the relationship between plants and their consumers could in some cases be interpreted as mutualism and that plants are not necessarily always on the defensive (Owen and Wiegert, 1976) brought together a variety of evidence supporting the idea but offered no experimental support. Subsequent publications (Owen, 1977, 1978, 1980; Owen and Wiegert, 1981, 1982a,b, 1983) have met with a mixed response, including sharp criticism, but at the time of writing there is still no hard evidence one way or the other, although several experiments have been attempted (Petelle, 1980; Choudhury, 1984).

By "mutualism" we mean the association of two, often unrelated organisms such that the relative fitness of both is higher than it would be if each existed by itself. Sometimes one or both partners in a mutualistic relationship cannot even exist without the other partner. In evolutionary terms, mutualism is reciprocal adaptive radiation. Any tendency of one organism to form a more beneficial relationship (in terms of fitness) results in the evolution of traits in the other that increase the advantage of the new tendency. Thus, natural selection works on both parties in a mutualistic relationship in such a way that their respective fitnesses are enhanced.

We acknowledge difficulty in defining, let alone measuring, relative fitness. Here we use "fitness" to mean the contribution of offspring of one genotype relative to the contribution of others. This leads, by natural selection, to a greater number of those genotypes with the highest relative fitness in a given environment at a given time. We perceive the fitness of an individual at a given age as comprising two components, one associated with immediate reproduction or prospects of immediate reproduction and the other associated with survival to achieve future reproduction. In perennial grasses, for example, immediate reproduction (flower and seed formation) may be inhibited by grazing, but grazing may prolong the life of the individual so that eventually it achieves a higher relative fitness than an ungrazed individual. Furthermore, if an individual increases its own fitness it will be selected against if by so doing it increases the fitness of others even more (altruism). Conversely, if an individual decreases its own fitness, it will be selected for provided that it decreases the fitness of others even more, a possibility that has been called "spite."

There is probably no statistical evidence of any plant–animal mutualism enhancing the fitness of the participants. Even in well-known, universally accepted mutualisms such as between pollinator and pollinated and between fruit and fruit eater, evidence of enhanced fitness for both parties is at best indirect and at worst nothing more than a guess, although, of course, what might be called mutual "benefits" are well described and documented. Hence, those that seek hard data for evidence of fitness enhancement in plant–animal mutualisms will probably seek in vain. We believe, nevertheless, that there is much suggestive evidence that mutualism is more widespread than is generally supposed and that, although plants are obviously able to regulate consumption by animals, the notion that they are always on the defensive is much exaggerated. Our ideas of leaf eating as mutualism are in a sense a special way of looking at coevolution between plants and animals. "Coevolution," a term invented by Ehrlich and Raven (1965), is reciprocal adaptive radiation and may or may not involve mutualism. The idea of coevolution, which has been implicit in evolutionary thinking since the time of Darwin, is that one organism does not remain evolutionarily constant or static while another that exploits it continues to evolve better means of exploitation. Ehrlich and Raven document coevolution by reference to ploy–counterploy adaptations between butterflies and the plants their caterpillars feed on. Most female butterflies lay eggs with great precision, selecting not only the correct potential food plant, but also the right place on that food plant. In contrast plants respond to the egg-laying activities of butterflies and have, according to Ehrlich and Raven, evolved adaptations that (1) restrict the number of species of butterfly that exploit a given species of plant and (2) restrict the number of places, or niches, open to a butterfly for egg laying. That Ehrlich and Raven did not regard plant–butterfly coevolutionary relationships as mutualism does not preclude them.

III. SOME EVOLUTIONARY DIMENSIONS

In pre-Cambrian times the first photosynthetic plants were presumably vulnerable to attack by other organisms, traces of which appear in the fossil record, although we have little idea of who was eating what. In the Cambrian itself most of the major invertebrate groups, except insects, made their appearance, and any plant feeders among them must have fed on algae since these are the only plants whose existence is indicated by the fossil record. Whatever plant–animal feeding relationships had developed must have been in water, and there were, perhaps, similarities with the trophic arrangements in today's oceans. Not until the Silurian did the first vascular plants appear, still in water. It was only with the advent of the Devonian that life on land began to consolidate and the first feeding relationships between terrestrial plants and animals developed. The first trees appeared in the Devonian: They were fern- or gymnospermlike and were almost certainly utilized by the first insects, which also appeared at this time. The emergence of trees and treelike plants is one of the most extraordinary aspects of plant evolution; photosynthetic plants were evidently able to develop hard structure and outstrip the impact of consumers on their productivity.

All the major groups of plants, except the angiosperms, had appeared by the Carboniferous, and we assume that leaf-eating insects had evolved, although no positive evidence (fossils) of chewed holes and chewed leaf edges occurs until the early Permian. The angiosperms originated in the late Jurassic or early Cretaceous and with them occurred an enormous radiation of plant-feeding insects and their predators and parasitoids. Most authors are enthusiastic about widespread reciprocal adaptive radiation between insects and plants. Most, however, view leaf eating as detrimental and find, or claim to find, much evidence of plants' defending themselves from leaf-eating insects and other animals.

IV. SOME ECOLOGICAL DIMENSIONS

There are about 360,000 known species of plant-eating insects in the world, and no doubt many thousands await discovery and description. This represents about a quarter of the known species of organisms, excluding fungi, algae, and microbes (Strong *et al.*, 1984). Plant feeders are effectively confined to nine orders of insects, with leaf eaters (as opposed to leaf and stem suckers) virtually confined to the Lepidoptera (nearly all species), Orthoptera (nearly all species), Hymenoptera (chiefly sawflies, ~11% of the total species in the order), Diptera (~30% of all species), and Coleoptera (~38%). In terms of number of species (~120,000), size, biomass, and structural diversity, the Lepidoptera constitute the most important group of leaf eaters. Moreover, it is this order whose evolution and radiation coincide so closely with the evolution and radiation of the angiosperms. There are also leaf-eating (grazing and browsing) mammals whose (local) impact may be ecologically and evolutionarily considerable, a few grazing reptiles (such as the Aldabra tortoises), and a variety of invertebrate groups (including mollusks) whose number and diversity cannot be compared with those of the leaf-eating insects. Very few extant birds eat leaves, but as discussed in Section VIII recently extinct forms like moas may have played a significant part in plant evolution.

There are about 308,000 species of green plants, most of them angiosperms. Their diversity is greatest in equatorial regions, where the highest diversities of leaf-eating insects are also found. Almost all Lepidoptera caterpillars feed on angiosperms; many are confined to a relatively narrow range of related species or to just one species, whereas many others are highly polyphagous. The number of feeding interactions between the Lepidoptera and their food plants is phenomenal and on the basis of present knowledge cannot possibly be estimated. In one well-known group, the tropical American heliconiine butterflies, the 66 species feed on 123 species of plants (all Passifloraceae); although there is considerable host specificity, 387 interactions between species of butterflies and species of plant have been documented (Gilbert, 1977), which gives some indication of the complexity of plant–insect relationships.

V. PRUNING, POLLARDING, COPPICING, AND MOWING

Ecologists, particularly those used to observing animals, often find it difficult to understand that an organism might actually benefit from losing parts of itself, and this, perhaps, is one reason why leaf eating is not usually viewed as a mutalistic process. There are several precedents, however, regarding initial incomprehension of evolutionary phenomena. Thus, the lateral line of fish could not at first be understood as an adaptation because there is no analogous structure in humans (Williams, 1966).

Gardeners, especially fruit and rose growers, know that pruning alters the shape of a tree or bush and that regrowth after pruning often results in more or better flowers and fruit. The literature on pruning is extensive, but most of it is advisory and authoritarian: It tells what to do to achieve the best results, and there is usually no evidence resulting from experimentation that would satisfy a scientist. Yet at the same time the practice of pruning is so widespread and so ancient that it has to be acknowledged as an undoubted means of increasing yield. There is an abundance of published information on how to prune to get the best results; what is missing is a proper comparative analysis involving the use of experimentals and controls.

Cutting the tops of trees, or pollarding, is also a widespread and ancient practice. It is done to make the tree safe by the removal big branches that might otherwise fall off and to promote regrowth, which in some species, such as *Salix fragilis,* is extremely vigorous. It also seems to prolong the life of the tree. Coppicing, or cutting the tree down at its base and allowing regrowth, is also widely practiced, and there is no question that this extends the individual's life expectancy. In fact, coppice stools seem to live indefinitely. In England, stools of *Ulmus glabra, Fraxinus excelsior, Castanea sativa, Tilia* sp., and *Quercus* spp. are commonly 500 years old and may be up to 1000 years old (Rackham, 1981). Clipping or mowing grass is also believed to prolong the life of individual grass plants; perhaps, too, it tends to make individuals grow much larger than those that are uncut.

The natural analogs of pruning, pollarding, coppicing, and mowing may be found in leaf-eating animals, especially leaf-eating insects. Young stems and shoots are cut off and eaten; the growing points of a plant may be removed by chewers; a plant may be cut down at its base by a noctuid caterpillar; while many species are grazers whose effect may not be obvious because of the nature of the structure of a grass plant. If pruning, pollarding, coppicing, and mowing are deemed beneficial to growing plants, is it not possible that leaf eating is likewise beneficial? And is it not further possible that plants, especially long-lived perennials, have largely escaped from a defensive strategy and instead regulate consumption and use consumers to their own advantage?

VI. EFFECTS OF LEAF CHEWING

Almost every piece of research attempting to show the effect of consumers on plants ends up by taking the measurement of production as a means of assessing consumer impact. From an evolutionary point of view this is almost a worthless measurement, because production may or may not be indicative of a consumer's effect on individual plant fitness. A better measurement would be flower, seed, or fruit numbers.

There are some investigations, however, that start with the assumption that leaf loss can lead to reduced reproduction. One piece of work (Rockwood, 1973) tests the hypothesis that "increased foliage losses lead to decreased reproduction. . . . Six . . . tree species were defoliated by hand." And what was the result? Defoliated trees produced fewer or no fruit, which is hardly astonishing, given that defoliation is so rare in nature and that normally leaf eating results in very little loss of leaf biomass. Nevertheless, Rockwood believes his results support the notion that much of the morphology and chemistry of plants is a consequence of herbivore consumption and that his data "support the view that physical and chemical defenses evolved by plants have played an important role in plant–herbivore co-evolution." We suggest that artificial defoliation or damage to leaves is a poor replica of natural leaf chewing by caterpillars, mammals, and other leaf eaters: The experimenter is unlikely to be able to mimic the precise way in which a leaf eater eats its leaf.

The assumption that leaf eaters invariably cause damage to plants is so pervasive that few investigators looking into the effects of leaf eating even consider possible benefits to the plant. Compensation and regrowth following leaf eating are often detected but interpreted more as a plant's "defensive" response than as a specific adaptation enhancing fitness. One piece of work (Islam and Crawley, 1983) seems to us to provide much of the evidence required for a "new look" at plant–animal feeding relationships and sets the scene for a reappraisal of the effects of leaf eating on fitness. Islam and Crawley do not appear to have thought of their results in terms of mutualism but instead take the conventional view that "more insects means more damage to the plant" (Lawton and McNeill, 1979).

Like many before them, Islam and Crawley (1983) investigated the impact of the cinnabar moth, *Tyria jacobaeae,* on ragwort, *Senecio jacobaea,* a noxious weed of pastures that has been introduced to many parts of the world. The cinnabar moth has been used as a means of biological control of ragwort, but there has always been the suspicion that feeding by its caterpillars increases the density of ragworts because several new rosettes develop from each eaten plant (Dempster, 1971). The new rosettes and the plants that develop from them are, of course, the same evolutionary individuals as the original plant, and so the question is whether increased rosette formation following feeding by cinnabar caterpillars increases the relative fitness of the plant. In this instance a comparison could be envisaged between individual plants not subjected to consumption by cinnabars, and hence with fewer rosettes, and those that have been eaten and consequently have more rosettes.

Islam and Crawley do not quite answer this question, but they do show that, since cinnabar caterpillars are strongly seasonal, once they have pupated the ragworts can easily compensate by regrowth and eventually by the formation of new and more rosettes. Within one season the number of seeds produced by cinnabar-free plants is higher than that produced by plants whose foliage has been eaten, but there is the suggestion that a plant that has been eaten lives longer. This is obviously something that requires a follow-up. The seeds of eaten plants may be produced 2 months later and are considerably smaller than those of uneaten plants. In Australia (where cinnabar and ragwort are aliens) regrowth seeds do not germinate as well (Bornemissza, 1966), but this result requires checking in an area where both species occur naturally. Islam and Crawley are intrigued by the small size of the regrowth seeds and ask why the uneaten plants do not produce more and smaller seeds, thus saving energy and enhancing potential fitness. But such a question arises only if the herbivore is assumed to be detrimental to the plant; it does not arise if it is assumed that there is a mutualism in which the plant makes an adaptive response to the presence of the herbivore. In other words, one way for ragwort to produce smaller seeds is to enlist the "help" of the cinnabar. As is well known, not all ragwort individuals support cinnabar caterpillars, and hence the possibility exists that there are two (polymorphic) forms of the plant: one more chemically attractive to cinnabars that produces smaller seeds later in the season as a consequence of the induction of regrowth and one less attractive that produces larger seeds earlier in the season. If this is so we could be dealing with a balanced polymorphism maintained by natural selection in a heterogeneous environment.

Along with most ecologists with interests in plant–insect feeding relationships, we have repeatedly observed that caterpillars frequently feed on the new, soft leaves and avoid older, tougher-looking leaves, presumably because they are physically and chemically unacceptable. Species after species of rain forest butterfly lay eggs only on new leaves of the potential food plant, and these new leaves typically appear for a short and restricted, although predictable, time of year, their appearance usually being triggered by some facet of the alternation of wet and dry seasons. As a result of this, growing points are frequently destroyed by catepillars, which results in compensation through regrowth and the production of additional (usually lateral) branches. This is exactly analogous to the response to pruning, and although we have been unable to find references to the effects of compensation on flower, seed, or fruit production, it seems to us likely that such effects are detectable and hence worthy of investigation. The conventional explanation for the concentration of leaf-eating caterpillars on new leaves is that these leaves have not yet elaborated sufficient secondary compounds to deter the caterpillars. Older leaves, it is claimed, are avoided because they are toxic or too tough, or both. We do not dispute this claim but question why so many plants (especially trees and bushes) are unable to protect the new leaves from caterpillars. Is it possible that there is a selective advantage for the new leaves to be relatively palatable? Perhaps the most effective way for a tree to grow, develop, and produce seed is to branch as repeatedly as possible, and the best way to facilitate branching is to encourage consumers to feed on new leaves and shoots at growing points. Plants clearly regulate consumption by herbivores, but is this regulation always a matter of "defense" and never a matter of attraction? Even plants that require insect pollinators, usually accepted as mutualism, regulate the amount of nectar given to the pollinating insect.

We do not mean to imply that more palatable plants or parts of plants should increase in frequency at the expense of less palatable plants or parts of plants. Rather, we envisage each plant as a mosaic of palatability, a mosaic that varies in space and time and in relation to levels of consumption. If, as we hypothesize, leaf eating is selectively advantageous to the plant and hence constitutes a mutualism between eater and eaten, it is essential that consumption be regulated within strict limits. This most land plants seem able to do: When they do not there are outbreaks of plant feeders and significant, destructive defoliations.

VII. THE SALIVA QUESTION

In one of our publications (Owen and Wiegert, 1981) we cite Dyer (1980), who found that an epidermal growth factor (EGF) extracted from the submaxillary glands of mice increased shoot elongation in the seedlings of *Sorghum bicolor*. At the time, Dyer's work seemed to confirm that saliva from plant feeders could stimulate plant growth, but later Dyer *et al.* (1982) could find no effect of EGF on the seedlings of corn, mustard, and tomato. The suggestion that animal saliva promotes growth dates from an article by Vittoria and Rendina (1960). Reardon *et al.* (1972) claim to have found that bovine saliva stimulates growth in the grass *Bouteloua curtipendula,* but in a later article (Reardon *et al.,* 1974) could not confirm the original results. Dyer and Bokhari (1976) report that grasshopper grazing stimulates tiller production in the grass *Bouteloua gracilis,* more so than mechanical cutting, and invoked the effect of saliva. Detling *et al.* (1980) could find no effect of saliva on any aspect of growth in *Bouteloua gracilis*. Thus far, then, the question of saliva affecting plant growth is an open one. We admit that we probably made too much of Dyer's results when they first appeared, and we can only add that the existence of growth-promoting saliva or other substance produced by browsers or grazers is not essential to our hypothesis. All the same it would be beneficial to have the results of more experiments on the effect of animal saliva on plant growth. Such experiments should ideally make use of a plant feeder on its natural food plant, not a laboratory mouse on a cultivar of an alien crop, as in Dyer's experiment.

VIII. DIVARICATING PLANTS OF NEW ZEALAND AND BROWSING BY MOAS

The size, structure, and diversity of trees, shrubs, and other (especially woody) plants is both striking and to a large extent evolutionarily inexplicable. The sudden, relatively recent origin and radiation of the angiosperms, what Darwin called an "abominable mystery," is a puzzle to this day. There was no mass extermination of gymnosperms or ferns by climatic or other unspecified events that might have made possible the explosive evolution of a major new line, as with, for example, the Early Tertiary radiation of mammals that immediately followed the extinction of dinosaurs in the Late Cretaceous. Moreover, the simultaneous diversification of insects in the Cretaceous is viewed as a consequence rather than a cause of angiosperm radiation (Doyle, 1978). This seems to us to be a simplification of what is likely to have happened.

The fossill record of Cretaceous leaf eaters and signs of leaf eating in plants is so fragmentary as to be virtually worthless. Hence, any attempt to postulate mutualistic coevolution to explain the apparent simultaneous radiation of plants and insects in the Cretaceous is fraught with difficulty. The picture becomes clearer in the Tertiary, by which time most of the important lines of plants and plant feeders had developed. Relating plant structure to past and present patterns of herbivory seems to us to be a useful, if neglected, line of inquiry; in the work of Greenwood and Atkinson (1977) many facets of plant structure are related to browsing by the extinct moas of New Zealand.

According to Greenwood and Atkinson (1977), nearly 10% of the woody species of plants in New Zealand are classed as "divaricating," a term used to describe branching at a wide angle, often of 90° or more. In all, 54 species of 20 genera and 16 families of New Zealand angiosperms and 1 family of gymnosperms possess the trait, which is rare or absent in other parts of the world. Divaricating plants are typically small-leaved, woody shrubs with long, interlacing branches, usually without spines or thorns. The interlacing results from reduced apical growth and pronounced lateral branching. A cut branch is described as being difficult to disentangle. The stems are also difficult to break, and in many species outer branches have larger internodes with fewer and smaller leaves than inner branches. The similarity in structure among so many unrelated plants is an example of convergent evolution, which must have been generated by selective forces common to all the species. Such selection must also be or have been confined or almost confined to New Zealand.

Various hypotheses have been put forward to explain the high frequency of divarication among the plant species in New Zealand. Most involve the supposed effects of climate, but as Greenwood and Atkinson (1977) point out, to assert that climate is the driving force it would be necessary to discover something peculiar about the climate of New Zealand. Such a discovery has not been made. A somewhat bizarre, nonadaptive explanation was proposed by Went (1971). He suggested that a chromosome segment carrying the genes for divarication was transferred asexually among species, genera, and families. This proposal seems to have arisen because no other explanation could be suggested. Went actually ruled out browsing as a cause on the grounds that New Zealand lacks indigenous browsing mammals. A link between browsing and divarication was made by Denny (1964), who though accepting the climatic explanation hinted of a "remote possibility" that moas might have been involved. In the 1970s several other authors entertained the possibility of moa browsing as a selective force in divarication, but the idea was not followed up until Greenwood and Atkinson (1977) assembled all the available evidence.

Moas, wingless ratite birds belonging to two families, the Dinornithidae and the Anomalopterigidae, became extinct about 300 years ago. Confined to New Zealand, there were between 14 and 20 species, and they varied in height between 0.5 and 3 m. They had long necks and long legs and were undoubtedly browsers, a mode of feeding that is extremely rare in extant birds. Everything points to their having been exterminated by immigrant Polynesians (who later became known as Maoris), who started to hunt them around 1350. When the first Europeans reached New Zealand all the moas had gone, but a large quantity of remains has been found, including bones, egg shells, feathers, skin, and even flesh and gizzard contents, making accurate reconstructions possible. Some of the finest reconstructions can be seen in the Zoological Museum, Tring, England.

Greenwood and Atkinson (1977) estimate that New Zealand trees were browsed by moas for about 65 million years. Lacking teeth, they could not have chewed or cut foliage; instead they must have tugged it with their blunt and horny bills. Unlike browsing mammals, which have soft noses easily repelled by a plant's spines, moas had hard bills, which would have enabled them to reach deep into plants. This may account for the lack of spines on divaricating plants. (There were, of course, no indigenous browsing mammals in New Zealand.)

Greenwood and Atkinson identify three main evolutionary trends in divaricating plants, each of them probably arising from a long browsing association with moas. The most obvious is the wide-angle, lateral branching resulting from the removal of the apical growing point and the marked tendency for branches to become entangled with one another. As Greenwood and Atkinson remark, most divaricating species show a genetically controlled absence of apical dominance, which resembles the loss of apical dominance when normal plants are browsed (or pruned). Another trend is that smaller leaves occur on the outermost branches, where there is often also an increase in the length of the internodes. The overall effect is of a more woody exterior and more leafy interior. The third trend is that the stems of divaricating plants are relatively tough and springy, making them difficult to break off.

Other apparent moa-induced adaptations can be identified, the most striking being divaricating juvenile and nondivaricating adults in nine species of trees. In these trees the divaricating growth form disappears once a height of about 4 m is reached.

If divarication is an evolutionary response to browsing by moas, the conventional explanation is that this is yet another example of plants "defending" themselves from consumers. Even put this way, however, we are led to the conclusion that the evolution of so many species with the divaricating trait is a direct response to browsing. The actual existence of these species in the form in which we now see them is a consequence of natural selection induced by the browsers. Is this really just a matter of the plants defending themselves? Is it not likely that they have adjusted to browsing by changing genetically in a manner that is also beneficial to them and that has tended to maximize their individual relative fitness? We think that loss of apical dominance and conspicuous lateral branching in particular are likely to benefit the plant, not only by regulating browsing but also by developing more branches and hence more seed.

The moas are gone; only the plants remain, and so there is no possibility for experimentation, even if an experiment could be devised. Nonetheless, similar associations exist, for example, in the acacia–giraffe relationship in Africa. Giraffes can eat more than 80% of leaf and shoot production of an acacia in a year. The acacias of the African savanna have a highly characteristic shape, which most giraffe watchers would agree is an evolved response to browsing by giraffes. This in effect means that an acacia could not have evolved to its present form unless it possessed a battery of giraffe-induced genes that make it what we call an acacia. Mammal and tree have coevolved, and we consider that both (not just the mammal) have benefited from their evolutionary association. Everything in the foregoing interpretation of the evolution of trees and their browsers can be explained by Darwinian selection; there is no need to invoke Lamarckism, even in the giraffe–acacia example.

IX. CONCLUSION

Natural selection should favor traits in plants that result in benefits from the effects of browsing and grazing. There is no need to assume that, in the coevolution between plants and leaf eaters, the plants are necessarily always on the defensive. The evolution of terrestrial plants, especially trees and woody shrubs, is intimately associated with the evolution of browsers and grazers. Although many adaptive features in plants can be viewed as defenses against browsing and grazing, many can equally well be viewed in terms of mutualistic coevolution.

Darwin's "abominable mystery" can be solved by postulating that the explosive radiation of the angiosperms was brought about by these plants "coming to terms" with consumers to an extent that the evolution of one group would not be possible without the other. In suggesting this we are rejecting the idea that the radiation of insects in the Cretaceous followed the radiation of plants; the two occurred simultaneously.

The (usually) low levels of primary consumption of living plant tissue suggest that plants have evolved mechanisms for regulating consumption. Outbreaks (of leaf-eating insects or of any other animal) in which a plant is killed are rare and can be explained in most cases by the plant growing in abnormal circumstances or by its having been genetically altered by humans.

Finally, if experiments are to be devised to test for the presence or absence of mutualism, let them be confined to a plant in its own environment with its natural consumer. There should be no repetition of the sorghum–mouse experiment.

ACKNOWLEDGMENTS

We thank all those who, since 1976, have provided published and unpublished criticism of our views. We would be the first to agree that there are gaps in, and alternative explanations to, many of our ideas. In particular, we thank R. M. Greenwood and I. A. E. Atkinson for information about moa–plant interactions.

REFERENCES

Bornemissza, G. F. (1966). An attempt to control ragwort in Australia with the cinnabar moth, *Callimorpha jacobaeae* (L.) (Arctiiae: Lepidoptera). *Aust. J. Zool.* **14,** 201–243.

Choudhury, D. (1984). Aphids and plant fitness: A test of Owen and Wiegert's hypothesis. *Oikos* **43,** 401–402.

Dempster, J. P. (1971). The population ecology of the cinnabar moth, *Tyria jacobaeae* L. (Lepidoptera: Arctiidae). *Oecologia* **7,** 26–67.

Denny, G. (1964). Habit heteroblastism of *Sophora microphylla* Ait. Unpublished M.Sc. Thesis, University of Canterbury (cited in Greenwood and Atkinson, 1977).

Detling, J. K., Dyer, M. I., Procter-Gregg, C., and Winn, D. T. (1980). Plant–herbivore interactions: Examination of potential effects of bison saliva on regrowth of *Bouteloua gracilis* (H.B.K.) Lag. *Oecologia* **45,** 26–31.

Doyle, J. A. (1978). Origin of angiosperms. *Annu. Rev. Ecol. Syst.* **9,** 365–392.

Dyer, M. I. (1980). Mammalian epidermal growth factor promotes plant growth. *Proc. Natl. Acad. Sci. U.S.A.* **77,** 4836–4837.

Dyer, M. I., and Bokhari, U. G. (1976). Plant–animal interactions: Studies of the effects of grasshopper grazing on blue grama grass. *Ecology* **57,** 762–772.

Dyer, M. I., Detling, J. K., Coleman, D. C., and Hilbert, D. W. (1982). The role of herbivores in grasslands. *In* "Grasses and Grasslands: Systematics and Ecology" (J. R. Estes, R. J. Tyrl, and J. N. Brunken, eds.), pp. 255–295. Univ. of Oklahoma Press, Norman.

Ehrlich, P. R., and Raven, P. H. (1965). Butterflies and plants: A study in coevolution. *Evolution (Lawrence, Kans.)* **18,** 586–608.

Fox, L. R., and Macauley, B. J. (1977). Insect grazing on *Eucalyptus* in response to variation in leaf tannins and nitrogen. *Oecologia* **29,** 145–162.

Gilbert, L. E. (1977). Development of theory in the analysis of insect–plant interactions. *In* "Analysis of Ecological Systems" (D. J. Horn, R. D. Mitchell, and G. R. Stairs, eds.), pp. 117–154. Ohio State Univ. Press, Columbus.

Greenwood, R. M., and Atkinson, I. A. E. (1977). Evolution of divaricating plants in New Zealand in relation to browsing. *Proc. N. Z. Ecol. Soc.* **24,** 21–33.
Hairston, N. G., Smith, F. E., and Slobodkin, L. B. (1960). Community structure, population control, and competition. *Am. Nat.* **94,** 421–425.
Islam, Z. and Crawley, M. J. (1983). Compensation and regrowth in ragwort (*Senecio jacobaea*) attacked by cinnabar moth (*Tyria jacobaeae*). *J. Ecol.* **71,** 829–843.
Lawton, J. H., and McNeill, S. (1979). Between the devil and the deep blue sea: On the problem of being a herbivore. *Symp. Br. Ecol. Soc.* **20,** 223–224.
Owen, D. F. (1977). Are aphids really plant pests? *New Sci.* **75,** 9–11.
Owen, D. F. (1978). Why do aphids synthesize melezitose? *Oikos* **31,** 264–267.
Owen, D. F. (1980). How plants may benefit from the animals that eat them. *Oikos* **35,** 230–235.
Owen, D. F., and Wiegert, R. G. (1976). Do consumers maximize plant fitness? *Oikos* **27,** 488–492.
Owen, D. F., and Wiegert, R. G. (1981). Mutualism between grasses and grazers: An evolutionary hypothesis. *Oikos* **36,** 376–378.
Owen, D. F., and Wiegert, R. G. (1982a). Grasses and grazers: Is there a mutualism? *Oikos* **38,** 258–259.
Owen, D. F., and Wiegert, R. G. (1982b). Beating the walnut tree: More on grass/grazer mutualism. *Oikos* **39,** 115.
Owen, D. F., and Wiegert, R. G. (1983). Darwin's "abominable mystery" and the need for speculation in evolutionary ecology. *Oikos* **41,** 154–155.
Petelle, M. (1980). Aphids and melezitose: A test of Owen's 1978 hypothesis. *Oikos* **35,** 127–128.
Rackham, O. (1981). Memories of our wildwood. *Nat. World,* pp. 26–29.
Reardon, P. O., Leinweber, C. L., and Merrill, L. B. (1972). The effect of bovine saliva on grasses. *J. Anim. Sci.* **34,** 897–898.
Reardon, P. O., Leinweber, C. L., and Merrill, L. B. (1974). Response of sideoats grama to animal saliva and thiamine. *J. Range Manage.* **27,** 400–401.
Rockwood, L. L. (1973). The effect of defoliation on seed production of six Costa Rican tree species. *Ecology* **54,** 1363–1369.
Strong, D. R., Lawton, J. H., and Southwood, R. (1984). "Insects on Plants." Blackwell, Oxford.
Vittoria, A., and Rendina, N. (1960). Fattori condizionati la funzionalita tiaminica in piante superiori e cenni sugli effetti del bocca dei ruminanti sull erb pascolative. *Acta Med. Vet.* **6,** 379–405.
Went, F. W. (1971). Parallel evolution. *Taxon* **20,** 197–226.
Whitham, T. G. (1981). Individual trees as heterogeneous environments: Adaptation to herbivory or epigenetic noise? *In* "Insect and Life History Patterns: Habitat and Geographic Variations" (R. F. Denno and H. Dingle, eds.), pp. 9–27. Springer-Verlag, Berlin and New York.
Whitham, T. G. (1983). Host manipulation of parasites: Within-plant variation as a defense against rapidly evolving pests. *In* "Variable Plants and Herbivores in Natural and Managed Systems" (R. F. Denno and M. S. McClure, eds.), pp. 15–41. Academic Press, New York.
Whitham, T. G., and Slobodchikoff, C. N. (1981). Evolution by individuals, plant–herbivore interactions, and mosaics of genetic variability: The adaptive significance of somatic mutations in plants. *Oecologia* **49,** 287–292.
Williams, G. C. (1966). "Adaptation and Natural Selection." Princeton Univ. Press, Princeton, New Jersey.

Part II

COMMUNITY STRUCTURE: NATURAL AND MANIPULATED ECOSYSTEMS

Chapter 5

Insect Outbreaks and Community Structure

ANDREW REDFEARN* and STUART L. PIMM*,†

* Graduate Program in Ecology
† Department of Zoology
University of Tennessee
Knoxville, Tennessee 37996

I. INTRODUCTION

A. Approaches and Purposes

An understanding of pest outbreaks is likely to come from a consideration of these phenomena at a number of levels. Certainly a detailed understanding of a particular outbreak must come from a study of the species involved. But the study of individual species is an abstraction: No species exists in isolation from all the other species in its community. Although it would be impossible to study all the species in a community in any detail, we can examine broad community- or ecosystem-level features for correlates of outbreaks. The information such examinations might yield would be quite general, but it still might be useful in suggesting strategies for particular systems.

In this chapter we take both theoretical and empirical approaches and examine the patterns of trophic interactions and their effects on community stability. In particular, we consider whether instability is promoted by the simplification in plant and animal communities caused by the planting of crops in near monoculture.

We have two purposes. First, we hope to eliminate the confusion stemming from the various meanings of the word "instability" and so bring the literature into clearer focus. Second, we hope to find theoretical approaches that might provide fresh insights. There is no guarantee that theory will prove useful, so we shall try to point to field studies that seem best explained by the theories that have been developed.

B. Meanings of "Instability"

Both Elton (1958) and MacArthur (1955) argued forcefully that there might be a relationship between a population's dynamics and the intrinsic properties of the species, as well as those of the community to which it belongs. Both attributed instability to system simplicity. Elton argued that pest outbreaks were just one of many manifestations of instability and that they were more likely to occur in simple, agricultural systems than in complex, natural systems. He relayed a conversation with some tropical foresters that (perhaps mistakenly) led him to believe that pest outbreaks are a feature of simple, temperate forests but not of complex, tropical ones.

MacArthur developed his ideas somewhat more formally. He defined instability in these terms: "Suppose, for some reason, that one species has an abnormal abundance, then we shall say that the community is unstable if the other species change markedly in abundance as a result of the first. The less effect this abnormal abundance has on the other species, the more stable the community." MacArthur defined the correlate of stability, complexity, as "the amount of choice of the energy in going through the [food] web."

It is tempting to be highly critical of these early studies. Elton's arguments are heterogeneous and often based on scant evidence. Agricultural systems differ in many ways from natural ones, and there are some remarkably simple natural systems that are not devastated annually by insect herbivores. (Examples include the large stands of bracken fern, *Pteridium,* studied extensively by J. H. Lawton. Strong *et al.,* 1984, provide a review.) MacArthur's argument is fine as it stands, but is incomplete; it considers changes in abundance of species at the base of food chains. Changing abundances of top predators might (and indeed do) have the opposite effect: The more complex the web, the more widely disturbances may propagate. Yet we consider these early studies to be particularly important. They argue that examining the characteristics of a single species is not enough. We must look at the system to which it belongs. They also point to a wide variety of possible meanings, not just of instability, but of community features that may correlate with instability.

Clearly, what we must do first is to look at the definitions of population stability. Then we must ask these questions: To what extent do the various kinds of instability correspond to pest outbreaks? How do these kinds of instability vary with the properties of the systems to which the species belong? Is there any evidence that outbreaks are more likely to occur in systems with certain trophic structures—simple ones, for example?

In reviewing the meanings of "stability," we have recognized five major ideas: stability (in the strict, mathematical sense), resilience, persistence, resistance, and variability (Pimm, 1984a).

Stability exists if and only if the species densities in a system tend to return to their equilibrium values following disturbances to the densities. In a variable, uncertain world, equilibrium levels may not be the population levels at which species remain; in such cases, equilibrium is better defined as the level below which the population tends to increase and above which the population tends to decrease (Tanner, 1966; Pimm, 1984b). *Resilience* is a measure of how fast a population returns to equilibrium. Resilience is measured in models by the characteristic return time—the time taken for the perturbation (equilibrium density minus the population density) to fall to $1/e$ (~37%) of its initial value. A resilient system has a short return time. *Persistence* measures the time a system lasts before it is changed to a different one—for example, how long a system may last before one equilibrium is replaced by another. *Resistance* is the tendency for a system to remain unchanged by a disturbance. *Variability* includes such measures as the variance, standard deviation, or coefficient of variation of population densities over time.

Stability (in the strict mathematical sense), resilience, and variability seem most relevant to the topic of pest outbreaks, and it is these we consider in the remainder of this chapter.

II. STABILITY

A. The Stability–Complexity Question

Stability is well defined mathematically, and most theoretical studies examine it alone. Early studies (such as Gardner and Ashby, 1970; May, 1972) found that a smaller proportion of models of multispecies systems were stable when there were more species, when a greater proportion of those species interacted (high connectance), and when the species interacted more strongly. This seemed so contradictory to the notions of Elton and MacArthur that considerable efforts were made to evaluate the many unrealistic assumptions these early models required. The patterns of the interactions were made more realistic, as were the parameters and even the form of the equations.

Some reviews of this literature are given by May (1973, 1979) and Pimm (1982, 1984a); to report them in detail here would be repetitious. The initial results, however, seem fairly robust. They can be reversed most easily by using models in which the predators have no effect on their prey's population growth rate (so-called donor-controlled models) (DeAngelis, 1975). This can happen if predators take those prey that are most likely to die from other causes—starvation, for example. There is a large body of literature on removing predatory species from communities or, as in the case of biological control, introducing them. The vast majority of these studies show that predators do have an impact on the densities of their prey (Pimm, 1980). For insects, this impact can be very large, with predators depressing prey populations to a fraction of 1% of the levels in the predator's absence (Beddington *et al.*, 1978). In short, the donor-controlled assumption does not seem to be a good one, and so we are faced with the conclusion that more complex systems are less likely to have a stable equilibrium than simple ones.

On closer inspection, this conclusion does not contradict the ideas of MacArthur and Elton as much as it might superficially appear (Pimm, 1982, 1984a). There is no difficulty with the idea that systems with stable equilibria are likely to persist. Those with unstable equilibria can have one of two initial fates: They can lose species and settle to a new stable equilibrium, or the populations may persist, oscillating probably in some complex manner. For many natural systems these two initial fates are really the same. Large-amplitude oscillations will eventually mean that populations will be driven to such low levels that they will not be able to recover. Thus, stable systems will persist, whereas unstable systems will tend to lose species and simplify to the point where they will contain a stable species assembly.

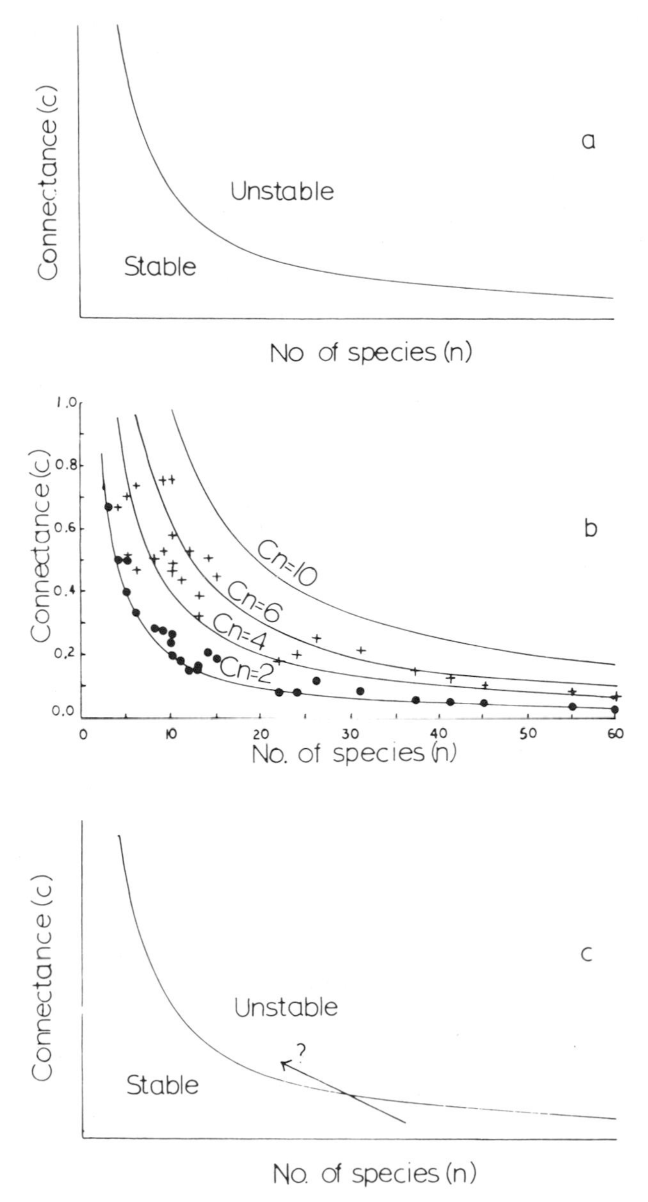

Fig. 1. (a) Values of connectance C and species number n that lead to stable and unstable systems in food web models. This result suggests that natural systems should have values of C and n bounded by a hyperbola. (b) Four hyperbolic approximations and the observed values of C and n for aphids, their plant hosts, and their parasitoids. Discovered trophic connectance (●) was calculated on the basis of only the discovered interactions, whereas potential total C (+) also includes all potential competitive interactions. After Rejmanek and Stary (1979). Reprinted by permission from *Nature*, Vol. 280, No. 5720, pp. 311–313. Copyright © 1979 Macmillan Journals Limited. (c) When species number is reduced, it is possible that the resultant system is unstable, even though, other things being equal, simple systems are more likely to be stable.

Although the quantitative results of stability analyses depend on the various assumptions made, there is a qualitative prediction that seems relatively robust: To retain stability, the product of two measures of community complexity, species number n and connectance C, should be smaller than a critical value (which depends on the strength of the species interactions). If the systems we observe in the real world are those that are stable, observed values of C and n should fall in a region *below* a hyperbolic function, as suggested by Fig. 1a. Data from a variety of communities, including the aphid-dominated systems shown in Fig. 1b, show this to be the case (Rejmanek and Stary, 1979). What this reveals is straightforward: The systems we observe in nature are relatively simple ones. There is now considerable evidence that the patterns of trophic interactions we observe in nature are simpler, in a variety of ways, than we would expect by chance (Pimm, 1982). There seems to be no reason to expect that simple crop ecosystems should not be stable.

B. Species Deletion Stability

1. *Introduction and Model Results*

What happens when we simplify a system, say by reducing the number of species present? How often are we likely to retain a stable system, and how often will the system become unstable (Fig. 1c)? Simple systems may often be stable, but there is no guarantee that we will produce stable systems from simplifying existing stable, complex ones. Answering these questions requires an examination of what we have called "species deletion stability" (Pimm, 1979a).

A system is deemed "species deletion stable" if, after the removal of a species, the remaining $n - 1$ species can coexist at a new, stable equilibrium (Pimm, 1979a, 1980). We can determine a system's species deletion stability repeatedly for the same species and the same model structure, but over ranges of interaction parameters designed to mimic those found in nature. This gives a probability of species deletion stability for that species' removal and for that particular web. These probabilities, averaged over all the species in a food web, vary in much the same way with connectance, species number, and interaction strength as does simple stability. The more complex the model community, the more likely it is that the loss of a species will cause further species losses. Most model systems with complexities anything near those observed in nature are not species deletion stable, nor are natural systems. The vast majority of natural systems cannot withstand species removals without changes in species composition (a review of this literature can be found in Pimm, 1980).

From these studies it might seem that we have an explanation of pest oubreaks in accord with the view of Elton and MacArthur: Instability is caused by simplification (rather than just by simplicity). This may be so, but on close inspection it is not anywhere near as clear-cut as it might seem. Species deletion stability varies markedly depending on which species are removed from a community (Pimm, 1980). The reviews of species removals mentioned earlier tend to focus on the removals of predators or top predators. For these we expect, and find, further species losses. For plant removals, however, particularly those of plants fed on by generalized herbivores, models predict fewer losses. Moreover, these losses should become increasingly less likely with more complex systems. There is less evidence to support this result, but one cannot help notice the lack of an effect of removing chestnut, *Castanea dentata* (Marsh.) Borkh, trees from eastern North America. Chestnuts occupied more than 40% of the canopy in some areas in the early twentieth century and have now almost totally disappeared (Krebs, 1978). Although the disappearance may have caused the loss of seven insect species that fed only on chestnuts, most insects that fed on chestnuts also fed on other tree species (Opler, 1978). There do not seem to have been any losses of vertebrate species.

The simplification practiced in agriculture leads us to ask, How often does the removal of an unwanted plant species cause the crop species' load of insect herbivores to increase? Thus, in reducing the competitors of the plants we wish to harvest, do we make the crop species more vulnerable to attack by insects? This can certainly happen theoretically. A simulation is shown in Fig. 2, where the removal of a plant causes the loss of a generalist predator on a specialist herbivore; the remaining plant species goes to a lower equilibrium than before. However, the models do not tell us the frequency of this occurrence in practice; it is certainly not inevitable, and it may be unlikely. Common sense dictates the conditions under which it will be a likely event—when we remove plant species essential to the survival of generalist predators that have a controlling effect on the herbivores feeding on the crop (just as in Fig. 2). Moreover, the loss due to these herbivores must be greater than the gain obtained from competitive release.

In short, taking an existing system and simplifying it by removing species will usually cause further species losses. It is far from certain, however, whether removing one plant species will cause a decrease in the other remaining plant species by increasing their vulnerability to insect

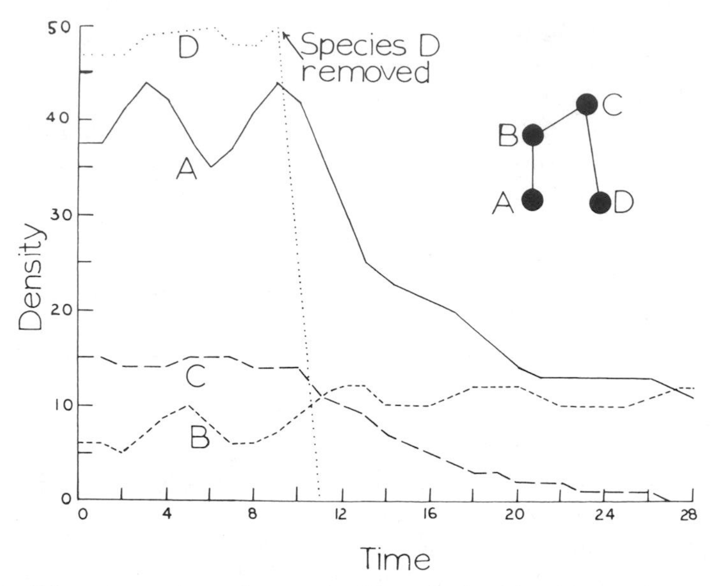

Fig. 2. Effect of removing a plant species (D) on the densities of the species remaining in the system. Note that the other plant species (A) may end with a lower density because of the increased attention of its herbivore (B).

herbivores. How do these results match our observations and intuition about the real world?

2. Some Field Studies

The kinds of studies that have tested the ideas about simplification have asked, How do a crop's insect numbers differ if that crop is grown singly or in a multispecies planting? Root's (1973) work is an early example of such a study. *Brassica* were grown in a single-species planting and also among many other plant species. In multispecies plantings, more species of insects were present throughout the planting, and on the *Brassica* itself, insect herbivores did not reach such high levels. From this, we might conclude that simplification caused a pest outbreak. But how general is this result, and exactly what is being simplified?

Answering these questions requires many other studies. More than 150 such studies have been compiled in a timely and important review by Risch *et al.* (1983). In a highly significant proportion of cases, insect herbivores were more likely to reach high densities in single-species plantings, but there were some important patterns of variation.

Risch *et al.* (1983) argued that increased density in single-species plantings might occur for one of two reasons. First, reduced predator diversity and impact might make herbivore outbreaks more likely. Second, on the basis of the phenomenon described in the "resource concentration" hypothesis (Root, 1973), the plants associated with the crop in a multispecies planting might have a direct effect on the ability of insects to find and utilize the crop. They argued that these associated plants might mask the herbivore's host-finding stimuli, generally reduce movement between individual plants, or in various other ways lower herbivore colonization rates.

The two hypotheses make different predictions about the effects of plant diversification on monophagous and polyphagous herbivores. Both groups might be expected to suffer from the increased attention of predators in multispecies plantings, if this is the cause of the reduced densities. Monophagous species, however, should decline far more than polyphagous species, if the "resource concentration" hypothesis is correct, because for polyphages the multispecies plantings will not represent such a dilution of resources.

The data support the "resource concentration" hypothesis. For monophagous species, 61% of studies showed a decrease in density with multispecies plantings, 10% showed an increase, and the rest were equivocal. For polyphagous species, 27% of the studies showed a decrease and 44% an increase. The differences were highly significant.

Risch *et al.* (1983) went on to consider the differences between herbivores on annual and perennial plant species. They argued that annual species might rely more heavily on escape in time from their herbivores, whereas perennials might rely on chemical defenses to slow herbivore growth. In the latter case, herbivores would be subject to longer periods of exposure to predation. If the reduced numbers in multispecies plantings were due to the effects of predators, we might expect differences between annuals and perennials. Risch *et al.* were unable to detect such an effect. Thus, monophagous herbivores were less abundant in diversified plantings of annuals in 58% of the cases and in 67% of the cases for perennials. For polyphagous species, the corresponding figures were 27 and 28%, respectively.

In short, Risch *et al.* (1983) make a persuasive case that complexity reduces herbivore densities, but the complexity is that of the plant species and the physical effects spacing plants might have. It seems to have little to do with the trophic structure of the insect communities.

C. Summary

Early studies suggested that pest outbreaks in agricultural systems might be due to simplification of the system. Later theoretical studies suggested that there is nothing inherently unstable about simple systems. Indeed, it is the sufficiently complex systems that should be unstable. We might expect such systems to become simplified through species losses. The result should be that the systems we observe in nature are relatively simple compared with what chance dictates (Fig. 1a). This seems to be the case.

Models show that the actual process of simplifying a system by removal of species from it can be expected to cause further species losses and changes in the densities of the remaining species. Removal of plant species can lead to increased herbivore levels on some of the remaining plant species, but this is not inevitable. There is now a large collection of studies that show the effects on insect herbivores of simplifying a system by removing plant species. The insect herbivores are generally more abundant on plant species grown in monoculture, but this seems to have far more to do with the difficulty of getting from host plant to host plant in the multispecies planting than to any trophic interactions.

III. RESILIENCE

A. Does Insecticide Use Increase Resilience?

How fast species' densities recover after a perturbation is a measure of stability that permits comparisons to be made among existing systems. We might usually equate resilience with "stability": The greater the resilience, the faster the system recovers following some reduction in number. There is an interesting twist to this argument for pest outbreaks, however. Let us consider a pest population increasing toward a high density that involves far more pests and far too few of its hosts for our liking. A species with a resilient population will reach unacceptably high levels faster, is more likely to oscillate, and is numerically more likely to get away from its predators than a species that is not resilient. Population resilience can thus be a measure of instability and also a measure of how often we may have to control the pest by chemical means.

In a general way, it is their high resilience that makes insects pests. Small size is coincident with high population growth rates (Southwood, 1981). Within insects, it is also clear why some groups cause a greater problem than others. Aphids, for example, have particularly high population growth rates. It is not, however, on such intrinsic factors that we wish to concentrate. Rather, we ask, How does trophic structure modify a species' rate of return to equilibrium? In the specific terms of insect outbreaks, this means asking how trophic structure affects the rate at which outbreaks reach unacceptable levels.

A major determinant of resilience appears to be food chain length. The longer the food chain, the lower is the resilience of the system's constituent species (Pimm and Lawton, 1977; Pimm, 1979b, 1982). Thus, removing trophic levels will mean that pest outbreaks will occur faster (Pimm, 1984a).

A way to evaluate this idea may also give some insight into why insecticide applications often induce outbreaks. Insecticides clearly kill many pests, but many predators are often killed as well. The predators, moreover, cannot recover until the pests come back. This suggests that there will be a "time window" after insecticide application in which pests have very few predators and so, according to the theory, should increase more rapidly. We pose this question: At comparable densities, do pests increase more rapidly after insecticide applications than do untreated controls?

We might add, parenthetically, that it has often been documented that insect pests can reach higher densities after spraying, but this does not necessarily mean that they are increasing faster at any particular density. Although this idea seems reasonable, it has not been tested experimentally. The tests we now present are new and must be explained in somewhat more detail than those elsewhere in this chapter.

B. Methods

1. Data Collection

We restricted our attention to phytophagous aphids and mites on the above-ground parts of cereal, vegetable, and orchard crops. Why choose these organisms? We must assume that the population changes are a reflection more of reproduction than of emergence and immigration.

Many arthropod pests of agricultural crops overwinter as immatures in the soil, before emerging in the spring. Our assumption would be invalid if we included species in which soil emergence were an important cause of within-season increases in numbers. Environmental conditions during the previous season would clearly influence our estimates of the rate of increase of populations. Immigration can also be a dominant force in causing the increases in numbers we observe. Since aphids can be winged in certain life stages, ideally we should include counts only of the wingless stages. Unfortunately, this would severely reduce our data set, because most studies did not distinguish among the different stages.

We chose aphids and mites because, although we cannot exclude emergence and immigration, the high reproductive capacity of these species, plus other features of their natural history, means that the changes in numbers we observe will more likely reflect the contribution of population growth.

We included studies on commercially farmed land as well as those using experimental plots. We grouped chemicals according to (1) whether they were applied as granules or liquids into the soil and then covered ("soil-applied" pesticides) or (2) whether they were applied as sprays, dusts, or foliage granules above ground ("above-ground" pesticides). It was assumed that pesticides in both categories would reduce the densities of herbivorous species, whereas those in the second category would also severely affect predatory species. Systemic pesticides applied above ground were considered to be in the second category.

The data we assembled were counts of individual species, groups of species, or all mites or aphids encountered. Samples involved one life stage or more. We extracted data from articles appearing in the *Journal of Economic Entomology* during the years 1957–1976. Some articles presented more than one set of data, for example, when more than one pest species, area, or year was studied. All data sets included counts in one untreated plot or field, but most sets contained separate measurements from more than one treated plot or field, each receiving a particular amount or type of chemical. In total, we made 112 comparisons between treated and untreated populations from 34 data sets and from 23 articles. From each data set we extracted pest densities on different dates for the control and for each treatment type. The recorded times represented the time period, in days, between the sampling date and the last treatment date.

We treated the data according to a number of rules, which satisfy three general requirements. First, population growth rates were calculated from density measurements that were based on samples taken over similar dates for the control and for the treated plots. Second, in order to make valid comparisons between control and treated populations, we restricted our attention to treated populations that eventually recovered densities similar to those of the corresponding control populations. Specifically, densities in treated plots had to reach the same order of magnitude as the maximum density measurement taken from the control plots as a requirement for inclusion. Third, only populations that were genuinely increasing were included. We excluded data involved in the drops in populations that often occurred at the end of the season. Details of the data we analyzed can be found in Table 1.

TABLE 1

Summary of Data Sources and Treatments Tested[a]

Type of pesticide, reference	Table from which data were extracted	Organisms counted	Type of crop	Treatment			
				Chemical	Amount applied[b] (lb/acre unless otherwise stated)	Method[c]	Reference no. assigned to treatment
Soil-applied pesticides							
Bacon *et al.* (1976)	1	Apterous green peach aphids	Potato	Disulfoton	3.0	A	1
				Phorate	3.0	A	2
	2	Apterous green peach aphids	Potato	Disulfoton	3.0	C	3
	3	Apterous green peach aphids	Potato	AC-92100	3.0	C	4
				Carbofuron	3.0	C	5
				Disulfoton	3.0	C	6
				Aldicarb	1.5	C	7
Cone (1975)	2	Two-spotted spider mites	Hops	HERC-17413	2.86	G	8
				HERC-18536	2.86	G	9
				Phorate	5.71	G	10
				Tartan	?	G	11
	3	Two-spotted spider mites	Hops	Aldicarb	2 ×2.86[d]	G	12
				Aldicarb	5.71 + 2.86[d]	G	13
				Disulfoton	2 × 2.86[d]	F	14
				Disulfoton	2 × 2.86[d]	E	15
				Disulfoton	2 ×5.71[d]	F	16
				Aldicarb	5.71	G	17
				Disulfoton	5.71	E	18
Daniels (1972)	2	Greenbugs	Irrigated sorghum	Disulfoton	0.15	A	19

(continued)

TABLE 1 (*Continued*)

Type of pesticide, reference	Table from which data were extracted	Organisms counted	Type of crop	Treatment			
				Chemical	Amount applied[b] (lb/acre unless otherwise stated)	Method[c]	Reference no. assigned to treatment
	6	Greenbugs	Dryland sorghum	Disulfoton	0.5	C	20
				Disulfoton	1.0	C	21
Depew (1971)	3	Greenbugs	Irrigated sorghum	Demeton	0.25	D	22
				Demeton	0.75	D	23
				Disulfoton	0.25	D	24
				Phorate	0.25	D	25
Depew (1972)	2	Greenbugs	Irrigated sorghum	Aldicarb	0.25	C	26
				Aldicarb	0.5	C	27
				Aldicarb	1.0	C	28
				Carbofuran	0.25	D	29
				Carbofuran	0.5	C	30
				Carbofuran	0.5	D	31
				Carbofuran	1.0	C	32
				Cyolane	0.5	C	33
				Cyolane	1.0	C	34
				Cytrolane	0.25	D	35
				Cytrolane	0.5	C	36
				Cytrolane	0.5	D	37
				Cytrolane	1.0	C	38
				Disulfoton	0.25	C	39
				Disulfoton	0.25	D	40
				Disulfoton	0.5	D	41
				Methomyl	0.5	D	42
				Methomyl	1.0	D	43

				Phorate	0.25	C	44
				Phorate	0.25	D	45
				Phorate	0.5	C	46
				Phorate	0.5	D	47
Hagel (1970)	3	Two-spotted spider mites	Field beans	Disulfoton	1.0	A	48
				Disulfoton	2.0	A	49
				Disulfoton	3.0	A	50
	4	Two-spotted spider mites	Field beans	Disulfoton	5.0	H	51
Savage and Harrison (1962)	2	Green peach aphids	Tobacco	Phorate	1.0	A	52
Above-ground-applied pesticides							
Anthon (1957)	3	Plum nursery mites	Cherry trees	Parathion (25% WP)[e]	1 lb/100 gal water	I	53
Daniels (1972)	4	Greenbugs	Irrigated sorghum	Disulfoton	0.5	I	54
	4	Corn leaf aphids	Irrigated sorghum	Diethyl parathion	0.5	I	55
Dickinson (1958)	3	Two-spotted spider mites	Apple trees	Aramite (15% WP)	1.5 lb/100 gal water	I	56
Forsythe *et al.* (1962)	1	Pea aphids	Alfalfa	Dibrom	0.5	I	57
				Dibrom	1.0	I	58
				Dylox	0.5	I	59
				Dylox	1.0	I	60
				Ethion	0.5	I	61
				Ethion	1.0	I	62
				Guthion	1.0	I	63
				Methoxychlor	1.0	I	64
				Telodrin	0.5	I	65
				Telodrin	1.0	I	66

(*continued*)

TABLE 1 (*Continued*)

Type of pesticide, reference	Table from which data were extracted	Organisms counted	Type of crop	Treatment			
				Chemical	Amount applied[b] (lb/acre unless otherwise stated)	Method[c]	Reference no. assigned to treatment
Hagel (1970)	4	Two-spotted spider mites	Field beans	Thiodan	1.0	I	67
				Trithion	1.0	I	68
				Aldicarb	2.0	K	69
				Dimethoate	2.0	I	70
				Disulfoton	2.0	K	71
Hale and Shorey (1971)	1 (exp. 2)	Apterous/nymph green peach aphids	Peppers	Carbofuran	1.0	I	72
	1 (exp. 3)	Apterous/nymph green peach aphids	Peppers	Endosulfan	0.75	I	73
	1 (exp. 4)	Apterous/nymph green peach aphids	Peppers	Dimethoate	0.5	I	74
		Apterous/nymph green peach aphids	Peppers	Dimethoate	1.0	I	75
				Endosulfan + cottonseed oil	1.0	I	76
				Endosulfan	1.0	I	77
Harding (1973)	2 (test 3)	Green peach aphids	Spinach	Biothion	0.5	I	78
				Dimethoate	0.5	I	79
				Pirimor	0.25	I	80
				Pirimor	0.5	I	81
	2 (test 4)	Green peach aphids	Spinach	Acephate	0.25	I	82
				Acephate	0.5	I	83
				Carbofuran	0.5	I	84

				Chlordimeform	0.5	I	85
				Formetanate	0.5	I	86
				MGK-Pyrocide growers	0.056	I	87
				NIA-26021	0.1	I	88
				Supracide	1.0	I	89
Johansen (1960)	5	Clover aphids	Red clover	Endrin (1 lb/gal emulsifiable)	0.2	I	90
				Phostex (8 lb/gal emulsifiable)	1.0	I	91
Landis and Schopp (1958)	2	Green peach aphids	Potato	Diazinon	0.5	I	92
					1.0	I	93
					2.0	I	94
Madsen and Bailey (1959)	1	European red mites	Apple trees	10% Phostex-oil miscible	3.6 gal/acre	I	95
Madsen *et al.* (1961)	3 (Sect. 1)	Apple aphids	Apple trees	Bayer 30911	1.5 pt 4 miscible/100 gal	I	96
				Dimethoate	1 pt 4 miscible/100 gal	I	97
				Stauffer-R2968	1 pt 4 miscible/100 gal	I	98
Nelson and Show (1975)	5	Adult two-spotted spider mites	Strawberries	Hexadecylcyclo-propropane 2.1 carboxylate		I	99
Randolph (1957)	2	Spotted alfalfa aphids	Alfalfa	Parathion	0.25	I	100
Reynolds *et al.* (1960)	1	Spider mite, *Tetranychus cinnabarinus*	Sugar beet	3% Thiodan	40.0	J	101
				3% Trithion	40.0	J	102
Savage and Harrison (1962)	2	Green peach aphids	Tobacco	Phosphamidon	0.25	I	103

(*continued*)

TABLE 1 (*Continued*)

Type of pesticide, reference	Table from which data were extracted	Organisms counted	Type of crop	Treatment			
				Chemical	Amount applied[b] (lb/acre unless otherwise stated)	Method[c]	Reference no. assigned to treatment
Stern and Reynolds (1957)	3 (test 5)	Apterous spotted alfalfa aphids	Alfalfa	Phosdrin	0.019	I	104
				Phosdrin	0.0375	I	105
Warren and King (1959)	3	Mite, *Tetranychus hicoriae*	Pecan trees	Malathion	Unknown	I	106
Wene (1957)	1	Cabbage aphids	Cabbage	Demeton	0.25	I	107
				Demeton	0.5	I	108
Westigard and Berry (1970)	4	Yellow spider mites	Pear trees	Carbophenothion (25% WP)	0.38 lb/100 gal water	I	109
	6	Yellow spider mites	Pear trees	Galecron (4 lb/gal emulsifiable concentrate)	0.25 lb/100 gal water	I	110
				GS-13005 (25% WP)	0.5 lb/100 gal water	I	111
				GS-19851 (25% WP)	0.5 lb/100 gal water	I	112

[a] A list of the scientific names and authors for each species mentioned in this table can be found in the Appendix.

[b] Units refer to the amount of *actual toxicant;* when studies failed to mention whether the amount referred to actual toxicant, we assumed that it did.

[c] A, Granules as sidedressing; B, liquid as sidedressing; C, granules applied with seed; D, liquid applied with seed; E, granules stirred into soil around plant crowns; F, liquid stirred into soil around plant crowns; G, stirred into soil around plant crowns; unclear whether liquid or granules used; H, granules applied before planting; I, spray; J, dust; K, foliage granules.

[d] A multiplication or addition sign indicates that chemicals were applied on more than one date.

[e] WP, Wettable powder.

2. *Data Analysis*

We used two methods:

1. We assumed exponential population growth and applied the following simple linear regression model:

$$\text{log density} = a + b(\text{time}) \quad (1)$$

where a is the intercept and b an estimate of r, the intrinsic rate of increase in a population.

2. If population growth is not simply exponential, that is, log density does not increase linearly with time, the addition of a quadratic term to Eq. (1) may improve the model. Therefore, we tested the model

$$\text{log density} = a + b(\text{time}) + c(\text{time})^2 \quad (2)$$

where a, b, and c are constants. Populations may behave in this fashion if, for example, resource limitation were restricting population growth at high densities.

C. Results

1. *Simple Linear Regression Model*

Estimates of the rate of population growth obtained by applying the simple linear regression model are given in Table 2. Values of r are directly related to the amount of time taken for the population to double in density. The units of density used, therefore, do not alter the estimate of r and so are not included in the table. Units of time, however, are not arbitrary, and we have used days as the unit throughout.

From 60 comparisons, 88% of populations treated with above-ground pesticides have higher r values than the corresponding control population. The different comparisons from a particular study may not have been independent, and so treatments were grouped according to the control with which they were compared. A one-tailed paired t test on the differences between the means for the treated populations in each such group and the corresponding control population was used. This test showed that, on average, the treated populations increased significantly faster than the controls ($.0005 < p < .005$) (see Table 3).

For populations treated with soil-applied pesticides, only 67% of 52 comparisons had greater r values than the corresponding control. The t test showed that these treated populations were not increasing at significantly higher rates than the controls ($p > .4$) (see Table 3).

It can be seen from Table 3 that, when aphids and mites were analyzed

TABLE 2

Rates of Population Growth *r*, Density Ranges over Which *r* Values Were Calculated, and Instantaneous Population Growth Rates r_{inst} for All Control and Treated Populations Studied[a]

	Control			Treated		
Ref. no. assigned to treatment[b]	r	Density range[c]	r_{inst}	r	Density range[c]	r_{inst}
Soil-applied pesticides						
	0.0375	10–224				
1			0.0459	0.0554	1–149	0.0359
2			0.0442	0.0504	10–671	0.0436
	0.0616	17–607				
3			0.0592	0.0547	3–288	0.0459
	0.0656	1–454				
4			0.0551	0.0640	1–449	0.0643
5			0.0585	0.0590	1–542	0.0590
6			0.0524	0.0617	1–624	0.0830
	0.0546	11–454				
7			0.1017	0.1183	4–181	0.1030
	0.0419	52–182				
8			0.0340	0.0302	81–574	0.0368
9			0.0350	0.0787	51–514	0.1753
10			0.0351	0.0105	124–312	0.0093
11			0.0352	0.0251	65–440	0.0128
	0.0771	14–313				
12			0.1143	0.0806	25–597	0.0954
13			0.1330	0.0598	9–128	0.1129
14			0.0952	0.0476	115–754	0.0362
15			0.1015	0.0354	106–312	0.0776
16			0.0868	0.0288	153–527	0.0716
	0.1059	1–313				
17			0.1154	0.0523	19–486	0.0468
18			0.1082	0.0585	24–320	0.1468
	0.0791	15–18,000				
19			0.1269	0.0956	12–14,000	0.1401
	0.0729	175–8,500				
20			0.1180	0.0714	65–3,000	0.0854
21			0.1273	0.0957	50–2,000	0.0888
	0.0686	13–305				
22			0.0687	0.0741	18–362	0.0761
23			0.0695	0.0919	11–253	0.1142
24			0.0804	0.1020	1–143	0.1055
25			0.0736	0.0992	3–291	0.0989
	0.0280	80.3–507.9				
26			0.0512	0.0579	10.7–629.1	0.0658
27			0.0597	0.0447	17.2–256.4	0.0389
28			0.0624	0.0501	5.7–171.5	0.0413

TABLE 2 (*Continued*)

Ref. no. assigned to treatment[b]	Control r	Control Density range[c]	Control r_{inst}	Treated r	Treated Density range[c]	Treated r_{inst}
29			0.0633	0.1030	7.4–204.9	0.0884
30			0.0640	0.0529	10.7–131.4	0.0244
31			0.0602	0.0363	16.4–160.9	0.0203
32			0.0694	0.0557	4.5–116.7	0.0372
33			0.0568	0.0454	10.6–258.9	0.0498
34			0.0670	0.0591	6.2–121.0	0.0233
35			0.0622	0.0521	8.6–209.8	0.0392
36			0.0673	0.0653	13.9–113.9	0.0921
37			0.0706	0.0600	2.1–134.5	0.0327
38			0.0682	0.0544	1.7–113.9	0.0386
39			—	0.0319	13.9–120.3	—
40			0.0615	0.0762	3.8–443.1	0.0689
41			0.0652	0.0779	2.8–317.7	0.0664
42			0.0578	0.0588	12.8–236.5	0.0585
43			0.0562	0.0425	13.9–261.3	0.0408
44			0.0453	0.0532	26.5–337.2	0.0470
45			0.0568	0.0542	10.0–361.2	0.0510
46			0.0548	0.0319	21.4–219.9	0.0369
47			0.0582	0.0514	10.9–281.7	0.0473
	0.0270	166–1,253				
48			0.0410	0.0250	168–1,037	0.0281
49			0.0384	0.0280	179–1,450	0.0385
50			0.0410	0.0324	100–1,131	0.0341
	0.0390	225–1,692				
51			0.0432	0.0305	429–2,170	0.0643
	0.0483	51–933				
52			0.0594	0.0048	23–300	0.0701
Above-ground-applied pesticides						
	0.1057	0.1–24.0				
53			0.1909	0.0779	0.3–19.0	0.1171
	0.1413	43–14,500				
54			0.1814	0.1267	150–12,250	0.2701
	0.0945	210–1,500				
55			0.1041	0.1472	100–1,950	0.2504
	0.0610	4.3–36.8				
56			—	0.1213	0.2–13.8	—
	0.0204	47.2–99.9				
57			0.0194	0.0452	11.1–63.2	0.0444
58			0.0192	0.0787	4.7–97.1	0.0857
59			0.0199	0.0313	21.1–80.1	0.0788
60			0.0204	0.0357	34.3–158.5	0.1098

(*continued*)

TABLE 2 (*Continued*)

Ref. no. assigned to treatment[b]	Control			Treated		
	r	Density range[c]	r_{inst}	r	Density range[c]	r_{inst}
61			0.0192	0.0612	6.1–65.9	0.0630
62			0.0186	0.0927	1.5–57.9	0.1040
63			0.0183	0.0897	1.4–42.5	0.0775
64			0.0206	0.0209	61.4–152.8	0.0796
65			0.0194	0.0470	13.3–70.2	0.0633
66			0.0179	0.0998	0.5–28.1	0.1139
67			0.0169	0.1055	0.3–13.5	0.1343
68			0.0177	0.1042	0.5–27.3	0.0780
	0.0390	225–1,692				
69			0.0433	0.0321	500–2,519	0.0351
70			0.0365	0.0883	22–1,932	0.0995
71			0.0437	0.0240	575–2,076	0.0422
	0.0291	3,451–8,826				
72			—	0.0381	409–3,770	—
	0.0487	2,238–10,767				
73			0.0529	0.1012	387–10,113	0.0963
	0.0184	5,490–9,930				
74			0.0519	0.1146	105–4,218	0.1683
75			0.0517	0.0971	42–3,452	0.0673
76			0.0622	0.1062	12–1,290	0.0926
77			0.0544	0.0817	51–1,853	0.0452
	0.0195	245–363				
78			0.0088	0.0412	311–753	0.0626
79			0.0401	0.0528	148–448	0.0682
80			0.0552	0.0637	81–310	0.1039
81			0.0680	0.0771	34–165	0.0315
	0.0369	422–895				
82			0.0841	0.0700	91–410	0.2028
83			0.0962	0.0484	56–158	0.1623
84			0.0626	0.1044	127–1,110	0.1074
85			0.0868	0.0393	95–218	0.0990
86			0.0668	0.0431	219–543	0.0753
87			0.0434	0.0382	522–1,190	0.0818
88			0.0199	0.0570	701–2,400	0.0998
89			0.0306	0.0742	466–2,240	0.0937
	0.0745	89–1,657				
90			0.0756	0.0849	36–1,121	0.0969
91			0.0750	0.0754	92–1,548	0.0855
	0.0718	2,199–11,725				
92			0.0884	0.1399	418–11,489	0.1624
93			0.0983	0.2180	68–12,004	0.2431
94			0.1036	0.2538	26–10,883	0.2844
	0.0244	1.3–13.4				
95			0.0116	0.0323	0.6–12.6	0.0413

TABLE 2 (*Continued*)

Ref. no. assigned to treatment[b]	Control			Treated		
	r	Density range[c]	r_{inst}	r	Density range[c]	r_{inst}
	0.0333	21.7–191.9				
96			0.0515	0.0483	7.1–202.1	0.1051
97			0.0682	0.0738	0.1–155.9	0.0853
98			0.0629	0.0661	0.3–174.9	0.0831
	0.0485	0.9–8.9				
99			0.1168	0.0187	1.1–2.5	0.0181
	0.0363	380.3–1,049.1				
100			0.0286	0.0906	56.3–1,081.3	0.3281
	0.0196	305–574				
101			0.0272	0.0305	102–273	0.0318
102			0.0312	0.0477	38–177	0.0309
	0.1318	145–933				
103			0.06964	0.0784	16–163	0.06968
	0.0751	1,251–4,198				
104			0.0761	0.1913	212–5,105	0.2068
105			0.0775	0.3426	2–1,038	0.6245
	0.0698	8–76				
106			0.0690	0.0378	3–29	0.0397
	0.0303	5.6–28.0				
107			0.0648	0.0571	1.1–21.3	0.0513
108			0.0765	0.0720	0.3–11.9	0.0563
	0.0404	98–398				
109			0.0364	0.0637	8–242	0.0663
	0.0106	2.24–4.48				
110			—	0.0352	0.08–1.68	—
111			—	0.0714	0.08–7.76	—
112			—	0.0254	0.60–3.12	—

[a] The r value for each treated population (as well as reference number) is placed on the lines below the r value for the corresponding control. The r_{inst} values for the control and treated populations are both placed on the same line as the appropriate reference number.
[b] See Table 1.
[c] See relevant paper for units.

separately, this situation held only for aphids. For mites, the rates of increase of control populations were not significantly different from those for populations treated with above-ground or soil-applied pesticides.

2. Model with a Quadratic Term

There was evidence that at least some of the populations studied were not increasing exponentially; an additional term in $(\text{time})^2$ made, in some cases, a significant improvement in the model. For many populations,

TABLE 3

Summary of Our Results[a]

Method	Soil-applied pesticides			Above-ground-applied pesticides		
	Overall	Aphids	Mites	Overall	Aphids	Mites
(a) Log density = $a + b$(time); frequency of times r is higher in treated populations than in the corresponding control population	$\frac{35}{52} = 67.31\%$	$\frac{31}{37} = 83.78\%$	$\frac{4}{15} = 26.67\%$	$\frac{53}{60} = 88.33\%$	$\frac{44}{46} = 95.65\%$	$\frac{9}{14} = 64.29\%$
t_r (from paired t test on control and treated populations' r values)	0.111 $p > .4$ (d.f. = 13)	1.173 $p > .1$ (d.f. = 8)	−1.924 $p > .05$ (d.f. = 4)	2.893 $.0005 < p < .005$ (d.f. = 23)	3.027 $.0005 < p < .005$ (d.f. = 14)	0.672 $p > .25$ (d.f. = 8)
(b) Log density = $a + b$(time) + c(time)^2; frequency of times r_{inst} is higher in treated populations than in the corresponding control populations	$\frac{21}{51} = 41.18\%$	$\frac{16}{36} = 44.44\%$	$\frac{5}{15} = 33.33\%$	$\frac{45}{55} = 81.82\%$	$\frac{41}{45} = 91.11\%$	$\frac{4}{10} = 40.00\%$
r_{inst} (from paired t test on control and treated populations' r_{inst} values)	−0.104 $p > .4$ (d.f. = 13)	−0.056 $p > .45$ (d.f. = 8)	−0.085 $p > .45$ (d.f. = 4)	2.369 $.025 < p < .01$ (d.f. = 20)	3.135 $.0005 < p < .005$ (d.f. = 13)	−0.894 $p > .2$ (d.f. = 6)

[a] Method (a) assumes exponential population growth; method (b) includes a quadratic term in the population growth model.

however, we could not judge whether the quadratic model was a statistical improvement because we had insufficient density measurements.

For populations receiving soil-applied pesticides, in 12 of the 43 cases with sufficient data, population growth was modeled more accurately by the quadratic equation [Eq. (2)]. This was true for 8 control populations of 16 and 1 above-ground-treated population of 17. In all cases where the quadratic model was a good fit, the coefficient c of $(\text{time})^2$ was negative. This result shows that at least some of the populations were progressively increasing more slowly at higher densities.

The nonlinear growth raises the following possibility. The application of above-ground pesticides may not have affected population growth rates directly. Our previous estimates of r for treated populations may have been higher than those for controls due to sampling at different points along a common curve of log density against time, namely, at lower densities. As we might expect, the density measurements used to estimate r values for treated populations were lower, on average, than those for the corresponding control. The lowest density measurement used for a given treated population was lower than that used for the control in more than 75% of the 112 comparisons made in our study.

3. Dealing with the Implications of Nonexponential Growth

To test whether our results are altered if one assumes that growth is nonexponential, one would need to compare growth rates of treated and untreated populations at equal densities. We compared instantaneous rates of increase (which we call r_{inst}) at a density y^* midway between the mean log densities of each corresponding treated and control population. This procedure was carried out for all data sets. The equation

$$T^* = \frac{-b + [b^2 - 4(a - y^*)c]^{1/2}}{2c} \tag{3}$$

gives the time T^* when the density y^* occurred in each treated and control population. The parameters a, b, and c were obtained from the quadratic statistical model [Eq. (2)]. Instantaneous population growth rates can then be found by substituting the result from Eq. (3) into

$$r_{\text{inst}} = b + 2c(T^*) \tag{4}$$

The estimate of r_{inst} for each treated population was thus compared with a unique estimate for the corresponding control at the same density y^*.

The results of this analysis are included in Table 2. In six cases (blanks in the table), Eq. (3) could not solved and r_{inst} could not be calculated. From 55 comparisons, however, 82% of populations treated with above-ground pesticides had higher instantaneous r values than the corresponding controls. A one-tailed paired *t* test compared the mean instantaneous *r* values obtained for treated populations averaged over each data set with the mean values for the controls. The test showed that the treated populations had significantly higher instantaneous *r* values than the controls ($.025 < p < .01$) (see Table 3).

For populations treated with soil-applied pesticides, only 41% of 51 comparisons had greater instantaneous *r* values than the control. The *t* test showed that there was no significant difference between control and treated populations in this respect ($p > .4$) (see Table 3). These results parallel those from the exponential model. Again, when mite populations were analyzed separately, their instantaneous *r* values were not significantly increased by either category of pesticide (see Table 3).

D. Summary

Above-ground pesticides accelerate population growth of aphids but not mites. This accelerated growth is not due entirely to the fact that these insecticides might release the aphid populations from some density-dependent factor, such as resource limitation, which affects the untreated populations at high densities. We reach this conclusion because our results were not altered by assuming density-dependent growth and comparing estimates of the instantaneous rates of increase of treated and untreated populations at equal densities. Furthermore, it is unlikely that the acclerated growth caused by above-ground pesticides was entirely a consequence of the removal of herbivorous species competing with the aphids. These would have been attacked by soil-applied pesticides, which had no significant effect on the population growth rates of the aphids. A more detailed study is required to determine why our hypothesis applies to aphids but not mites. Aphids and mites may suffer from different degrees of predation, and their capacities to increase—either through reproduction or by colonization—are likely to differ.

IV. VARIABILITY

A. Introduction

Since Elton (1958) clearly equated variability with instability, a number of studies have examined the variability of insect populations. Variability, on a year-to-year basis, is not difficult to calculate. Rather more difficult are attempts to understand and predict what the trends in variability should be. How much a population varies may depend on the trophic structure of the community to which a species belongs, but it is also likely to depend on the variability of such extrinsic factors as the weather. Moreover, there is a general tendency in biological data for variability to increase with the mean, although by the use of such measures as the coefficient of variation (standard deviation/mean) it may be possible to avoid such a trend. Simply, variability might be among the most difficult kinds of "stability" in which to detect trends.

B. Two Alternatives

We might expect a close tie between variability and resilience. Resilient populations return quickly to equilibrium and thus may be expected to stay closer to that equilibrium and vary less than populations that are not resilient. Turelli (1978, 1986) has pointed out, however, that this is not the only argument that can be made. In fact, there is an argument that leads to a diametrically opposite result. Resilient populations may chase continually fluctuating resources, whereas less resilient ones may average out such variations. For bird (Pimm, 1984b) and fish (May *et al.*, 1978) populations, the first argument seems to be the better description of nature: More resilient populations vary less than others. For noctuid moths, Spitzer *et al.* (1984) have shown the opposite effect. Species with the greatest fecundity and potential rates of increase were the most variable.

C. Empirical Studies

Looking at variability directly has produced some interesting studies. Wolda (1978) evaluated Elton's (1958) contention that tropical populations vary less than temperate ones. By comparing a large number of populations studies, Wolda showed that this was not the case. However, populations did appear to be more variable in systems, such as deserts, that we might expect to have a less predictable climate.

Watt's (1965) study is more closely related to trophic structure. He analyzed data collected by the Canadian Forest Insect Survey (McGugan, 1958; Prentice, 1962, 1963) on 552 species of forest Macrolepidoptera. Watt (1965) separated gregarious larval feeders from solitary larval feeders, since he had already observed that the former were generally more abundant and more variable. For both of these broad categories, he grouped species of Macrolepidoptera according to their number of tree host species. He then calculated the standard error of the logarithms of the yearly counts for each species and found the mean of this measure of variability for each group. He concluded that groups of moths that had many host trees were generally more variable than specialized groups. This result was obtained by combining data from all regions of Canada.

Watt (1965) suggested that the amount of "available habitat" was the central determinant of variability. Thus, trophically generalized species would have more available habitat than specialized ones and would be able to reach a higher number during periods of favorable conditions. Watt went on to note the spruce budworm, *Choristonera fumiferana* (Clemens), as a species that, though trophically specialized, is highly variable but occurs over large areas of almost pure stands of its food supply (and available habitat).

We have performed a similar analysis for aphid counts from suction traps located at Silwood Park, England. Counts for the years 1969–1983 were obtained from the Rothamsted Experimental Station Reports (Taylor and French, 1970, 1971, 1972, 1973, 1974, 1975, 1976, 1977, 1978, 1979, 1980, 1981; Taylor *et al.*, 1982, 1983, 1984). We calculated the standard deviation of the logarithms of the yearly counts of 28 aphid species and found the total number of British plant genera fed on by each species from Tatchell *et al.* (1983).

We found that the measure of variability was significantly *negatively* correlated with the number of host plant genera ($F = 4.25$; $p = .0495$). Highly polyphagous aphid species were less variable, on a year-to-year basis, than specialized species (see Fig. 3).

The results of Watt's analysis and our own are apparently contradictory, and both data sets require further attention before any definite explanations can be presented. However, one difference between the two analyses stands out. Watt studied Macrolepidoptera inhabiting perennial forests in Canada, whereas we examined aphids living in far more ephemeral, disturbed habitats, namely, agroecosystems. Herbivorous insects living in undisturbed habitats may be more effectively controlled by predators than are those of disturbed habitats [Risch *et al.* (1983) discuss the arguments for the view that control by predators may be more important for herbivores on perennial plants than for those on annual plants.] Thus, perhaps Watt's more polyphagous species were more variable because they were able to reach high numbers during those few times when predator densities were low. In contrast, changes in aphid population densities in British agroecosystems may largely reflect changes in resource availability rather than changes in predators. Polyphagous aphids have a greater choice of alternative food plant species and so may be less susceptible to the loss of one species (and thus be less variable) than specialized species, which may fall to lower population densities during unfavorable years. [This is essentially MacArthur's (1955) argument, discussed earlier.]

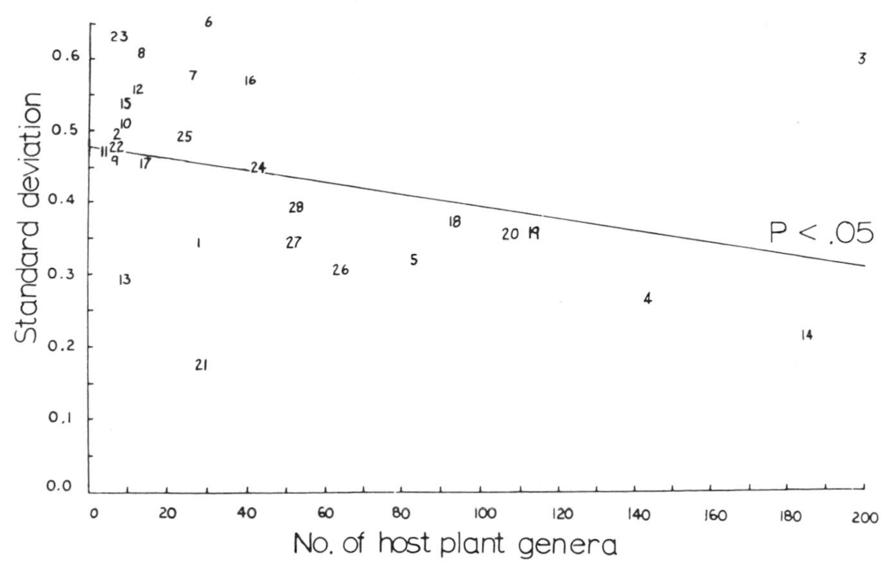

Fig. 3. Standard deviation of the logarithms of yearly counts of different aphid species at Silwood Park, England, versus the number of host plant genera for each of those species. The aphid species are 1, *Acyrthosiphon pisum* (Harris); 2, *Amphorophora rubi* (Kaltenbach); 3, *Aphis fabae* Scopoli; 4, *Aulacorthum solani* (Kaltenbach); 5, *Brachycaudus helichrysi* (Kaltenbach); 6, *Brevicoryne brassicae* (Linnaeus); 7, *Cavariella aegopodii* (Scopoli); 8, *Drepanosiphum platanoidis* (Schrank); 9, *Dysaphis plantaginae* (Passerini), 10, *Elatobium abietinum* (Walker); 11, *Eriosoma ulmi* (Linnaeus); 12, *Hyalopterus pruni* (Geoffroy); 13, *Hyperomyzus lactucae* (Linnaeus); 14, *Macrosiphum euphorbiae* (Thomas); 15, *Megoura viciae* Buckton; 16, *Metopolophium dirhodum* (Walker); 17, *Metopolophium festucae* (Theobald); 18, *Myzus ascalonicus* Doncaster; 19, *Myzus certus* (Walker) and *Myzus persicae* (Sulzer); 20, *Myzus ornatus* Laing; 21, *Nasonovia ribisnigri* (Mosley); 22, *Phorodon humuli* (Schrank); 23, *Phyllaphis fagi* (Linnaeus); 24, *Rhopalosiphum insertum* (Walker); 25, *Rhopalosiphum maidis* (Fitch); 26, *Rhopalosiphum padi* (Linnaeus); 27, *Sitobion avenae* (Fabricius); 28, *Sitobion fragariae* (Walker).

Watt (1965) used the same data set to show that species of Macrolepidoptera that shared their host plant species with many other insect species were, on average, less variable than Macrolepidoptera species that had few potential competitors. This result is perplexing, and "potential" is the key word: Strong *et al.* (1984) and Lawton and Strong (1981) make a persuasive case that competition for resources is not a pervasive feature of herbivorous insect communities. They do show, however, that the number of insect species per plant species is closely correlated with the size of the plant species' geographic range—particularly for plant species with similar physical structure. (All of Watt's data are for trees.) This would seem to change the interpretation of Watt's second result, as follows: Populations of insects on more widely distributed plants are less likely to be variable, a result that is not immediately reconcilable with his "available habitat" idea.

V. SUMMARY AND DISCUSSION

To develop our ideas we have noted several definitions of stability, three of which seem particularly relevant to studies of insect pests:

1. *Stability*. In the strict sense this refers to a population returning (or not, if unstable) to some equilibrium value.
2. *Resilience*. A population may have a stable equilibrium at a level we find unacceptable. Resilience is a measure of how fast the population reaches that level.
3. *Variability*. A highly variable population will sporadically reach unacceptable levels.

The relationships between trophic structure and these kinds of instability are varied. Despite early suggestions, there seems to be nothing inherently unstable about simple communities. Theory predicts that communities should be simple relative to what they could be (Fig. 1a), and data show this to be the case (Fig. 1b). There are some natural stable communities whose floristic simplicity approaches that of agricultural monocultures. Simply, pest outbreaks in agricultural systems do not appear to be an ecologically inevitable consequence of trophic simplicity.

The actual process of simplifying a trophic structure will almost certainly cause changes in the densities of the remaining species, but it is far from obvious how and if plant removal will predispose the remaining herbivores to become pests. Actual studies comparing herbivores in monocultures with multispecies plantings show that many species do increase in density in monocultures. Detailed studies, however, refute the idea that this is related to changes in the control effected by the predators.

Populations do appear to become more resilient as the system's trophic structure becomes simpler.

Temperate population densities seem no more variable than tropical ones. The relationship between variability and the degree of polyphagy can be either negative or positive. It is possible that in predator-limited groups, polyphagous species can attain unusually high densities when their predators are unusually rare. Monophagous species will not reach such high densities and thus will not vary as much. For resource-limited groups, polyphagous species may vary less than monophagous species, because the polyphages are less susceptible to the failure of one of their food plant species.

We have presented the thesis that insect pest outbreaks may be considered "dynamic instabilities" and instability may be related not just to the intrinsic properties of the species, but to the trophic structure of the community to which the species belongs. Though the results are varied, the thesis seems correct and it argues persuasively for our recognizing the trophic structure of the community in which an outbreak occurs.

ACKNOWLEDGMENTS

We thank Dr. David Andow, Dr. John Lawton, Dr. Donald Strong, Jr., Dr. Jack C. Schultz, and Dr. Pedro Barbosa for valuable discussions.

REFERENCES

Anthon, E. W. (1957). Control of the plum nursery mite on sweet cherries. *J. Econ. Entomol.* **50,** 564–566.

Bacon, O. G., Burton, V. E., Mclean, D. L., James, R. H., Riley, W. D., Baghott, K. G., and Kinsey, M. G. (1976). Control of the green peach aphid and its effect on the incidence of potato leaf roll virus. *J. Econ. Entomol.* **69**, 410–414.

Beddington, J. R., Free, C. A., and Lawton, J. H. (1978). Modelling biological control: On the characteristics of successful natural enemies. *Nature* (*London*) **273,** 513–519.

Cone, W. W. (1975). Crown-applied systemic acaricides for control of the two-spotted spider mite and hop aphid on hops. *J. Econ. Entomol.* **68,** 684–686.

Daniels, N. E. (1972). Insecticidal control of greenbugs on grain sorghum. *J. Econ. Entomol.* **65,** 235–240.

DeAngelis, D. L. (1975). Stability and connectance in food web models. *Ecology* **56,** 238–243.

Depew, L. J. (1971). Evaluation of foliar and soil treatments for greenbug control on sorghum. *J. Econ. Entomol.* **64,** 169–172.

Depew, L. J. (1972). Further evaluation of insecticides for greenbug control on grain sorghum in Kansas. *J. Econ. Entomol.* **65,** 1095–1098.

Dickinson, B. C. (1958). Ethion, a promising new acaricide and insecticide. *J. Econ. Entomol.* **51,** 354–357.

Elton, C. S. (1958). "The Ecology and Invasions by Animals and Plants." Chapman & Hall, London.

Forsythe, H. Y., Jr., Hardee, D. D., and Gyrisco, G. G. (1962). Field tests for the control of certain alfalfa insect pests in New York. *J. Econ. Entomol.* **55,** 828–830.

Gardner, M. R., and Ashby, W. R. (1970). Connectance of large, dynamical (cybernetic) systems: Critical values for stability. *Nature* (*London*) **228,** 784.

Hagel, G. T. (1970). Systemic insecticides and control of insects and mites on beans. *J. Econ. Entomol.* **63,** 1486–1489.

Hale, R. L., and Shorey, H. H. (1971). Effect of foliar sprays on the green peach aphid on peppers in southern California. *J. Econ. Entomol.* **64,** 547–549.

Harding, J. A. (1973). Green peach aphid: Field trials with newer insecticides on cabbage and spinach. *J. Econ. Entomol.* **66,** 459–460.

Johansen, C. (1960). Bee poisoning versus clover aphid control in red clover grown for seed. *J. Econ. Entomol.* **53,** 1012–1015.

Krebs, C. J. (1978). "Ecology: The Experimental Analysis of Distribution and Abundance." Harper & Row, New York.

Landis, B. J., and Schopp, R. (1958). Further studies with systemic insecticides against the green peach aphid on potatoes. *J. Econ. Entomol.* **51,** 138–140.

Lawton, J. H., and Strong, D. R., Jr. (1981). Community patterns and competition in folivorous insects. *Am. Nat.* **118,** 317–338.

MacArthur, R. H. (1955). Fluctuations of animal populations and a measure of community stability. *Ecology* **36,** 533–536.

McGugan, B. M. (1958). "Forest Lepidoptera of Canada," Vol. 1, Publ. 1034. Can. Dep. Agric., Ottawa.

Madsen, H. F., and Bailey, J. B. (1959). Control of the apple aphid and the rosy apple aphid with new spray chemicals. *J. Econ. Entomol.* **52,** 493–496.

Madsen, H. F., Westigard, P. H., and Falcon, L. A. (1961). Evaluation of insecticides and sampling methods against the apple aphid, *Aphis pomi*. *J. Econ. Entomol.* **54,** 892–894.

May, R. M. (1972). Will a large complex system be stable? *Nature* (*London*) **238,** 413–414.

May, R. M. (1973). "Stability and Complexity in Model Ecosystems." Princeton Univ. Press, Princeton, New Jersey.

May, R. M. (1979). The structure and dynamics of ecological communities. *Symp. Br. Ecol. Soc.* **20,** 385–407.

May, R. M., Beddington, J. R., Horwood, J. W., and Shepherd, J. F. (1978). Exploiting natural populations in an uncertain world. *Math. Biosci.* **42,** 219–252.

Nelson, R. D., and Show, E. D. (1975). A novel group of miticides containing the cyclopropane moiety: Laboratory experiments and field studies in strawberries on the two-spotted spider mite. *J. Econ. Entomol.* **68,** 261–266.

Opler, P. A. (1978). Insects of American Chestnut: Possible importance and conservation concern. *Proc. Amer. Chestnut Symp.* 83–84.

Pimm, S. L. (1979a). Complexity and stability: Another look at MacArthur's original hypothesis. *Oikos* **33,** 351–357.

Pimm, S. L. (1979b). The structure of food webs. *Theor. Popul. Biol.* **16,** 144–158.

Pimm, S. L. (1980). Food web design and the effects of species deletion. *Oikos* **35,** 139–149.

Pimm, S. L. (1982). "Food Webs." Chapman & Hall, London.

Pimm, S. L. (1984a). The complexity and stability of ecosystems. *Nature* (*London*) **307,** 321–326.

Pimm, S. L. (1984b). Food chains and return times. *In* "Community Ecology: Conceptual Issues and the Evidence" (D. R. Strong, Jr., D. Simberloff, L. G. Abele, and A. B. Thistle, eds.), pp. 397–412. Princeton Univ. Press, Princeton, New Jersey.

Pimm, S. L., and Lawton, J. H. (1977). The number of trophic levels in ecological communities. *Nature* (*London*) **268,** 329–331.

Prentice, R. M. (1962). "Forest Lepidoptera in Canada," Vol. 2, Bull. 128. Can. Dep. For., Ottawa.

Prentice, R. M. (1963). "Forest Lepidoptera in Canada," Vol. 3, Publ. 1013. Can. Dep. For., Ottawa.

Randolph, N. M. (1957). Control of the spotted alfalfa aphid on alfalfa. *J. Econ. Entomol.* **50,** 124–126.

Rejmanek, M., and Stary, P. (1979). Connectance in real biotic communities and critical values for stability in model ecosystems. *Nature* (*London*) **280,** 311–313.

Reynolds, H. T., Fukuto, T. R., and Peterson, G. D., Jr. (1960). Effect of topical applications of granulated systemic insecticides and of conventional applications of other insecticides on control of insects and spider mites on sugar beet plants. *J. Econ. Entomol.* **53,** 725–729.

Risch, S. J., Andow, D., and Altieri, M. (1983). Agroecosystem diversity and pest control: Data, tentative conclusions and new research directions. *Environ. Entomol.* **12,** 625–629.

Root, R. B. (1973). Organization of a plant arthropod association in simple and diverse habitats: The fauna of collards, *Brassica oleracea. Ecol. Monogr.* **43,** 95–124.

Savage, L. B., and Harrison, F. P. (1962). Control of the green peach aphid on tobacco with systemic insecticides. *J. Econ. Entomol.* **55,** 623–626.

Southwood, T. R. E. (1981). Bionomic strategies and populations parameters. *In* "Theoretical Ecology: Principles and Applications" (R. M. May, ed.), pp. 30–52. Blackwell, Oxford.

Spitzer, K., Rejmanek, M., and Soldan, T. (1984). The fecundity and long-term variability in abundance of noctuid moths (Lepidoptera, Noctuidae). *Oecologia* **62,** 91–93.

Stern, V. M., and Reynolds, H. T. (1957). Developments in chemical control of the spotted alfalfa aphid in California, 1955–56. *J. Econ. Entomol.* **50,** 817–821.

Strong, D. R., Jr., Lawton, J. H., and Southwood, T. R. E. (1984). "Insects on Plants." Blackwell, Oxford.

Tanner, J. T. (1966). Effects of population density on the growth rates of animal populations. *Ecology 47,* 433–445.

Tatchell, G. M., Parker, S. J., and Woiwood, I. P. (1983). Synoptic monitoring of migrant insect pests in Great Britain and Western Europe. IV. Host plants and their distribution for pest aphids in Great Britain. *In* "Rothamsted Experimental Station Report for 1982," Part 2, pp. 45–159. Bartholomew Press, Dorking.

Taylor, L. R., and French, R. A. (1970). Rothamsted insect survey. *In* "Rothamsted Experimental Station Report for 1969," Part 2, pp. 168–185. Bartholomew Press, Dorking.

Taylor, L. R., and French, R. A. (1971). Rothamsted insect survey. *In* "Rothamsted Experimental Station Report for 1970," Part 2, pp. 237–253. Bartholomew Press, Dorking.

Taylor, L. R., and French, R. A. (1972). Rothamsted insect survey. *In* "Rothamsted Experimental Station Report for 1971," Part 2, pp. 181–199. Bartholomew Press, Dorking.

Taylor, L. R., and French, R. A. (1973). Rothamsted insect survey. Fourth Report. *In* "Rothamsted Experimental Station Report for 1972," Part 2, pp. 182–211. Bartholomew Press, Dorking.

Taylor, L. R., and French, R. A. (1974). Rothamsted insect survey. Fifth annual summary. *In* "Rothamsted Experimental Station Report for 1973," Part 2, pp. 240–269. Bartholomew Press, Dorking.

Taylor, L. R., and French, R. A. (1975). Rothamsted insect survey. Sixth annual summary. *In* "Rothamsted Experimental Station Report for 1974," Part 2, pp. 201–229. Bartholomew Press, Dorking.

Taylor, L. R., and French, R. A. (1976). Rothamsted insect survey. Seventh annual summary. *In* "Rothamsted Experimental Station Report for 1975," Part 2, pp. 97–128. Bartholomew Press, Dorking.

Taylor, L. R., and French, R. A. (1977). Rothamsted insect survey. Eighth annual summary. *In* "Rothamsted Experimental Station Report for 1976," Part 2, pp. 195–227. Bartholomew Press, Dorking.

Taylor, L. R., and French, R. A. (1978). Rothamsted insect survey. Ninth annual summary. *In* "Rothamsted Experimental Station Report for 1977," Part 2, pp. 79–112. Bartholomew Press, Dorking.

Taylor, L. R., and French, R. A. (1979). Rothamsted insect survey. Tenth annual summary. *In* "Rothamsted Experimental Station Report for 1978," Part 2, pp. 137–173. Bartholomew Press, Dorking.

Taylor, L. R., and French, R. A. (1980). Rothamsted insect survey. Eleventh annual summary. *In* "Rothamsted Experimental Station Report for 1979," Part 2, pp. 111–137. Bartholomew Press, Dorking.

Taylor, L. R., and French, R. A. (1981). Rothamsted insect survey. Twelfth annual summary. *In* "Rothamsted Experimental Station Report for 1980," Part 2, pp. 123–151. Bartholomew Press, Dorking.

Taylor, L. R., Macaulay, E. D. M., Dupoch, M. J., and Nicklen, J. (1982). Rothamsted insect survey. Thirteenth annual summary. *In* "Rothamsted Experimental Station Report for 1981," Part 2, pp. 129–157. Bartholomew Press, Dorking.

Taylor, L. R., Woiwood, I. P., Macaulay, E. D. M., Dupuch, M. J., and Nicklen, J. (1983). Rothamsted insect survey. Fourteenth annual summary. *In* "Rothamsted Experimental Station Report for 1982," Part 2, pp. 169–202. Bartholomew Press, Dorking.

Taylor, L. R., Woiwood, I. P., Macaulay, E. D. M., Dupuch, M. J., and Nicklen, J. (1984). Rothamsted insect survey. Fifteenth annual summary. *In* "Rothamsted Experimental Station Report for 1983," Part 2, pp. 301–331. Bartholomew Press, Dorking.

Turelli, M. (1978). A reexamination of stability in randomly versus deterministic environments with some comments on the stochastic theory of limiting similarity. *Theor. Popul. Biol.* **13,** 244–267.

Turelli, M. (1986). Stochastic community theory: A partially guided tour. *In* "Mathematical Ecology: An Introduction" (T. G. Hallam and S. A. Levin, eds.). Springer-Verlag, Berlin and New York, pp. 321–339.

Warren, F. W., III, and King, D. R. (1959). The biotic effect of insecticides on populations of aphids and mites of pecans. *J. Econ. Entomol.* **52,** 163–165.

Watt, K. E. F. (1965). Community stability and the strategy of biological control. *Can. Entomol.* **97,** 887–895.

Wene, G. P. (1957). Cabbage aphid control. *J. Econ. Entomol.* **50,** 576–577.

Westigard, P. H., and Berry, D. W. (1970). Life history and control of the yellow spider mite on pear in southern Oregon. *J. Econ. Entomol.* **63,** 1433–1437.

Wolda, H. (1978). Fluctuations in abundance of tropical insects. *Am. Nat.* **112,** 1017–1045.

APPENDIX: SCIENTIFIC NAMES AND AUTHORS OF SPECIES MENTIONED IN TABLE 1

Common name	Scientific name and author
Apple aphid	*Aphis pomi* De Geer
Cabbage aphid	*Brevicoryne brassicae* (Linnaeus)
Clover aphid	*Anuraphis bakeri* (Cowen)
Corn leaf aphid	*Rhopalosiphum maidis* (Fitch)
European red mite	*Panonychus ulmi* (Koch)
Greenbug	*Schizaphis graminum* (Rondani)
Green peach aphid	*Myzus persicae* (Sulzer)
Pea aphid	*Macrosiphum pisi* (Harris)
Plum nursery mite	*Vasates fockeui* (Nal.)

APPENDIX (*Continued*)

Common name	Scientific name and author
Spotted alfalfa aphid	*Pterocallidium* sp. (in Randolph, 1957)
	Therioaphis maculata (Buckton) (in Stern and Reynolds, 1957)
Two-spotted spider mite	*Tetranychus urticae* Koch (in Cone, 1975; Hagel, 1970; Nelson and Show, 1975)
	Tetranychus telarius (Linnaeus) (in Dickinson, 1958)
Unspecified spider mite	*Tetranychus cinnabarinus* (Boisduval)
Unspecified mite	*Tetranychus hicoriae* McGregor
Yellow spider mite	*Eotetranychus carpini borealis* (Ewing)

Chapter 6

Community Structure and Folivorous Insect Outbreaks: The Roles of Vertical and Horizontal Interactions

STANLEY H. FAETH

Department of Zoology
Arizona State University
Tempe, Arizona 85287

I. INTRODUCTION

Outbreaks of folivorous (foliage-feeding) insects are rare events considering the myriad insect species that feed on plants. At least one-fourth of the earth's macroscopic species are phytophagous insects (Southwood, 1973; Strong, 1983), and a large fraction of these are folivorous insects on trees. Yet relatively few species undergo violent oscillations in population size such that trees are severely defoliated (see below). The rarity of

INSECT OUTBREAKS

outbreaks implies that most folivorous insect species are under continuous and rigorous control. The goal of community ecology is to determine whether controls are governed by internal organization of insect communities or by extrinsic factors. If the rarity of outbreaks is explained largely by insect community structure, occasional outbreaks can be viewed simply as symptomatic of some breakdown of internal organization. Outbreaks can then be considered within the framework of community ecology theory, and the intensity and frequency of outbreaks can be viewed as another pattern resulting from community processes.

This is certainly an optimistic viewpoint. Problems and controversies abound in determining factors responsible for insect outbreaks, not the least of which is reconciling community ecology with patterns of insect outbreaks. The purpose of this chapter is to evaluate these factors, particularly interactions, within the framework of community ecology and to suggest modifications of current theory. The focus is on folivorous insects that feed on forest trees; these represent systems in which the paucity of insect outbreaks is especially puzzling.

Outbreaks of Insects on Forest Trees: An Anomaly

Schultz (1983a) argued that the rarity of insect outbreaks on forest trees is paradoxical. Trees are particularly long-lived and certainly available to many potential defoliators during their long life span. For example, more than 120 lepidopteran species feed on *Quercus robur,* English oak (Feeny, 1970). Considering the high fecundity of insects and their potential to evolve counteradaptations to defenses of trees that exhibit long life spans, one expects population explosions and resulting defoliation to be commonplace. Yet this clearly is not the case. For example, Schultz (1983a) stated that <10% of lepidopteran species on Canadian trees show periodic or occasional outbreaks. Of the vast number of insect species that feed on western North American trees, only 31 species are considered major pests and undergo outbreaks (Furniss and Carolin, 1977). Of these, only 16 species are defoliators; the remaining 15 species are bark and wood feeders. Three of the 16 defoliator species are introduced pests, like the gypsy moth, a major defoliator in northeastern forests. Subtracting these introduced species reduces the number of outbreak species to a paltry fraction of the pool of leaf-feeding species.

One might expect that all species of forest insect folivores collectively would remove a considerable amount of leaf area, even though few species exhibit high population densities. A tree species may have 100 or more potential leaf feeders (e.g., 120 lepidopteran species on *Q. robur*). If each species consumed only 1% of the total leaf area, complete defoliation would result. This usually does not occur. Folivorous insects collectively remove less than 10% of primary production in most forests (Wiegert and Owen, 1971; Golley, 1972; Mattson and Addy, 1975). Typically, the annual amount of leaf area removed by leaf consumers is surprisingly low, especially in temperate forests (Table 1). Even in tropical forests, where more stable climatic conditions seem favorable to insect growth and survivorship, leaf area consumed by phytophages is only a small fraction of total production (Table 1). A possible exception to this rule is Australian *Eucalyptus* and *Jarrah* forests, where chronic consumption can range from 10 to 50% of total leaf area (Fox and McCauley, 1977; Morrow and LaMarche, 1978). However, more recent estimates of insect consumption in Australian forests (Ohmart *et al.*, 1983; Ohmart, 1984) are much lower (2–3%) and generally similar to those in other forests (but see Fox and Morrow, 1986). Despite fairly extensive evidence that insects consume very small fractions of leaf area, little attention has been focused on the percentage of leaves grazed. Edwards and Wratten (1983) contend that a large percentage of leaves on several tree species are damaged by insects, but little leaf area is actually removed. This is a very important, yet largely overlooked point. Only two studies in Table 1 present data on percentage of leaves damaged. Renaud (1987) reports that 90 and 70% of leaves of *Juglans arizonica* in desert and montane riparian zones, respectively, suffer herbivory. Leaf chewers on *Quercus emoryi* removed >5% of the total leaf area, but leaves with >10% leaf area removed by insects comprise at least 30% of all leaves (Faeth, 1985a). The relation of this pattern to horizontal interactions among folivorous insects is discussed later in this chapter.

One must conclude, therefore, that the majority of insect species feeding on forest trees are extremely rare and consume minuscule fractions of leaf area, since most tree species potentially are fed on by tens or hundreds of species. Yet overall annual consumption of leaf area is typically low. These patterns of amazingly low levels of foliage consumption suggest very strong constraints on population expansion of folivorous insect species.

Alternatively, infrequent, yet violent outbreaks (see Chapter 2, this volume, for definition of "outbreak") of certain insect species must involve an escape from one or more of these constraints. I now examine potential factors controlling insect populations on trees within the framework of classical community ecology. I then argue that conventional community ecology theories are inadequate to explain the rarity of outbreaks. Finally, I propose modifications to current theory that are needed to account for patterns of insect outbreaks.

TABLE 1

Annual Amount of Leaf Area of Forest Trees Consumed by Herbivores

Tree species	Locality	Habitat	Geographic area	Area consumed by herbivores (%)	Reference
Trichilia cipo	Panama	Rain forest	Tropical	11	Coley (1982, 1983a)
13 persistent spp.	Panama	Rain forest	Tropical	21	Coley (1982)
38 shrub spp.	Puerto Rico	Rain forest	Tropical	7	Odum and Ruiz-Reyes (1970)
Rain forest	Barro Colorado Island	Rain forest (dry season)	Tropical	10	Leigh and Smythe (1978)
		Rain forest (wet season)	Tropical	7	Leigh and Smythe (1978)
Eucalyptus spp.	Australia	Unknown	Unknown	20–50	Fox and McCauley (1977), Morrow and LaMarche (1978)
	Australia	Subalpine	Temperate	2–3	Ohmart (1984), Ohmart *et al.* (1983)
Beech forest	Russia	Mesophytic	Temperate	8	Bukovskii (1936, from Pimentel *et al.*, 1975)
Quercus forest	Tennessee	Mesophytic	Temperate	6.7	Rothacher *et al.* (1954)
Prunus–Quercus forest	Southern Ontario	Mesophytic	Temperate	11.7	Bray (1961)
Quercus forest	Southern Ontario	Xeric upland	Temperate	10.6	Bray (1961, 1964)
Acer–Fagus forest	Southern Ontario	Moist lowland	Temperate	5.9	Bray (1961, 1964)
Liriodendron forest	Tennessee	Mesophytic	Temperate	7.7	Reichle *et al.* (1973)
Several tree species	Unknown	Mesophytic	Temperate	4–11	Edwards and Wratten (1983)
Quercus emoryi	Arizona	Riparian (chaparral)	Temperate	<5	S. H. Faeth (unpublished data)
Juglans arizonica	Arizona	Riparian (montane)	Temperate	6	Renaud (1987)
	Arizona	Riparian (desert)	Temperate	40	Renaud (1987)
Quercus rubra	Missouri	Mesophytic	Temperate	22	Linit *et al.* (1986)

II. COMMUNITY ECOLOGY AND FOLIVOROUS INSECTS

Lawton and Strong (1981) stated that "for community ecology, the primary null hypothesis is that species coexist independently, without effective interaction, and without evolution having made any combinations of species more compatible." Their review focused on folivorous insect communities and especially competitive interactions among folivorous insect species. Obviously, the alternative hypothesis is that folivorous insect species are not independent of one another and that their interactions serve as the primary foundation for community structure. With regard to insect outbreaks on trees, the relevant questions are the following: (1) Is the infrequency of outbreaks determined by constraints imposed by interactions of folivorous insect species? (2) When outbreaks do occur, are they the result of a breakdown of these constraints?

If species interactions are important determinants of folivorous insect community structure and resulting patterns, insect outbreaks and community theory should be inseparable. Insect outbreaks could then be examined under the umbrella of community ecology theory rather than as unrelated and disjunct events. If, however, folivorous insect communities are noninteractive random assemblages, community ecology would fail to embrace insect outbreaks within its framework.

Few ecologists think that assemblages of folivorous insects are strictly random. For example, Lawton and Strong (1981) and Strong *et al.* (1984) concluded that vertical interactions (interactions among trophic levels) and autecological factors are primarily responsible for most community patterns observed in folivorous insect communities. Vertical interactions include those with the host plant or with natural enemies. Autecological factors are influences of climate, such as changing temperature and humidity regima. In short, Lawton and Strong (1981), Strong *et al.* (1984), and Lawton and MacGarvin (1986) discount horizontal interactions (interactions within a trophic level, e.g., interspecific competition among folivorous insect species) as important factors in community organization of folivorous insects. In terms of insect outbreaks, or lack thereof, this proposal means that horizontal interactions play little or no role.

I consider to be premature the view that within-trophic-level interactions are insignificant in community structure and in insect outbreaks on trees. In the following sections, I examine the role of autecological factors and between- and within-trophic-level interactions in patterns of community structure and folivorous insect outbreaks on trees. I propose that these factors very likely interact to serve as very effective constraints on folivorous insect population growth, thus accounting for the rarity of outbreaks. Folivorous insects face conflicting selective pressures, such that evolution in one direction is opposed by counterpressure from other factors. I suggest that any theory that explains outbreak within a framework of community ecology will necessarily be complex and multifaceted.

III. AUTECOLOGICAL FACTORS

I consider autecological factors to include primarily general climatic conditions and, more specifically, localized weather factors such as changing humidity and temperature. This view is quite different from that of Lawton and Strong (1981), who considered host plant factors (i.e., plant toxins, nutrition, phenology, and distribution) to be autecological factors. Although the distinction is not critical, since these plant factors can be modified in ecological and evolutionary time by insect feeding (discussed later) I think it more logical to include host plant factors as between-trophic-level interactions (vertical interactions).

Certainly, harsh and variable climatic conditions can have major impacts on population densities of folivorous insects (Andrewartha and Birch, 1954, 1984). Harsh weather can reduce population densities such that interactions among insect species are unlikely to occur (McClure and Price, 1975, 1976). However, it seems that climatic conditions alone are inadequate to explain the rarity of insect outbreaks, for two reasons. First, if harsh climatic conditions played the major role in limiting outbreaks, then outbreaks of folivorous insects would be more frequent in relatively benign climates, such as the tropics. Certainly, tropical forests are not entirely constant and exhibit regular wet and dry seasons (Wolda, 1978). Nevertheless, compared with the radical fluctuations in climate in temperate and subarctic forests, conditions in tropical forests are much more conducive to insect growth, survival, and reproduction. Yet levels of defoliation are very comparable in temperate and tropical forests (Odum and Ruiz-Reyes, 1970; see also Table 1). Second, human-made monocultures in both temperate and tropical systems are very vulnerable to insect outbreaks (Goodman, 1975). If weather were the sole determinant of frequencies of insect outbreaks, one would expect the occurrence of insect outbreaks in monocultures to be comparable to that in surrounding forests. Since outbreaks are much higher in monocultures of trees in both tropical and temperate areas (Goodman, 1975), weather factors alone are not responsible for outbreaks. Aspects of weather may be correlated with insect abundance, but the pattern often varies radically for individual insect species within the same forest or among forests in the same region (Kendeigh, 1979). Other mortality factors (e.g., intensity of predation) are often collinear with weather variables (Kendeigh, 1979). Climatic factors, however, can interact with other factors, such as plant chemistry and nutrition (White, 1978), and this aspect of climate is discussed in Section IV.

IV. INTERACTIONS

Folivorous insects potentially interact with species at three different trophic levels: the host plant, other folivores, and natural enemies. I argue, as have others (Lawton and MacNeil, 1979; Price *et al.,* 1980; Whitham, 1981, 1983; Schultz, 1983a,b), that these three sources of interaction impose conflicting selective pressures on insect growth, survival, and reproduction, such that evolution to minimize negative interactions at one level may increase selective pressures at one or both of the other levels. Therefore, few if any interactions at one level are mutually exclusive of the others. Furthermore, multiple within-level interactions may exhibit similar opposing selective pressures. I believe this intricate system of counteracting forces is the key to understanding insect outbreaks and, generally, organization of folivorous insect communities. This view necessitates a highly complex and often very subtle structure of folivorous insect communities.

A. Vertical Interactions: Natural Enemies

Natural enemies of folivorous insects include parasites, predators, and pathogens. Pathogens include a large array of disease-causing organisms such as viruses, bacteria, protozoans, and fungi. Technically, pathogens are simply microscopic parasites (May, 1983; Toft, 1986). Macroscopic invertebrate parasites of folivorous insects are usually mites and insects, especially in the orders Hymenoptera and Diptera. Most macroscopic species directly kill their insect hosts and are therefore termed parasitoids (Price, 1975). Predators can be roughly divided into invertebrate and vertebrate species.

Pathogens, parasites, and predators can undoubtedly have a major impact on populations of folivorous insects. In fact, community ecology of folivorous insects has seen a resurgence of Hairston *et al.*'s (1960) hypothesis that insect populations are controlled primarily by interactions with natural enemies rather than competitive interactions among folivorous insect species (Lawton and Strong, 1981; Schoener, 1983; Strong *et al.,* 1984; but see Connell, 1983; Faeth, 1985b, 1986; Karban, 1986; Harrison and Karban, 1986).

1. Pathogens

Anderson and May (1980) considered pathogens (microparasites) to be primary regulators of populations of forest insects, especially of insect species prone to periodic outbreak cycles. However, models of control via pathogens are based on simplistic parasite–host equations (Anderson and May, 1980) that are constrained by unrealistic assumptions (Heck, 1976). Furthermore, these models ignore other regulatory factors such as interactions with the host plant, other folivorous species, and other natural enemies. Parasitoids and predators may transmit pathogens (Tinsley, 1979; Otvos, 1979). Changes in host plant defenses, nutrition, and phenology could conceivably alter the growth and development of insects, thus increasing susceptibility to pathogenic infections.

2. Parasitoids

Macroscopic parasite species attack egg, larval or nymphal, pupal, and adult stages of folivorous insects. Parasite species are highly diverse and often very specialized, restricting attack to a single species or even a single stage of a given species. Most folivorous insect species, and phytophagous insects in general, are attacked by parasitoids (but not all, see Janzen, 1975). Some folivorous insect species suffer high rates of parasitism. For example, parasites of larval leaf miners account for more than 50% of mortality (Faeth and Simberloff, 1981). Because of their commonness, enormous diversity, and specialization, parasitoids have been considered a major force in keeping folivorous insects species at low densities and thus preventing competition among folivores from occurring (Hairston *et al.*, 1960; Faeth and Simberloff, 1981; Lawton and Strong, 1981). Parasitoids are also potential candidates for the regulation of folivorous insects because of their relatively rapid numerical response. However, accumulating evidence indicates that parasitoids may not be effective regulators of folivorous insect populations (Klomp, 1966; Varley and Gradwell, 1968; Dempster, 1975; Baltensweiler, 1971; van Der Meijden, 1980; but see Murdoch *et al.*, 1984).

The reasons for lack of control by parasitoids parallel the basic tenet of this chapter regarding folivorous insects: Parasitoids face multiple selective pressures. Parasitoid effectiveness may be limited by competition within the parasitoid guild via chemical interference of other species, direct cannibalism, or intra- and interspecific competition by larvae within the host (Force, 1970; Vinson, 1976; Vinson and Iwantsch, 1980; but see Dean and Ricklefs, 1979). Pathogens may reduce host numbers and limit parasitoid species (Janzen, 1975). Parasitoids may actually transmit certain competing pathogens, such as viruses, to host insects (Tinsley, 1979).

Alternatively, transmission of a virus by a parasitoid may be necessary for survival of the parasitoid egg, because the virus suppresses encapsulation response by the insect host (Edson *et al.*, 1981). Hyperparasites and pathogens (among-trophic-level interactions) may attack parasitoids and reduce their number (Vinson and Iwantsch, 1980). Finally, parasitoids may be limited by aspects of their insect hosts, including phenology (size, shape, and age), location (patchiness of insect hosts and their host plants and chemical and physical cues), and host defenses (physiological responses). I suggest that regulation of folivorous insect species and parasitoid species is remarkably similar. Both are affected by the same array of selective pressures—the host plant, species within their trophic level, and natural enemies—the effects of which are simply removed by one trophic level.

Experimental evidence for the role of parasitoids in preventing outbreaks is scarce. In exclosure experiments, Faeth and Simberloff (1981) found that exclusion of parasitoids results in a 10-fold increase in a species of leaf miner on oak. However, endemic levels of the leaf miner were initially low (<2 mines per 100 leaves), so even after parasitoids were excluded, 80% of leaves still were not mined. Faeth and Simberloff (1981) concluded that other factors (e.g., lack of suitable leaves and overwintering sites) prevented further increases in leaf miner densities.

Correlative and experimental evidence suggests that parasitoids alone cannot account for rarity of folivorous insect outbreaks, probably because parasitoid populations themselves are restricted by aspects of their hosts, other parasitoid species, and their own natural enemies. In fact, parasitoids tend to be more effective regulators of insect populations in biocontrol situations, where one or more of these three selective pressures are minimized (van der Meijden, 1980; Murdoch *et al.*, 1984).

3. Vertebrate Predators

Avian predators seem to be less likely candidates than parasitoids for keeping folivorous insect populations under control, because birds have generally lower diversity and abundances and longer lags in numerical response (Holmes and Sturges, 1975). However, strong avian functional responses, at least at low to moderate densities of insect prey, ability of birds to immigrate into areas of high insect density, and switching behavior to more numerous insect species may counteract these effects (Franz, 1961). It has been suggested that low populations of insectivorous birds are responsible for the pattern of higher levels of defoliation in *Eucalyptus* forests in Australia (P. A. Morrow, personal communication, in Strong *et al.*, 1984). Otvos (1979) reviewed examples of the impact of avian predators on several important defoliator species on forest trees and showed that the amount of insects consumed can be quite variable. It is generally agreed that insectivorous birds may have a large impact on insect populations when insects are at low or moderate densities, in outbreak declines, or in very localized outbreaks, but usually do not control insect populations during an outbreak (Holmes *et al.*, 1979; Otvos, 1979). Bird predation may have little effect at even low or moderate densities of insect species that are defended by toxins or physical features, because these insects are generally avoided by birds (Otvos, 1979; Robinson and Holmes, 1982; but see Heinrich and Collins, 1983).

Bird exclosure experiments are important ways to examine the role of avian predation on insect populations. The exclusion of birds generally increased the number of insects within cages (Solomon *et al.*, 1976; Holmes *et al.*, 1979; Gradwohl and Greenberg, 1982). However, these experiments were short term (two seasons or less) and less is known of the long-term effects of bird removal on insects. Although insect numbers were higher within exclosures, their numbers did not increase to levels characteristic of outbreak situations. This suggests that limiting factors in addition to bird predation account for the rarity of outbreaks.

As with parasitoids, the interactions of avian predators with other factors affecting folivorous insects can be significant. If birds consume adult parasites of folivorous insects or prefer parasitized larvae, pupae, or adults, avian predators could reduce the effectiveness of parasites as controlling agents. Usually birds do not consume significant numbers of adult parasites (Otvos, 1979) but may prefer more susceptible parasitized larvae (Betts, 1955; but see Sloan and Coppel, 1968; MacClellan, 1970; Sloan and Simmons, 1973). Birds may also distribute insect pathogens such as nuclear polyhedrosis virus by consuming infected insects and then defecating in noninfected localities (Entwhistle *et al.*, 1977). Birds may also spread plant diseases (Otvos, 1979), influencing both the quantity and quality of host plants for folivorous insects. Finally, birds can alter microhabitats of folivorous insects and thus increase the susceptibility of insects to pathogens, parasites, or other predators. For example, birds open leaf mines without removing the larvae (S. H. Faeth, personal observation), but in doing so may allow the entry of fungi or bacteria or cause desiccation of the larvae.

4. Invertebrate Predators

Less is known of the role of invertebrate predators relative to vertebrate predators in maintaining low levels of folivorous insects on forest trees. Probably the most widespread insect predators are ants, which may have a significant impact on the population sizes of folivorous insects (Jeanne, 1979; Faeth, 1980; Laine and Niemela, 1980; Skinner, 1980;

Skinner and Whittaker, 1981). Furthermore, ant–plant mutualisms are common where trees provide shelter, food bodies, or nectar in return for protection from herbivorous insects (Janzen, 1966; Thompson, 1982). Some ant species tend aphids, lycaenids, or membracids and repel other herbivores (e.g., Messina, 1981; Skinner and Whittaker, 1981); thus ants cause an increase in tended herbivorous species but a decrease in other herbivores on plants.

B. Vertical Interactions: Host Trees

Host trees exhibit interspecific, intertree, and within-tree variability in phenology, distribution, density, or leaf chemical defenses and nutrition (see references in Denno and McClure, 1983, for a general discussion). Furthermore, these factors may vary temporally, both within and among growing seasons. This enormous temporal and spatial variability in host trees represents a formidable array of selective pressures to insects and has been implicated in the coevolution of insects and host trees, organization of insect communities, and rarity of insect outbreaks (Schultz, 1983a,b). In this section, I examine some of these host tree factors in relation to the organization of insect communities and its offshoot, insect outbreaks.

1. *Seasonal Variation in Quantitative Defenses and Nutrition*

Feeny (1970, 1976) and Rhoades and Cates (1976) attempted to account for the rarity of defoliation on long-lived plants, such as forest trees, by suggesting that such apparent plants had evolved quantitative defenses. Quantitative defenses act in a dosage-dependent fashion. Tannins were viewed as ideal examples of quantitative defense because they purportedly bind plant proteins and insect proteolytic enzymes and thus render protein, a limiting nutrient (Southwood, 1973; Mattson, 1980), less available to insects. This type of defense would be particularly difficult for insects to overcome in ecological time (Feeny, 1976). Indeed, Feeny (1970) showed that both the diversity and abundance of lepidopteran insects are greater when the content of tannins and that of protein are low and high, respectively.

More recently, however, the notion that quantitative defenses, particularly tannins in trees, are responsible for both overall amount and seasonal patterns of folivory has been challenged (Bernays, 1978, 1981; Fox, 1981; Berenbaum, 1983; Coley, 1983b; Zucker, 1983; Faeth, 1985a). Bernays (1981) concluded that tannins have variable effects on insects, ranging from adverse, neutral, to beneficial (Bernays *et al.*, 1980; Bernays and

Woodhead, 1982). Bernays (1981) contended that insects that usually feed on high-tannin plants are less likely than naïve insects to be adversely affected by high tannin content. Thus, tannins could have a negative effect on population densities of insects unaccustomed to feeding on trees with tannins but may have little impact on the vast array of insect species that ordinarily use these trees as food sources. Fox (1981) showed that *Eucalyptus* trees suffer a large amount of folivory despite high tannin levels. Coley (1983b) found little support for the notion that higher levels of phenols are related to lower amounts of folivory in tropical trees. I have shown that both tannin and protein content are poor predictors of degree of folivory on *Quercus emoryi* (Faeth, 1985a). Finally, Zucker (1983) argued that the notion of tannins as generalized, quantitative defenses may be erroneous since tannins are very specific in complexing proteins.

It is unlikely that tannins present insurmountable obstacles to insects on trees and thus prevent insect outbreaks (Schultz, 1983a). A large number of species feed on high-tannin trees (Feeny, 1970; Schultz, 1983a; Faeth, 1985a); some of these species feed when tannin content is highest (Faeth *et al.,* 1981a; Feeny, 1970), and some seem to benefit, at least indirectly, from high tannin levels (Taper *et al.,* 1985; Faeth and Bultman, 1986). These species do not appear to exhibit outbreaks with any lower or higher frequency than other folivore species. The pattern of early feeding on some tree species can also be explained by intensity of bird predation (Holmes *et al.,* 1979), high water content, low fiber content, and low toughness of newly emerged leaves (Feeny, 1970; Coley, 1983b; Faeth, 1985a), intolerance of high temperatures in summer months by insects (Faeth, 1986), and avoidance of leaves damaged previously by other folivores (Edwards and Wratten, 1983; West, 1985; Faeth, 1986; see Section IV,B,2).

2. Variable Plant Chemistry and Natural Enemies

Schultz (1983a) and Whitham (1983) proposed not only that folivorous insects must contend with enormous variation in host defenses on spatial and temporal scales, but also that evolutionary response to chemical variation is resisted by interaction with natural enemies. Schultz (1983a) argued that variable plant chemistry increases the vulnerability of insects to parasites and predators in several ways. First, insects may be restricted to feeding on certain leaves or tissues, which would make them more predictable to natural enemies. For example, Whitham (1981, 1983) showed that predation pressure was greater on galling aphids that are clumped on preferred leaves (large leaves with low phenolic content). Second, variable leaf quality may cause an increase in the movement of insects to locate suitable leaves, thus rendering insects more obvious to natural enemies (Schultz, 1983a,b). Greater movement may also enhance the likelihood of contact with pathogens. However, others (Bergelson *et al.*, 1986; Fowler and MacGarvin, 1986) found no relationship between increased movement due to physical and chemical changes in leaves and increased parasitism of leaf-feeding caterpillars.

Changing plant chemistry may not be exclusively detrimental to insect survival. In fact, I discuss in later sections how changes in plant chemistry may actually enhance the survival of insects. First, however, I discuss phenological aspects of host trees that create further selective pressures on insect growth and survival.

3. Variable Host Tree Phenology

a. Leaf Flush Forest trees differ considerably in timing of initial leaf flush, even within species. For example, *Q. emoryi* trees in the same riparian locale may vary by as much as 2 weeks in timing of initial leaf flush (S. H. Faeth, unpublished data). Since most leaf-chewing insects feed on new foliage of *Q. emoryi* (Faeth, 1986), the majority of overwintering insects must synchronize emergence to this initial budbreak. When insects do emerge, many trees either may not yet have broken bud or may have flushed earlier such that leaves are no longer suitable for the development of insects. Thus, asynchrony of leaf flush may impose strong limitations on insect distribution and growth and population increases. Phenological variation with trees may impose similar selective pressures. For instance, Rhomberg (1984) proposed that the distribution of galling aphids on cottonwood trees is determined by differential susceptibility of unfurling leaves rather than habitat selection, as suggested by Whitham (1978, 1980).

Niemelä and Haukioja (1982) have demonstrated an important correlation between phenological characteristics of trees and peaks of lepidopteran diversity. In trees such as *Quercus* that usually produce a single flush of leaves and stop growing early in the growing season, the diversity of lepidopterous caterpillars peaks early in summer. For trees that continue to grow throughout the growing season such as *Betula* and *Alnus*, lepidopteran richness peaks in the fall. These results suggest that the availability of preferred foliage strongly affects timing and extent of feeding by lepidopterans. There are many examples of insects that develop poorly if feeding is even slightly out of phase with the phenology of host plants (see references in Strong *et al.*, 1984).

Variability in the secondary flush of leaves on trees can profoundly affect insect dynamics. Rockwood (1974) demonstrated experimentally that hand defoliation of calabash trees resulted in a sudden secondary flush of leaves that were devastated by the flea beetle, *Oedionychus*, whereas mature leaves on control trees showed little damage. His results suggest that flea beetle population sizes are limited under normal circumstances by a short interval of availability of new leaves.

Auerbach and Simberloff (1985) showed similar limitations on leaf miner densities on *Quercus nigra* in Florida. They demonstrated experimentally an interaction between fertilization and drought stress, leaf production, and density of leaf miners. Two species of leaf miners are restricted to feeding on supple, secondary leaves flushed in August. The density of these leaf miners is positively correlated with the amount of secondary leaf production. Fertilization and drought stress caused atypical patterns of secondary leaf production in some trees. Atypical reflushing was followed by marked increases in the leaf-mining species. However, if reflushing occurred too early, there was no noticeable effect on densities. Auerbach and Simberloff's (1985) study showed that leaf miner densities were restricted by changes in the phenology of trees and that phenology, in turn, is affected by nutrients and water stress. Reflushing of leaves after defoliation can have a positive or negative effect, or no effect, on probabilities of insect outbreaks depending on timing of defoliation, chemistry and phenology of refoliated leaves, and preference of foliage type of insects that feed after refoliation. These variable effects of defoliation and refoliation are discussed in Section IV,C.

b. Interaction of Phenology and Weather Stress White (1969, 1974, 1976, 1984) hypothesized that outbreaks of insects are often associated with episodes of physical stress, particularly droughts and flooding, because stress causes increases in free amino acid concentration and in the level of soluble nitrogen in leaves or needles. These increases result in greater survivorship and fecundity of folivorous insects. Rhoades (1979, 1983b) extended this hypothesis by suggesting that physical stress leads to a breakdown in plant defensive chemistry. White's hypothesis (1969, 1974, 1976, 1984) and Rhoades's modification seem confirmed by the frequent correlation of physical stress and propensity of insects to reach outbreak levels. White (1969, 1974, 1976) gives several examples of periods of flooding and drought stress followed by insect outbreaks. Even within a forest, trees under physical stress (e.g., on ridgetops, where wind and poor soils contribute to stress) are more susceptible to defoliation than conspecifics under less stressful conditions (Kulman, 1971). In Arizona, walnut trees suffer greater folivory at low elevation than at high-elevation sites, where soil moisture and nutrients are greater (Renaud, 1987). However, the relation between stress and insect outbreaks may have causal components other than change in nutrients and defensive chemistry.

Two factors that have generally been overlooked are alterations in the leaf production of trees under stress and changes in natural enemies at stress locales. Drought, flooding, and nutrient stress (either lack or excess of nitrogen) can cause atypical leaf production patterns (Auerbach and Simberloff, 1985). Drought stress and excessive nitrogen can cause leaves to abscise irregularly followed by reflushing of leaves. Since many folivorous insects prefer or are restricted to feeding on new leaves (Feeny, 1970; Rockwood, 1974; Schweitzer, 1979; Cates, 1980; Fowler and Lawton, 1985; Strong *et al.,* 1984; Faeth, 1985a), these bursts of new secondary leaves could lead to rapid colonization, development, and high fecundity of insects. For example, Washburn and Cornell (1981) showed that the density of cynipid gall wasps is correlated with the availability of new growth at certain times of the year. When fire (a physical stress) enhanced new growth, gall wasp densities increased. Auerbach and Simberloff (1985) observed that leaf miners that specialized on secondary new leaves also increased in density after refoliation of oaks following a notodontid defoliaton. Defoliators of *Quercus robur* (Silva-Bohorquez, 1986) and *Q. agrifolia* (Opler, 1974) also seem to increase resources for aphids and leaf miners, respectively. Thus, previous insect defoliation may itself be a stress resulting in altered phenology and higher densities of subsequent feeders.

It is possible that the natural-enemy complex of folivorous insects is altered in areas of physical stress. Drought, wind, or flooding could eliminate certain parasitoids, predators, or pathogens and thus contribute to insect outbreaks in these areas. The reduction of parasite or predator pressure associated with physical stress has received little attention. In Arizona, a leaf-mining species, *Cameraria* sp. nov., reaches high densities on *Q. emoryi* trees that grow in areas of low soil moisture and nutrients and high exposure to winds. Densities of the same leaf miner on nearby trees in protected arroyos with higher soil moisture are 10 times lower. Parasitism by hymenopterans is much lower on miners feeding on exposed trees and may contribute to high densities (T. Bultman, personal observation). A similar situation exists for *Q. geminata* trees in Florida, where outbreak levels of a leaf miner, *Stilbosis quadracustatella,* persist on trees that grow on coastal dunes, where soils are poor and predators may be depauperate (Mopper *et al.,* 1984). It is plausible that sites where trees are under physical stress generally exhibit lower intensities of parasitism and predation.

c. Leaf Morphology and Abscission Leaves within forest trees vary in morphology and timing of abscission. Both of these factors can impose selective pressures and have resulted in counteradaptations by folivorous insects. Morphological variations include trichomes, spines, and hairiness, which may deter insects from colonizing or feeding on leaves

(Rhoades, 1979). Another aspect of morphology, leaf size, may be particularly important to sedentary insects, such as gall formers and leaf miners, because these insects generally cannot move to other leaves and must complete their development on leaves where adult females have oviposted (Faeth, 1985b). For example, Whitham (1978, 1980, 1981, 1983) has shown that galling aphid stem mothers select large leaves in high-density situations, and survivorship and fecundity are higher on larger leaves. Larger leaves have a smaller amount of phenolic compounds (Zucker, 1982), which may account for both selection and increased survivorship on these leaves. Whitham (1981, 1983) reported that predation pressure is higher on clumped aphid galls on large leaves, a selective pressure that might counteract the selection of large leaves.

In the leaf miner–*Quercus* system, we have shown that the selection of leaves based on size depends on the life histories of individual species. For example, *Stilbosis juvantis,* a leaf miner on Emory oak that consumes less than 20% of an individual leaf surface, does not select leaves on the basis of size (Faeth, 1985b). Mining on either small or large leaves does not influence the mortality of this miner. However, leaf miners such as *Cameraria* sp. nov. that consume large fractions of leaves (80%) select large leaves, and pupal weight (and probably fecundity) is positively correlated with leaf size (Bultman and Faeth, 1986c). Leaf size should be less critical for more mobile insects that can move to feed on several leaves, although risks may increase during leaf-to-leaf movements (Schultz, 1983b).

Leaf abscission is a phenological variable that affects folivorous insects (Faeth *et al.,* 1981b; Williams and Whitham, 1986). Both temperate and tropical forests usually exhibit regular patterns of peak leaf falls, but some leaves are abscised at irregular intervals throughout the growing season. Some of this leaf fall is associated with weather stress, such as storms, or is a response to insect damage (Faeth *et al.,* 1981b). The consequences for folivorous insects, particularly sedentary ones, can be severe. Dislodgement from the tree usually results in death from starvation or desiccation. Askew and Shaw (1978) reported that large fractions of second-generation leaf miners are killed because they fail to complete development before leaves are abscised. Williams and Whitham (1986) indicated that premature leaf abscission can reduce galling aphid populations by as much as 53%. Apparently, leaf abscission has been a significant selective pressure such that insects have evolved counteradaptations. Engelbrecht *et al.* (1969), Kahn and Cornell (1983), and Faeth (1985b) documented cases of the "green island effect," whereby leaf mining larvae exude cytokinins that continue photosynthesis in the leaf after abscission occurs. Williams and Whitham (1986) proposed that sedentary galling aphids are more likely to shift hosts than nongalling species because of selective pressures from premature leaf abscission.

Bultman and Faeth (1986b) showed that variable leaf abscission within trees affects selection of leaves by a leaf-mining insect. *Cameraria* sp. nov. feeds on *Q. emoryi* as a larval miner for about 10 months. This miner is especially susceptible to desiccation from leaf abscission, because completion of larval and pupal stages occurs near the time leaves are abscised from trees. We found that this miner prefers central over peripheral leaves on trees because peripheral leaves are abscised sooner than central ones. Through behavioral selection by ovipositing, offspring of this leaf miner avoid death from abscission and ensuing desiccation. However, central leaves are less active photosynthetically and receive less solar radiation (the larvae continue feeding through winter months). Parasitism and the probability of sharing leaves is higher for miners clumped on central leaves. These results suggest counteracting selective pressures. Central leaves are superior in terms of slower abscission but are inferior in terms of photosynthesis, temperature, attack by parasites, and probability of co-occurrence with a competing leaf miner. This example of opposing selective pressures is a recurring pattern among folivorous insects. In terms of survival and fecundity, all leaves within a tree may be suboptimal.

I have chosen not to discuss chemical or nutritional variability among plants, nor variability in the distribution of plants in space and time (patchiness), discussions of which can be found in reviews by Kareiva (1983, 1986), Denno (1983), Krischik and Denno (1983), and Strong *et al.* (1984). These aspects of variability present additional selective pressures on folivore survival and reproduction.

The next section examines horizontal interactions (interactions among folivorous species) as the third major selective force, which can operate in directions opposite to selection by natural enemies and the host tree.

C. Horizontal Interactions

Horizontal interactions are interactions among species of folivores. They may be negative (e.g., competition or amensalism) or positive (e.g., mutualism or commensalism). The role of horizontal interactions in the community organization of folivorous insects and a related pattern, the rarity of outbreaks, has undergone a cyclic history in the literature. Hairston *et al.* (1960) proposed that herbivores are generally not resource-limited but instead are controlled by predation and parasitism and therefore are unlikely to compete. During the heyday of belief in interspecific competition as a primary organizing factor in animal communities, Janzen

(1973) proposed that insects on plants "automatically compete with all other species" on a given host plant. The predominant view has now definitely reverted to that of Hairston *et al.* (1960). More recent reviews by Lawton and Strong (1981), Strong (1983), Price (1983), and Strong *et al.* (1984) conclude that resource-based competition is either a weak or nonexistent factor in organizing folivorous insect communities on plants. This view is now held even by those who consider interspecific competition to be a primary organizing force in other ecological communities (Schoener, 1983). A return to the hypothesis of Hairston *et al.* (1960) was precipitated by a general lack of evidence for interspecific competition in both experimental and observation studies (for examples, see Lawton and Strong, 1981; Strong *et al.*, 1984), although experimental studies of interspecific competition among folivorous insects are generally few in proportion to the diversity of these species (Connell, 1983; Schoener, 1983).

Although classical resource-based competition is probably an infrequent and weak factor in folivorous insect communities, I consider species interactions (both negative and positive) via changes in host plant to be important influences on insect communities and the rarity of insect outbreaks. These interactions are often indirect and subtle and often defy conventional expectations of community ecology theory. Two patterns related to insect–plant interactions are critical to the argument that indirect interactions are prevalent and important: (1) recent evidence that physical and induced chemical changes in leaves are widespread and (2) the pattern of leaf feeding exhibited by most folivorous insects. In the following section, I describe these patterns and then experimental studies that exemplify the importance of indirect interactions.

1. Induced Responses to Folivory

The induction of chemical defenses in plants as a response to folivory or mechanical damage is now well documented in a variety of trees (fir: Puritch and Nijholt, 1974; birch: Haukioja and Niemela, 1976, 1979; Niemela *et al.*, 1979; oak: Schultz and Baldwin, 1982; Faeth, 1986; alder: Rhoades, 1983a; maple and poplar: Baldwin and Schultz, 1983) as well as other plants (see Rhoades, 1983b, for a review). These chemical responses may be short term, lasting only a few hours (Schultz and Baldwin, 1982), or persist for years (Haukioja, 1980). Induced tree responses seem the rule rather than the exception, but the defensive nature of such responses has been questioned (Myers and Williams, 1984; Fowler and Lawton, 1985). Not all plant responses to herbivory are detrimental to future folivores (Fowler and Lawton, 1985), especially when trees are completely defoliated (see Section IV,C,4).

The generality of induced plant responses suggests that folivores can affect subsequent folivorous insects. This scenario has been proposed to explain yearly fluctuations in the autumnal moth, *Oporinia autumnata*, which feeds on birch (Haukioja, 1980). Not only may folivores alter plant chemistry, but damage to leaves may also decrease nutritional content (Faeth, 1986) and indirectly alter attack by parasites and predators (Schultz, 1983a,b; Faeth, 1985b, 1986) or decrease food or refugia for later feeders (Karban, 1986). Even frass from insect defoliators in soils beneath trees may reduce larval survival and fecundity of subsequent feeders (Haukioja *et al.*, 1985).

2. *Patterns of Leaf Damage by Folivores*

As discussed in Section I,B, folivorous insects usually consume less than 10% of the leaf area of forest trees (Table 1) but damage considerably larger fractions of the total number of leaves. I measured cumulative fractions of leaves that had 10% or more leaf area removed on six trees of *Quercus emoryi* in central Arizona over three growing seasons. *Quercus emoryi* is an evergreen oak with leaves persisting from budbreak in mid-April to early May to the following budbreak. The cumulative fraction of leaves with 10% or more of leaf area removed is at least 30% in the first two seasons and at least 18% in the third season (Fig. 1). These fractions increase substantially if leaves with smaller amounts of insect grazing are

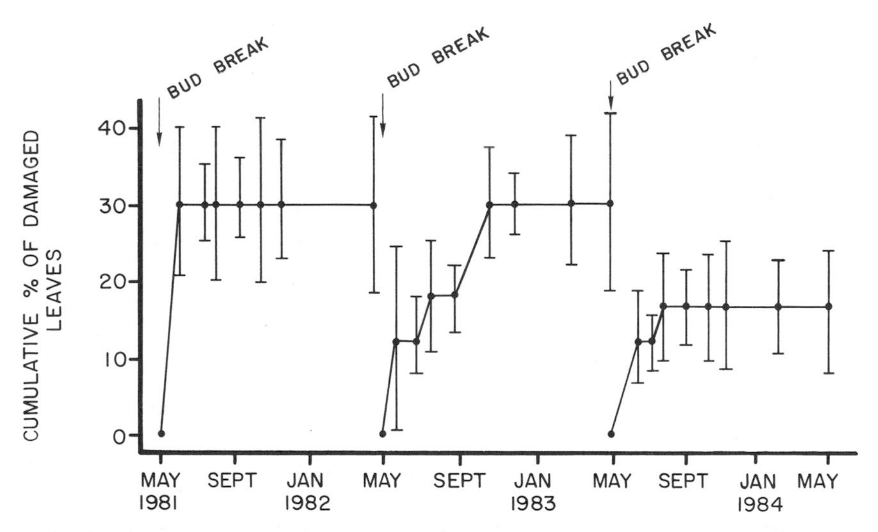

Fig. 1. Cumulative percentage of damaged leaves on *Quercus emoryi* from 1981 to 1984. Each point is the mean cumulative percentage of damaged leaves for six trees; bars are standard errors of means at each sample date.

included (>60%). However, in all three growing seasons, less than 5% of total leaf area is consumed on each tree.

Edwards and Wratten (1983) showed similar results for several species of forest trees and herbaceous plants. Not only were large fractions of leaves damaged (40–80%) with little leaf area consumed (4.3–11%); damage within leaves tended to be either regularly spaced or overdispersed, but not clumped (however, these patterns were not statistically tested). Edwards and Wratten (1983) proposed that localized, wound-induced changes in plant chemistry limited the total amount of leaf area that is grazed and caused overdispersion of grazing damage within leaves (but see Fowler and Lawton, 1985). Similarly, Coley (1982) reported that only 2.7% of leaves of *Trichilia cipo* are completely consumed by herbivores. Heinrich (1979) and Heinrich and Collins (1983) reported similar patterns for many forest caterpillar species, but dichotomized feeding patterns and behavior by caterpillars based on susceptibility to bird predation. Palatable caterpillars tended to minimize damage on leaves, whereas unpalatable caterpillars often consumed large fractions of leaves. Heinrich and Collins (1983) showed that some bird species can be cued by physical damage, and they hypothesized that bird predation was the primary reason for the minimal damage to individual leaves caused by palatable species. They discounted parasitism, although parasites may respond to damaged leaves, because, they reasoned, parasitism should be equivalent for palatable and unpalatable caterpillars. However, variable rates of parasitism were not measured. Therefore, they expected both types to minimize leaf damage if parasitism associated with damaged leaves is increased for palatable and unpalatable caterpillars. Heinrich and Collins (1983) did not evaluate the role of inducible plant chemistry on grazing patterns of caterpillars.

The feeding patterns of folivorous insects suggest that insect species can interact either through effects of inducible plant chemistry or indirectly by altering attack by natural enemies. Although these interactions are quite different from strict resource-based competition and may be only unidirectional (amensalistic), they nevertheless can affect the growth and survival of subsequent folivores. Furthermore, the effects could conceivably occur at low consumption levels of leaf area, provided that sufficient fractions of leaves are damaged or remaining intact leaves are inferior in terms of chemistry, nutrition, or attack by natural enemies (Faeth, 1985). In the next two sections, I describe experiments that have tested the effects of early-feeding folivores on later feeders via alterations in the host tree when individual leaves are only partially damaged and when whole leaves are removed from trees.

3. Interactions between Folivorous Guilds: Partial Consumption of Leaves

Leaf chewers on *Q. emoryi* tend to feed early in the growing season (Fig. 1), whereas adult leaf miners oviposit and larval miners begin feeding when most herbivory by leaf chewers has ceased (Faeth, 1985a,b, 1986). I tested the hypothesis that early-feeding, leaf-chewing guilds affect the distribution, density, and survivorship of late-feeding leaf miners on the shared host plant by partially damaging leaves. The test involved monitoring the distribution and survivorship of leaf miners on six control trees in which 25–30% of leaves were damaged by chewers and on six experimental trees in which similar fractions of leaves were damaged by chewers and an additional 50% of leaves were manually damaged (total fraction of leaves damaged, 75–80%). All insect-damaged and manually damaged leaves had about 10–50% of area removed. Seasonal changes in tannin (defensive compounds) and protein (nutrition) content were monitored for intact, insect-damaged, and manually damaged leaves. This study included two complete growing seasons (1981–1982 and 1982–1983).

On both control and experimental trees, leaf miners occurred more frequently than by chance on intact leaves in both seasons. However, because leaf miner densities did not differ between control trees (25–30% of leaves damaged) and experimental trees (75–80% of leaves damaged), selection of leaves occurs within but not among trees (Faeth, 1986).

Leaf miners experienced significantly greater survivorship if they were in intact rather than damaged leaves in both growing seasons (Table 2). Survivorship was lower in damaged leaves because rates of parasitism were increased on miners in damaged leaves. However, the incidence of death from other causes was lower for miners on damaged leaves, although this positive effect did not compensate for the negative effect of increased parasitism (i.e., overall survivorship was lower for miners on intact leaves; Table 2).

Phytochemistry changed in damaged leaves (Fig. 2). Both leaf chewing and manual damage induced increases in condensed tannin content (Fig. 2). Damaged leaves also had lower protein content. These chemical changes were sustained throughout each growing season and therefore potentially affected the leaf-mining guild that feeds from July to April. Furthermore, the chemical changes were localized on damaged leaves within trees, since overall levels of condensed tannin and protein in intact leaves did not differ between control and experimental trees.

Parasitism was increased on damaged leaves either because parasitoids

TABLE 2

Percentage of Survivorship and Mortality of Leaf Miners on Intact and Damaged Leaves of *Quercus emoryi* in Two Growing Seasons[a]

Category	1981–1982 Intact leaves (%)		1981–1982 Damaged leaves (%)	1982–1983 Intact leaves (%)		1982–1983 Damaged leaves (%)
Survived	46.7	*	27.8	36.6	*	23.0
Did not survive[b]	53.2	*	72.2	63.4	*	76.0
Parasitized	21.1	*	44.2	20.1	*	56.6
Preyed upon	12.3		11.5	9.7		2.6
Abscission death	17.5		19.2	24.7		15.8
Other death[c]	49.1	*	25.0	45.5	*	25.0

[a] Damaged leaves were both insect-damaged and manually damaged. Asterisks indicate significant difference ($p < .05$) between categories (chi-square tests).

[b] "Did not survive" category is divided into respective categories of mortality, which sum to 100% of total mortality.

[c] This category includes death from bacterial, fungal, or viral attack, plant chemistry, or weather factors. The last-named are probably unimportant because leaf miners within plant tissues are buffered from humidity and temperature changes (Faeth and Simberloff, 1981).

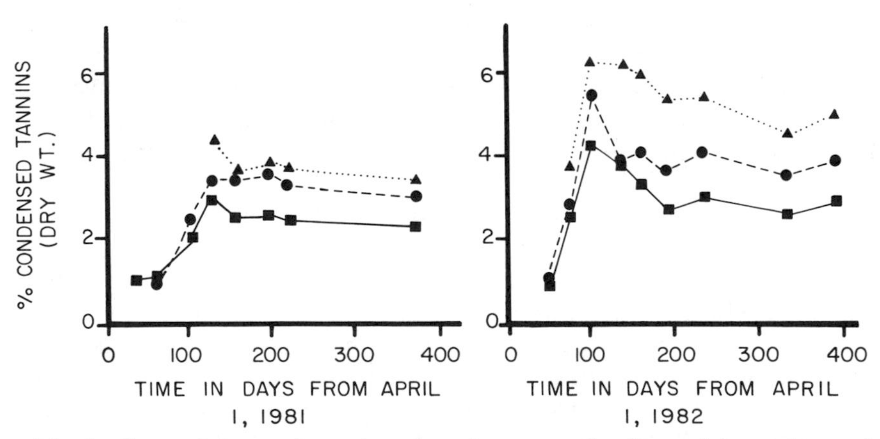

Fig. 2. Seasonal changes in condensed tannin content of undamaged, insect-damaged, and manually damaged leaves of *Quercus emoryi* in 1981–1982 and 1982–1983. Each point is the mean of six trees (■, undamaged; ●, insect-damaged; ▲, manually damaged).

cued to physical damage or chemical changes in leaves or because leaf miners on damaged leaves were more vulnerable to parasitoid attack because of prolonged development caused by chemical changes in leaves. We have shown in another experiment that increased tannins alone increase parasitism (Faeth and Bultman, 1986), possibly by acting as contact attractants (Vinson, 1976) or as long-range attractants via their volatile degradation products. Preliminary results indicate that physical damage independent of chemical changes also contributes to the attraction of parasitoids (S. H. Faeth, unpublished data). Predation rates did not differ for miners on intact and damaged leaves, contradicting Heinrich and Collins's (1983) hypothesis, but predation rates were nonetheless very low on these leaf miners (Table 2).

It is interesting that previous damage may actually decrease certain categories of mortality (Table 2). We hypothesized that increased tannin content in damaged leaves protected leaf miners from bacterial and fungal attack, owing to the well-known bactericidal and fungicidal properties of tannins (Swain, 1979; Faeth and Bultman, 1986). Leaf miners may be particularly susceptible to fungi and bacteria because they constantly encounter 100% humdidity within the mine. Coating mines with purified tannins seemed to confirm this hypothesis, since this category of mortality was significantly decreased (Faeth and Bultman, 1986). These results suggest that the effects of earlier-feeding folivores on later-feeding folivores need not always be negative. As discussed previously, notodontid defoliations on oaks increased secondary flushing of leaves, thus increasing the densities of two leaf-mining species (a positive effect) that specialize on secondary growth (Auerbach, 1982; Auerbach and Simberloff, 1984).

The negative effect of leaf chewers on leaf miners appeared to be inversely dependent on density. Leaf miners in damaged leaves when few leaves within a tree were damaged experienced greater survival and less parasitism than leaf miners in damaged leaves when a large fraction of a tree's leaves were damaged (Faeth, 1986a,b). This may occur because parasitoid search for leaf miners becomes less effective when many damaged leaves are present.

This experimental study suggests that interactions between temporally separated guilds may be important in the distribution and survivorship of folivores. West (1985) found similar interactions between chewers and miners on *Q. robur*. Leaf miners avoid damaged leaves (C. West, personal communication), and survivorship is lower on damaged than on intact leaves. Thus, horizontal interactions may influence both organization of folivore communities, at least at certain spatial scales, and probabilities of insect outbreaks. "Interaction" may be an inappropriate term, since leaf miners are likely to have no effect on leaf chewers; leaf miners usually occur at very low densities, consume small fractions of leaves, and feed on leaves that will be abscised before leaf chewers begin feeding again. Instead, such effects are probably amensalistic and are likely common among insect species (Lawton and Hassell, 1981). Leaf-chewing species may interact, since most species feed within a narrow window of time (Strong *et al.*, 1984) and avoid leaves damaged by other leaf chewers (Edwards and Wratten, 1983; Faeth, 1985a).

4. Interactions among Folivorous Guilds: Complete Defoliation

Further experiments were performed to test the effect of complete defoliation (all leaves removed from trees) on late-feeding leaf chewers and leaf miners (S. H. Faeth, unpubished data). Six trees of *Q. emoryi* were manually defoliated in two growing seasons at different times in each season. Leaf miner densities and amount of herbivory by leaf chewers were monitored on the six trees after refoliation and compared with those on mature leaves of six control trees. Changes in tannin and protein content were also determined on refoliated and mature leaves.

When trees were defoliated early in the growing season (6 weeks after budbreak) refoliation occurred relatively quickly (18 ± 3 days) after defoliation. When trees were defoliated slightly later in the growing season (11 weeks after budbreak), releafing was delayed (32 ± 5 days). In both cases the phytochemistry of refoliated leaves resembled that of primary new leaves produced at spring budbreak: Leaves were low in condensed tannins, high in hydrolyzable tannins (but quickly declining), and similar in protein content. The only difference was that the condensed tannin content of reflushed leaves remained low for the remainder of the growing season but quickly increased in primary new leaves (Fig. 3). Therefore, the quality of refoliated leaves should at least be equivalent (similar water protein and hydrolyzable content) and probably superior (lower condensed tannin content) to that of new spring leaves. At the very least, this result suggests that defoliation need not always induce chemical defenses and lower leaf quality for subsequent folivores.

It is interesting that late-season leaf chewers did not respond to refoliated leaves that appeared to be higher in quality in either season. In fact, herbivory by these insects on mature and reflushed foliage was equivalent. I propose two explanations for this pattern: First, late-season chewers prefer mature foliage since secondary flushing of leaves on *Q. emoryi* is rare in nature. However, if this were true, one would expect mature foliage to have suffered greater herbivory losses than reflushed foliage. This did not happen. Second, late-season chewers are generalists and

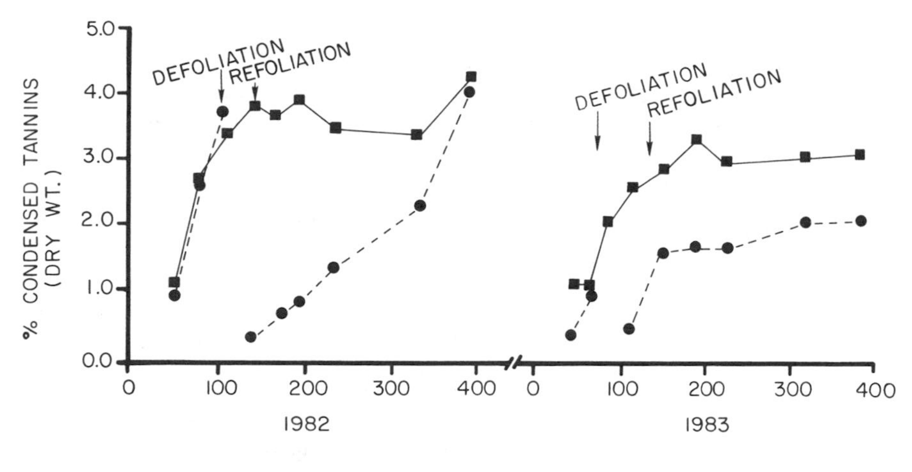

Fig. 3. Seasonal changes in condensed tannin content of control and experimental (defoliated) trees in 1982–1983 and 1983–1984. Each point is the mean of six trees (■, control; ●, experimental).

prefer neither new nor mature foliage (Niemelä, 1983). This explanation seems to hold. Late-season herbivory appears to be related simply to the availability of leaf chewers rather than plant chemistry; seasonal patterns in herbivory on refoliated and mature leaves coincided almost perfectly.

Unlike the densities of leaf chewers, leaf miner densities differed on refoliated and mature leaves. When defoliation occurred early and refoliation was rapid, leaf miner densities were greater on refoliated than on mature leaves. Possibly, leaf miners selected refoliated leaves because these leaves were chemically superior or simply because the leaves sustained less cumulative damage from leaf chewers. (Although mature and refoliated leaves had similar amounts of leaf-chewing herbivory after refoliation, mature leaves had much greater cumulative damage because of herbivory early in the spring.) When defoliation occurred late in the growing season and refoliation was delayed, leaf miner densities were lower on refoliated than on mature leaves. However, this decrease was due to the absence of one dominant, univoltine leaf miner that oviposits when leaves are not present on defoliated trees (before refoliation).

In terms of potential for insect outbreaks, it is clear that the timing of interaction is critical. If defoliation occurs early in the growing season, the interaction might be positive and leaf miner densities increase. Alternatively, if defoliation occurs late and refoliation is delayed, the interaction is negative because windows of oviposition are missed. The same could hold for leaf chewers; early defoliation and refoliation could prolong the availability of new foliage for specialized spring feeders, increasing survivorship and adding generations of insect species that are normally doomed because of maturation of primary spring leaves. It is clear that previous defoliation need not reduce leaf quality and quantity for later feeding insects.

In sum, the two experiments (Sections IV,C,3 and 4) suggest that the direction and mechanisms of horizontal interactions at the intra- and interleaf levels may be very different. Possible influences of leaf chewers on leaf miners at these levels are shown in Table 3. Some of the effects are hypothetical or are supported by only circumstantial evidence. However, the effects of leaf chewers on leaf miners via the host plant have been shown to be complex, subtle, and variable (Faeth, 1985a,b, 1986; West, 1985), and future studies with leaf miner–leaf chewer and other insect species may reveal that the list in Table 3 is incomplete. At the very least, the proposed effects underscore the conflicting selective pressures exerted on leaf miners by the action of leaf chewers on the same host plant.

5. *Implications for Folivorous Insect Communities*

The experimental studies discussed in the preceding sections have several important implications for community theory and the cognate pattern of rarity of insect outbreaks:

1. Horizontal (within-trophic-level) interactions can be important for the distribution of folivorous insects, as well as for their survivorship and mortality. This is in contrast to prevailing views of folivorous insect communities as loosely structured by among-insect-species interactions because such interactions are rare (Lawton and Strong, 1981; Price, 1983; Strong *et al.*, 1984).
2. Horizontal interactions are often subtle and indirect and mediated by morphological, phenological, and chemical changes in the host plant that, in turn, affect natural enemies.
3. Horizontal interactions may occur at low levels of folivory, provided that sufficient fractions of leaves are grazed or remaining leaves are of lower quality. The latter seems likely given the large amount of variability found within and among trees (Whitham, 1983). In fact, interactions among folivorous insects may occur *only* at low folivory levels, since increasing numbers of damaged leaves may swamp physical and chemical cues used by parasitoids and predators. Furthermore, partial herbivory of individual leaves may result in very different consequences for later phytophagous insects than partial or complete defoliation of whole leaves (Table 3).
4. The direction (positive or negative) and intensity of horizontal inter-

TABLE 3

Direct and Indirect Effects of Leaf Chewers on Leaf Miners via the Host Plant at Intra- and Interleaf Spatial Scales[a]

Level of interaction	Plant effect	Direct effect on leaf miners[b]	Indirect effect on leaf miners[b]
Intraleaf (partial damage to individual leaves)	Reduced area of individual leaves	Increased co-occurrence on individual leaves because of adult avoidance of damaged leaves (−)	Increased larval competition/cannibalism (−) Increased parasitism (−)
	Premature leaf abscission	Increased abscission death (−)	Reduced parasitism after leaf abscission (+)
	Increase in tannins, decrease in protein	Slower development; reduced body size (−) Decreased survivorship (−)	Increased parasitism (−) Failure to complete larval stages before abscission (−) Reduced fecundity (−) Protection from fungal and bacterial attack (+)
Interleaf (partial or complete defoliation of whole leaves)	Lag in refoliation or no refoliation at all	No leaves available for oviposition (−)	—
	Refoliated leaves are smaller and fewer in number	Increased co-occurrence on individual leaves (−)	Increase larval competition/cannibalism (−)
	Refoliated leaves are less prone to abscission	Less abscission death (+)	—
	Decreased condensed tannins and toughness, increased water content	Faster development, larger body size (+) Increased survivorship (+)	Decreased parasitism (+) Decreased abscission death (+) Increased fecundity (+)
	Refoliated leaves have little or no damage	Decreased co-occurrence on individual leaves; few damaged leaves to avoid (+)	Decreased competition/cannibalism (+) Decreased parasitism (+)

[a] Some effects are hypothetical.

[b] (+) and (−) indicate positive and negative effects on leaf miners, respectively, and therefore interactions that are likely to increase (+) or reduce (−) the probability of leaf miner outbreaks.

actions should be temporally variable. For example, defoliation very early in the growing season could result in rapid refoliation of new leaves, thus providing a prolonged resource for insects specializing on new leaves. Therefore, early-season defoliation could increase the probability of outbreak of these insects. Late-season defoliations, however, may delay refoliations, reduce the availability of leaves, and thus decrease the probabilities of outbreaks. Furthermore, reflushing of new leaves late in the growing season may reduce resources for late-feeding insects that may prefer mature foliage.

5. Given items 2 through 4, it is not surprising that little evidence exists for direct resource-based competition among folivorous insect species (Lawton and Strong, 1981; Connell, 1983; Price, 1983; Schoener, 1983; Strong *et al.,* 1984). Theories of community ecology grounded on resource-based competition are unlikely to apply to folivorous insect communities. Any theory of community organization as related to insect outbreaks will have to incorporate these indirect, subtle, and possibly inversely density-dependent interactions.

6. If indirect horizontal interactions among folivorous insects are common and intense, such interactions may either lessen or increase chances of outbreaks, depending on whether the interactions are positive or negative (Table 3).

V. FUTURE DIRECTIONS

I have presented a complex picture of community organization of folivorous insects and the control of insect outbreaks based on autocological factors and ecological interactions at vertical (natural enemies and host trees) and horizontal (among folivorous species) levels. Furthermore, few, if any, of these factors or interactions are mutually exclusive, since interactions between any two levels often affect other between-level or within-level interactions. The picture is even more complex if one incorporates short-term genetic and long-term evolutionary changes in the participating organisms. No doubt the enormous complexity of the structure of folivorous insect communities is the major reason entomologists and insect ecologists have been frustrated in their attempts to derive a general theory of insect outbreaks, especially theories involving only simple population models (see Rhoades, 1983b, for discussion and references). Certainly, community theories that are founded on resource-based interspecific competition have fared no better (Lawton and Strong, 1981; Strong *et al.,* 1984). As have others (Price *et al.,* 1980; Whitham, 1981, 1983; Schultz, 1983a,b), I advocate a holistic approach to understanding community organization of folivorous insects and insect outbreaks, with equal emphasis on horizontal interactions. The prevailing view that horizontal interactions are insignificant may only delay the understanding of community structure and its kindred pattern, insect outbreaks.

As a first step in examining horizontal interactions and their relation to community organization and insect outbreaks, ecologists must abandon conventional notions of where and when such interactions may occur. Classical community theory suggests that interactions should be most prevalent among closely related species within guilds and among species that feed contemporaneously (Root, 1967; Pianka, 1983). Recent evidence suggests that interactions occur among distantly related species of different guilds that feed at different times. Previous failures to discover horizontal interactions may stem from examining the wrong species at the wrong times and ignoring indirect effects.

The approach of examining multiple levels of interactions is very broad indeed. I am not suggesting that it replace any existing population or community theory; it is simply a conceptualization. It is unlikely that any one theory will be applicable to all folivorous insects because of the idiosyncrasies of each folivorous insect–host-tree–natural-enemy system.

To date, very few studies have incorporated a comprehensive approach to insect outbreaks. Most of the evidence I have provided for multiple-level interactions is piecemeal, most studies having examined only one or a few possible factors influencing insect outbreaks. However, these studies considered collectively provide strong inferential evidence that trilevel interactions are important in community structure and insect outbreaks.

Where should we begin? I suggest that simple systems in which only one or a few species of folivorous insect, host tree, or natural enemy are involved may be a good starting point. Studies involving sedentary insects (e.g., gall formers and leaf miners) have been particularly fruitful because feeding is restricted to single leaves within one host plant, and survivorship and fecundity are fairly easy to ascertain. Finally, I urge an experimental approach to community structure and insect outbreaks in which one or more interactions can be controlled to determine their relative importance.

VI. SUMMARY

In this chapter, I have examined the role of autocological factors and vertical (among-trophic-level) and horizontal (within-trophic-level) interactions in folivorous insect outbreaks from a community standpoint.

None of these factors considered alone is adequate to explain the rarity of insect outbreaks. Instead, interactions at any one level are likely to influence other interactions. Horizontal interactions among folivorous insect species may be as important as vertical interactions in community structure and the control of outbreaks, contradicting recent opinions. Horizontal interactions are often subtle and indirect, but nevertheless important. Finally, I have discussed why current population and community theories fail to explain frequencies of folivorous insect outbreaks. A comprehensive, multiple-level interaction approach to insect outbreaks and community structure is needed.

ACKNOWLEDGMENTS

T. L. Bultman and the editors of this volume made helpful comments on the manuscript. P. Sanchez efficiently typed the manuscript. M. Axelrod performed phytochemical analyses and prepared the figures. This work was supported by National Science Foundation Grants BSR 8110832 and BSR 8415616.

REFERENCES

Anderson, R. M., and May, R. M. (1980). Infectious diseases and population cycles of forest insects. *Science* **210,** 658–661.

Andrewartha, H. G., and Birch, L. C. (1954). "The Distribution and Abundance of Animals." Univ. of Chicago Press, Chicago, Illinois.

Andrewartha, H. G., and Birch, L. C. (1984). "The Ecological Web." Univ. of Chicago Press, Chicago, Illinois.

Askew, R. R., and Shaw, M. R. (1978). Mortality factors affecting the leaf-mining stages of *Phyllonorycter* (Lepidoptera: Gracillariidae) on oak and birch. 1. Analysis of mortality factors. *J. Linn. Soc. London, Zool.* **67,** 31–49.

Auerbach, M. J. (1982). Population biology and community ecology of leaf-mining insects on native and introduced oaks and chestnuts. Ph.D. Dissertation, Florida State University, Tallahassee.

Auerbach, M., and Simberloff, D. (1985). Responses of leaf miners to atypical leaf production patterns. *Ecol. Entomol.* **9,** 361–367.

Baldwin, I. T., and Schultz, J. C. (1983). Rapid changes in tree leaf chemistry induced by damage: Evidence for communication between plants. *Science* **221,** 277–279.

Baltensweiler, W. (1971). The relevance of changes in the composition of larch bud moth population. *In* "Dynamics Numbers Population" (P. J. den Boer and G. R. Gradwell, eds.), pp. 208–219. Pudoc, Wageningen.

Berenbaum, M. R. (1983). Effects of tannins on growth and digestion in two species of papilionids. *Entomol. Exp. Appl.* **34,** 245–250.

Bergelson, J., Fowler, S., and Hartley, S. S. (1986). The effects of foliage damage on casebearing moth larvae, *Coleophora serratella,* feeding on birch. *Ecol. Entomol.* **11,** 241–250.

Bernays, E. A. (1978). Tannins: An alternative viewpoint. *Entomol. Exp. Appl.* **24,** 244–253.

Bernays, E. A. (1981). Plant tannins and insect herbivores: An appraisal. *Ecol. Entomol.* **6,** 353–360.

Bernays, E. A., and Woodhead, S. (1982). Plant phenols utilized as nutrients by a phytophagous insect. *Science* **216,** 201–202.

Bernays, E. A., Chamberlain, D. J., and McCarthy, P. (1980). The differential effects of ingested tannin acid on different species of acridoidea. *Entomol. Exp. Appl.* **28,** 158–166.

Betts, M. M. (1955). The food of titmice in oak woodland. *J. Anim. Ecol.* **24,** 282–323.

Bray, J. R. (1961). Measurement of leaf utilization as an index of minimum level of primary consumption. *Oikos* **12,** 70–74.

Bray, J. R. (1964). Primary consumption in three forest canopies. *Ecology* **45,** 165–167.

Bultman, T. L., and Faeth, S. H. (1986a). Leaf size selection by leaf-mining insects on *Quercus emoryi: (Fagaceae). Oikos* **46,** 311–316.

Bultman, T. L., and Faeth, S. H. (1986b). Selective oviposition by a leaf miner in response to temporal variation in abscission. *Oecologia (Berlin)* **69,** 117–120.

Bultman, T. L., and Faeth, S. H. (1986c). Effect of within-leaf density and leaf size on pupal weight of the leaf-mining insect, *Cameraria* sp. nov. (Lepidoptera: Gracillaridae). *Southwest. Nat.* **31,** 201–206.

Cates, R. G. (1980). Feeding patterns of monophagous, oligophagous, and polyphagous insect herbivores: The effect of resource abundance and chemistry. *Oecologia (Berlin)* **46,** 22–31.

Coley, P. D. (1982). Rates of herbivory on different tropical trees. *In* "The Ecology of Tropical Forests: Seasonal Rhythms and Long Term Changes" (E. G. Lee, A. S. Rand, and D. M. Windsor, eds.), pp. 123–132. Smithsonian Inst. Press, Washington, D.C.

Coley, P. D. (1983a). Intraspecific variation in herbivory on two tropical tree species. *Ecology* **64,** 426–433.

Coley, P. D. (1983b). Herbivory and defensive characteristics of tree species in a lowland tropical forest. *Ecol. Monogr.* **53,** 209–233.

Connell, J. H. (1983). On the prevalence and relative importance of interspecific competition: Evidence from field experiments. *Am. Nat.* **122,** 661–696.

Dean, J. M., and Ricklefs, R. E. (1979). Do parasites of Lepidoptera larvae compete for hosts? No! *Am. Nat.* **113,** 302–306.

Dempster, J. P. (1975). "Animal Population Ecology." Academic Press, London.

Denno, R. F. (1983). Tracking variable host plants in space and time. *In* "Variable Plants and Herbivores in Natural and Managed Systems" (R. F. Denno and M. S. McClure, eds.), pp. 291–341. Academic Press, New York.

Denno, R. F., and McClure, M. S., eds. (1983). "Variable Plants and Herbivores in Natural and Managed Systems." Academic Press, New York.

Edson, K. M., Vinson, S. B., Stoltz, D. B., and Summers, M. D. (1981). Virus in a parasitoid wasp: Suppression of the cellular immune response in the parasitoid's host. *Science* **211,** 582–583.

Edwards, P. J., and Wratten, S. D. (1983). Wound-induced defenses in plants and their consequences for patterns of insect grazing. *Oecologia* **59,** 88–93.

Engelbrecht, L., Orban, U., and Heese, W. (1969). Leaf-miner caterpillars and cytokinins in the "green islands" of autumn leaves. *Nature (London)* **223,** 319–321.

Entwhistle, P., Adams, P., and Evans, H. (1977). Epizootiology of a nuclear-polyhedrosis virus in European spruce sawfly (*Gilpinia hercyniae*): The status of birds as dispersal agents of the virus during the larval season. *J. Invertebr. Pathol.* **29,** 354–360.

Faeth, S. H. (1980). Invertebrate predation of leaf miners at low densities. *Ecol. Entomol.* **5,** 111–114.

Faeth, S. H. (1985a). Quantitative defense theory and early feeding by insects on oaks. *Oecologia* **68,** 34–40.

Faeth, S. H. (1985b). Host leaf selection by a leaf miner: Interactions among three trophic levels. *Ecology* **66,** 870–875.

Faeth, S. H. (1986). Indirect interactions between temporally-separated herbivores mediated by the host plant. *Ecology* **67,** 479–494.

Faeth, S. H., and Bultman, T. L. (1986). Interacting effects of increased tannin levels on leaf-mining insects. *Entomol. Exp. Appl.* **40,** 297–300.

Faeth, S. H., and Simberloff, D. (1981). Population regulation of a leaf-mining insect, *Cameraria* sp. nov., at increased field densities. *Ecology* **62,** 620–624.

Faeth, S. H., Mopper, S., and Simberloff, D. (1981a). Abundance and diversity of leaf-mining insects on three oak species: Effects of host-plant phenology and nitrogen content of leaves. *Oikos* **37,** 238–251.

Faeth, S. H., Conner, E. F., and Simberloff, D. (1981b). Early leaf abscission: A neglected source of mortality for folivores. *Am. Nat.* **117,** 409–415.

Feeny, P. P. (1970). Seasonal changes in oak leaf tannins and nutrients as a cause of spring feeding by winter moth caterpillars. *Ecology* **51,** 565–581.

Feeny, P. P. (1976). Plant apparency and chemical defenses. *Recent Adv. Phytochem.* **10,** 1–40.

Force, D. C. (1970). Competition among four hymenopterous parasites of an endemic insect host. *Ann. Entomol. Soc. Am.* **63,** 1675–1688.

Fowler, S. V., and Lawton, J. H. (1985). Rapidly induced defense and talking tree: The devil's advocate position. *Am. Nat.* **126,** 181–195.

Fowler, S., and MacGarvin, M. (1986). The effects of leaf damage on the performance of insect herbivores on birch, *Betula pubescens*. *Oecologia (Berlin)* **55,** 565–573.

Fox, L. R. (1981). Defense and dynamics in plant–herbivore systems. *Am. Zool.* **21,** 853–864.

Fox, L. R., and McCauley, B. J. (1977). Insect grazing on *Eucalyptus* in response to variation in leaf tannins and nitrogen. *Oecologia* **29,** 145–162.

Fox, L. R., and Morrow, P. A. (1986). On comparing herbivore damage in Australia and north temperate systems. *Aust. J. Ecol.* **11,** 387–393.

Franz, J. M. (1961). Biological control of pest insects in Europe. *Annu. Rev. Entomol.* **6,** 183–200.

Furniss, R. L., and Carolin, V. M. (1977). Western forest insects. *Misc. Publ.—U.S., Dep. Agric.* **1339.**

Golley, F. B. (1972). Energy flux in ecosystems. *In* "Ecosystem: Structure and Function" (J. A. Wiens, ed.), pp. 69–90. Oregon State Univ. Press, Corvallis.

Goodman, D. (1975). The theory of diversity-stability relationships in ecology. *Q. Rev Biol.* **50,** 237–266.

Gradwohl, J., and Greenberg, R. (1982). The effect of a single species of avian predator on the arthropods of aerial leaf litter. *Ecology* **63,** 581–583.

Hairston, N. G., Smith, F. E., and Slobodkin, L. B. (1960). Community structure, population control, and competition. *Am. Nat.* **94,** 421–425.

Harrison, S., and Karban, R. (1986). Effects of an early-season folivorous moth on the success of a later-season species mediated by a change in the quality of the shared host, *Lupinus arboreus* Sims. *Oecologia* (*Berlin*) **69,** 354–359.

Haukioja, E. (1980). On the role of plant defenses in the fluctuation of herbivore populations. *Oikos* **35,** 202–213.

Haukioja, E., and Niemelä, P. (1976). Does birch defend itself actively against herbivores? *Rep. Kevo Subarct. Res. Stn.* **13,** 44–47.

Haukioja, E., and Niemelä, P. (1979). Birch leaves as a resource for herbivores. Seasonal occurrence of increased resistance in foliage after mechanical damage of adjacent leaves. *Oecologia (Berlin)* **39,** 151–159.

Haukioja, E., Suomela, J., and Neuvonen, S. (1985). Long-term inducible resistance in birch foliage: Triggering cues and efficacy on a defoliator. *Oecologia (Berlin)* **65,** 363–369.

Heck, K. L., Jr. (1976). Some critical considerations of the theory of species packing. *Evol. Theory* **1,** 247–258.

Heinrich, B. (1979). Foraging strategies of caterpillars: Leaf damage and possible predator avoidance strategies. *Oecologia* **42,** 325–337.

Heinrich, B., and Collins, S. L. (1983). Caterpillar leaf damage and the game of hide-and-seek with birds. *Ecology* **64,** 592–602.

Holmes, R. T., and Sturges, F. W. (1975). Bird community dynamics and energetics in a northern hardwoods ecosystem. *J. Anim. Ecol.* **44,** 175–200.

Holmes, R. T., Schultz, J. C., and Nothnagle, P. (1979). Bird predation on forest insects: An enclosure experiment. *Science* **206,** 462–463.

Janzen, D. H. (1966). Coevolution of mutualism between ants and acacias in Central America. *Evolution (Lawrence, Kans.)* **20,** 249–275.

Janzen, D. H. (1973). Host plants as islands. II. Competition in evolutionary and contemporary time. *Am. Nat.* **107,** 786–790.

Janzen, D. H. (1975). Interactions of seeds and their insect predators parasitoids in a tropical deciduous forest. *In* "Evolutionary Strategies of Parasitic Insects and Mites" (P. W. Price, ed.), pp. 154–186. Plenum, New York.

Jeanne, R. L. (1979). A latitudinal gradient in rates of ant predation. *Ecology* **60,** 1211–1224.

Kahn, D. M., and Cornell, H. V. (1983). Early leaf abscission and folivores: Comments and consideration. *Am. Nat.* **122,** 428–432.

Karban, R. (1986). Interspecific competition between folivorous insects on *Erigeron glaucus*. *Ecology* **67,** 1063–1072.

Kareiva, P. (1983). Influence of vegetation texture on herbivore populations: Resource concentration and herbivore movement. *In* "Variable Plants and Herbivores in Natural and Managed Systems" (R. F. Denno and M. S. McClure, eds.), pp. 259–289. Academic Press, New York.

Kareiva, P. (1986). Patchiness, dispersal, and species interactions: Consequences for communities of herbivorous insects. In "Community Ecology" (J. Diamond and T. J. Case, eds.), pp. 192–206. Harper & Row, New York.

Kendeigh, S. C. (1979). "Invertebrate Populations of the Deciduous Forest: Fluctuations and Relations to Weather," Ill. Biol. Monogr. No. 50. Univ. of Illinois Press, Urbana.

Klomp, H. (1966). The dynamics of a field population of the Pine looper (*Bupalus piniarius* L.) (Lepidoptera: Geometridae). *Adv. Ecol. Res.* **3,** 207–305.

Krischik, V. A., and Denno, R. F. (1983). Individual, population and geographic patterns in plant defense. *In* "Variable Plants and Herbivores in Natural and Managed Systems" (R. F. Denno and M. S. McClure, eds.), pp. 463–512. Academic Press, New York.

Kulman, H. M. (1971). Effects of insect defoliation on growth and mortality of trees. *Annu. Rev. Entomol.* **16,** 289–324.

Laine, K. J., and Niemelä, P. (1980). The influence of ants on the survival of mountain birches during an *Oporinia autumnata* (Lep., Geometridae) outbreak. *Oecologia (Berlin)* **47,** 39–42.

Lawton, J. H., and MacGarvin, M. (1986). The organization of herbivore communities. In "Community Ecology: Pattern and Process" (J. Kikkawa and D. J. Anderson, eds.). Blackwell, London.

Lawton, J. H., and Hassell, M. P. (1981). Asymmetrical competition in insects. *Nature (London)* **289,** 793–795.

Lawton, J. H., and MacNeil, S. (1979). Between the devil and the deep blue sea: On the problem of being an herbivore. *Symp. Br. Ecol. Soc.* **20,** 223–244.

Lawton, J. H., and Strong, D. R., Jr. (1981). Community patterns and competition in folivorous insects. *Am. Nat.* **118,** 317–333.

Leigh, E. G., Jr., and Smythe, N. (1978). Leaf production, leaf consumption, and the regulation of folivory on Barro Colorado Island. *In* "The Ecology of Arboreal Folivores" (G. G. Montgomery, ed.), pp. 33–50. Symp. Zool. Parks, Smithsonian Inst., Washington, D.C.

Linit, M. J., Johnson, P. S., McKinney, R. A., and Kearby, W. H. (1986). Insects and leaf area losses of planted northern red oak seedlings in an Ozark forest. *For. Sci.* **32,** 11–20.

MacClellan, C. R. (1970). Woodpecker ecology in the apple orchard environment. *Proc. Tall Timbers Conf. Ecol. Anim. Control Habitat Manage., 1970,* pp. 273–284.

McClure, M. S., and Price, P. W. (1975). Competition among sympatric *Erythroneura* leaf hoppers (Homoptera: Cicadellidae) on sycamore. *Ecology* **57,** 928–940.

McClure, M. S., and Price, P. W. (1976). Ecotype characteristics of coexisting *Erythroneura* leaf hoppers (Homoptera: Cicadellidae) on sycamore. *Ecology* **57,** 928–940.

Mattson, W. J. (1980). Herbivory in relation to plant nitrogen content. *Annu. Rev. Ecol. Syst.* **11,** 119–161.

Mattson, W. J., and Addy, N. D. (1975). Phytophagous insects as regulators of forest primary production. *Science* **190,** 515–522.

May, R. M. (1983). Parasitic infections as regulators of animal populations. *Am. Sci.* **71,** 36–45.

Messina, F. J. (1981). Plant protection as a consequence of an ant–membracid mutualism: Interactions of goldenrod (*Solidago* sp.). *Ecology* **62,** 1433–1440.

Mopper, S., Faeth, S. H., Boecklen, W. J., and Simberloff, D. S. (1984). Host specific variation in leaf miner population dynamics: Effects on density, natural enemies, and behavior of *Stilbosis quadricustatella* (Cham.) (Lepidoptera: Cosmopterigidae). *Ecol. Entomol.* **9,** 169–177.

Morrow, P. A., and LaMarche, V. C. (1978). Tree ring evidence for chronic insect suppression of productivity in subalpine Eucalyptus. *Science* **201,** 1244–1246.

Murdoch, W. M., Reeve, J. D., Huffaker, C. B., and Kennett, C. E. (1984). Biological control of olive scale and its relevance to ecological theory. *Am. Nat.* **123,** 371–392.

Myers, J. H., and Williams, K. S. (1984). Does tent caterpillar attack reduce the food quality of red alder foliage? *Oecologia* **62,** 74–79.

Niemelä, P. (1983). Seasonal patterns in the incidence of specialism: Macrolepidopteran larvae on Finnish trees. *Ann. Zool. Fenn.* **20,** 199–202.

Niemelä, P., and Haukioja, E. (1982). Seasonal patterns in species richness of herbivores: Macrolepidopteran larvae on Finnish deciduous trees. *Ecol. Entomol.* **7,** 169–175.

Niemelä, P., Aro, A. M., and Haukioja, E. (1979). Birch leaves as a resource for herbivores. Damaged-induced increase in leaf phenols with trypsin-inhibiting effects. *Rep. Kevo Subactic Res. Sta.* **15,** 37–40.

Odum, H. T., and Ruiz-Reyes, J. (1970). Holes in leaves and grazing control mechanism. *In* "A Tropical Rain Forest" (H. T. Odum and R. Pigeon, eds.), pp. I-69 to I-80. Div. Tech. Inf., U.S. At. Energy Commission, Oak Ridge, Tennessee.

Ohmart, C. P. (1984). Is insect defoliation in eucalypt forests greater than that in other temperate forests? *Aust. J. Ecol.* **9,** 413–418.

Ohmart, C. P., Stewart, L. G., and Thomas, J. R. (1983). Leaf consumption in three

Eucalyptus forest types in southeastern Australia and their role in short-term nutrient cycling. *Oecologia* **59,** 322–330.
Opler, P. A. (1974). Biology, ecology, and host specificity of Microlepi doptera associated with *Quereus agrifolia* (Fagaceae). Univ. of California Press, Berkeley.
Otvos, I. S. (1979). The effects of insectivorous bird activities in forest ecosystems: An evaluation. *In* "The Role of Insectivorous Birds in Forest Ecosystems" (J. G. Dickson, R. N. Connor, J. C. Kroll, R. R. Fleet, and J. A. Jackson, eds.), pp. 341–374. Academic Press, New York.
Pianka, E. R. (1983). "Evolutionary Ecology." Harper & Row, New York.
Pimentel, D., Levin, S. A., and Soans, A. B. (1975). On the evolution of energy balance in some exploiter–victim system. *Ecology* **56,** 381–390.
Price, P. W. (1975). The parasitic way of life and its consequences. *In* "Evolutionary Strategies of Parasitic Insects and Mites" (P. W. Price, ed.), pp. 1–13. Plenum, New York.
Price, P. W. (1983). Hypotheses on organization and evolution in herbivores insect communities. *In* "Variable Plants and Herbivores in Natural and Managed Systems" (R. F. Denno and M. S. McClure, eds.), pp. 559–596. Academic Press, New York.
Price, P. W., Bouton, C. E., Gross, P., McPheron, B. A., Thompson, J. N., and Weis, A. E. (1980). Interactions among three trophic levels: Influence of plants on interactions between insect herbivores and natural enemies. *Annu. Rev. Ecol. Syst.* **11,** 41–65.
Puritch, G. S., and Nijholt, W. W. (1974). Occurrence of juvabione-related compounds in grand firs and pacific silver fir infested by balsam wooly aphid. *Can. J. Bot.* **52,** 585–587.
Reichle, D. E., Goldstein, R. A., Van Hook, R. I., Jr., and Dodson, G. J. (1973). Analysis of insect consumption in a forest canopy. *Ecology* **54,** 1076–1084.
Renaud, D. E. S. (1987). Impact of herbivory on defensive and reproductive allocation of Arizona walnuts. Master's thesis, Arizona State University.
Rhoades, D. F. (1979). Evolution of chemical defenses against herbivores. *In* "Herbivores: Their Interaction with Secondary Plant Metabolites" (G. A. Rosenthal and D. H. Janzen, eds.), pp. 3–54. Academic Press, New York.
Rhoades, D. F. (1983a). Responses of alder and willow to attack by tent caterpillars and webworms: Evidence for pheromonal sensitivity of willows. *Amer. Chem. Soc. ACS Symp. Ser.* **208,** 55–68.
Rhoades, D. F. (1983b). Herbivore population dynamics and plant chemistry. *In* "Variable Plants and Herbivores in Natural and Managed Systems" (R. F. Denno and M. S. McClure, eds.), pp. 155–220. Academic Press, New York.
Rhoades, D. F., and Cates, R. G. (1976). Towards a general theory of plant antiherbivore chemistry. *Recent Adv. Phytochem.* **10,** 168–213.
Rhomberg, L. (1984). Inferring habitat selection by aphids from dispersion of their galls over the tree. *Am. Nat.* **124,** 751–756.
Robinson, S. K., and Holmes, R. T. (1982). Foraging behavior of forest birds: The relationship among search tactics, diet, and habitat structure. *Ecology* **63,** 1918–1931.
Rockwood, L. L. (1974). Seasonal changes in the susceptibility of *Crescentia alata* leaves to the flea beetle, *Oedionychus* sp. *Ecology* **55,** 142–148.
Root, R. B. (1967). The niche exploitation pattern of the blue-gray gnatcatcher. *Ecol. Monogr.* **37,** 317–350.
Rothacher, J. S., Blow, F. E., and Potts, S. M. (1954). Estimating the quantity of tree foliage in oak stands in the Tennessee Valley. *J. For.* **52,** 33–51.
Schoener, T. W. (1983). Field experiments on interspecific competition. *Am. Nat.* **122,** 240–285.

Schultz, J. C. (1983a). Impact of variable plant defensive chemistry on susceptibility of insects to natural enemies. *ACS Symp. Ser.* **208,** 37–54.

Schultz, J. C. (1983b). Habitat selection and foraging tactics of caterpillars in heterogenous trees. *In* "Variable Plants and Herbivores in Natural and Managed Systems" (R. F. Denno and M. S. McClure, eds.), pp. 61–90. Academic Press, New York.

Schultz, J. C., and Baldwin, I. T. (1982). Oak leaf quality declines in response to defoliation by gypsy moth larvae. *Science* **217,** 149–151.

Schweitzer, D. F. (1979). Effects of foliage age on body weight and survival in the tribe Lithophanine (Lepidoptera: Noctuidae). *Oikos* **32,** 403–408.

Silva-Bohorquez, I. (1986). Interspecific interactions between insects on oak trees, with special reference to defoliators and the oak aphid. D. Phil. Thesis, Univ. of Oxford.

Skinner, G. J. (1980). The feeding habits of the wood-ant, *Formica rufa* (Hymenoptera: Formicidae), in limestone woodland in Northwest England. *J. Anim. Ecol.* **49,** 417–433.

Skinner, G. J., and Whittaker, J. B. (1981). An experimental investigation of inter-relationships between the wood-ant (*Formica rufa*) and some tree-canopy herbivores. *J. Anim. Ecol.* **50,** 313–326.

Sloan, N. F., and Coppel, H. (1968). Ecological implications of bird predators on the larch casebearer in Wisconsin. *J. Econ. Entomol.* **61,** 1067–1070.

Sloan, N. F., and Simmons, G. A. (1973). Foraging behavior of the chipping sparrow in response to high populations of Jack Pine budworm. *Am. Midl. Nat.* **90,** 210–215.

Solomon, M. E., Glen, D. M., Kendall, D. A., and Milsom, N. F. (1976). Predation of overwintering larvae of codling moth (*Cydia pomonella* (L.)) by birds. *J. Appl. Ecol.* **13,** 341–352.

Southwood, T. R. E. (1973). The insect/plant relationship—an evolutionary perspective. *Symp. Entomol. Soc. London* **6,** 3–30.

Strong, D. R., Jr. (1983). Natural variability and the manifold mechanisms of ecological communities. *Am. Nat.* **122,** 636–660.

Strong, D. R., Jr., Lawton, J. H., and Southwood, R. (1984). "Insects on Plants: Community Patterns and Mechanisms." Blackwell, Oxford.

Swain, T. (1979). Tannins and lignins. *In* "Herbivores: Their Interaction with Secondary Plant Metabolites" (G. A. Rosenthal and D. H. Janzen, eds.), pp. 657–682. Academic Press, New York.

Taper, M. L., Zimmerman, E. M., and Case, T. J. (1985). Sources of mortality for a cynipid gall-wasp (*Dryocomus dubiosus* (Hymenoptera: Cynipidae)): The importance of the tannin/fungus interaction. *Oecologia (Berlin)* **68,** 437–445.

Thompson, J. N. (1982). "Interaction and Coevolution." Wiley, New York.

Tinsley, T. W. (1979). The potential of insect pathogenic viruses as pesticidal agents. *Annu. Rev. Entomol.* **24,** 63–87.

Toft, C. A. (1986). Communities of species with parasitic life-styles. In "Community Ecology" (J. Diamond and T. J. Case, eds.), pp. 445–463. Harper & Row, New York.

van der Meijden, E. (1980). Can hosts escape from their parasitoids? The effects of food shortage on the braconid parasitoid *Apanteles popularis* and its host, *Tyria jacobaeae*. *Neth. J. Zool.* **30,** 382–392.

Varley, G. C., and Gradwell, G. R. (1968). Population models for the winter moth. *In* "Insect Abundance" (T. R. E. Southwood, ed.), pp. 132–142. Blackwell, Oxford.

Vinson, S. B. (1976). Host selection of insect parasitoids. *Annu. Rev. Entomol.* **21,** 109–133.

Vinson, S. B., and Iwantsch, G. F. (1980). Host suitability for insect parasitoids. *Annu. Rev. Entomol.* **25,** 397–419.

Washburn, J. O., and Cornell, H. V. (1981). Parasitoids, patches, and phenology: Their possible role in the local extinction of a cynipid gall wasp population. *Ecology* **62,** 1597–1607.

West, C. (1985). Factors underlying the late seasonal appearance of the lepidopterous leaf-mining guild on oak. *Ecol. Entomol.* **10,** 111–120.

White, T. C. R. (1969). An index to measure weather-induced stress of trees associated with outbreaks of psyllids in Australia. *Ecology* **50,** 905–909.

White, T. C. R. (1974). A hypothesis to explain outbreaks of looper caterpillars with special reference to populations of *Seliosema suavis* in a plantation of *Pinus radiata* in New Zealand. *Oecologia* **16,** 279–301.

White, T. C. R. (1976). Weather, food and plagues of locusts. *Oecologia* **22,** 119–134.

White, T. C. R. (1978). The importance of a relative shortage of food in animal ecology. *Oecologia* **33,** 71–86.

White, T. C. R. (1984). The abundance of invertebrate herbivory in relation to the availability of nitrogen in stressed food plants. *Oecologia* **63,** 90–105.

Whitham, T. G. (1978). Habitat selection by *Pemphigus* aphids in response to resource limitation and competition. *Ecology* **59,** 1164–1176.

Whitham, T. G. (1980). The theory of habitat selection examined and extended using *Pemphigus* aphids. *Am. Nat.* **115,** 449–466.

Whitham, T. G. (1981). Individual trees as heterogenous environments. Adaptation to herbivory or epigenetic noise. *In* "Insect and Life History Patterns: Habitat and Geographic Variations" (R. F. Denno and H. Dingle, eds.), pp. 9–27. Springer-Verlag, Berlin, New York.

Whitham, T. G. (1983). Host manipulation of parasites: Within-plant variation as a defense against rapidly evolving pests. *In* "Variable Plants and Herbivores in Natural and Managed Systems" (R. F. Denno and M. S. McClure, eds.), pp. 15–41. Academic Press, New York.

Wiegert, R. G., and Owen, D. F. (1971). Trophic structure, available resources, and population density in terrestrial vs. aquatic ecosystems. *J. Theor. Biol.* **30,** 69–81.

Williams, A. G., and Whitham, T. G. (1986). Premature leaf abscission: An induced plant defense against galling aphids. *Ecology* **67,** 1619–1627.

Wolda, H. (1978). Fluctuations in the abundance of tropical insects. *Am. Nat.* **112,** 1017–1045.

Zucker, W. V. (1982). How aphids choose leaves: The role of phenolics in host selection by a galling aphid. *Ecology* **63,** 972–981.

Zucker, W. V. (1983). Tannins: Does structure determine function? An ecological perspective. *Am. Nat.* **121,** 335–365.

Chapter 7

Population Outbreaks of Introduced Insects: Lessons from the Biological Control of Weeds

JUDITH H. MYERS

The Ecology Group
Departments of Zoology and Plant Science
University of British Columbia
Vancouver 8, Canada V6T 2A9

I. INTRODUCTION

Why do populations of some insects show periodic outbreaks to high density whereas others maintain low, relatively stable densities? Insect population ecologists have attempted to answer this question largely through descriptive studies, life table analysis, or theoretical machinations. Although these studies have yielded interesting stories, conclusive answers are rare. An experimental approach to the study of insect populations provides an alternative that may yield more conclusive interpretations.

The introduction of insects to exotic areas for the biological control of weeds provides a unique experimental situation. The insects are freed from predators, parasites, competitors, and diseases and are introduced to areas where their food plants are abundant. In this chapter I review information on insects introduced for the biological control of weeds in an attempt to answer the question, Why do some introduced insects erupt into outbreak densities whereas others maintain low densities or fail to become established?

II. HYPOTHESES TO EXPLAIN THE FAILURE OF INTRODUCED INSECTS TO ATTAIN OUTBREAK DENSITIES

Several hypotheses can account for the failure of some introduced insects to become established or reach high numbers whereas the populations of other herbivorous insects erupt to outbreak densities after their introduction to new areas:

1. Climatic differences between native and exotic areas prevent outbreak of the introduced insect populations.
2. Food plant quality or taxonomic differences in host plants between exotic and native areas prevent outbreaks of introduced populations.
3. Predators or parasites in the exotic area attack the introduced insects and prevent population outbreak.
4. Reduced genetic variability in the introduced population reduces the vigor of individuals and prevents population outbreak.
5. Insects that are competitively superior among the native guild of insects attacking a particular plant species or those that suffer little natural mortality are less likely to expand to outbreak densities after their introduction to areas lacking competitors, parasites, and diseases than are competitively inferior species or those that have evolved with high natural mortality (Zwölfer, 1973).

In the following sections I review case studies from the literature on the introduction of insects for the biological control of weeds to determine whether they are consistent with these hypotheses. The surveys of Julien (1982) and Maw (1984) provide much of the basic information. In most cases the only information available concerns the outbreak of the insect after its introduction. In a few examples information is also available on subsequent fluctuations of the insect populations.

A. Climatic Compatibility and the Outbreak of Insect Populations

The first rule of thumb in selecting biological control organisms is that the insects selected for release be collected from areas most similar in climate to the area of the weed problem. This is not always possible since sometimes areas of climatic equivalency are not accessible for political or practical reasons. The failure of an introduced insect can rarely be attributed to climatic differences with certainty. However, cool summers and/or cold winters have been blamed for some failures of biological control agents in Canada (Harris, 1984a), and in Australia drought has caused the disappearance of both weeds and introduced insects (Maw, 1984).

A delay in population outbreak until several generations after the establishment of an introduced insect may indicate a period of adaptation to a new climate. DeBach (1965) states that adaptation after establishment and before outbreak is rare for entomophagous insects: The insects are either well adapted and increase immediately, or they do not become established or remain rare. Herbivorous insects do not necessarily follow the same pattern. *Chrysolina quadrigemina* (Suffr.) "disappeared" for four years after its introduction to Australia and then became abundant (Wilson, 1965). In western Canada, populations of *C. quadrigemina* did not increase until 7 to 11 years after introduction and then reached outbreak densities (Harris *et al.*, 1969). However, *C. quadrigemina,* which begins laying eggs in the autumn (Peschken, 1972), has not adapted to areas with severe winter weather that kills overwintering eggs and larvae. In these areas, *C. quadrigemina* remains rare, and *C. hyperici* (Forester), which delays oviposition, is common (J. H. Myers, personal observation; Williams, 1986).

Metzneria paucipunctella Zell., a seed-feeding moth introduced from Switzerland to British Columbia, attacks spotted knapweed. The population trend after introduction suggests postintroduction adaptation (Fig. 1). Approximately 200 individuals were released in both 1973 and 1974, but an outbreak did not occur until 1978. Winter survival of the *Metzneria* was monitored from 1976 through 1980. Over the winter of 1976–1977 only 5% of larvae survived. The next year, however, 70% of larvae survived and the population increased approximately 10 times the next summer. In the next 2 years winter survival of *Metzneria* larvae was about 30%. No environmental correlate with the improved winter survival was observed. The population may have, at least partially, adapted to the new

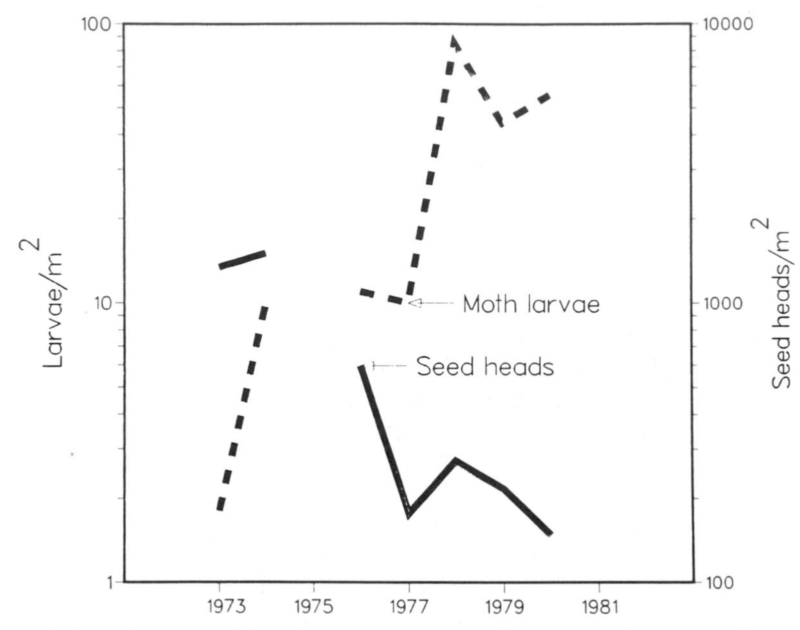

Fig. 1. Population trend for *Metzneria paucipunctella* introduced on spotted knapweed in British Columbia. Data from J. H. Myers and P. Harris (unpublished).

climate before the outbreak (J. H. Myers, unpublished data; Harris and Myers, 1984).

The weevil *Erytenna consputa* Pascoe was collected from a variety of sites in Australia and introduced to South Africa in locations judged to be climatically similar to the collection sites. However, the success of the introductions was influenced more by the strain of host plant, *Hakae sericea* Schrader, than by the match between the climates of the collection and release sites (Neser and Kluge, 1986). Either the weevils were better able to adapt to climatic differences than they were to host strain differences, or they were preadapted to a wide range of climates.

Two cases have been recorded of phenological shifts by introduced insects adapting to new climatic conditions. Mild winters in California are a likely explanation for a shift in temperature threshold (a genetically controlled trait) of introduced cinnabar moth, *Tyria jacobaeae* (L.). A higher temperature threshold for pupal development prevents the premature emergence of adults in the spring (Myers, 1979; Richards and Myers, 1980). Although it was not recorded, strong selection against individuals with low temperature thresholds must have occurred in the first years after introduction. Murray (1982, 1986) found a shift in incubation times from oviposition to larval hatching among populations of *Cactoblastis cactorum* (Berg.) in different climatic regions of Australia.

Whether the insects will adapt to climatic differences depends on the impact and degree of difference between the native and exotic areas. For example, the establishment of first-instar larvae of the root-boring beetle, *Sphenoptera jugoslavica* Obenb., is dependent on diffuse knapweed plants being dormant during summer drought (Zwölfer, 1976). Slight changes in rainfall can determine the growth status of the plants and therefore the ability of the larvae to penetrate the plants. It may be much more difficult for insects to adapt to this type of situation than to modify temperature thresholds that are directly exposed to selection for phenological adjustments.

In conclusion, severe climatic incompatibility may explain the failure of some insect introductions, particularly in areas of extreme cold or drought. However, introduced insects do adapt to shifts in climate. Although the outbreak of an introduced insect may be delayed by a necessary period of adaptation, several documented cases suggest that insects frequently carry sufficient genetic or physiological variability to allow them to adjust to new environments.

B. Host Plant Quality and Taxonomic Status and the Outbreak of Populations of Introduced Insects

The quality of food plants is a crucial determinant of the success of a herbivorous insect in becoming established and increasing to outbreak densities. Two aspects of food quality are the nutritional condition of the plant and the secondary chemicals and physical features that characterize taxonomic entities.

1. Nutritional Quality

In the classical case of the control of prickly pear cactus in Australia by the moth *Cactoblastis cactorum,* successful control was achieved in an area with poor soil only after nitrogen fertilizer improved the nutritional quality of the plants (Dodd, 1940). A similar technique has been used in the control of the aquatic fern *Salvinia molesta* D. S. Mitchell by the weevil *Cyrtobagous salviniae* Hustache (Thomas and Room, 1986). The weevils were successfully established in the field only when urea was used to raise the nitrogen content of *Salvinia* at the original release site. As the beetle population slowly expanded, the previously attacked *Salvinia* plants were noticeably darker green than the undamaged plants. The

TABLE 1

Crude Protein of Defoliated and Control *Hypericum perforatum*[a]

Defoliation (%)	Water treatment every 3 days[b]		
	120 ml	160 ml	200 ml
0	22.4	16.8	14.1
100	29.0	19.8	17.1

[a] Values (percent dry weight) are for combined foliage of four plants. Data from A. Chow and J. H. Myers (unpublished).

[b] Plants maintained in the greenhouse for 54 days.

regenerating plants had higher nitrogen content. In this situation the original increase in the nitrogen levels of the *Salvinia* allowed the weevils to become established. Subsequently, their feeding improved the nitrogen content of surrounding plants and permitted the beetle population to expand.

Artifical defoliation and/or water stress of *Hypericum perforatum* L. increased the nitrogen levels of regrowth foliage (Table 1). This was probably the result of defoliated plant shunting more nitrogen to leaves and less to flowers. Initial insect attack on trees sometimes improves foliage quality for insects (Myers and Williams, 1984; Williams and Myers, 1984). These examples suggest that insects can improve conditions for themselves and thus pave the way for their own population outbreak.

Do introduced insects frequently expand to outbreak densities following plant quality improvement? Little information is available on this interesting possibility, but in several of the situations in which data have been collected to quantify the population trends of introduced insects, an apparent decline in the plant population occurred before the outbreak of the insect (Figs. 1–3). This suggests that environmental conditions stressing the plants may reduce the plant density and improve the quality of the plants for the insects (White, 1984).

The dynamics of cinnabar moth, introduced to North America as a biological control agent on tansy ragwort, are strongly influenced by plant nitrogen levels (Myers and Post, 1981). The fecundity of moths, survival of larvae, and amplitude of fluctuations of introduced moth populations were correlated to the nitrogen content of the foliage. Populations living in areas with plants of higher quality had greater potential for overexploitation, and this magnified the fluctuations of the population (Myers, 1976).

The nutritional quality of food plants is very important for the success and subsequent outbreak of introduced insects. Fertilization may be a necessary tool for the establishment of biological control agents in some cases. Experimental investigation of the interactions between insect herbivores and the quality of their food plants should be pursued to determine whether low levels of herbivory prime the pump for insect outbreaks.

2. Taxonomic Status

The species or strain of host plant can also influence the potential for outbreak of introduced insects. Although one would expect that plants for which biological control was being attempted would be properly identified, this is not always the case. Misidentification has sometimes resulted in the introduction of insects to the wrong species of food plant. Larvae of a root-boring moth from European leafy spurge introduced to Canada became paralyzed by the latex of the Canadian spurge (Harris, 1984c). As mentioned previously, weevils introduced from Australia to South Africa to control *Hakea* were influenced more by the host plant strain than by the climate (Neser and Kluge, 1986). The cerambycid beetle *Aledion cereicola* Fisher readily attained outbreak densities on *Eriocereus martinii* (Labouret) Riccobona in Australia but remained rare on *Eriocereus tortuosus* (Forbes). However, the mealy bug, *Hypogeococcus festerianus* (Lizer y Trelles), rapidly increased on both species of Harrisia cactus (Julien, 1982). Host plant specificity can be critical to the success of introduced insects (Ehler and Andres, 1983; Sands and Harley, 1981), but it is not necessarily so.

A lack of specificity is shown by the moth *Cactoblastis cactorum* and the cochineal insect *Dactylopius opuntiae* (Cockerell). These insects have been introduced successfully to 9 and 12 species of cactus, respectively, and have reached outbreak densities on most. Eleven of 21 insects introduced on cactus around the world have become established on nonnative host plants (Moran and Zimmerman, 1984).

Rhinocyllus conicus Forelich is another insect that has become established and has reached high population densities on several plant species, in this case all thistles in the genus *Carduus*. Thus far, however, it has successfully controlled only *Carduus nutans* L. (Goeden and Richer,

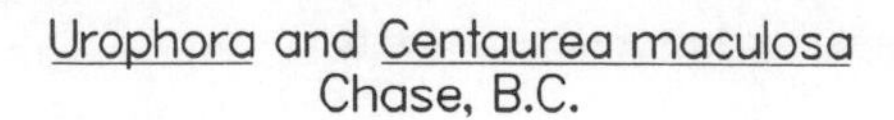

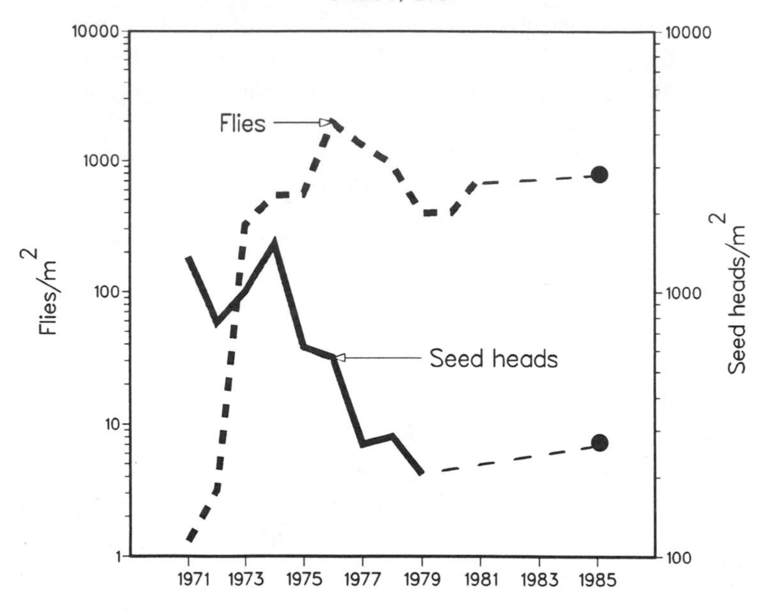

Urophora and Centaurea maculosa
Walachin, B.C.

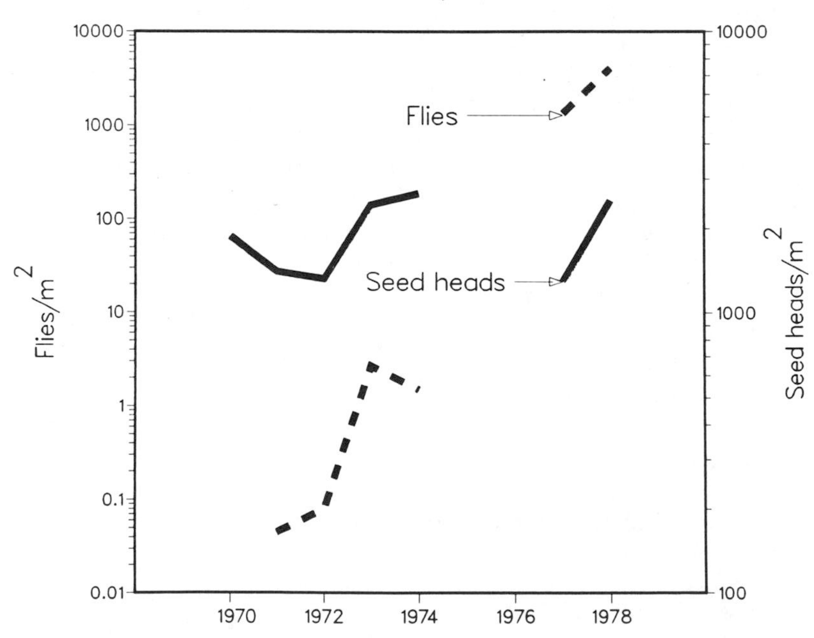

Fig. 2. Combined population trends for *Urophora affinis* and *U. quadrifasciata* introduced to British Columbia on spotted knapweed. Data from Harris (1980).

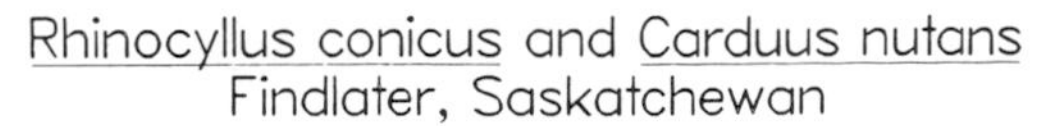

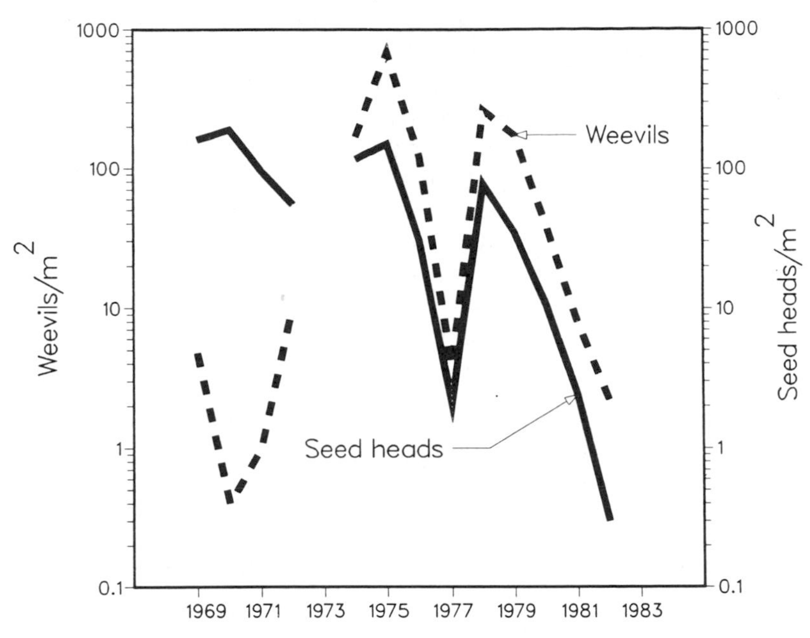

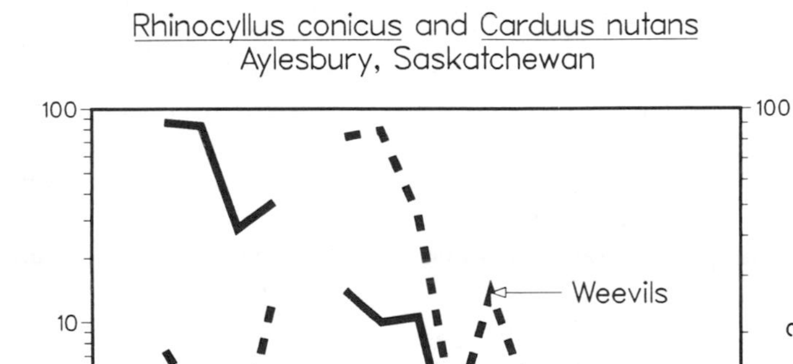

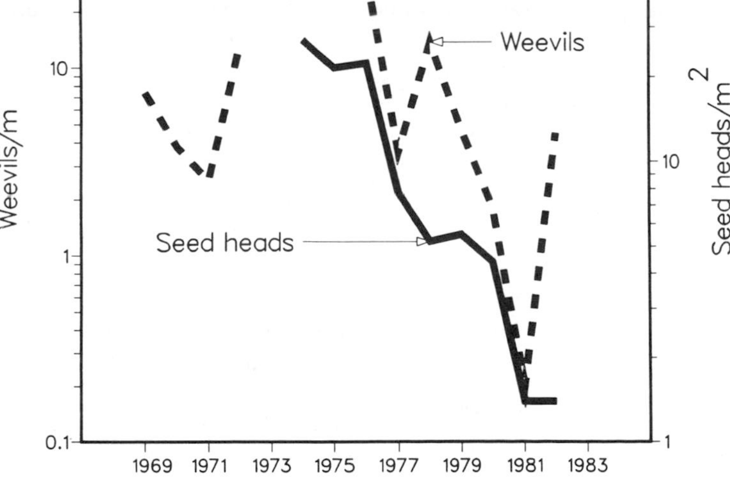

Fig. 3. Population trends for *Rhinocyllus conicus* introduced on nodding thistle to Saskatchewan. Data from Harris (1984c).

1978; Julien, 1982). *Urophora affinis* Frauenfeld, a gall fly, has attained outbreak densities on three species of knapweed, *Centaurea,* in Canada (P. Harris and J. H. Myers, 1984; also personal observation).

The specificity of insects to host plant strains varies among groups. Current biological control practices demand careful screening of insects before introduction, and only feeding specialists pass the test. This probably selects against species capable of establishing on a range of varieties even within the same taxonomic group. Even so, 67–70% of insects introduced for the biological control of weeds became established (Ehler and Andres, 1983; Harris, 1984a; Myers, 1986). This high success rate is undoubtedly due, in part, to the matching of food plant and insect species.

C. Native Predators and Parasites and the Outbreak of Introduced Insects

Biological control insects are released without their natural enemies. Therefore, any population suppression by parasites or predators must be due to the adaptation of native insects to the exotics. The native parasite community in Australia seems particularly adept at suppressing insects introduced for the biological control of weeds (McFadyen, 1981). For example, a trypetid, gall-forming fly, *Procecidochares utilis* Stone, maintains sufficient densities to control crofton weed, *Eupatorium adenophorum* Sprengel, in Hawaii, but within 2 years of establishment and population explosion in Australia, native parasites suppressed the population (Haesler, 1981). Because the phenologies of the gall flies and the parasites are not well synchronized seasonally, populations of the host insects have enormous fluctuations and sometimes reach high levels. In this example, interactions between hosts and parasites result in periodic outbreaks of the introduced gall-forming fly. Native parasites also reduced populations of *E. adenophum* in India and New Zealand (Julien, 1982). In contrast, *Cactoblastis cactorum* is parasitized by 20 species of wasps in Australia, but this did not suppress its population outbreak (Mann, 1970).

Of 41 cases of introduction for the control of a variety of cactus species around the world, success was reduced by predators in 25% and by parasites in 7% of the cases (Moran and Zimmerman, 1984). In one example, the densities of an imported cochineal insect, *Dactylopius opuntiae,* were raised by spraying with DDT to control predaceous coccinellid beetles (Hassell and Moran, 1976). Predators are frequently blamed for the failure of biological control introductions, but experimental testing of this interpretation is rarely carried out (Goeden and Louda, 1976; see also Chapter 12, this volume). Generalist predators are more likely to adapt rapidly to introduced prey than are more specialized parasites and to respond both functionally and numerically to augmented prey densities (Nuessly and Goeden, 1983).

D. Genetic Variability and the Outbreak of Introduced Insects

Increased genetic variation has been shown to be associated with outbreaks of laboratory populations of insects (Ayala, 1968; Carson, 1961; Myers,; see Table 2), and population outbreaks can be associated with increased phenotypic and perhaps also genetic variation (Ford and Ford, 1930). A number of studies on a variety of organisms have shown that growth and survival rates are higher for more heterozygous individuals (Mitton and Grant, 1984), and this could explain the relationship between genetic variation and population increase. One might expect that biological control introductions started from a small number of individuals would have reduced genetic variability and reduced potential for outbreak.

Some extreme examples of successful biological control introductions with few individuals are the introduction of the offspring of a single female *Dactylopius opuntiae* from Ceylon to Mauritius, the introduction of five female *D. austrinus* De Lotto to South Africa (Moran and Zimmerman, 1984), and the introduction of the offspring of three mealybugs, *Hypogeococcus festerianus,* to Australia (W. H. Haesler, personal communication). Many biological control introductions involve 50–100 individuals from a single native population.

The observed outbreaks following introductions of a small number of individuals demonstrate either that these "bottlenecks" do not seriously reduce genetic variability or that genetic variability is not necessary for outbreaks. Lewontin (1965) pointed out that much of the genetic variance of a population is carried by a single fertilized female, even though rare alleles are lost at some polymorphic loci. Particularly if the population

TABLE 2

Influence of the Addition of New Genetic Material to Laboratory Populations of Onion Root Maggots[a]

Proportion of wild males	0.00	0.20	0.50
Number of flies[b]	349.5 (35)	546.5 (37)	635.0 (9)
Pupal weight (mg)[c]	12.25 (0.1)*	12.9 (0)*	12.45 (0.3)

[a] Asterisks indicate significantly different, $p < .01$.

[b] Two populations were started with 30 males and females for each treatment, and mean and standard error (in parentheses) values are recorded.

[c] Means and standard error (in parentheses) of average pupal weights for two populations.

expands after introduction, reductions in heterozygosity and additive genetic variance are minor (Nei *et al.,* 1975; Franklin, 1980; Frankel and Soule, 1981). Myers and Sabath (1981) reviewed the evidence on the genetic variability of introduced organisms based on electrophoretic markers. Introduced organisms seem not to lose genetic variability.

Genetic transiliencies following introduction and expansion have been discussed by Carson (1968), Templeton (1980), Carson and Templeton (1984), and Murray (1986). Little evidence exists, however, for major genetic shifts in populations of introduced insects. Acceptance of new plant species for oviposition and larval development might indicate a major genetic change in introduced insects. Biotypes of *Rhinocyllus conicus* collected from Italian thistle, *Carduus pycnocephalus* L., and milk thistle, *Silybum marianum* L., have been introduced to California. The Italian thistle biotype of beetle began to oviposit on milk thistle after reaching explosive population densities, but larvae were not able to develop on that plant species. Therefore, this does not represent a genetic shift. The milk thistle biotype of the beetle, however, increasingly transferred to Italian thistle under field conditions as the population increased (Goeden *et al.,* 1986). The beetles attacked Italian thistle before outbreak, so again this does not represent a genetic shift.

Perhaps the best way to look for genetic change after an outbreak is to perform test crosses with introduced and native populations. Crosses made between North American and European cinnabar moth populations indicated no reproductive incompatibility, but sample sizes were small (Myers, 1978).

Introductions from previously established populations more frequently result in outbreak than do those from native population (Kok, 1978) or from laboratory cultures (Schroeder, 1983). Newly established populations could be better adapted to the new environment, or they might contain combinations of genotypes resulting from reduced selection during the outbreak phase that make them more vigorous. Myers and Sabath (1981) recommended collecting insects for introduction from increasing and expanding populations in native habitats.

Introduced insects have sometimes declined in number after the original outbreak and have never attained outbreak densities again, or new introductions from the original established population were not successful. The introduction of cinnabar moth from Europe to Nanaimo, British Columbia, exemplifies this. The original release consisted of 20,000 larvae, but after the first outbreak, the population declined and moths remain rare. Only one of many introductions of this population resulted in outbreak (F. Wilkinson, personal communication), and there was little spread from the population, which eventually became extinct (J. H.

Myers, personal observation). Similarly, after its first introduction to Queensland, *Trirhabda baccharidis* (Weber) readily became established and the population has persisted. Subsequent populations, however, either did not become established or did not develop into outbreak densities (McFadyen, cited in Julien, 1982). The explanation for these population trends is not apparent, but they could suggest genetic deterioration following a genetic flush during the original outbreak (Carson, 1968). Population eruption and decline is typical of animal introductions (Caughley, 1970; Elton, 1958; Myers and Iyer, 1981).

Many of the ecologically relevant shifts in phenotype that occur with outbreaks of biological control insects have a large environmental component. For example, although the heritability of body size of cinnabar moths can be measured (Richards and Myers, 1980), most of the shift in pupal weight and fecundity after introduction is due to crowding and food limitation (Table 3). This phenotypic plasticity allows rapid adjustment to changing conditions.

To summarize, although the introduction of biological control insects frequently starts from a small number of individuals, this has not suppressed the potential for outbreak in many species. There is little evidence that major genetic shifts follow outbreaks of insects introduced as biological control agents. The postintroduction adaptations that have been observed indicate that introduced populations have sufficient genetic variation to respond to selection. Phenotypic variability is also important in allowing insects to adjust to new conditions. Because the genotype is difficult to measure, little is known about the genetics of insect outbreaks (see Chapter 19).

TABLE 3

Relationship between Density and Pupal Weight for a Cinnabar Moth Population Introduced near Cultus Lake, British Columbia

Year	Density (eggs/g food plant)	Pupal weight (mg)
1975	1.03	119
1976	0.07	157
1977	0.15	167
1978	0.16	166
1979	0.84	149

E. Population Density and Competitive Status of Native Populations as Predictors of the Potential for Outbreak of Introduced Insects

In his study of the competition and coexistence of insects attacking flower heads of the thistle *Carduus nutans,* Zwölfer (1973) concluded that the insect species whose growth and survival are most strongly reduced by other insects feeding on the same plants will survive by virtue of its ability to lay more eggs than its superior competitors. Inferior competitors are the best insects to introduce for biological control because they have the potential for outbreak. The outbreaks of introduced populations of *Rhinocyllus conicus* support this. In Europe, the number of *Rhinocyllus* individuals is reduced in thistle flower heads that also contain other insects. The weevils are also heavily parasitized. In spite of this competition and parasitization, *Rhinocyllus* persists by laying eggs on many flower heads. In North America, *Rhinocyllus* populations without competitors and parasites have become established, have rapidly increased, and have controlled thistle (Fig. 3; also Harris, 1984b).

Cinnabar moth is the predominant herbivore on tansy ragwort in many parts of Europe, but introduced populations were difficult to establish in Canada and Australia and were not permanent after introduction. This may represent another example of a species that has not evolved under conditions of extreme competition or high mortality and has been a poor invader.

The mealybug, *Hypogeococcus festerianus,* immediately became established and rapidly increased following its introduction on Harrisia cactus in Australia. In its native Argentina, predators and parasites suppress the density of the mealybug, and it is found only in scattered patches (McFadyen and Tomley, 1981). From this limited damage done to the cactus by the insects in Argentina, Rachel McFadyen did not expect it to be a successful biological control agent. However, the ability of the mealybugs to rapidly reach and maintain high densities in the absence of predators has resulted in a dramatic reduction in the Harrisia cactus in Queensland. A second insect imported from Argentina to Australia for the control of Harrisia cactus, *Alcidion cereicola,* also increased rapidly after introduction. This species has few parasites in Argentina, but many young larvae die during wet months, when the turgor pressure of cactus is high. This species has therefore evolved under a regime of high larval mortality. In contrast, a species of *Cactoblastis* that is common and widely distributed on Harrisia cactus in Argentina did not thrive in Australia.

Outbreaks of exotic insects demonstrate experimentally that herbivorous insects that are suppressed in their native habitats have the potential to erupt if released from predators, parasites, and competitors and that herbivorous insects that are common in their native habitats are sometimes difficult to introduce or are not persistent in exotic habitats (Force, 1972). Information about the insects in their new and native habitats should be integrated to determine the generality of the relationship between competitive dominance and the potential for outbreak after introduction. Biological control introductions can provide unique experimental information on basic ecological questions of competition, predation, and parasitization.

III. AFTER THE OUTBREAK

After the initial outbreak of successfully established biological control agents, a variety of population trends are possible. The desired trend is for the plant population to decline and the insects to remain sufficiently common to maintain pressure on the plants. Although few case histories have been studied, the population trends in Figs. 1–3 are typical; both the plant density and insect density decline. In several studies of insects that attack the seed heads of plants, the resources available to insects at peak density can be compared with those after the decline (Table 4). Although the number of flower or seed heads was reduced after peak density in

TABLE 4

Mean Number of Insects per Seed Head during the Peak Years of the Outbreak and after the Population Declined[a]

Species	Peak year(s)			Postoutbreak		
	$\bar{x}$	SE	N	$\bar{x}$	SE	N
Rhinocyllus conicus	3.1	0.6	4	2.2	0.7	5
Urophora (on spotted knapweed)	4.99	—	1	1.95	—	1
Urophora (on diffuse knapweed)	1.32	0.2	4	0.72	0.1	2

[a] Means are of data from several years or several populations. Data are from Harris (1980, 1984b).

these examples, the number of insects per head was even lower. The insects were not using the food as efficiently after the outbreak. This could have been because the food quality declined or could result from larval competition (Harris, 1980).

In successful biological control situations the plants are reduced to scattered patches. In the most successful example of biological weed control, the control of prickly pear cactus by *Cactoblastis,* the equilibrium was thought to represent a hide-and-seek between plants and insects (Andrewartha and Birch, 1954). However, Monro found that the populations of both *Opuntia* and *Cactoblastis* are persistent but fluctuating (Monro, 1967; Myers *et al.,* 1981; Osmond and Monro, 1981). Both plants and insects increase during years with sufficient moisture and crash during periods of drought. Similarly, in drier areas of British Columbia, St. John's wort and *Chrysolina* beetles maintain persistent, patchy populations, particularly along roadsides or previously disturbed sites. In moister areas the plants periodically reach a high density, but the *Chrysolina* beetles rapidly expand in number; they do not become extinct when plant density is low (J. H. Myers, personal observation).

The population dynamics of cinnabar moths in Europe and North America are very similar. Moth densities fluctuate because these insects overexploit their food supply, which results in larval dispersal and starvation, reduced size and fecundity of moths, and reduced size of plants after defoliation (Dempster, 1982; Myers, 1980). The plants recover very quickly after the insects crash, and outbreaks occur every 3–4 years. Moth populations have not been persistent in North America, however, and many have become nearly extinct.

The *Urophora–Centaurea* system is similar in Europe and North America. Varley (1947) studied the gall fly, *Urophora jacenana,* feeding on black knapweed, *Centaurea nemoralis*. He predicted that without parasites the *Urophora* population would increase 16 times. The introduced *Urophora* in British Columbia lack parasites and have "stabilized" at densities 12–18 times that of the European populations (Myers and Harris, 1980). In both Europe and Canada larval mortality of *Urophora* increases with density and intraspecific competition, which may be the main regulating factor (Harris, 1980). Interactions among weather, phenology of bud development, and fly attack are important determinants of the variation in number in both areas (Berube, 1980).

To summarize, the population dynamics of native and introduced insect populations are similar in the few examples that make comparisons possible. Although the average densities may differ, the factors that cause fluctuations are similar.

IV. CONCLUSIONS

It is not possible to reject the general hypotheses investigated here. Examples can be cited that do and do not support each of them. Clearly, some climatic differences are within the range of variability that insects can adapt to and others are not. Host plant specificity is of greater importance in some insect–plant systems than in others. There is no evidence that herbivorous insects are better able to control species or strains of weeds different from that on which they evolved (Room *et al.,* 1981; Harris, 1986). In fact, for some insects outbreak is highly dependent on host plant matching.

Host plant quality, measured by nutritional status, seems to be a generally important characteristic and one that can be investigated experimentally. The fertilization of weed plants to facilitate the initial introduction of insects may be an important tool for biological control.

Predators and parasites undoubtedly suppress the outbreak of some introduced insects. Generalist predators are more likely to respond rapidly to the presence of new prey than are more specific parasites. Whether predators are responsible for suppressing the outbreak of as many introduced insects as they are given credit for can be evaluated only through experimental manipulations.

The information available so far does not suggest that introduced insects are inhibited by low genetic variability following introduction "bottlenecks." The adaptation of some species to new climatic conditions suggests that they have maintained sufficient genetic variance to allow adaptation. However, major genetic shifts, as might be expected following the outbreak of an introduced insect, have not been recorded.

A relationship between high reproductive potential and the outbreak of introduced insects is to be expected. Whether high reproductive potential evolved in response to heavy natural mortality or low competitive status remains an interesting idea to be further tested.

It is often difficult to evaluate the biological control literature because mechanisms for success, failure, outbreak, or suppression are frequently surmised rather than investigated. Most studies involve species that were successfully introduced and have reached outbreak densities. No clear-cut patterns have emerged to indicate that particular taxonomic groups or life styles are more successful. More comparisons of native and introduced insects will reveal the details of those factors that control the dynamics of insect population fluctuations. The introduction of insects as biological control agents provides a powerful experimental tool for the study of insect and plant ecology.

ACKNOWLEDGMENTS

I thank the following people for their helpful comments on an earlier draft of the manuscript: Jamie Smith, Rachel McFadyen, Paul McFadyen, Peter Harris, Richard Goeden and Fred Wilkinson. Previously unpublished work cited here was supported by Natural Science and Engineering Research Council of Canada, Agricultural Canada, British Columbia Ministry of Agriculture and Food, and British Columbia Ministry of Forests.

REFERENCES

Andrewartha, H. G., and Birch, L. C. (1954). "The Distribution and Abundance of Animals." Univ. of Chicago Press, Chicago, Illinois.

Ayala, F. J. (1968). Genotype, environment, and population numbers. *Science* **162,** 1453–1459.

Berube, D. E. (1980). Interspecific competition between *Urophora affinis* and *U. quadrifasciata* (Diptera: Tephritidae) for ovipositional sites on diffuse knapweed (*Centaurea diffusa* Compositae). *Z. Angew. Entomol.* **90,** 299–306.

Carson, H. L. (1961). Heterosis and fitness in experimental populations of *Drosophila melanogaster. Evolution (Lawrence, Kans.)* **15,** 496–509.

Carson, H. L. (1968). The population flush and its genetic consequences. *In* "Population Biology and Evolution" (R. C. Lewontin, ed.), pp. 123–137. Syracuse Univ. Press, Syracuse, New York.

Carson, H. L., and Templeton, A. R. (1984). Genetic revolutions in relation to speciation phenomena: The founding of new populations. *Annu. Rev. Ecol. Syst.* **15,** 97–131.

Caughley, G. (1970). Eruption of ungulate populations, with emphasis on Himalayan Thar in New Zealand. *Ecology* **51,** 53–72.

DeBach, P. (1965). Some biological and ecological phenomena associated with colonizing entomophagous insects. *In* "The Genetics of Colonizing Species" (H. G. Baker and G. L. Stebbins, eds.), pp. 287–303. Academic Press, New York.

Dempster, J. P. (1982). The ecology of the cinnabar moth, *Tyria jacobaeae* L. (Lepidoptera: Arctiidae). *Adv. Ecol. Res.* **12,** 1–36.

Dodd, A. P. (1940). "The Biological Campaign against Prickly Pear." Commonw. Prickly Pear Board, Brisbane, Australia.

Ehler, E. H., and Andres, L. A. (1983). Biological Control: Exotic natural enemies to control exotic pests. *In* "Exotic Plant Pests and North American Agriculture" (C. L. Wilson and C. L. Graham, eds.), pp. 396–419. Academic Press, New York.

Elton, C. S. (1958). "The Ecology of Invasions by Animals and Plants." Methuen, London.

Force, D. C. (1972). *r*- and *k*-strategists in endemic host–parasitoid communities. *Bull. Entomol. Soc. Am.* **18,** 135–137.

Ford, H. D., and Ford, E. B. (1930). Fluctuation in numbers and its influence on variation in *Melitaea aurinia. Trans. R. Entomol. Soc. London* **78,** 345–351.

Frankel, O. H., and Soule, M. E. (1981). "Conservation Ecology." Cambridge Univ. Press, London and New York.

Franklin, I. R. (1980). Evolutionary change in small populations. *In* "Conservation Biology" (M. E. Soule and B. A. Wilcox, eds.), pp. 135–149. Sinauer Assoc., Sunderland, Massachusetts.

Goeden, R. D., and Louda, S. M. (1976). Biotic interference with insects imported for weed control. *Annu. Rev. Entomol.* **21,** 325–342.
Goeden, R. D., and Richer, D. W. (1978). Establishment of *Rhinocyllus conicus* (Col.: Curculionidae) on Italian thistle in southern California. *Environ. Entomol.* **7,** 787–789.
Goeden, R. D., Richer, D. W., and Hawkins, B. A. (1986). Ethological and genetic differences among three biotypes of *Rhinocyllus conicus* (Coleoptera: Curculionidae) introduced into North America for the biological control of asteraceous thistles. *Proc. Int. Symp. Biol. Control Weeds, 6th, 1984,* pp. 181–190. Agriculture Canada, Ottawa.
Haesler, W. H. (1981). Lessons from early attempts at biological control of weeds in Queensland. *Proc. Int. Symp. Biol. Control Weeds, 5th, 1980,* pp. 3–10.
Harris, P. (1980). Establishment of *Urophora affinis* Frfld. and *U. quadrifasciata* (Meig.) (Diptera: Tephritidae) in Canada for the biological control of diffuse and spotted knapweek. *Z. Angew. Entomol.* **89,** *504–515.*
Harris, P. (1984a). Current approaches to biological control of weeds. *In* "Biological Control Programmes Against Insects and Weeds in Canada 1969–1980" (J. S. Kelleher and M. A. Hulme, eds.), pp. 95–104. Commonw. Agric. Bur., England.
Harris, P. (1984b). *Carduus nutans* L., nodding thistle and *C. acanthiodes* L., plumeless thistle (Compositae). *In* "Biological Control Programmes Against Insects and Weeds in Canada 1969-1980" (J. S. Kelleher and M. A. Hulme, eds.), pp. 115–121. Commonw. Agric. Bur., England.
Harris, P. (1984c). *Euphorbia esula–virgata* complex, leafy spurge, and *E. cyparissias* L., cypress spurge (Euphorbiaceae). *In* "Biological Control Programmes Against Insects and Weeds in Canada 1969-1980" (J. S. Kelleher and M. A. Hulme, eds.), pp. 159–169. Commonw. Agric. Bur., England.
Harris, P. (1986). Biological control of weeds. "Fortsch. der Zoologie," Bd. 32. Franz (Hrsg): "Biological Plant and Health Protection." Fischer Verlag, New York.
Harris, P., and Myers, J. H. (1984). *Centaurea diffusa* Lam. and *C. maculosa* Lam. s. lat., diffuse and spotted knapweed (Compositae). *In* "Biological Control Programmes Against Insects and Weeds in Canada 1969–1980" (J. S. Kelleher and M. A. Hulme, eds.), pp. 127–137. Commonw. Agric. Bur., England.
Harris, P., Peschken, K., and Milroy, J. (1969). The status of biological control of the weed *Hypericum perforatum* in British Columbia. *Can Entomol.* **101,** 1–15.
Hassell, M. P., and Moran, V. C. (1976). Equilibrium levels and biological control. *J. Entomol. Soc. South. Afr.* **39,** 357–366.
Julien, M., ed. (1982). "Biological Control of Weeds: A World Catalogue of Agents and Their Target Weeds." Commonw. Agric. Bur., England.
Kok, L. T. (1978). Biological control of Carduus thistles in northeastern USA. *Proc. Int. Symp. Biol. Control Weeds, 4th, 1976,* pp. 101–104.
Lewontin, R. (1965). Discussion of paper by W. E. Howard. *In* "The Genetics of Colonizing Species" (H. G. Baker and G. L. Stebbins, eds.), p. 481. Academic Press, New York.
McFadyen, P. J. (1981). Hyperparasites: An option for biological control of weeds. *Proc. Int. Symp. Biol. Control Weeds, 5th, 1980,* pp. 201–204.
McFadyen, R. E., and Tomley, A. J. (1981). Biological control of Harrisia cactus, *Eriocereus martinii,* in Queensland by the mealybug, *Hypogeococcus festerianus. Proc. Int. Symp. Biol. Control Weeds, 5th, 1980,* pp. 589–594.
Mann, J. (1970). "Cacti Naturalized in Australia and Their Control." Dept. of Lands, Brisbane, Australia.
Maw, M. G., ed. (1984). "An Update of Julien's World Catalogue of Agents and Their Target Weeds." Agriculture Canada, Regina.

Mitton, J. B., and Grant, M. C. (1984). Associations among protein heterozygosity, growth rate, and developmental homeostasis. *Annu. Rev. Ecol. Syst.* **15,** 479–499.

Monro, J. (1967). The exploitation and conservation of resources by populations of insects. *J. Anim. Ecol.* **36,** 531–547.

Moran, V. C., and Zimmerman, H. G. (1984). The biological control of cactus weeds: Achievement and prospects. *Biocontrol News Inf.* **5,** 297–320.

Murray, N. D. (1982). Ecology and evolution of the *Opuntia–Cactoblastis* ecosystem in Australia. *In* "Ecological Genetics and Evolution: The Cactus-Yeast- *Drosophila* Model System" (J. S. F. Barker and W. T. Starmer, eds.), pp. 17–30. Academic Press, New York.

Murray, N. D. (1986). Rates of change in introduced organisms. *Proc. Int. Symp. Biol. Control Weeds, 6th, 1984,* pp. 191–202. Agriculture Canada, Ottawa.

Myers, J. H. (1976). Distribution and dispersal in populations. A simulation model. *Oecologia* **23,** 255–269.

Myers, J. H. (1978). Biological control introductions as grandiose field experiments: Adaptation of the cinnabar moth to new surroundings. *Proc. Int. Symp. Biol. Control Weeds, 4th, 1976,* pp. 181–188.

Myers, J. H. (1979). The effects of food quantity and quality on emergence time in the cinnabar moth. *Can. J. Zool.* **57,** 1150–1156.

Myers, J. H. (1980). Is the insect or the plant the driving force in the cinnabar moth–tansy ragwort system? *Oecologia* **47,** 16–21.

Myers, J. H. (1986). How many insect species are necessary for successful bio-control of weeds? *Proc. Int. Symp. Biol. Control Weeds, 6th, 1984,* pp. 77–82. Agriculture Canada, Ottawa.

Myers, J. H., and Harris, P. (1980). Distribution of *Urophora* galls in flower heads of diffuse and spotted knapweed in British Columbia. *J. Appl. Ecol.* **17,** 359–367.

Myers, J. H., and Iyer, R. (1981). Phenotypic and genetic characteristics of the European cranefly following its introduction and spread in western North America. *J. Anim. Ecol.* **50,** 519–532.

Myers, J. H., and Post, B. J. (1981). Plant nitrogen and fluctuations of insect populations: A test with the cinnabar moth–tansy ragwort system. *Oecologia* **48,** 151–156.

Myers, J. H., and Sabath, M. D. (1981). Genetic and phenotypic variability, genetic variance and the success of establishment of insect introductions for the biological control of weeds. *Proc. Int. Symp. Biol. Control Weeds, 5th, 1980,* pp. 91–102.

Myers, J. H., and Williams, K. S. (1984). Does tent caterpillar attack reduce the food quality of red alder foliage? *Oecologia* **62,** 74–79.

Myers, J. H., Monro, J., and Murray, N. (1981). Egg clumping, host plant selection and population regulation in *Cactoblastis cactorum* (Lepidoptera). *Oecologia* **51,** 7–13.

Nei, M., Maruyama, T., and Chakraborty, R. (1975). The bottleneck effect and genetic variability in populations. *Evolution(Lawrence, Kans.)* **29,** 1–10.

Neser, S., and Kluge, R. L. (1986). A seed-feeding insect showing promise in the control of a woody, invasive plant: The weevil *Erytenna consputa* on *Hakea sericea* (Proteaceae) in South Africa. *Proc. Int. Symp. Biol. Control Weeds, 6th, 1984,* pp. 805–810. Agriculture Canada, Ottawa.

Nuessly, G. S., and Goeden, R. D. (1983). Spider predation on *Cokophora parthenica* (Lepidoptera: Coleophoridae), a moth imported for the biological control of Russian thistle. *Environ. Entomol.* **12,** 1433–1438.

Osmond, C. B., and Monro, J. (1981). Prickly pear. *In* "Plants and Man in Australia" (D. J. Carr and S. G. M. Carr, eds.), pp. 194–222. Academic Press, New York.

Peschken, D. P. (1972). *Chrysolina quadrigemina* (Coleoptera: Chrysomelidae) introduced

from California to British Columbia against the weed *Hypericum perforatum:* Comparison of behaviour, physiology and colour in association with post colonization adaptation. *Can. Entomol.* **104,** 1689–1698.
Richards, L. J., and Myers, J. H. (1980). Maternal influences on size and emergence time of the cinnabar moth. *Can. J. Zool.* **58,** 1452–1457.
Room, P. M. (1981). Biogeography, apparency and exploration for biological control agents in exotic ranges of weeds. *Proc. Int. Symp. Bio. Control Weeds, 5th, 1980,* pp. 113–124.
Sands, D. P. A., and Harley, K. L. S. (1981). Importance of geographic variation in agents selected for biological control of weeds. *Proc. Int. Symp. Biol. Control Weeds, 5th, 1980,* pp. 81–89.
Schroeder, D. (1983). "Biological Control of Weeds. Recent Advances in Weed Research," pp. 42–78. Commonw. Agric. Bur., England.
Templeton, A. R. (1980). The theory of speciation via the founder principle. *Genetics* **94,** 1011–1038.
Thomas, P. A., and Room, P. M. (1986). Towards biological control of *Salvinia* in Papua New Guinea. *Proc. Int. Symp. Biol. Control Weeds, 6th, 1984,* pp. 567–576. Agriculture Canada, Ottawa.
Varley, G. D. (1947). The natural control of population balance in the knapweed gall-fly (*Urophora jaceana*). *J. Anim. Ecol.* **16,** 139–187.
White, T. C. R. (1984). The abundance of invertebrate herbivores in relation to the availability of nitrogen in stressed food plants. *Oecologia* **63,** 90–105.
Williams, K. S. (1986). Climatic influences on weeds and their herbivores: Biological control of St. John's wort in British Columbia. *Proc. Int. Symp. Biol. Control Weeds, 6th, 1984,* pp. 127–134. Agriculture Canada, Ottawa.
Williams, K. S., and Myers, J. H. (1984). Previous herbivore attack of red alder may improve food quality for fall webworm larvae. *Oecologia* **63,** 166-170.
Wilson, F. (1965). Biological control and the genetics of colonizing species. *In* "The Genetics of Colonizing Species" (H. G. Baker and G. L. Stebbins, eds.), pp. 307–330. Academic Press, New York.
Zwölfer, H. (1973). Competition and coexistence in phytophagous insects attacking the heads of *Carduus nutans* L. *Proc. Int. Symp. Biol. Control Weeds, 2nd, 1971,* pp. 74–80.
Zwölfer, H. (1976). Investigations on *Sphenoptera* (*Chilostetha*) *jugoslavica* Oben. (Col. Buprestidae), a possible biocontrol agent of the weed *Centaurea diffusa* Lam. (Compositae) in Canada. *Z. Angew. Entomol.* **80,** 170–190.

Chapter 8

Insect Pest Outbreaks in Agroecosystems

DAVID N. FERRO

Department of Entomology
University of Massachusetts
Amherst, Massachusetts 01003

I. INTRODUCTION

Most of the chapters in this book are concerned with environmental and biological factors that allow insect population outbreaks in natural ecosystems. Although many of these factors are driving forces of pest outbreaks in agroecosystems, other contributory factors are unique to agroecosystems. Although there may be little that is "natural" about agricultural cropping systems, the crop environment is an ecological unit. In agricultural systems, soil, plant, and animal interactions are rarely persistent enough, in time and space, to provide the ecological stability or equilibrium characteristic of nonagricultural systems. Nevertheless, if these interactions are understood and properly managed, it may be possible to reduce insect pest outbreaks.

Early efforts to control pests were based on innate plant resistance, cultural practices, and natural enemies. As new cultivars of plants were selected, it became apparent that higher yields could be attained if pest control was maximal. Farmers began using a wide range of pesticides, which initially had tremendously positive effects on the yield and quality of product. Under the assumption that pesticides could control all pest problems, plant breeders became complacent, focusing their efforts on developing plant cultivars that produced higher yields with better storage qualities, higher levels of starch, oil, or protein, that were easier to harvest, and so on, but almost totally disregarding plant pests.

There are many factors that, alone or in combination, allow insect populations to reach densities that are of outbreak proportions. Although some are not unique to agroecosystems or to species that undergo outbreaks, it is important to be cognizant of all these factors and to recognize that it is the interaction of all or some of them that allows for outbreaks. This chapter concentrates on the important influence of the rate of colonization, weather, insecticides, natural enemies, and plant cultivars on pest population outbreaks.

II. COLONIZATION

The rate of colonization of a crop within an agroecosystem determines the initial phase leading to pest population outbreaks. The mechanisms that govern this process are basically the same for herbivores and entomophagous insects. Because most agricultural crops are grown as monocultures, the probability of a pest and its natural enemies colonizing an acceptable host crop is much greater than in a diversified plant community. The rate of colonization is dependent on the ability of the insect to locate a suitable host, cultural practices, natural enemies, reproductive potential, and environmental limitations.

A. Location of a Suitable Host

A series of steps is completed as an insect colonizes a crop, and the amount of time devoted to this is influenced by the characteristics of the habitat of neighboring crops and species diversity of this habitat. The summation of time allocated to these steps and the number of colonizers determines the rate of colonization. These steps are host-habitat selection, host location, host recognition, host acceptance, and host suitability.

1. Host-Habitat Location

The habitat of the crop is generally located through phototactic, anemotactic, and geotactic responses by the insect. This initial step toward colonization is extremely important for migratory insect pests, especially those with a limited host range. Most aphid species in temperate areas overwinter as eggs on primary hosts. Aphids seeking acceptable secondary host plants respond to different environmental cues than do aphids searching for primary host plants on which to overwinter. In the spring, alates (winged forms) are produced that leave the primary host, become windborne, and may travel hundreds of miles before descending. Many aphid species that colonize cultivated plants are attracted to objects that have a peak reflectance of wavelengths of about 550 nm (530–570 nm), which is the same as for many cultivated plants, weeds, and herbaceous plants (Johnson, 1969).

2. Host Location

Once within the crop habitat, the insect must find its host. It does so by using several sensorial mechanisms. Color, shape, and odor are important cues for the cabbage maggot in orienting to its host. Dapsis and Ferro (1983) showed that cabbage maggot, *Delia radicum* (L.), is highly attracted to "federal safety yellow"-painted stakes (peak reflectance at 540 nm). Tuttle (1985) showed that when allyl isothiocyanate and some other mustard oils, found in cruciferous plants, were used with yellow stakes, the number of adult cabbage maggots trapped was higher than when yellow stakes were used alone. Furthermore, in addition to color and odor, cabbage maggot adults used shape as a cue for host orientation. Once on the plant, the insect uses olfactory (kairomone) and tactile (pubescence, texture) cues to assess the quality of its host.

3. Host Recognition

This step is closely linked to host location and host suitability and is generally accomplished through chemical and tactile cues. Plant chemicals may be detected by olfaction through gustatory activity or palpation of the plant surface. Host cues used in plant recognition may also be chemicals emitted by microbes associated with the plant; for example, microbes coinhabit larval mines with onion maggots, *Delia antigua* (Meigen) (Dindonis and Miller, 1980). In this study it was found that decomposing plants infested with onion maggots were most attractive to male and female flies. The primary host odor, *n*-dipropyl sulfide and other volatiles, appears to be very important for host recognition by this species. Tactile cues are used by codling moth, *Cydia pomonella* (L.), adults, which prefer to oviposit on the waxy surface of leaves and apples and generally avoid the bark and stems.

4. Host Acceptance

Tactile cues, odor, and ingestion of plant material are all important for finding acceptable hosts. For example, aphids generally must probe tissue and ingest plant protoplasm before accepting the host (Garrett, 1973).

5. Host Suitability

Although all sensorial responses by the insect may indicate that it has colonized an acceptable host, the host still may not be suitable for survival and reproduction. Nutritionally, the plant may lack essential amino acids, it may have low concentrations of carbohydrates, or there may be an imbalance of these nutrients. The plant may contain an antibiotic (i.e., toxin) that kills the colonizer or prevents the normal development of offspring. If the host is suitable for sustained population growth and reproduction, the pest has successfully colonized the crop. The probability of colonizers successfully mating and placing their offspring in the right environment is enhanced in crops grown as monocultures.

B. Reproductive Potentials and Environmental Limitations

Life history parameters determine, to a large extent, how rapidly an insect colonizes a crop and how quickly the population builds up. A migratory pest that is incapable of overwintering must have a high reproductive potential, colonize in large numbers, or have a short life cycle if it is to reach outbreak numbers. This type of pest is represented by both the fall armyworm, *Spodoptera frugiperda* (J. E. Smith), and the corn earworm, *Heliothis zea* (Boddie), which migrate each year from the southern United States to the Northeast via northerly moving weather fronts. Neither of these species overwinters in the Northeast and they seldom arrive before mid-July, yet they routinely reach outbreak proportions (D. N. Ferro, unpublished). Female moths of these species are capable of laying about 1500 eggs, and as they move into the Northeast, virtually free of any predators or parasites, they can reach high population densities within one generation. The converse situation occurs with pests that do not have a high reproductive potential but have a short life cycle and routinely reach outbreak numbers. Pests like mites and aphids represent this group.

Death due to climate or natural enemies can regulate the size of the colonizing population. *Pediobius foveolatus,* a parasite of the Mexican bean beetle, *Epilachna varivestis* Mulsant, when released in an inundative manner, caused high mortality to last-generation Mexican bean beetle in the fall. The parasites, caused sufficient mortality to reduce the colonizing population to a level that prevented it from reaching outbreak levels the following season (Forrester, 1982). In addition, low winter temperatures can inflict high mortality on overwintering populations, effectively reducing the number of colonizers. An extreme example of this is represented by the fall armyworm, which cannot overwinter north of 30° latitude and does not survive temperatures below 15.6°C (Barfield *et al.*, 1978).

C. Cultural Practices

1. Crop Rotation

Crop rotation is one of many cultural practices that may enhance or retard colonization. On Long Island, New York, potatoes planted in fields previously containing potatoes had up to 40 times more colonizing adult Colorado potato beetles than when planted in fields previously containing grains like winter rye and wheat (Wright, 1984). Lashomb and Ng (1984) showed that Colorado potato beetle population buildup, in fields rotated into potato from winter wheat, was delayed by 2 to 3 weeks, which in some areas is enough time to reduce the number of generations. If host plants are not in the immediate vicinity of overwintering sites, the beetle must find host plants by walking or flight, effectively retarding the rate of colonization of fields newly planted with potato or another acceptable host crop.

2. Planting and Harvest Dates

In areas where crops are planted as monocultures, insect population densities on these crops can reach outbreak proportions and are the major source of pest inoculum the following season. If crops are harvested before the insect enters diapause or planted the following season so that the crop phenology is not in synchrony with the insects' phenology, the rate of colonization can be retarded. This is the key to preventing the pink bollworm, *Pectinophora gossypiella* (Saunders), from becoming a major pest of cotton in Texas. If pink bollworm larvae successfully enter diapause and a large number overwinter, the population will reach outbreak numbers during the next growing season. Diapause is controlled by a photoperiod of less than 13 hr, which occurs during early September (Adkisson, 1964). The use of early-maturing cultivars that can be planted later and harvested earlier significantly reduces the development of most of the overwintering population of pink bollworms. This is accomplished by applying defoliators and dessicants to the cotton crop in late August or early September before the onset of diapause and by planting so that cotton fruit is not available for oviposition by moths emerging in the spring (Adkisson, 1972).

Harvesting a crop can actually accelerate the rate of colonization. The lygus bug, *Lygus hesperus* Knight, is considered a key pest of cotton in California. A major habitat of the bugs is provided by alfalfa fields, which are generally interspersed throughout the cotton-producing areas. When alfalfa is cut, lygus bugs leave the alfalfa, colonizing adjacent cotton fields in numbers that often exceed the economic threshold level. In this case, the influx of the insect population is so great that it is at an outbreak density during the colonization phase. This can be prevented if the alfalfa is cut in alternate strips in order to slow migration (Stern *et al.*, 1967) or if strips of alfalfa are interplanted within cotton fields to attract colonizers (Stern, 1969).

3. Natural Enemies

The spatial and temporal occurrence of host crops and nonhosts of herbivores determines to a large extent the natural-enemy complex and the importance of this complex in preventing pest outbreaks. Barfield and Stimac (1981) present a likely scenario of the importance of a multiple-crop system in outbreaks of the fall armyworm and the velvet bean cater-piller, *Anticarsia gemmatalis* Hübner. Outbreaks of these species occur in the corn–peanut–soybean cropping system in North Florida and other areas of the southeastern United States. Fall armyworm is a perennial pest of corn; velvet bean caterpillar is a perennial pest of soybean, and both pests occur in the intermediate crop peanut. The complex of predators (Buschman *et al.*, 1977) and parasites (Ashley, 1979) is essentially the same for these species. The initial predator numbers in corn are probably the result of a density-dependent functional response of predators in non-crop host plant communities preceding or coexisting with corn. Populations of the fall armyworm immigrate to the corn–peanut–soybean production area and infest corn. One or two generations are completed on corn, after which a substantial number of females move into peanut to oviposit. Natural enemies respond and, provided that the infestation in peanuts is sufficient, build up in peanuts where the velvet bean caterpillar arrives later. If the timing is right, the natural-enemy complex, present as a result of a response to fall armyworm densities, now attacks the velvet bean caterpillar. If the timing or number of natural enemies, either entering the corn field initially or that build up on fall armyworm in corn, is not right, outbreaks of these pests occur. C. S. Barfield (personal communication) has indicated that additional data have been collected to substantiate this scenario. Although the phenomenon is not well documented, it illustrates the need to study pests and their natural enemies as part of a complex cropping system in conjunction with noncrop hosts and, that only through such studies will we be able to manage insect pests without the use of insecticides.

D. Weather

Weather and climate influence insect populations directly by affecting the survival, physiology, and behavior of insects and indirectly by affecting the food supply, habitat, and natural enemies. Only direct effects are discussed in this chapter. "Climate" refers to the prevailing or average weather conditions that, in many cases, define the geographic range and distribution of insect species, whereas "weather" includes atmospheric conditions like temperature, humidity, wind, rain, and radiation that at any one time directly influence the physiology and behavior of an insect.

1. Temperature

Insects are poikilotherms; that is, chemical reactions that take place within insects are directly regulated by temperature. Lewontin (1965) stated that developmental rate is the most important variable influencing the intrinsic rate of increase of colonizing species. Although insects may have a wide geographic distribution, it is generally in those areas where optimal temperatures occur that insects reach outbreak numbers.

The developmental rate of an insect is determined by the accumulation of heat units or degree-days above a threshold temperature. The threshold temperature is that temperature below which development ceases. Above the threshold temperature there is a geometric increase in rate of development with increasing temperature, until the maximum rate of development occurs. At temperatures above the maximum rate of increase, the developmental rate rapidly decreases, until the thermal maximum or death is reached. Some insects can develop and survive over a wide range of temperatures, whereas others tolerate only a narrower temperature range. For example, the Banks grass mite, *Oligonychus pratensis* (Banks), is characterized by sporadic outbreaks and rapid population increases. In the semiarid high plains of the southwestern United States, the Banks grass mite reaches outbreak numbers. It is able to survive the cold winters of the Plains states and to develop at temperatures from 10 to 43°C, maximum growth occurring at about 37°C (Perring *et al.*, 1984).

Another insect that sporadically reaches outbreak numbers from southern Florida to New Hampshire is the fall armyworm. In the southeastern states it is a pest of field and sweet corn, pasture grass, and peanut, whereas in the northeastern states it is primarily a pest of sweet corn. This pest has a much narrower temperature range for development, about 18–35°C, with 100% mortality at 38°C (Barfield *et al.,* 1978). There is a rapid decline in the rate of development between 35 and 38°C.

Another insect that reaches outbreak numbers in defined areas of its distribution is the Colorado potato beetle, *Leptinotarsa decemlineata* (Say). This pest overwinters as far north as Alberta, Canada, as far west as eastern Washington, as far south as southern Mexico, and as far east as Maine. Yet only in the North Central region and Northeast region south to North Carolina do populations in a single season reach high densities. This pest also has a narrow temperature range for development, ranging from 12 to 33°C, with maximum development occurring at 28°C (Ferro *et al.,* 1985). Development virtually ceases at temperatures above 30°C, and high summer daytime temperatures coupled with cool nights most likely limit the capacity of this insect to reach higher population levels in many parts of its distribution. In the Northeast, especially on Long Island, New York, where the average daytime temperatures are 24–29°C, the Colorado potato beetle can complete three generations per year, whereas in the Northwest, where average daytime temperatures routinely reach 35°C and night time temperatures are often below 15°C, the beetle completes one to two generations per year.

In addition to growth and development, temperature may regulate feeding rates and behavior. This becomes important when a herbivore develops faster than its predators and parasites or the herbivore disperses more rapidly and colonizes the host crop, or parts of the crop, earlier than its natural enemies, allowing it to reach outbreak numbers.

The favorable temperature zone for insects is delimited by temperatures above and below which growth and multiplication cease, even though the insect may survive for a long time at these temperatures. The ability of an insect population to survive cold winters is dependent on the ability of some life stage to tolerate cold temperatures. Conversely, for an insect population to survive extremely hot conditions, some stage must be able to tolerate or to avoid these high temperatures. In general, high temperature in nature is not as important a regulating mechanism of insect populations as cold temperature. Insects are capable in most situations of avoiding extremely high temperatures but are not as capable of avoiding cold temperatures. In North America and Europe, the influence of lethal low winter temperatures on the survival rate of an insect can have a dramatic effect on the resulting population size the following summer. The life stage of an insect that is adapted to survive exposure to extreme cold is said to be "cold-hardy." There are several physiological and biochemical mechanisms that allow for cold-hardiness. Last-instar larvae of the codling moth can be cooled to less than −20°C and survive (Andrewartha and Birch, 1954, p. 194). This survival mechanism for these pests enables them to survive the winter in high enough numbers to reach outbreak levels routinely in most areas where they occur.

2. Humidity and Moisture

Insects can tolerate minor fluctuations in the amount of body water. They are able to maintain a balance between the water taken in orally or through the integument and the water lost by excretion and transpiration through the integument or through respiration. However, under extreme conditions, either excessive or inadequate water can be fatal.

Insects generally have a preferred humidity range, and the ability to recognize and respond to slight differences in humidity enables an insect to move into a regime of preferred humidity. Larvae of the wireworm, *Agriotes* sp., avoided dry air and moved toward the moist end of a humidity gradient, provided that the relative humidity exceeded 70%, and in air dryer than 70% the larvae seemed unresponsive to a humidity gradient (Lees, 1943). Humidity may affect other aspects of insect behavior that allow for outbreaks. The calling behavior, hence the ability to mate and propagate, of the European corn borer is affected by humidity (Webster and Cardé, 1982).

Mukerji and Gage (1978) examined the effects of soil moisture on embryonic development and egg mortality of a nondiapause strain of the migratory grasshopper *Melanoplus sanguinipes* (F.). A minimum soil moisture of about 13.5% was necessary for eggs to complete development. This soil moisture was absolutely critical before blastokinesis. Outbreaks of this species are dependent on an early spring hatch, which is dependent on advanced embryonic development from the preceding fall. If fall moisture levels are too low, embryogenesis ceases until the following spring and delays oviposition until later in the season.

Too much moisture may be lethal either directly [e.g., when heavy precipitation dislodges early-instar Colorado potato beetle (Harcourt, 1971), causing up to 30% mortality] or indirectly, by encouraging the development and spread of pathogenic microorganisms. In Saskatchewan, Canada, in 1963, unusually humid weather with above-average rainfall provided ideal conditions for the pathogenic fungus *Entomophthora grylli* as it underwent a widespread epizootic, reducing the clear-winged grasshopper *Campula pellucida* (Scudder) to 7% of the species complex from a high of 64%.

Xeric conditions may also influence survival. Survival can be expressed as a function of the rate at which the insect loses water into unsaturated air and the extent to which it can withstand desiccation of its tissues. Ferro and Chapman (1979) showed that the percentage of egg hatch of the two-spotted spider mite, *Tetranychus telarius* (*urticae*) Koch, over a preferred temperature range was independent of temperature but was greatly influenced by humidity. Survival dropped markedly at saturation deficits greater than 25 mm Hg.

3. Wind

Wind is the primary mode of transportation by which many insects disperse and colonize crops. This can occur over very short distances, such as from one crop to another, as occurs with the two-spotted spider mite moving from field corn to peanuts (Brandenburg and Kennedy, 1982). This influx into peanuts accounted for about 750,000 mites blown through a 1-m^2 window per week. A subsample of these mites showed that 86% were mated colonizing females. In the case of migratory pests, such as the fall armyworm and corn earworm, wind may be the main mechanism by which insect pests move from one region to another, covering hundreds of miles. These pests do not overwinter in the northeast, yet each year they arrive in sufficient numbers to infest up to 100% of sweet corn plants. Studies in eastern Massachusetts (D. N. Ferro, unpublished) have shown that pheromone trap catches of the corn earworm change from zero per week to 500 or more moths per trap within 24 hr of a storm originating in the South. The drought in the southeast in the summer of 1986 apparently reduced the oviposition sites for the corn earworm, and extremely large numbers dispersed from that region to invade Massachusetts, where over 2000 moths were captured in a single pheromone trap in one week! Most often these pests arrive without their natural enemies or far enough in advance of them that their population growth is unchecked, reaching outbreak numbers within one generation.

III. PESTICIDES

Insecticides can induce insect outbreaks by reducing the number of natural enemies of the insect or by selecting for pest populations resistant to insecticides. The most dramatic outbreaks are likely to occur when a pest develops resistance to a wide range of insecticides and its natural enemies are killed by insecticides.

A. Resurgence and Outbreaks of Secondary Pests

The widespread use of highly toxic, persistent insecticides determines to a large extent the pest complex in an agroecosystem. The cottony-cushion scale, *Icerya purchasi* Maskell, was a primary pest of citrus in

California and was controlled for almost 60 years by the predatory beetle *Rodolia cardinalis* (Mulsant). In the late 1940s, the cottony-cushion scale again became a major pest of citrus following the widescale use of DDT, which was highly toxic to *R. cardinalis*. In the Imperial Valley of California an obscure insect, the cotton leaf perforator, *Bucculatrix thurberiella* Busck, became a major cotton pest following the local use of carbaryl. Carbaryl was used to control the pink bollworm and at the same time virtually eliminated the natural enemies of the cotton leaf perforator (Smith, 1970). Luck *et al.* (1977) reported that 24 of the 25 most destructive insect pests of California agriculture attained this status due either to secondary pest outbreaks or to resurgences induced by the use of insecticides (but see Chapter 5, this volume).

Another, more subtle way that insecticides influence the natural control of pests is by disrupting food chains. For example, if aphids, spider mites, and thrips (the usual prey of abundant, omnivorous predators such as *Chrysopa* spp., *Nabis* spp., *Geocoris* spp., and *Orius* spp.) are eliminated from the cotton agroecosystem early in the season by insecticides, predators will eventually starve, emigrate, or cease to reproduce. Later, when migratory pests such as *Spodoptera* spp. and *Heliothis* spp. invade the treated fields, they are essentially free of predator attack, and explosive outbreaks of these pests occur (Reynolds *et al.*, 1982).

It is difficult to quantify precisely the effect of parasites and predators and other beneficial organisms in preventing insect pest outbreaks, except in the simplest systems (see Chapter 12). For this reason, it is difficult to assess the importance of the killing of natural enemies by pesticides in releasing pests from predation and parasitism. Whitcomb and Bell (1964) identified more than 600 species of predators and parasites in cotton, which effectively kept most pests from reaching outbreak numbers. A situation such as this is so complex that, when all environmental and cultural factors are included, it is virtually impossible to assess the effects of natural control within this pest complex. Only through simulation modeling, whereby the biology of pests and natural enemies can be examined in an interactive manner, can we understand these complex interactions and the effects of insecticides on the crop system.

B. Resistance

Many insect pests have a high reproductive potential, and if left uncontrolled by natural enemies or pesticides will reach outbreak numbers in one to two generations. When an insect has developed resistance to an insecticide or a group of insecticides, repeated applications of the insecticide actually exacerbate the situation because they destroy most natural enemies and exert little control on the target insect. In most cases, it may be the interaction between the development of resistance and the destruction of natural enemies that allows for insect outbreaks in an agroecosystem, rather than just insecticide resistance. This can be illustrated best by the example of two insect species that have well-documented histories of insecticide resistance but different biological systems and different mortalities caused by natural enemies.

The Colorado potato beetle is the major pest of Irish potato, *Solanum tuberosum,* in many potato-growing areas of the United States and Europe and is capable of completely defoliating a crop before the end of the growing season (Ferro *et al.,* 1983). In the northeastern United States, it has developed resistance to all currently registered insecticides (Forgash, 1985). It has developed cross-resistance to the carbamates (such as carbaryl, carbofuran, oxamyl, and aldicarb), cross-resistance to the organophosphates (such as azinphos-methyl, malathion, methidathion, phosmet, and chlorfenvinphos) (Hare, 1980; Harris and Svec, 1981), and multiple resistance to DDT and two synthetic pyrethroids, fenvalerate and permethrin (Forgash, 1985). The development of resistance by the Colorado potato beetle to insecticides has been very rapid in some cases. On Long Island, New York, the Colorado potato beetle went from being 20 times resistant to fenvalerate in the spring of 1980 to 600 times resistant by the fall of 1982, and 13 times resistant to aldicarb in the first generation of 1980 to 60 times resistant to the second generation of 1982, when compared with a susceptible population (Forgash, 1985).

The Colorado potato beetle is thought to be indigenous from southern Mexico to the Rocky Mountains of Colorado (Hsiao, 1985). The occurrence of Colorado potato beetle in these regions coincides with the distribution of its natural hosts, *Solanum rostratum* Dun. and *S. angustifolium* Lam. (Whalen, 1979). The eastward expansion of the beetle was swift and appears to have been closely linked to the widespread cultivation of potatoes. There are a number of predators, including *Chrysopa* spp., several species of coccinelids [especially *Coleomegilla maculata* (Degeer)], and pentatomids, like *Perillus bioculatus* (F.) and *Podisus maculiventris* (Say), and several species of parasites, including the tachinid parasite *Doryphorophaga doryphorae* (Riley), that attack the beetle. Apparently, as the beetle shifted from its native host plants to the cultivated potato, its adaptation to this new host so changed its biology, relative to its natural enemies, that these beneficial species have not been able to prevent the beetle from reaching outbreak numbers, even in the absence of insecticides (Harcourt, 1971). The Colorado potato beetle is a native insect that has a high reproductive potential (about 500–1000 eggs per female), exists in an environment that has optimum temperatures (25–28°C) for its growth during the growing season (Ferro *et al.*, 1985) and ideal soil structure for pupation (sandy or sandy loam) and overwinter survival, has a widespread preferred host plant, experiences essentially no control by natural enemies in commercial fields, and is resistant to all registered insecticides. The combined effect is that the beetle begins the growing season at a population density of less than one adult per square meter and by the second generation reaches about 500 pupae per square meter, resulting in devastating outbreaks (R. H. Voss and D. N. Ferro, unpublished). Serious outbreaks of this pest threaten the potato industry of Long Island and areas of southern New England.

Once the Colorado potato beetle shifted hosts and colonized potato, it was a recognized pest; however, the tobacco budworm, *Heliothis virescens* (F.), was not always a recognized pest in the cotton-growing areas of the southeast and southwest United States. Before the boll weevil, *Anthonomus grandis* Boheman, invaded the United States and northern Mexico in 1892, there was relatively little damage to cotton by insects; there were no key pests. Insecticides applied to control the boll weevil soon changed the insect complex and created pests out of nonpest species. Although insecticides such as lead and calcium arsenate and nicotine sulfate were being used to control the boll weevil as early as the 1920s, it was not until widespread use of the broad-spectrum organochlorine insecticides like DDT, BHC, and toxaphene in the 1950s and the shift to the organophosphorus and carbamate compounds like methyl parathion, azionphos-methyl, ethyl parathion, malathion, EPN, and carbaryl in the 1960s that there was a dramatic change in the pest status of the budworm. In the early 1960s the budworm became a more important pest than the boll weevil. By the late 1960s, there was a drastic change in the pest status of the tobacco budworm in the lower Rio Grande Valley of Texas and northeastern Mexico because it had become resistant to organophosphorus insecticides. Many growers were treating their fields 15–18 times with methyl parathion and still not preventing outbreaks.

Larvae of a budworm population collected from Monte, Mexico, in 1970 were 201 times more resistant to methyl parathion than was a susceptible population (Wolfenbarger *et al.*, 1973). As a result of this high level of resistance only 1200 acres of cotton were planted in this area in 1970 compared with 710,715 acres in 1960 (Adkisson, 1971). The history of cotton production and control of insect pests of cotton in the Southwest has been closely linked to the development of new insecticides. There was a lack of new pesticide chemistry from the late 1960s until the mid-1970s, and *H. virescens* was resistant to the major groups of insecticides (organochlorine, carbamate, and organophosphorus). During this time, growers in many areas were forced to adhere to integrated pest management (IPM) strategies, relying on natural control of the budworm. By the mid-1970s, a new group of insecticides, synthetic pyrethroids, was being used in Mexico and cotton-growing areas of the Southwest. The application of these insecticides has now become the accepted method of controlling the budworm in many parts of the United States. A dependency on these insecticides could be short-lived, since Sparks (1981) pointed out that the budworm could exhibit cross-resistance to DDT and the pyrethroids. In fact, there is evidence that resistance to the synthetic pyrethroids is already occurring in the Imperial Valley of California (Martinez-Carillo and Reynolds, 1983).

It is well documented that, when general predators and parasites of the budworm are eliminated from cotton fields, there are tremendous outbreaks of this pest (Reynolds *et al.*, 1982). Some of the known parasites of *H. virescens* include *Chelonus texanus* Cresson, *Cardiochiles nigriceps* (Vier.), *Eucelatoria armigera* (Coqvillet), *Campeletis sonorensis* (Cameron), *Microplitis croceipes* (Cresson), *Cotesia marginiventris* (Cresson), and *Trichogramma* spp. (Roach *et al.*, 1979). Some of the known predators include *Orius insidiosus* Say, *Coleomegilla maculata* (Degeer), *Chrysopa oculata* Say and *C. carnea* Stephens, *Hippodamia convergens* Guerin-Meneville, *Collops balteata* Leconte, *Scymnus creperus* Mulsant, *Lebia analis* Dej., *Podisus maculiventris* Say, *Geocoris* spp., and several species of ants, carabids, and spiders (Roach *et al.*, 1979). The budworm has been studied as well as any major pest in the United States, yet because of the complexity of its natural enemies, it has not been possible to assess quantitatively the role of each of these beneficial species in the regulation of budworm populations. Even though the budworm has become resistant to a wide range of insecticides, it has been possible through integrated pest management programs to prevent outbreaks of this pest by preserving its natural enemies (Reynolds *et al.*, 1982), unlike the situation with the Colorado potato beetle.

C. Induced Resistance

Little attention has been given to the area of induced resistance, that is, altering the biochemistry and physiology of herbivores, rendering them less susceptible to pesticides, without changing the genetic makeup of the population. This form of resistance has the potential to lead to pest outbreaks and to reduce the overall effectiveness of the pest management systems based on host plant resistance. The same enzymes that are involved in the metabolism or detoxification of pesticides could be involved in the metabolism of natural products or allelochemicals (Brattsten, 1979). The group of enzymes most often associated with this phenomenon includes the mixed-function oxidases (MFO) and glutathione transferase (Brattsten *et al.*, 1977; Gould and Hodgson, 1980). Brattsten *et al.* (1977)

showed that the southern armyworm, *Spodoptera eridania* (Cramer), when administered a diet containing the secondary plant substance (+)-α-pinene, had accelerated MFO activity within 2 hr of ingestion and that previous exposure to this substance doubled the LD_{50} for the insecticide nicotine when compared with the control. This means that the induction process could begin within hours of feeding, making it more difficult to control larvae feeding on plants containing this secondary plant substance. Gould *et al.* (1982) showed that a susceptible subpopulation of two-spotted spider mite, *Tetranychus telarius,* when reared on a resistant cultivar of cucumber, *Cucumis sativus* L., a marginally acceptable host, had greater resistance to three of four pesticides than a subpopulation reared on lima bean, *Phaseolus limensis* MacFadyen, a favorable host. Kennedy (1984) demonstrated that when the secondary plant substance 2-tridecanone was selected in tomato for resistance to the tobacco hornworm, *Manduca sexta* (L.), and the Colorado potato beetle, this substance altered the physiology of the tomato fruitworm, *Heliothis zea* (Boddie), making it more difficult to kill with the insecticide carbaryl. He showed that the allelochemical 2-tridecanone, the resistance factor in the wild tomato, *Lycopersicon hirsutum* f. *glabratum* Luckwill, induced resistance by *Heliothis zea* to the carbamate insecticide carbaryl.

As long as insecticides play a significant role in pest management, care must be taken to select plant cultivars that do not ultimately create insect outbreaks through induced resistance. Another aspect of induced resistance, hypothesized by Gould *et al.* (1982) and others, is the reverse of that just discussed; that is, where herbivores have developed resistance to a wide range of insecticides, they may be able to survive on some host plants that are toxic to insecticide-susceptible populations. Gould *et al.* tested this hypothesis with the two-spotted spider mite but were unable to support it. However, it would be better to study an insect pest that is already resistant to a wide range of insecticides and that feeds on plants containing a variety of toxic secondary plant substances. The Colorado potato beetle would be a good candidate for research, since it meets these criteria.

D. Hormologosis

Insecticides can stimulate the reproductive physiology of some insects, causing outbreaks. Carbaryl induces two-spotted spider mite outbreaks when applied to host crops. Although it has been generally accepted that this is due to the elimination of predators (Putman, 1963), it has now been shown by Dittrich *et al.* (1974) and more clearly by Brandenburg and Kennedy (1983) that carbaryl can actually alter the physiology of mites, increasing fecundity.

E. Fungicides and Herbicides

Natural epizootics of entomopathogenic fungi often prevent insect outbreaks, especially late in the growing season. This insidious form of mortality frequently goes unnoticed. Many of these pathogenic fungi are sensitive to fungicides and herbicides used in agricultural production.

The two-spotted spider mite, *Tetranychus telarius,* is an annual pest of many cultivated crops that often reaches pest population levels following pesticide applications. Carner and Canerday (1970) found that epizootics of *Neozygites floridana* Weiser and Muma (=*Entomophthora floridana*) were associated with declines in *T. telarius* populations in Alabama cotton, while Boykin *et al.* (1984) found that fungicides benomyl and mancozeb to suppress the incidence of *N. floridana* infected mites, resulting in *T. telarius* outbreaks.

The fungus *Nomuraea rileyi* (Farlow) Samson is frequently observed in soybean ecosystems as a natural epizootic of at least 6 major lepidopterous pests (Puttler *et al.,* 1976). *In vitro* studies by Ignoffo *et al.* (1975) showed *N. rileyi* to be sensitive to many fungicides, even at one-tenth the recommended field rates, and to the herbicide dinoseb. When *Trichoplusia ni* (Hübner) larvae were fed leaflets treated with *N. rileyi* conidia, 79.6% mortality occurred compared to 21.8% for conidia + benomyl-treated leaflets. Similarly, conidia-treated leaflets killed 31.8% of the larvae compared to 5.7% for larvae fed leaflets treated with conidia + dinoseb.

Beauvaria bassiana (Balsamo) Vuillemin has a wide insect host range, and the natural incidence of this fungus is greatest in soil habitats or habitats with microclimatic conditions similar to those of soil (Ferron, 1980). The persistence of this fungus is then dependent on survival of infective conidia in the environment and mycelial growth. Gardner and Storey (1985) found a wide range of herbicides to reduce germination of conidia and retard mycelial growth. When *S. frugiperda* larvae were exposed to soil containing *B. bassiana* conidia with and without the herbicide alachlor there was a 50% reduction in larval mortality in the alachlor treatment. *Beauvaria bassiana* can cause high mortality to soil inhabiting *L. decemlineata* pupae, especially late in the season (Watt and LeBrun, 1984). However, this fungus is sensitive to the fungicides chlorothalonil and mancozeb which are used to control the spread of late blight, *Phytophthora infestans* (Montagne). *In vitro* studies by Clark *et al.* (1982) showed chlorothalonil and mancozeb to cause 80% and 100% reduction in *B. bassiana* development, respectively. In small field plot studies conidia were blended with mineral oil and an emulsifying agent and then applied to the soil. At 2 and 9 days postinoculation, mancozeb was applied to the soil. Commercial growers use a 5–7-day fungicide schedule over a 60–90-day period. Larval mortality due to *B. bassiana* infection ranged from 60 to 80% in nonfungicide-treated plots and from 18 to 40% in mancozeb-treated plots.

Fungicides can act in a more indirect manner by disrupting the normal soil microflora. Ovipositing seedcorn maggots, *Delia platura* (Meigen), oviposit near germinating seeds because of stimulatory substances produced by certain microorganisms proliferating in exudates from germinating seeds (Eckenrode *et al.*, 1975). Harman *et al.* (1978) found the fungus *Chaetomium globosum* Kze. to reduce oviposition by *D. platura* by inhibiting microbial activity. When seeds were treated with the fungicide thiram in the presence of *C. globosum*, the thiram-treated seeds suffered 2.5 times as much damage as untreated seeds due to increased ovipositional activity as a result of the effects of thiram on *C. globosum*.

IV. PLANT RESISTANCE

In addition to induced resistance to insecticides by insects feeding on resistant plant cultivars, resistant plant cultivars can contribute to insect outbreaks of nontarget herbivores. The exudates of the glandular trichomes of the wild potato, *Solanum berthaultii* Hawkes, contain secondary plant substances, including the glycoalkaloids. These trichomes confer resistance on the plant to the green peach aphid, *Myzus persicae* (Sulzer), to the potato leafhopper, *Empoasca fabae* (Harris) (Tingey and Laubengayer, 1981), and to the Colorado potato beetle (Dimock and Tingey, 1985). *Solanum berthaultii* is being used to develop potato cultivars resistant to these pests; however, recent evidence indicates that this type of resistance is not necessarily compatible with other noninsecticidal control strategies. Obrycki *et al.* (1985) found that the adults of the Colorado potato beetle egg parasite, *Edovum puttleri* Grissell, became entangled in the sticky exudates of the trichomes and were unable to parasitize Colorado potato beetle eggs. Campbell and Duffey (1979) found that a parasite of the tomato fruitworm, *H. zea,* could be poisoned by α-tomatine, an antibiotic substance bred for tomato host plant resistance. Low levels of α-tomatine were found to prolong larval development, reduce pupal eclosion, produce smaller individuals, and reduce adult longevity of the parasite *Hyposoter exiguae* (Viereck). These combined effects could render this parasite ineffective in controlling tomato fruitworm populations when tomato cultivars containing high levels of α-tomatine are used. This reveals the importance of examining the interactions of different control strategies, no matter how innocuous they may seem.

V. CONCLUSIONS

Insect outbreaks in an agroecosystem can be due to a single factor, whether biotic or abiotic; generally, however, a complex interaction of factors allows for such outbreaks. In many cases, it is our lack of knowledge of the biology and behavior of these pests that keeps us from preventing these outbreaks. Only through detailed studies on the interactions of all factors will we be able to prevent such outbreaks, and only through simulation modeling and appropriate field experiments will we be able to examine these factors in an interactive manner. Because of their interaction, it is difficult to predict which of the factors discussed in this chapter play the most important role in insect outbreaks. For this reason, it is important to be cognizant of all of them when attempting to study insect outbreaks in an agroecosystem.

ACKNOWLEDGMENTS

I thank the following colleagues for providing me with valuable information: Carl S. Barfield (University of Florida), Raymond E. Frisbie (Texas A & M University), Kerry F. Harris (Texas A & M University), George G. Kennedy (North Carolina State University), and Richard T. Roush (Cornell University). I am especially grateful to Craig S. Hollingsworth (University of Massachusetts) and Ronald J. Prokopy (University of Massachusetts) for reviewing the manuscript.

REFERENCES

Adkisson, P. L. (1964). Action of the photoperiod in controlling insect diapause. *Am. Nat.* **98**, 357–374.

Adkisson, P. L. (1971). Objective uses of insecticides in agriculture. *In* "Agricultural Chemicals—Harmony or Discord for Food, People, Environment" (J. E. Swift, ed.), pp. 43–51. University of California, Div. Agric. Sci., Sacramento.

Adkisson, P. L. (1972). The integrated control of the insect pests of cotton. *Proc. Tall Timbers Conf. Ecol. Anim. Control Habitat Manage.*, Vol. 4, pp. 175–188.

Andrewartha, H. G., and Birch, L. C. (1954). "The Distribution and Abundance of Animals." Univ. of Chicago Press, Chicago, Illinois.

Ashley, T. R. (1979). Classification and distribution of fall armyworm parasites. *Fla. Entomol.* **62,** 114–123.

Barfield, C. S., and Stimac, J. L. (1981). Understanding the dynamics of polyphagous, highly mobile insects. *Proc. Int. Congr. Plant Prot.*, Vol. 1, pp. 43–46.

Barfield, C. S., Mitchell, E. R., and Poe, S. L. (1978). A temperature-dependent model for fall armyworm development. *Ann. Entomol. Soc. Am.* **71,** 70–74.

Boykin, L. S., Campbell, W. V., and Beute, M. K. (1984). Effect of pesticides on *Neozygites floridana* (Entomophthorales: Entomophtheraceae) and arthropod predators attacking the two-spotted spider mite (Acari: Tetranychidae) in North Carolina plant fields. *J. Econ. Entomol.* **77,** 969–975.

Bradenburg, R. L., and Kennedy, G. G. (1982). Intercrop relationships and spider mite dispersal in a corn/peanut agro-ecosystem. *Entomol. Exp. Appl.* **32,** 269–276.

Brandenburg, R. L., and Kennedy, G. G. (1983). Interactive effects of selected parasites on the two-spotted spider mite and its fungal pathogen *Neozygites floridana. Entomol. Exp. Appl.* **34,** 240–244.

Brattsten, L. B. (1979). Biochemical defense mechanisms in herbivores against plant allelochemicals. *In* "Herbivores: Their Interaction with Secondary Plant Metabolites" (G. A. Rosenthal and D. H. Janzen, eds.), pp. 200–270. Academic Press, New York.

Brattsten, L. B., Wilkinson, C. F., and Eisner, T. (1977). Herbivore–plant interactions: Mixed-function oxidase and secondary plant substances. *Science* **196,** 1349–1352.

Buschman, L. L., Whitcomb, W. H., Heal, T. M., and Mays, D. L. (1977). Winter survival and hosts of the velvetbean caterpillar in Florida. *Fla. Entomol.* **60,** 267–273.

Campbell, B. C., and Duffey, S. S. (1979). Tomatine and parasitic wasps: Potential and incompatibility of plant antibiosis with biological control. *Science* **205,** 700–702.

Carner, G. R., and Canerday, T. O. (1970). *Entomophthora* spp. as a factor in the regulation of the two-spotted spider mite on cotton. *J. Econ. Entomol.* **63,** 638–640.

Clark, R. A., Casagrande, R. A., and Wallace, D. B. (1982). Influence of pesticides on *Beauvaria bassiana,* a pathogen of the Colorado potato beetle. *Environ. Entomol.* **11,** 67–70.

Dapsis, L. J., and Ferro, D. N. (1983). Effectiveness of baited cone traps and colored sticky traps for monitoring adult cabbage maggots: With notes on female ovarian development. *Entomol. Exp. Appl.* **33,** 35–42.

Dimock, M. B., and Tingey, W. M. (1985). Resistance in *Solanum* spp. to the Colorado potato beetle: Mechanisms, genetic resources and potential. *In* "Proceedings from the Symposium on the Colorado Potato Beetle, 17th International Congress of Entomology" (D. N. Ferro and R. H. Voss, eds.), Agric. Exp. Stn. Bull. No. 704. University of Massachusetts, Amherst.

Dindonis, L. L., and Miller, J. R. (1980). Host-finding responses of onion and seedcorn flies to healthy and decomposing onions and several synthetic constituents of onions. *Environ. Entomol.* **9,** 467–472.

Dittrich, V., Streibert, P., and Bathe, P. A. (1974). An old case reopened: Mite stimulation by insecticide residue. *Environ. Entomol.* **3,** 534–540.

Eckenrode, C. J., Harman, G. E., and Webb, D. R. (1975). Seed-borne microorganisms stimulate seedcorn maggot egg laying. *Nature (London)* **256,** 487–488.

Ferro, D. N., and Chapman, R. B. (1979). Effects of different constant humidities and temperatures on two-spotted spider mite egg hatch. *Environ. Entomol.* **8,** 701–705.

Ferro, D. N., Morzuch, B. J., and Margolies, D. (1983). Crop loss assessment of the Colorado potato beetle on potatoes in western Massachusetts. *J. Econ. Entomol.* **76,** 349–356.

Ferro, D. N., Logan, J. A., Voss, R. H., and Elkinton, J. S. (1985). Colorado potato beetle temperature dependent growth and feeding rates. *Environ. Entomol.* **14,** 343–348.

Ferron, P. (1980). Pest control by the fungi *Beauvaria* and *Metarhizium. In* "Microbial Control of Pests and Plant Diseases, 1970–1980" (H. D. Burges, ed.), pp. 463–482. Academic Press, London.

Forgash, A. J. (1985). Insecticide resistance of the Colorado potato beetle, *Leptinotarsa decemlineata* (Say). *In* "Proceedings from the Symposium on the Colorado Potato Beetle, 17th International Congress of Entomology" (D. N. Ferro and R. H. Voss, eds.), Agric. Expt. Stn. Bull. No. 704. University of Massachusetts, Amherst.

Forrester, T. O. (1982). "Control of Mexican Bean Beetle in Truck Farm and Small Plantings of Green Beans with *Pediobius foveolatus*, Lab. Rep., pp. 197–199. USDA/APHIS—Otis Methods Development Center, Massachusetts.

Gardner, W. A., and Storey, G. K. (1985). Sensitivity of *Beauvaria bassiana* to selected herbicides. *J. Econ. Entomol* **78,** 1275–1279.

Garrett, R. G. (1973). Non-persistent aphid-borne viruses. *In* "Viruses and Invertebrates" (A. J. Gibbs, ed.), pp. 476–492. Am. Elsevier, New York.

Gould, F., and Hodgson, E. (1980). Mixed-function oxidase and glutathione transferase activity in last instar *Heliothis virescens* larvae. *Pestic. Physiol.* **13,** 34–40.

Gould, F., Carroll, C. R., and Futuyma, D. J. (1982). Cross-resistance to pesticides and plant defenses: A study of the two-spotted spider mite. *Entomol. Exp. Appl.* **31,** 175–180.

Harcourt, D. G. (1971). Population dynamics of *Leptinotarsa decemlineata* (Say) in Eastern Ontario. III. Major population processes. *Can. Entomol.* **103,** 1049–1061.

Hare, J. D. (1980). Contact toxicities of ten insecticides to Connecticut populations of the Colorado potato beetle. *J. Econ. Entomol.* **73,** 230–231.

Harman, G. E., Eckenrode, C. J., and Webb, D. R. (1978). Alteration of spermosphere ecosystems affecting oviposition by the bean seed fly and attack by soilborne fungi on germinating seeds. *Ann. Appl. Biol.* **90,** 1–6.

Harris, C. R., and Svec, J. H. (1981). Colorado potato beetle resistance to carbofuran and several other insecticides in Quebec. *J. Econ. Entomol.* **74,** 421–424.

Hsiao, I. H. (1985). Ecophysiological and genetic aspects of geographic variations of the Colorado potato beetle. *In* "Proceedings from the Symposium on the Colorado Potato Beetle, 17th International Congress of Entomology" (D. N. Ferro and R. H. Voss, eds.), Agric. Exp. Stn. Bull. No. 704. University of Massachusetts, Amherst.

Ignoffo, C. M., Hostetter, O. L., Garcia, C., and Pinnell, R. E. (1975). Sensitivity of the entomopathogenic fungus *Nomuraea rileyi* to chemical pesticides used on soybeans. *Environ. Entomol.* **4,** 765–768.

Johnson, C. G. (1969). "Migration and Dispersal of Insects by Flight." Methuen, London.

Kennedy, G. G. (1984). 2-Tridecanone, tomatoes and *Heliothis zea*: Potential incompatibility of plant antibiosis with insecticidal control. *Entomol. Exp. Appl.* **35,** 305–311.

Lashomb, J. H., and Ng, Y.-S. (1984). Colonization of Colorado potato beetles, *Leptinotarsa decemlineata* (Say) (Coleoptera: Chrysomelidae), in rotated and nonrotated fields. *Environ. Entomol.* **13,** 1352–1356.

Lees, A. D. (1943). On the behavior of wireworms of the genus *Agriotes* Esch. (Coleoptera: Elateridae). I. Reactions to humidity. *J. Exp. Biol.* **20,** 43–53.

Lewontin, R. C. (1965). Selection for colonizing ability. *In* "The Genetics of Colonizing Species" (H. G. Baker and G. L. Stebbins, eds.), pp. 77–91. Academic Press, New York.

Luck, R. F., Van den Bosch, R., and Garcia, R. (1977). Chemical insect control—a troubled strategy. *BioScience* **27,** 606–611.

Martinez-Carillo, J. L., and Reynolds, H. T. (1983). Dosage–mortality studies with pyrethroids and other insecticides on the tobacco budworm (Lepidoptera: Noctuidae) from the Imperial Valley, California. *J. Econ. Entomol.* **76,** 983–986.

Mukerji, M. K., and Gage, S. H. (1978). A model for estimating hatch and mortality of grasshopper egg populations based on soil moisture and heat. *Ann. Entomol. Soc. Am.* **71,** 183–190.

Obrycki, J. J., Tauber, M. J., Tauber, C. A., and Bollands, B. (1985). *Edovum puttleri* (Hymenoptera: Eulophidae), an exotic egg parasitoid of the Colorado potato beetle (Coleoptera: Chrysomelidae): Responses to temperate zone conditions and resistant potato plants. *Environ. Entomol.* **14,** 48–54.

Perring, T. M., Holtzer, T. O., Toole, J. L., Norman, J. M., and Myers, G. L. (1984). Influences of temperature and humidity on pre-adult development of the Banks grass mite (Acari: Tetranychidae). *Environ. Entomol.* **13,** 338–343.

Putman, W. L. (1963). Lack of effect of DDT on fecundity and dispersion of the European red mite, *Panonychus ulmi* (Koch) (Acarina: Tetranychidae), in peach orchards. *Can. J. Zool.* **41,** 603–610.

Puttler, B., Ignoffo, C. M., and Hostetter, D. L. (1976). Relative susceptibility of nine caterpillar species to the fungus *Nomuraea rileyi. J. Invert. Pathol.* **27,** 269–270.

Reynolds, H. T., Adkisson, P. L., Smith, R. F., and Frisbie, R. E. (1982). Cotton insect pest management. *In* "Introduction to Insect Pest Management" (R. L. Metcalf and W. H. Luckman, eds.), pp. 375–441. Wiley, New York.

Roach, S. H., Smith, J. W., Vinson, S. B., Graham, H. M., and Harding, J. A. (1979). Sampling predators and parasites of *Heliothis* species on crops and native host plants. *South. Coop. Ser. Bull.* **231,** 132–145.

Smith, R. F. (1970). Pesticides: Their use and limitations in pest management. *In* "Concepts of Pest Management" (R. L. Rabb and F. E. Guthrie, eds.), pp. 103–113. North Carolina State University, Raleigh.

Sparks, T. C. (1981). Development of insecticide resistance in *Heliothis zea* and *Heliothis virescens* in North America. *Bull. Entomol. Soc. Am.* **27,** 186–192.

Stern, V. M. (1969). Interplanting alfalfa in cotton to control lygus bugs and other insect pests. *Proc. Tall Timbers Conf. Ecol. Anim. Control Habitat Manage.,* Vol. 1, pp. 55–69.

Stern, V. M., van den Bosch, R., Leigh, T. F., McCutcheon, O. D., Sallee, W. R., Houston, C. E., and Garber, M. J. (1967). Lygus control by strip-cutting alfalfa. *Univ. Calif. Ext. Serv.* **AXT-241,** 1–13.

Tingey, W. M., and Laubengayer, J. E. (1981). Defense against the green peach aphid and potato leafhopper by glandular trichomes of *Solanum berthaultii. J. Econ. Entomol.* **71,** 721–725.

Tuttle, A. F. (1985). Attraction of adult cabbage maggots to visual and olfactory traps. M.S. Thesis, University of Massachusetts, Amherst.

Watt, B. A., and LeBrun, R. A. (1984). Soil effects of *Beauvaria bassiana* on pupal populations of the Colorado potato beetle (Coleoptera: Chrysomelidae). *Environ. Entomol.* **13,** 15–18.

Webster, R. P., and Cardé, R. T. (1982). Influence of relative humidity on calling behavior of female European corn borer moth (*Ostrinia nubilalis*). *Entomol. Exp. Appl.* **32,** 181–185.

Whalen, M. J. (1979). Taxonomy of *Solanum* section *Androceras. Gentes Herb.* **11,** 359–426.

Whitcomb, W. H., and Bell, K. (1964). Predaceous insects, spiders and mites of Arkansas cotton fields. *Arkansas, Agric. Exp. Stn. Bull.* **690.**

Wolfenbarger, D. A., Lukefahr, M. J., and Graham, H. M. (1973). LD_{50} values of methyl parathion and endrin to tobacco budworms and bollworms collected in the Americas and hypothesis on the spread of resistance in these lepidopterans to these insecticides. *J. Econ. Entomol.* **6,** 211–216.

Wright, R. (1984). Evaluation of crop rotation for control of Colorado potato beetles (Coleoptera: Chrysomelidae) in commercial potato fields on Long Island. *J. Econ. Entomol.* **77**, 1254–1259.

Chapter 9

Agricultural Ecology and Insect Outbreaks

STEPHEN J. RISCH

Division of Biological Control
University of California
Berkeley, California 94720

I. INTRODUCTION

There are two groups of questions concerning insect outbreaks in agroecosystems. First, what distinguishes agroecosystems from so-called natural systems in terms of the general occurrence of insect outbreaks? Are outbreaks more or less common in agroecosystems and why? Second, within the category of agroecosystems, what accounts for the large fluctuations in insect numbers from generation to generation and from year to year? What triggers an insect outbreak in agricultural systems? Are some kinds of agroecosystems more prone to experience outbreaks and why? The answers to these questions are of considerable interest to both the basic ecologist and the applied entomologist. From the basic ecologist's perspective, by discovering what accounts for differences between natural systems and agroecosystems we can learn much about the underlying ecological processes that create the observed patterns of distribution and abundance in nature, one of the principal goals of the discipline of ecology. Furthermore, the careful ecological studies on agroecosystems that seek to examine the underlying causes of insect outbreaks usually cannot be carried out in more natural ecosystems. Because agroecosystems are simpler, more manipulable, and more easily replicated, it is often much easier to test basic ecological hypotheses in these systems.

From the applied perspective, the study of insect outbreaks is obviously important because of the management goal. Presumably a more complete understanding of the processes causing outbreaks will generate control strategies. Yet between basic knowledge and its application via technology stand tremendous political, economic, and philosophical variables that act as selective filters. Our knowledge of what causes insect outbreaks in agroecosystems far outstrips our ability to reach a consensus on how to act on it (Risch, 1984). Although in this chapter I emphasize the basic question of what accounts for insect outbreaks in agroecosystems, I also incorporate some conclusions on management strategies where appropriate.

II. COMPARING NATURAL SYSTEMS AND AGROECOSYSTEMS

A. Morphology and Chemistry of Plants

The invention of agriculture some 10,000 years ago and its gradual evolution have created fundamental differences between natural and agricultural systems. From the perspective of insects and their tendency to experience outbreaks in the two systems, there are several major differences.

First, humans tended to select plants with fewer morphological and chemical defenses. This occurred for several reasons. In those cases in which humans consumed particular plant parts, selection resulted in the reduction of defenses in those parts so that the crop would be more palatable. Evidence for this includes the fact that, when botanical insecticides were extracted and used in traditional agriculture, the chemicals tended to come from nonconsumptive-type plants and plant parts (Secoy and Smith, 1983). In addition, overall intense human selection for fast growth and high reproductive output resulted in a general lowering of the plants' allocation to defense. Of course, significant amounts of toxic secondary compounds remain in many edible crops (Ames, 1983), but the general trend has been the gradual reduction of those chemicals and morphological features that protected plants from arthropod and vertebrate herbivores (Rhoades, 1979). This often left the plants more vulnerable than their wild relatives, and it largely explains the widespread belief that there are more outbreaks of insects in agroecosystems than natural ecosystems (Elton, 1958; Pimentel, 1961a,b; Margalef, 1968).

Although the elimination of morphological protective devices such as thorns, trichomes, and silicaceous particles has nearly always increased crop plant vulnerability, there are some interesting exceptions in the case of chemical defenses. For example, in the family Curcurbitaceae, a group of bitter substances known as cucurbitacins are repellent to most generalist herbivores, but several specialists have overcome this defense, and these substances now act as specific arrestants and feeding stimulants (DaCosta and Jones, 1971). Chrysomelid beetles in the genera *Diabrotica* and *Acalymma* accumulate in extremely large numbers on plants that are very bitter. Since breeding programs have deliberately selected nonbitter varieties to suit human taste, these specialist herbivores are more attracted to some of the wild *Cucurbita* species than the domesticated ones. Generalist herbivores, however, show the reverse trend, being more attracted to the domesticated varieties (Carroll and Hoffman, 1980).

The same pattern can be seen in some species of the Cruciferae. Breeding programs have tended to select varieties of cabbage and its relatives that have much lower concentrations of mustard oils, which apparently are repellent to many generalist herbivores but extremely attractive to the specialist flea beetles in the genus *Phyllotreta* (Feeny *et al.,* 1970). Some wild mustard weeds, such as *Brassica kaber,* when grown in the presence of collards significantly lower flea beetle abundance on the collards by acting as a trap plant (B. Platts, personal communication). Thus, in those cases where specialist herbivores have evolved strategies to detoxify qualitative defenses (*sensu* Feeny, 1976), and even use them as specific feeding attractants, the agricultural plant may under some circumstances be subject to lower herbivore loads than the "natural" wild relative.

B. Landscape Diversity and Insect Outbreaks

The invention and evolution of agriculture have resulted in tremendous changes in landscape diversity. There has been a consistent trend toward simplification that entails (1) the enlargement of fields, (2) the aggregation of fields, (3) an increase in the density of crop plants, (4) an increase in the uniformity of crop population age structure and physical quality, and (5) a decrease in inter- and intraspecific diversity within the planted field.

Although these trends appear to exist worldwide, they are more apparent, and certainly best documented, in industrialized countries. For instance, in the United States, a USDA Task Force on Spatial Heterogeneity in Agricultural Landscapes indicated that the amount of crop diversity per unit of arable land has markedly decreased and that crop lands have shown a tendency toward concentration [U.S. Department of Agriculture (USDA), 1973]. These trends are particularly evident in the Corn Belt, Mississippi Delta, Red River Valley, Texas High Plains, California irrigation areas, southern Florida, and the Kansas–Oklahoma Winter Wheat Belt.

Although these changes in landscape diversity have not always led to more insect outbreaks, it seems generally agreed that this is the overall trend (Altieri and Letourneau, 1984). This is best documented in the case of within-field diversity. Ecologists have long claimed that less taxonomically and structurally diverse systems are more subject to outbreaks of specialist insects (Pimentel, 1961a,b). The idea was stated most strongly in the post-World War II writings of several ecologists. MacArthur (1955) and Elton (1958) held that a food web of interactions between trophic levels acts to resist change in the abundance of individual species more effectively than do single food chains. Margalef (1968), Odum (1971), and others developed and popularized the idea, and it was widely quoted by agricultural entomologists (Smith and van den Bosch, 1967).

In the 1970s a series of articles seriously challenged the generality of this idea on theoretical and logical grounds (May, 1973; Goodman, 1975; Murdoch, 1975). Root's (1973) seminal paper raised the issue yet again, but this time in a much more specific and testable way by posing two clearly stated hypotheses. The first, called the enemies hypothesis, predicts that natural enemies will be augmented in diversified systems and thereby control the herbivores more effectively. This may occur for several reasons related to the interaction between food availability and natural-enemy populations in these diverse systems (Risch, 1981): (1) a greater temporal dispersion and variety of types of pollen and nectar resources, both of which can augment the longevity and fecundity of natural enemies; (2) more ground cover, which is often very important for certain predators; and (3) a greater species richness of herbivores that can act as alternative prey/hosts when the preferred hosts/prey are locally extinct or at inappropriate stages of the life cycle.

The second hypothesis, usually called the resource concentration hypothesis, focuses on the movement and reproductive behavior of the adult herbivores. It predicts that specialist herbivores will build up in concentrated stands of their host plants and will be much less common in diverse vegetation due to (1) higher immigration rates into less diverse areas, (2) lower emigration rates out of these areas, and/or (3) greater tendency to reproduce in these areas.

Several reviews have been published documenting the effects of within-habitat diversity on insects (Andow, 1983; Cromartie, 1981; Altieri and Letourneau, 1984; Risch *et al.*, 1983). In the review by Risch *et al.*, 150 published studies of the effect of diversifying an agroecosystem on insect pest abundance of the effect of diversifying an agroecosystem on insect pest abundance were summarized; 198 total herbivore species were examined in these studies. Fifty-three percent of these species were found to be less abundant in the more diversified system, 18% were more abundant in the diversified system, 9% showed no difference, and 20% showed a variable response. Two major drawbacks of the studies were that (1) proper experiments were not conducted to show that differences in pest abundance between more and less diverse systems resulted in yield differences, and (2) the ecological mechanisms accounting for the differences in pest abundance were carefully examined in only a few of the studies. A fundamental understanding of these ecological mechanisms is critical to the development of a predictive theory of how agricultural diversification affects insect pests.

Despite these problems, both empirical data and theoretical arguments suggest that differences in pest abundance between diverse and simple annual cropping systems can best be explained by the movement and reproductive behavior of herbivores and not be the activities of natural enemies. Thus, the resource concentration hypothesis rather than the enemies hypothesis best explains insect outbreaks in simplified systems. In the case of *perennial* cropping systems, however, the role of natural enemies (i.e., the enemies hypothesis) seems to be more important in explaining pest outbreaks in simplified systems.

These conclusions should be treated as very tentative since there have been so few detailed experimental studies in which the relative importance of two hypotheses has been compared. For logistical reasons most of these have involved annual systems (Risch, 1980, 1981; Bach, 1980a,b, 1981). A number of recent studies have examined in some detail the effect of crop habitat simplification either on enemies or on herbivore movement (for a review, see Altieri and Letourneau, 1984), but from the perspective of developing a general predictive theory of how agroecosystem simplification causes insect outbreaks, this is not adequate. We must refine these hypotheses, develop new ones if necessary, and compare their relative explanatory power within particular cropping systems.

The agricultural simplification that has taken place on a more regional level (enlargement and aggregation of fields) is also claimed to increase the probability of insect outbreaks (Andow, 1983), but for obvious reasons there are many fewer experimental data that clearly document it. There are two main issues here. The first is colonization of crop "islands" by insects. In the case of annual crops, insects must colonize from the borders each season, and the larger the field, the greater is the distance that must be covered. Several studies suggest (not surprisingly) that natural enemies tend to colonize after their hosts/prey (Tepedino and Stanton, 1976; Schoener, 1974) and that the lag time between the arrival of pest and natural enemy increases with distance from border (source pool). Price (1976), for instance, found that the first occurrence of an herbivore and that of a predatory mite in a soybean field were separated by 1 week on the edge versus a 3-week lag in the center. To the extent that this is a general phenomenon, increased field size should lead to more frequent insect outbreaks. Yet we have very few data concerning the effect of field size on this lag time in other agricultural systems, nor do we know how the lag time is affected by specialist versus generalist natural enemies—a critical issue when one is considering potential biological control of the pest.

A second major issue is the relative value of hedgerows, second-growth windbreaks, and other noncrop habitats bordering planted fields. As field aggregation increases, these tend to disappear. Although these habitats can serve as refugia for both pests and natural enemies, it is often assumed that on balance they are beneficial, and therefore the gradual loss of these habitats is thought to contribute to insect outbreaks in agroecosystems. The data, however, are very mixed.

In a review of the effects of windbreaks, hedgerows, ditches, and other noncrop habitats, Altieri and Letourneau (1984) list these advantages and disadvantages:

1. Disadvantages

1. Uncultivated land as a source of pests for adjacent crops. These habitats often harbor alternative food sources for many pest insect species, and they may encourage insect outbreaks in neighboring agroecosystems, especially if the pests tend to move relatively short distances (van Emden, 1981). Pests with seasonal host variation such as heteroecious aphids are particularly good candidates, and such aphids comprise 42% of pest species (Eastop, 1981). It is also possible that the presence of wild crop relatives in nearby habitats could result in the maintenance of genetic diversity in local, relatively nonvagile insect populations, leading to the emergence of new biotypes that could overcome host plant resistance.

2. Direct physical effects. Uncultivated land such as windbreaks may have direct physical effects, such as providing overwintering sites for pests or halting flight and thereby depositing insects on the leeward side of such obstructions. Litter under shelterbelts has been implicated in the facilitation of boll weevil outbreaks by providing overwinter sites (Jones and Sterling, 1978). Outbreaks of aphids in lettuce and thrips in cereals

have apparently been facilitated by the passive deposition of pests by windbreaks (Lewis, 1965, 1970).

2. *Advantages*

1. Adjacent habitats as trap crops. The planting of a small area of highly attractive and disposable crop or weed near a valuable crop can suppress insect outbreaks by keeping the pest on the so-called trap plants. Some plants are usually sown before the main crop and often destroyed after the pest has colonized it. Numerous variations on this theme are documented by Altieri and Letourneau (1984). Such techniques must be managed carefully, as demonstrated by the case of wild grasses growing adjacent to wheat fields (Watt, 1981). Although there is a dilution effect of aphid damage early in the season, the scattered phenologies of the surrounding grasses probably allow the buildup of aphid populations.

2. Uncultivated land and natural enemies. Hedgerows, windbreaks, and woodlands adjacent to crops can provide extremely important refugia for natural enemies as well as sources of nectar, pollen, and hosts/alternate prey. The importance of local extinctions of natural enemies in triggering insect pest outbreaks is well known, and these uncultivated areas can prevent such extinctions. However, the documentation of many examples of this benefit (see Altieri and Letourneau, 1984; Bugg, 1985) should not lead one to infer that it always occurs, as is sometimes suggested in the popular and even academic literature.

There is not now, nor does it seem there could be, a general theory that will predict the overall value of uncultivated land in preventing insect outbreaks. We know something, however, of the ecological mechanisms that underly the various advantages and disadvantages of such habitats, and for the time being the overall value of uncultivated land will have to be judged on a case-by-case basis.

C. Escape in Space of Crop Pests from Their Natural Enemies

Crops are not grown only in their site of origin; humans have transplanted them over the entire earth. Maize, for instance, which originated in what is now central Mexico, is grown in nearly every country on earth. While transporting a crop, people often inadvertently carried the plant's associated pest insects but usually not the pest's natural enemies. This would account for many insect outbreaks in agroecosystems. The widely successful program of classical biological control, in which the insect pest's natural enemies are collected at or near the site of pest origin and released where the pest has newly migrated (Huffaker and Messenger, 1976), often brings insect outbreaks under control. Not all insect pest outbreaks, however, are caused by accidentally introduced exotics. Pimentel (1985) estimated that 60% of insect pests are native insects that have moved onto crops that were imported. Thus, the absence of coevolved natural enemies on these pests does not explain why these herbivores sometimes cause outbreaks on crops.

III. FACTORS TRIGGERING INSECT OUTBREAKS IN AGROECOSYSTEMS

The previous section has addressed two interrelated questions: (1) why insect outbreaks are more common in agroecosystems than in natural systems, and (2) how habitat diversity affects the potential for insect outbreaks. This section focuses on mechanisms that trigger large temporal fluctuations in pest insects in agroecosystems; that is, it examines what causes an insect to reach outbreak levels within a particular agroecosystem.

A. Pesticide-Induced Insect Outbreaks

Pesticides are, of course, applied to prevent or "cure" insect outbreaks, but in fact they often cause them. There are several ways in which this can happen.

1. Reduction of Natural Enemies

The biological control literature is replete with examples of pest resurgence (i.e., outbreaks) following insecticide application (Huffaker and Messenger, 1976). The factors that contribute to this resurgence include the following:

1. Natural enemies often experience higher mortality than herbivores following a given spray. This is due, in part, to the greater mobility of many natural enemies, which exposes them to more insecticide per unit time following a spray (Hagen *et al.*, 1976; Vinson, 1981; Weseloh, 1981). This especially applies to holometabolous natural enemies requiring a diversity of secondary resources (nectar, pollen, etc.) to reproduce.

2. Natural enemies appear to evolve resistance to insecticides much more slowly than herbivores. This results from a smaller probability that some individuals in populations of natural enemies will have genes for insecticide resistance. This in turn is due to the much smaller size of the

natural-enemy population relative to the pest population and the different evolutionary history of natural enemies and herbivores. The coevolutionary history of many herbivore groups with host plants that contain toxic secondary compounds has resulted in a set of metabolic pathways more easily adjusted to produce insecticide resistance (Croft and Morse, 1979; Croft and Strickler, 1983).

3. The killing of a large proportion of pests in a particular area may cause a parasite or predator to become locally extinct, thus allowing for pest resurgence, which may last until the natural-enemy population becomes reestablished. This may also in part explain the underlying mechanism of secondary pest outbreaks, the outbreak of an insect that had not been a pest until the insecticide was applied and killed off its natural enemies (DeBach, 1974).

2. Removal of Competitive Species

In theory, an insect outbreak could be caused by the insecticidal removal of competitors of a potential pest insect, thus allowing for ecological release of the new pest. This is probably quite rare, compared with the incidence of insecticide-induced pest resurgence and secondary pest outbreaks. Although ecologists have long been enamored of the importance of competition in structuring ecological communities, the forces championing competition have more recently been in retreat (Strong, 1983, 1984; Simberloff, 1983; Price *et al.,* 1984). It seems that especially in phytophagous insect communities there is relatively little evidence that competition plays a dominant organizing role (Strong *et al.,* 1984). This should apply particularly to highly disturbed habitats such as agroecosystems (Risch and Carroll, 1986). There are, however, several interesting exceptions to this apparent general trend. Root and Skelsey (1969) found that population outbreaks of aphids occurred when the insecticide carbaryl was applied on collards in New York. Decreases of predators and parasites in the sprayed plots could not account for the aphid increases. Circumstantial evidence suggested that decreased interspecific competition from other herbivores killed by the pesticide, especially flea beetles, may have accounted for the aphid outbreak.

More recently, Stimac and Buren (1982) documented a dramatic resurgence of the imported fire ant (*Solenopsis invicta*) following single applications of the ant insecticides Amdro® and Mirex®. (Amdro and Mirex were specifically designed for the control of the imported fire ant, and considerable amounts of federal and state tax money have subsidized the distribution of these chemicals.) Several years after insecticide treatment of fields that resulted in the removal of most of the ant species, fire ant populations were 2–10 times higher than in untreated controls. The cause of the resurgence has not been determined rigorously but appears to involve (1) the great capacity of the fire ant to colonize relative to native ants due to frequency and size of mating flights, (2) the removal of several ants that act as important predators on nest-founding fire ant queens, and (3) the removal of native ants that act as effective competitors, thus reducing the number and eventual size of nests of native ants that became established.

3. Direct Stimulation of Pest Reproduction (Hormoligosis)

The physiology of some arthropods is apparently directly stimulated by insecticides, even to the point of causing outbreaks. One of the best documented example is the case of mites (Dittrich *et al.,* 1974). More recently, work in the Philippines has shown that populations of the brown planthopper, *Nilaparvata lugens,* increase dramatically following the use of a variety of insecticides. In fact, when outbreaks of the hopper are desired for demonstration or experimental purposes at the International Rice Research Institute (IRRI), insecticides are used to create them. Decreased populations of natural enemies cannot completely account for these outbreaks, and preliminary work has shown that the insecticides directly stimulate the growth and development of the brown planthopper (Heinrichs *et al.,* 1982; Chelliah and Heinrichs, 1980; Reissig *et al.,* 1982). Although low-level stimulatory effects may be more common than was previously thought, it appears unlikely that they often cause significant outbreaks.

4. Indirect Stimulation of Reproduction of Pests by Altering the Nutritional Biochemistry of the Plant (Trophobiosis)

Insecticides, fungicides, and herbicides can significantly alter the nutritional biochemistry of the plant by changing the concentrations of nitrogen, phosphorus, and potassium, by influencing the production of sugars, free amino acids, and proteins, and by influencing the aging process, which affects surface hardness, drying, and wax deposition (Chabbonseau, 1966a,b; Maxwell and Harwood, 1960; Oka and Pimentel, 1974, 1976; Rodriguez *et al.,* 1957, 1960). These effects are very widespread, but there is still relatively little work demonstrating how these changes in plant chemistry contribute to insect outbreaks. In part this is due to severe methodological problems of experimentally differentiating the various effects of the chemicals and in part because there seems to be less concern with these effects than with the more obvious and straightforward ways in which pesticides contribute to insect pest resurgence.

B. Weather-Induced Insect Outbreaks

Weather can trigger insect outbreaks, and some authors argue that this may be the most important overall cause of dramatic changes in pest abundance in agroecosystems (see Chapter 10, this volume). For example, Miyashita (1963), in reviewing the dynamics of seven of the most serious insect pests in Japanese crops, concluded that weather was the principal cause of the outbreaks in each case. There are several ways in which weather can trigger insect outbreaks. Perhaps the most straightforward mechanism is direct stimulation of the insect and/or host plant physiology. The development and widespread use of degree-day models to predict outbreaks of particular pests and appropriate control strategies are an indication of the importance of the linkage between temperature and growth and the development of herbivorous insects and their host plants (Gordon, 1984; Hughes *et al.,* 1984). For example, Gutierrez *et al.* (1971, 1974a,b) and Gutierrez and Yaninek (1983) have shown that weather plays the major role in the development of cowpea aphid populations in southeast Australia. In this case, a series of climatic events favor complex changes in aphid physiological development, migration, and dispersal in such a way as to cause localized outbreaks. Agroecosystems are probably much more vulnerable than are natural ecosystems to these climate-induced insect outbreaks because the crop plants themselves tend to be more responsive to small fluctuations in temperature and precipitation.

Insect movement is often affected by weather, and changes in temperature, humidity, and/or wind patterns can trigger outbreaks by directly stimulating insect movement (Wellington and Trimble, 1984). For example, dramatic local concentrations of gypsy moths (Cameron *et al.,* 1979; Mason and McManus, 1981) and locusts (Rankin and Singer, 1984) can be accounted for by convergent wind fields that deposit a large number of dispersing individuals in a small area. Although movement in direct response to weather may sometimes cause outbreaks of this kind, it appears more common that movement, and the resultant pest outbreak, are caused indirectly by weather via changes in insect density or host plant quality.

The effect of changes in host plant quality on insect outbreaks in agriculture has received much attention (White, 1974, 1976, 1978, 1984). White originally postulated that, when plants are stressed by certain changes in weather patterns, they become a better source of food for insects because this stress causes an increase in the amount of nitrogen available in their tissues for young herbivores feeding on them. One of the best examples of the weather-induced phenomenon comes from locusts.

White (1976) suggests that for locusts, as for many phytophagous insects in agriculture, there is often a relative shortage of nitrogenous food for rapidly growing young. Species adapt to this "inadequate environment" by producing large "surpluses" of young. When chance combinations of weather stress the food plants, making them a richer source of nitrogen, there is a greatly increased rate of survival of the very young, rapidly leading to an explosive increase in abundance—to an outbreak.

Originally the focus was on weather fluctuations as the factor that triggered increases in the amount of readily available and assimilable nitrogen. However, it is now known that many environmental factors can cause such a change and thereby stimulate insect outbreaks in agroecosystems. White (1984) discusses these and cites examples of insect outbreaks in agriculture caused by each of them:

1. Infection by disease organisms
2. Normal plant senescence
3. Deficiency of soil nutrients
4. Pesticides
5. Physical damage to plants
6. Airborne pollutants

Most workers have thus far emphasized how these environmental factors influence nitrogen availability, and although nitrogen may be the most important, there are other nutrients and feeding stimulants and deterrents that can be affected by environmental factors. For example, outbreaks of the squash beetle pest, *Acalymma themei,* were observed in Costa Rica on squash during periods of high humidity (Risch, 1981). Beetles strongly preferred squash leaves that were infected with powdery mildew, and local beetle outbreaks were apparently caused by the weather-induced disease epidemic. However, no difference in available nitrogen was found between diseased and nondiseased leaves, although the diseased leaves did have higher concentrations of the feeding stimulant cucurbitacin (unpublished data). When insect outbreaks are observed on stressed agricultural plants, one should not automatically assume that increased nitrogen availability is the proximate cause (Rhoades, 1979, 1983).

Weather can also cause insect outbreaks by decoupling predator–prey or pathogen–prey relations. This can occur when temperature and/or moisture have differential effects on the behavior, mortality, or metabolic efficiency of predators and prey. For example, as temperature decreases, the behavioral efficiency of some predators apparently suffers much more than the metabolic efficiency of the prey. This may be especially true for certain ground-dwelling generalist predators, such as ants and carabid beetles. This is thought to account for outbreaks of the flannel moth (*Panolis flammea*) in Poland. In this case low spring temperatures significantly decrease the foraging efficiency of *Formica* ant predators on first-instar moth larvae but have relatively little effect on moth larval development. By the time temperatures are warm enough to stimulate active *Formica* foraging, moth larvae have reached the larger instars, which are not efficiently attacked by the ant; an outbreak of the moth then occurs (Sitowski, 1924).

In a similar example, it was shown that low temperature decoupled the ant predation of tent caterpillars in Canada (Ayre and Hitchon, 1968). The caterpillars develop at temperatures below 50°F, but their major predators, the imported ant (*Formica obscuripes*) and the native ant (*Myrmica americana*), do not forage in this cool weather. Cool springs thereby increase the probability of a tent caterpillar outbreak.

Temperature and moisture conditions also have very significant effects on the development and reproduction of insect pathogens, and changes in weather can thus affect the mortality of pest insects. In northern California the blue and pea aphids (*Acyrthosiphon kondoi* and *A. pisum*) are important pests of alfalfa. The pea aphid, but not the blue aphid, is very susceptible to a fungus, *Erynia nouryi*. During the dry winters the fungus develops and disperses poorly, and pea aphid populations increase dramatically (J. Pickering and A. P. Gutierrez, unpublished).

C. Biotechnology and Insect Outbreaks in Agriculture

Currently there is much discussion in the popular and academic literature of biotechnology and its probable impact on agriculture. Although the claims being made concerning the potential beneficial effects of biotechnology are extravagant and probably cannot be substantiated, there is little doubt that the new technologies will have a significant impact on all areas of agriculture, including pest management [Office of Technology Assessment (OTA), 1984]. I discuss here several ways in which biotechnology could affect the probability of an insect outbreak in agriculture.

1. Biotechnology and Habitat Diversity

Techniques of plant genetic engineering allow for much more rapid development and proliferation of particular crop varieties with desirable traits [National Academy Press (NAS) 1984]. If disease and insect resistance prove easy enough to engineer into the plant, it may be argued that it is most economical to grow the single, best-adapted variety over a large area and change varieties as insect pests adapt to them. Thus, it is possible that the same trends toward decreased agricultural habitat diversity that have occurred over the past 100 years will now only be accelerated. The best genotype for a particular area will be developed and grown in monoculture stands with many of the same problematic consequences discussed earlier.

Yet genetic engineering technology also makes it easier to develop diverse multilines that could be grown in appropriate spatial and temporal patterns to avoid insect outbreaks. There is already an abundant theoretical and empirical literature documenting how multilines can economically decrease plant pathogen populations (see, e.g., Heybroek, 1982; Trenbath, 1977), but additional ecological studies are needed to explore the impact of multilines on herbivorous insect population dynamics. The literature on plant genetic engineering unfortunately makes no mention of the development of varieties for use in multilines, and it appears that most thinking is along the lines of monocultures of single, best-adapted varieties (NAS, 1984).

2. *Biotechnology and Insect–Microbe–Plant Interactions*

Because of the extreme importance of insect–microbe associations in the ecology of herbivorous insects, the release of genetically engineered microorganisms has the potential to influence the population dynamics of insects, in some cases affecting the probability of outbreaks. Jones (1984) discussed the importance of insect–microbe associations in the ecology of herbivorous insects and demonstrated the extreme functional diversity in these associations.

Microorganisms can have a beneficial impact on growth, development (molting, sexual maturation), survivorship, and fecundity (egg production). In addition, they can physically break down food, inhibit pathogens or parasitoids, produce oviposition stimulants and host attractants, stimulate the production of defensive compounds in plants, and even synthesize insect defenses themselves (Jones, 1984). At the trophic, metabolic level these activities include (1) serving as the sole food source in obligate predatory and mutualistic interactions (insect provisions microorganism from plant); (2) serving as a supplemental food source in predatory and mutualistic, obligate, and facultative associations; (3) provisioning macro- and micronutrients by biochemical and physiochemical activities. The last-named activities include concentration, synthesis of absent or deficient components, enzymatic conversion of refractile or conjugated components, and detoxification of interfering components.

Although the production and deliberate release of genetically engineered microbes in agricultural habitats are a relatively recent technological development, there are already several examples of the way such releases could potentially lead to insect outbreaks. For example, genetically engineered "ice minus" bacteria have been sprayed onto strawberry and potato plants in California. The engineered bacteria, unlike the wild type, do not enhance ice formation by acting as nucleating agents, and they thereby increase the number of frost-free days for crop production. It is not known whether insects that are sprayed or that consume the bacteria also develop increased temperature tolerance. In addition, it is not known whether there could be horizontal genetic transmission of this trait between the engineered bacteria and microbes associated with important potato insect pests. If either situation existed, the insect pest, but perhaps not its natural enemies, could undergo a range shift, leading to potential outbreaks in the new area.

In other planned experiments, the gene for one of the *Bacillus thuringiensis* toxins is being inserted into a common plant surface bacterium (*Pseudomonas syringae*) and into several species of common free-living soil bacteria, which will then be sprayed onto crop plants and the soil. A herbivore consuming the crop or coming into contact with the soil bacteria with the *Bt* toxin gene would presumably die. Although at present most *Bt* toxins are rather specific to particular insect groups and many are not very toxic, part of the biotechnology effort in this area is devoted to finding isolates with much stronger toxins and wider host ranges. When bacteria with the genes for producing these toxins are sprayed in an agricultural habitat, there could be dramatic and hard to predict effects on the ecological interaction among host plants, herbivores, symbionts, and natural enemies. Even assuming (1) no horizontal genetic transmission and (2) stability of the engineered genetic change, one can describe various complex effects that could result in insect pest outbreaks in nearby unsprayed agricultural habitats. In addition, if the bacteria successfully colonized important weeds that had previously been controlled by herbivores (a situation more common in pastures than annual cropping systems), there could be an increase in the abundance of particular weed species. Studies of the movement and fate of bioengineered organisms are important in determining possible hazards associated with the deliberate release of such organisms, but such work is only in the beginning stages (Andow, 1984).

In addition to the insertion of the *Bt* gene for toxin into microbes, a considerable effort is being directed at inserting the gene directly into crop plants. This might be an ecologically safer route, since crops are by nature much more contained; they do not spread through the environment as microbes do. For some crops, such as cotton, with economically serious lepidopterous pests, if the crop itself were to produce the *Bt* toxin this might significantly reduce pest damage. However, it is important to point out that the insecticidal activity of *Bt* results from a simple, special chemical toxin produced by Bt, just like a pesticide, and the toxicity of this compound probably has no greater evolutionary longevity than a pesticide.

The general point here is that insect pests are involved in complex associations with microbes that influence the crop–herbivore–natural-enemy interaction. It has long been known that host and range shifts of herbivores can lead to outbreaks. Such shifts could be unwittingly facilitated by the release of genetically engineered microbes in or near agricultural habitats. Because of the extremely important role of microbes in mediating herbivorous insect–host plant interactions, care should be exercised in the deliberate application of genetically engineered organisms in agricultural habitats.

The recent increase in the number of planned releases of biotechnology products and concern over the possible ecological consequences have resulted in several conferences and preliminary reviews that deal with the issue (Rissler, 1984; Sharples, 1982; Gillet *et al.*, 1984; Brown *et al.*, 1984; Regal, 1985). Some of their recommendations for basic research, if followed, could significantly decrease the probability that the release of bioengineered products, especailly microorganisms, would lead to insect outbreaks in agroecosystems. In general, they conclude that more research is needed in the following areas, and that before a particular release is planned, special attention be given to the following possible adverse effects:

Genetic

1. Rate and nature of horizontal (infections) genetic transmission among microbes and between microbes and higher organisms
2. Stability of the engineered genetic change(s) (role of movable genetic elements)

Evolutionary

1. Likelihood and nature of host range shifts
2. Likelihood of unregulated propagation
3. Likelihood of changes in the virulence of parasites and pathogens

Ecological

1. Effects on competitors
2. Effects on prey, hosts, and/or symbionts
3. Effects on predators, parasites, and/or pathogens

IV. SUMMARY

Although there is no literature review demonstrating unequivocally that insect outbreaks are more common in agroecosystems than in natural ecosystems, it is a belief widely held by both ecologists and applied entomologists. To the extent that it is true, the most important contributing factors are (1) the reduced chemical and physical defenses of crop plants compared with those of closely related wild species and (2) the "simplification" of agroecosystems compared with natural ecosystems (including a decrease in inter- and intraspecific diversity within the planted field and an increase in the uniformity of crop population age structure and physical quality). Recent changes in agriculture that have contributed to an increase of outbreaks in agricultural systems include the enlargement and aggregation of fields and an increase in plant density. The factors that appear to trigger insect outbreaks in particular planted fields mainly include conditions caused directly or indirectly by human intervention (spraying of pesticides, introduction of exotics, particular irrigation and fertilization regimes, release of airborne pollutants, etc.). Recent advances in biotechnology have greatly enhanced our ability to manipulate genetically the organisms associated with agroecosystems. Some of these manipulations could increase the probability of insect outbreaks. Weather, which remains largely outside human control, is considered by many to be the most significant factor induding insect outbreaks. It may play a greater role in agroecosystems than in natural systems because agroecosystems, especially annual ones, are often less buffered against climate change, and crop plants have been specifically bred so that small changes in temperature and precipitation can be reflected in the growth and development of the plants.

ACKNOWLEDGMENTS

I thank R. Carroll, K. S. Hagen, E. A. Bernays, A. P. Gutierrez, M. A. Altieri, D. Andow, and R. Colwell for help in preparing the manuscript.

REFERENCES

Altieri, M. A., and Letourneau, D. K. (1984). Vegetation diversity and outbreaks of insect pests. *CRC Crit. Rev. Plant Sci.* **2,** 131–169.

Ames, B. N. (1983). Dietary carcinogens and carcinogens. *Science* **221,** 1256–1264.

Andow, D. (1983). The extent of monoculture and its effects on insect pest populations with particular reference to wheat and cotton. *Agric., Ecosyst. Environ.* **9,** 25–35.

Andow, D. (1984). Dispersal of microorganisms with novel genotypes. *In* "Potential Impacts of Environmental Release of Biotechnology Products: Assessments, Regulation and Research Needs" (J. Gillet, S. A. Levin, M. A. Harwell, M. Alexander, D. A. Andow, and A. M. Stern, eds.), pp. 63–184. Ecosyst. Res. Cent. Publ., Cornell University, Ithaca, New York.

Ayre, G. L., and Hitchon, D. E. (1968). The predation of tent caterpillars, *Malacosoma americana,* (Lepidoptera: Lasiocampidae) by ants. *Can. Entomol.* **100,** 823–826.

Bach, C. E. (1980a). Effects of plant density and diversity on the population dynamics of a specialist herbivore, the striped cucumber beetle, *Acalymma vitata* (Fab.). *Ecology* **61,** 1515–1530.

Bach, C. E. (1980b). Effects of plant diversity and time of colonization on an herbivore–plant interaction. *Oecologia* **44,** 319–326.

Bach, C. E. (1981). Host plant growth form and diversity: Effects on abundance and feeding preference of a specialist herbivore, *Acalymina vittata* (Coleoptera: Chrysomelidae). *Oecologia* **50,** 370–375.

Brown, J. H., Colwell, R. K., Lenski, R. E., Levin, B. R., Lloyd, M., Regal, P. L., and Simberloff, D. (1984). Report on workshop on possible ecological and evolutionary impacts of bioengineered organisms released into the environment. *Bull. Ecol. Soc. Am.* **65,** 436–439.

Bugg, R. L. (1985). The use of a nectar-bearing weed to enhance biological control by big-eyed bugs (*Geocoris* spp.), important natural enemies of major agricultural pests. Ph.D. Dissertation, Dep. Entomol., Univeristy of California, Davis (unpublished).

Cameron, E. A., McManus, M. L., and Mason, C. J. (1979). Dispersal and its impact on the population dynamics of the gypsy moth in the USA. *Bull. Soc. Entomol. Suisse* **52,** 169–179.

Carroll, C. R., and Hoffman, C. (1980). Chemical feeding deterent mobilized in response to insect herbivory and counter adaptation by *Epilachna tredecimnota. Science* **209,** 414–416.

Chabbonseau, F. (1966a). Die Vermehrung der Milben als Foge der Verwendung von Pflanzenschutzmitteln und de biochische Veranderungen, die diese auf die Pflanze ausuben. *Z. Angew. Zool.* **53,** 257–276.

Chabbonseau, F. (1966b). Noeaux apects de la phytiatrie et de la phytopharmacie, le phenoène de la trophobiose. *Proc. FAO-Symp. Integr. Pest Control, 1965,* Vol. 1, pp. 33–61.

Chelliah, S., and Heinrichs, E. A. (1980). Factors affecting insecticide-induced resurgence of the brown planthopper, *Nilarvata lugens,* on rice. *Environ. Entomol.* **9,** 773–777.

Croft, B. A., and Morse, J. G. (1979). Research advances on pesticide resistance in natural enemies. *Entomophaga* **24,** 3–11.

Croft, B. A., and Strickler, K. (1983). Natural enemy resistance to pesticides: Documentation, characerization, theory and application. *In* "Pest Resistance to Pesticides" (G. P. Georghion and T. Saito, eds.), pp. 669–702. Plenum, New York.

Cromartie, W. J. (1981). Environmental control of insects using crop diversity. *In* "CRC Handbook of Pest Management" (D. Pimentel, ed.), Chapter 48. CRC Press, Boca Raton, Florida.

DaCosta, C. P., and Jones, C. M. (1971). Cucumber beetle resistance and mite susceptibility controlled by the bitter gene in *Cucumis sativus* L. *Science* **172,** 1145–1146.

DeBach, P. (1974). "Biological Control by Natural Enemies." Cambridge Univ. Press, London and New York.

Dittrich, V., Streibert, P., and Bathe, P. A. (1974). An old case reopened: Mite stimulation of insecticide residues. *Environ. Entomol.* **3,** 534–540.

Eastop, V. F. (1981). The wild hosts of aphid pests. *In* "Pests, Pathogens and Vegetation" (J. M. Thresh, ed.), pp. 285–298. Pitman, London.

Elton, C. S. (1958). "The Ecology of Invasions by Animals and Plants." Metheun, London.

Feeny, P. P. (1976). Plant apparency and chemical defense. *Recent Adv. Phytochem.* **10,** 1–40.

Feeny, P. P., Paauwe, R. L., and Demong, N. J. (1970). Flea beetles and mustard oils: Host plant specificity of *Phyllotreta cruciferae* and *P. striolata* adults (Coleoptera: Chrysomelidae). *Ann. Entomol. Soc. Am.* **63,** 832–841.

Gillet, J., Levin, S. A., Harwell, M. A., Alexander, M., Andow, D. A., and Stern, M. A., eds. (1984). Potential Impacts of Environmental Release of Biotechnology Products: Assessment, Regulation and Research Needs." Ecosyst. Res. Cent. Publ., Cornell University, Ithaca, New York.

Goodman, D. (1975). The theory of diversity–stability relationships in ecology. *Q. Rev. Biol.* **50,** 237–266.

Gordon, H. T. (1984). Growth and development of insects. *In* "Ecological Entomology" (C. B. Huffaker and R. L. Rabb, eds.), pp. 53–77. Wiley (Interscience), New York.

Gutierrez, A. P., and Yaninek, S. J. (1983). Responses to weather of eight aphid species commonly found in pastures of southeastern Australia. *Can. Entomol.* **115,** 1359–1364.

Gutierrez, A. P., Morgan, D. J., and Havenstein, D. E. (1971). The ecology of *Aphis craccivora* Koch and subterranean clover stunt virus. I. The phenology of aphid populations and the epidemiology of virus in pastures in S.E. Australia. *J. Appl. Ecol.* **8,** 699–721.

Gutierrez, A. P., Havenstein, D. E., Nix, H. A., and Moore, P. A. (1974a). The ecology of *Aphis craccivora* Koch and subterranean clover stunt virus. II. A model of cow pea aphid populations in temperate pastures. *J. Appl. Ecol.* **11,** 1–20.

Gutierrez, A. P., Havenstein, D. E., Nix, H. A., and Moore, P. A. (1974b). The ecology of *Aphis craccivora* Koch and subterranean clover stunt virus in southeastern Australia. III. A regional perspective on the phenology and migration of the cowpea aphid. *J. Appl. Ecol.* **11,** 21–35.

Hagen, K. S., Bambosch, S., and McMurtry, J. A. (1976). The biology and impact of predators. *In* "Theory and Practice of Biological Control" (C. B. Huffaker and P. S. Messenger, eds.), pp. 93–130. Academic Press, New York.

Heinrichs, E. A., Aquino, G. B., Valencia, S. L., and Reissig, W. H. (1982). Resurgence of *Nilaparvata lugens* (Stal) populations as influenced by method and timing of insecticide applications in lowland rice. *Environ. Entomol.* **11,** 788–84.

Heybroek, H. M. (1982). Monoculture versus mixture: Interaction between susceptible and resistant trees in a mixed stand. *In* "Resistance to Disease and Pests in Forest Trees" (H. M. Heybroek, B. R. Stephan, and K. von Weissenery, eds.), Proc. 3rd Int. Workshop on the Benefits of Host-Parasite Interaction in Forestry. Pudoc, Wageningen.

Huffaker, C. B., and Messenger, P. S., eds. (1976). "Theory and Practice of Biological Control." Academic Press, New York.

Hughes, R. D., Jones, R. E., and Gutierrez, A. P. (1984). Short-term patterns of population change: The life system approach to their study. *In* "Ecological Entomology" (C. B. Huffaker and R. L. Rabb, eds.), pp. 309–357. Wiley (Interscience), New York.

Jones, C. G. (1984). Microorganisms as mediators of plant resource exploitation by insect herbivores. *In* "A New Ecology: Novel Approaches to Interactive Systems" (P. W. Price, C. N. Slobodchikoff, and W. S. Good, eds.), pp. 54–84. Wiley, New York.

Jones, C. G., and Sterling, W. L. (1978). Locomotory activity and distribution of wintering boll weevils in east Texas litter. *Southwest. Entomol.* **3,** 315.

Lewis, T. (1965). The effect of an artificial windbreak on the distribution of aphids in a lettuce crop. *Ann. Appl. Biol.* **38,** 513.

Lewis, T. (1970). Patterns of distribution of insects near a windbreak of tall trees. *Ann. Appl. Biol.* **65,** 213.

MacArthur, R. H. (1955). Fluctuations of animal populations, and a measure of community stability. *Ecology* **36,** 533–536.

Margalef, R. (1968). "Perspectives in Ecological Theory." Univ. of Chicago Press, Chicago, Illinois.

Mason, C. J., and McManus, M. L. (1981). Larval dispersal of the gypsy moth. *U.S., Dep. Agric., Tech. Bull.* **1584,** 161–202.

Maxwell, R. C., and Harwood, R. F. (1960). Increased reproduction of pea aphids on broad beans treated with 2,4-D. *Ann. Entomol. Soc. Am.* **53,** 199–205.

May, R. M. (1973). "Stability and Complexity in Model Ecosystems." Princeton Univ. Press, Princeton, New Jersey.

Miyashita, K. (1963). Outbreaks and population fluctuations of insects, with special reference to agricultural insect pests in Japan. *Bull. Natl. Inst. Agric. Sci. Ser.* **C15,** 99–170.

Murdoch, W. W. (1975). Diversity, complexity, stability, and pest control. *J. Appl. Ecol.* **12,** 795–807.

National Academy Press (NAS) (1984). "Genetic Engineering of Plants: Agricultural Research Opportunities and Policy Concerns." NAS, Washington, D.C.

Odum, E. P. (1971). "Fundamentals of Ecology," 3rd ed. Saunders, Philadelphia, Pennsylvania.

Office of Technology Assessment (OTA) (1984). "Impacts of Applied Genetics: Microorganisms, Plants and Animals." OTA, Washington, D.C.

Oka, I. N., and Pimentel, D. (1974). Corn susceptibility to corn leaf aphids and common corn smut after herbicide treatment. *Environ. Entomol.* **3,** 911–915.

Oka, I. N., and Pementel, D. (1976). Herbicide (2,4-D) increases insect and pathogen pests on corn. *Science* **193,** 239–240.

Pimentel, D. (1961a). The influence of plant spatial patterns on insect populations. *Ann. Entomol. Soc. Am.* **54,** 61–69.

Pimentel, D. (1961b). Species diversity and insect population outbreaks. *Ann. Entomol. Soc. Am.* **54,** 76–86.

Pimentel, D. (1985). Native and exotic insect pests as invaders. *In* "Ecological Consequences of Biological Invasions" (H. Mooney, ed.). Springer-Verlag, Berlin and New York.

Price, P. W. (1976). Colonization of crops by arthropods: Non-equilibrium communities in soybean fields. *Environ. Entomol.* **5,** 605.

Price, P. W., Slobodchikoff, C. N., and Good, W. S., eds. (1984). "A New Ecology: Novel Approaches to Interactive Systems." Wiley, New York.

Rankin, M. A., and Singer, M. C. (1984). Insect movement: Mechanism and effects. *In* "Ecological Entomology" (C. B. Huffaker and R. L. Rabb, eds.), pp. 185–216. Wiley (Interscience), New York.

Regal, P. J. (1985). Potential ecological impact of genetically engineered organisms. *In* "Ecological Consequences of Biological Invasions" (H. Mooney, ed.). Springer-Verlag, Berlin and New York.

Reissig, W. H., Heinrichs, G. A., and Valencia, S. L. (1982). Effects of insecticides on *Nilaparvata lugens* and its predators: Spiders, *Microvelia atrolineata* and *Cyrtohinus lividipennis*. *Environ. Entomol.* **11,** 193–199.

Rhoades, D. F. (1979). Evolution of plant chemical defense against herbivores. *In* "Herbivores: Their Interaction with Secondary Plant Metabolites" (G. A. Rosenthal and D. H. Janzen, eds.), pp. 4–48. Academic Press, New York.

Rhoades, D. F. (1983). Herbivore population dynamics and plant chemistry. *In* "Variable Plants and Herbivores in Natural and Managed Systems" (R. F. Denno and M. S. McClure, eds.), pp. 155–204. Academic Press, New York.

Risch, S. J. (1980). The population dynamics of several herbivorous beetles in a tropical agroecosystem: The effect of intercropping corn, beans and squash in Costa Rica. *J. Appl. Ecol.* **17,** 593–612.

Risch, S. J. (1981). Insect herbivore abundance in tropical monocultures and polycultures: An experimental test of two hypothesis. *Ecology* **62,** 1325–1340.

Risch, S. J. (1984). The biological control alternative: Social impediments to faster development and implementation. *In* "Agriculture, Change, and Human Values" (R. Haynes and R. Lanier, eds.), Vol. 2, pp. 569–580. Humanities and Agriculture Program, Gainesville, Florida.

Risch, S. J., and Carroll, C. R. (1986). Effects of seed predation by a tropical ant, *Solonopsis geminata,* on competition among weeds. *Ecology* (in press).

Risch, S. J., Andow, D., and Altieri, M. (1983). Agroecosystem diversity and pest control: Data, tentative conclusions and new research directions. *Environ. Entomol.* **12,** 625–629.

Rissler, J. F. (1984). Research needs for biotic environmental effects of genetically-engineered microorganisms. *Recomb. DNA Tech. Bull.* **7,** 20–30.

Rodriguez, J. G., Chen, H. H., and Smith, W. T., Jr. (1957). Effects of soil insecticides on beans, soybeans and cotton and resulting effect on mite nutrition. *J. Econ. Entomol.* **50,** 587–593.

Rodriquez, J. G., Maynard, D. E., and Smith, W. T., Jr. (1960). Effects of soil insecticides and absorbents on plant sugars and resulting effect on mite nutrition. *J. Econ. Entomol.* **53,** 491–495.

Root, R. B. (1973). Organization of a plant–arthropod association in simple and diverse habitats: The fauna of collards (*Brassica oleracea*). *Ecol. Monogr.* **43,** 94–125.

Root, R. B., and Skelsey, J. J. (1969). Biotic factors involved in crucifer aphid outbreaks following insecticide application. *J. Econ. Entomol.* **62,** 223–232.

Schoener, A. (1974). Experimental zoogeography: Colonization of marine mini-islands. *Am. Nat.* **108,** 715.

Secoy, D. M., and Smith, A. E. (1983). Use of plants in control of agricultural and domestic pests. *Econ. Bot.* **37,** 28–57.

Sharples, F. E. (1982). Spread of organisms with novel genotypes: Thoughts from an ecological perspective. *Oak Ridge Natl. Lab.* [*Rep.*] *ORNL-TM* (*U.S.*) **ORNL-TM-8473,** Publ. No. 2040, 1–47.

Simberloff, D. (1983). Competition theory, hypothesis testing and other community ecological buzzwords. *Am. Nat.* **122,** 626–635.

Sitowski, L. (1924). Strzygonia choinowka (*Panolis flammea* Schiff) i jej pasorzyty na ziemiach polskich. Czesc II. *Rocz. Nauk Roln. Lesn.* **12,** 18.

Smith, R. F., and van den Bosch, R. (1967). Integrated control. *In* "Pest Control: Biological, Physical, and Selected Chemical Methods" (W. W. Kilgore and R. I. Doutt, eds.), pp. 295–340. Academic Press, New York.

Stimac, J. C., and Buren, W. L. (1982). "Population Resurgence of the Imported Fire Ant, *Solenopsis invicta* Buren," CRIS Rep. No. 0084603. USDA-CSRS, Washington, D.C.

Strong, D. R. (1983). Natural variability and the manifold mechanisms of ecological communities. *Am. Nat.* **122,** 636–660.

Strong, D. R. (1984). Exorcising the ghost of competition past from insect communities. *In* "Ecological Communities: Conceptual Issues and the Evidence" (D. R. Strong, D. Simberloff, and L. G. Abele, eds.), pp. 28–41. Princeton Univ. Press, Princeton, New Jersey.

Strong, D. R., Lawton, J. H., and Southwood, R. (1984). "Insects on Plants: Community Patterns and Mechanisms." Harvard Univ. Press, Cambridge, Massachusetts.

Tepedino, V. J., and Stanton, N. L. (1976). Cushion plants as islands. *Oecologia* **25,** 243.

Trenbath, B. R. (1977). Interactions among diverse hosts and diverse parasites. *Ann. N.Y. Acad. Sci.* **287,** 124–150.

U.S. Department of Agriculture (USDA) (1973). "Monoculture in Agriculture, Causes and Problems," Report of the Task Force on Spatial Heterogeneity in Agricultural Landscapes and Enterprises. U.S. Govt. Printing Office, Washington, D. C.

van Emden, H. G. (1981). Wilds plants in the ecology of insect pests. *In* "Pests, Pathogens and Vegetation" (J. M. Thresh, ed.), p. 251. Pitman, London.

Vinson, S. B. (1981). Habitat location. *In* "Semiochemicals" (D. A. Nordlund, R. Jones, and W. J. Lewis, eds.), pp. 52–68. Wiley, New York.

Watt, A. D. (1981). Wild grass and the grain aphid (*Sitobion avenae*). *In* "Pests, Pathogens, and Vegetation" (J. M. Thresh, ed.), p. 299. Pitman, London.

Wellington, W. G., and Trimble, R. M. (1984). Weather. *In* "Ecological Entomology" (C. B. Huffaker and R. L. Rabb, eds.), pp. 399–425. Wiley (Interscience), New York.

Weseloh, R. M. (1981). Host location by parasitoids. *In* "Semiochemicals" (D. A. Nordlund, R. Jones, and W. J. Lewis, eds.), pp. 79–90. Wiley, New York.

White, T. C. R. (1974). A hypothesis to explain outbreaks of looper caterpillars, with special reference to populations of *Selidosema suavis* in a plantation of *Pinus radiata* in New Zealand. *Oecologia* **16,** 279–301.

White, T. C. R. (1976). Weather, food and plagues of locusts. *Oecologia* **22,** 119-134.

White, T. C. R. (1978). The importance of a relative shortage of food in animal ecology. *Oecologia* **33,** 71–86.

White, T. C. R. (1984). The abundance of invertebrate herbivores in relation to the availability of nitrogen in stressed food plants. *Oecologia* **63,** 90–105.

Part III

BIOTIC AND ABIOTIC FACTORS IN INSECT OUTBREAKS

Chapter 10

The Role of Climatic Variation and Weather in Forest Insect Outbreaks

PETER J. MARTINAT[1]

Department of Entomology
University of Maryland
College Park, Maryland 20740

I. INTRODUCTION

This chapter examines the theory that outbreaks of forest insects are triggered by climatic anomalies. In the 40 years since the theory of climatic release was first postulated, important advances have been made in our understanding of the atmosphere as a dynamic system. I ask whether the theory is consistent with modern climatology and testable. Prior to a

[1] Present address: Chemical and Agricultural Product Division, Abbott Laboratories, North Chicago, Illinois 60064.

critical examination of the theory, a historical review summarizes the place that weather and climate have had in the development of insect population dynamics theory. An approach to developing hypotheses based on outbreak characteristics concludes the chapter.

Meteorology textbooks define "weather" as short-term variation of the atmosphere or as the state of the atmosphere at a given time with respect to temperature, pressure, wind, moisture, cloudiness, and precipitation. "Climate" is usually defined as the statistical collective of the weather of a specified area during a specific interval of time or as the prevailing or average weather conditions of an area over a long period. In this chapter, a "weather variable" is considered to be precipitation, temperature, humidity, and so on, as recorded at a weather station. For a given weather variable a "mean" (for temperature, humidity, etc.) or "total" (for precipitation) specifies a time of year, duration, and place (e.g., "New England July mean temperature" or "total summer precipitation"). "Climate" similarly refers to the combined effects of weather variables at a given time period and place (e.g., "the July climate in New England" or "the summer climate in New England"). A deviation from a long-term condition (if it is climatic) or a mean (if it is a weather variable) is referred to as an "anomaly." A part of the year and a duration are specified. Thus, a "5-year July temperature anomaly" is a departure in the same direction from the mean July temperature lasting five consecutive years. A 1- to 10-year deviation from a normal climatic condition is referred to as a "climatic anomaly." A "climatic trend" is a deviation lasting more than 10 years.

"Infestation" refers to the sudden appearance of visible damage (i.e., defoliation or dead timber) in a small continuous area caused by a high population of a forest insect herbivore. "Outbreak" refers to the simultaneous appearance of two or more disjunct infestations.

II. HISTORICAL BACKGROUND

The role of weather and climate in the control of insect abundance entered the "density-dependent" versus "density-independent" debate at an early stage. Uvarov (1931), the first to undertake a comprehensive review of the subject, rejected the theory that insect populations fluctuate around a stable equilibrium. He also rejected the idea that the principal controlling factors of a population are density-dependent natural enemies and competition for limited resources. Instead, he believed that "the key to the problem of balance in nature is to be looked for in the influence of climatic factors . . . which cause a regular elimination of an enormous percentage of individuals (even) under so-called normal conditions which are such that . . . insects survive them not because they are perfectly adapted to them, but only owing to their often fantastically high reproductive abilities.'' Andrewartha and Birch (1954) hypothesized that insects are limited by a shortage of time during which the weather is favorable enough to allow for population increase, and therefore the carrying capacity of the environment is never reached. These views contrast with those of Nicholson (1933, 1947), Smith (1935), Elton (1949), and Klomp (1962), who believed that climate and weather influence the distribution of animals (i.e., which habitats are occupied) but do not regulate or control the population within suitable habitats.

The controversy between the two points of view may never be resolved, but some reviews suggest that differences are primarily semantic (Thompson, 1956; Richards, 1961; Morris, 1964; Ives, 1981). Ehrlich and Birch (1967) rejected altogether the concept of ''balance'' as a fiction in nature.

Theory of Climatic Release

A general theory that relates outbreaks of forest insect herbivores to climatic anomalies was first suggested by Graham (1939) and later developed by Wellington (1954a,b) and Greenbank (1956, 1963). The theory postulates that the fluctuations in the abundance of many forest insect herbivores are under long-term climatic control. If a climatic anomaly that favors an increase in fecundity and/or survival persists over several consecutive generations, its effects on the population may be multiplicative. After 3 or 4 years of a continuous increase in density, the ''released'' population may cause noticeable defoliation. The anomaly responsible for the favorable conditions would thus occur before defoliation (Fig. 1). Graham (1939) conceptualized a favorable climatic anomaly as a ''temporary lowering of environmental resistance,'' which allows a forest insect to realize its tremendous reproductive ability. Wellington (1954b) believed that climatic anomalies are predictable and orderly and are therefore a strong and consistent signal to which forest insect populations will respond.

The concept of population ''release'' apparently originated with Solomon (1949), who believed that under certain circumstances natural control may become ''disorganized.'' This would allow a population temporary escape from its important controlling agents, or ''key factors,'' leading to an increase in number. This suggested another possibility, which later researchers, such as Morris (1963), exploited—namely, that release may be the result of conditions that affect the key factors. Implicit in this model is the bimodal population consisting of four distinct phases:

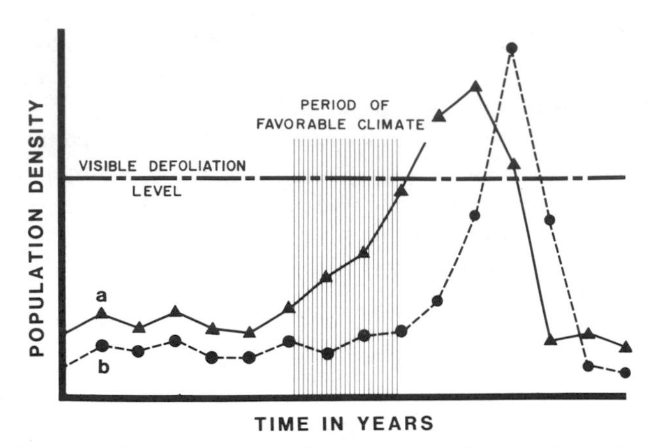

Fig. 1. Diagrammatic representation of the theory of climatic release. Both time scale and population density scale are arbitrary. (a) Defoliator population; (b) density-dependent population regulator.

an endemic or innocuous phase, during which numbers are low and stable; a release phase, during which favorable conditions allow population increase; an outbreak phase, in which damaging defoliation occurs; and a population collapse phase. It is a pattern that exhibits regular, often predictable cycles and is typical of many forest insects (Campbell and Sloan, 1978; Mason and Luck, 1978; Berryman and Stark, 1985).

III. MECHANISMS BY WHICH WEATHER CAUSES CHANGES IN FOREST INSECT ABUNDANCE

Weather has both direct and indirect effects on phytophagous forest insect populations (Fig. 2A). Direct effects of weather on behavior and physiology (Fig. 2B) are well documented, and by now there exists a vast literature (Uvarov, 1931, 1957; Bursell, 1974a,b; Tauber and Tauber, 1978). Most population studies examining direct effects are restricted to a single stage or generation in the life history of the insect. Few attempt to relate weather conditions to changes in density between generations. It is nevertheless generally believed that atypical or anomalous weather is directly responsible for widespread changes in the abundance of many forest insects, although the mechanisms are rarely understood in detail.

More recent hypotheses suggest that indirect effects on insects through effects on host plants and natural enemies may be more important than direct effects. Weather may influence the level of stress in the host plant, which in turn may alter its nutritional quality, chemical defenses, or di-

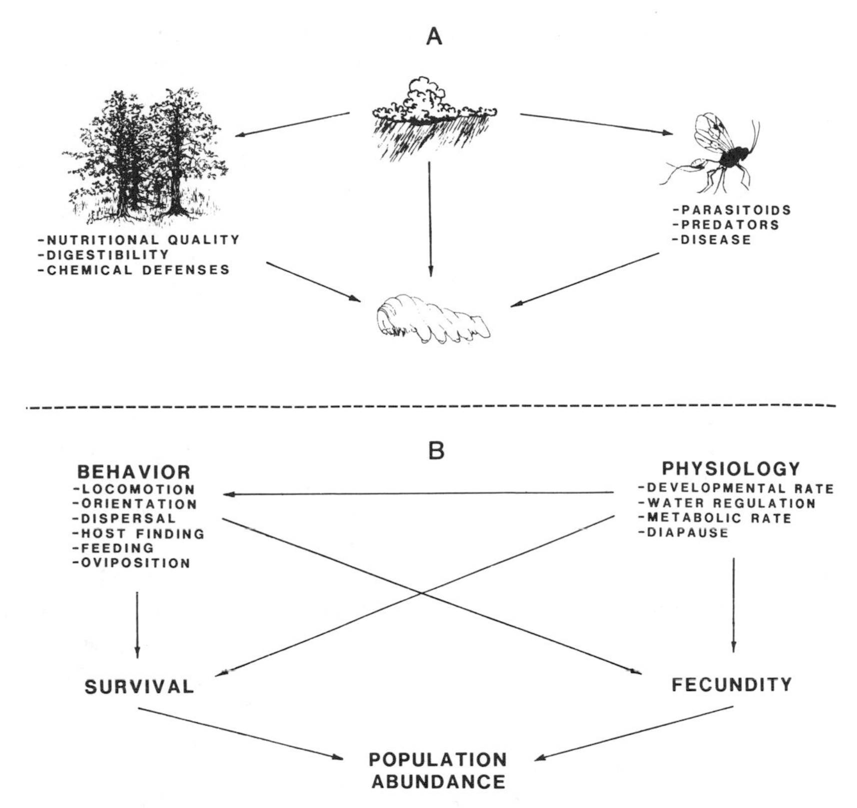

Fig. 2. Summary of the ways weather can influence the abundance of a forest insect herbivore. A, Weather can act either directly on the herbivore or indirectly through its effects on natural enemies or host plants. B, Both direct and indirect effects alter either the herbivore's behavior or its physiology and, ultimately, alter population abundance through its effect on survival or fecundity.

gestibility. White (1969) found that outbreaks of *Cardiaspina densitexta* and other psyllids in Australia were correlated with moisture-induced stress in host plants. Similarly, outbreaks of several species of loopers in New Zealand, South Africa, the Netherlands, and North America (White, 1974) and of desert locusts (White, 1976) seemed to be related to a pattern of rainfall that could have stressed the attacked plants. White hypothesized that defoliator populations may typically suffer high mortality or poor fecundity due to an insufficiency of nitrogen in their food. However, when plants are physiologically stressed due to an insufficiency or excess of water, there may be a drop in protein synthesis, which in turn may lead to an increase of available nitrogen in their aerial parts. If plants occasionally become a richer nitrogen source during periods of drought or excess precipitation, a greater number of herbivores may survive and an outbreak may result. Stress might also affect the ability of a plant to produce defensive compounds such as tannins and resins (Green and Ryan, 1972; Haukioja and Niemela, 1977; Rhoades, 1979). Other causes of plant stress that have been associated with insect outbreaks are discussed in Knight and Heikkenen (1980).

The fact that defoliation may induce the production of host plant defensive compounds suggests a mechanism whereby host plant induction interacts with climate to control the long-term abundance of a forest insect defoliator. Such induction may render the plant unpalatable (Walker-Simmons and Ryan, 1977; Wallner and Walton, 1979; Haukioja, 1980; Schultz and Baldwin, 1982). A decline in food quality due to defoliation not only might contribute to outbreak collapse, but might continue to suppress herbivore populations for many years until the host plant recovers full health or ceases to produce defensive compounds (see Chapter 16, this volume). At this point, the herbivore population might begin another outbreak cycle when a season of favorable weather occurs. Thus, outbreaks may occur when weather perturbs the subtle and complex balance between defensive and offensive mechanisms of both host plants and herbivores (Rhoades, 1983).

Other mechanisms by which forest insect herbivores might be indirectly affected by weather have received less attention. These include effects on natural enemies and disease epizootics. Parasitoid behavior and oviposition rates are affected by temperature, humidity, and wind (Juillet, 1964; Barney *et al.*, 1979; Barbosa and Frongillo, 1979; Ahmed *et al.*, 1982; Jackson and Butler, 1984). If a particular stage of a forest insect herbivore is vulnerable to parasitoid attack for a short duration, it may be possible for weather to control herbivore density by acting on its parasitoids. For example, egg parasitism can by very high, even in herbivore populations of relatively low density (Allen, 1972; Ticehurst and Allen, 1973; Kaya and Anderson, 1972, 1974; Surgeoner, 1976). If egg parasitoids are sensitive to inclement weather during the short period in which eggs are available for attack, parasitism rates may be drastically reduced.

The role of disease epizootics in the collapse of outbreaks is well documented (Tanada, 1959; Stairs, 1972; Anderson and May, 1980); and epizootics may be triggered by factors related to weather. However, little is known about the extent to which pathogenic microorganisms might contribute to the release of insect herbivore populations that are at an endemic level. To a certain extent, the virulence of a microorganism is controlled by weather conditions. For example, high humidity is optimal for the germination of fungal spores, temperature sometimes playing an important role (see references cited in Tanada, 1959). Microclimatic conditions unfavorable to the microorganism thus may suppress an important agent of herbivore mortality.

Many microorganisms persist in the insect in a latent state and are transmitted to the next generation through the eggs. Although latent, they may have sublethal effects, such as retardation of development, lowered fecundity, and reduced viability of eggs (Thompson, 1960), or cause morphological changes (Magnoler, 1974). The extent to which these sublethal effects are influenced by weather has not been adequately investigated.

IV. THREE CASE HISTORIES

Table 1 summarizes recent studies that explain outbreaks of one or more species of forest insect by means of a climatic release hypothesis. Uvarov (1931) reviewed many older papers. As examples of how researchers have attempted to use data from field studies to support climate release hypotheses, two well-documented cases will be reviewed: the spruce budworm, *Choristoneura fumiferana* (Clemens) (Lepidoptera: Tortricidae), and the forest tent caterpillar, *Malacosoma disstria* (Hubner). The review of work on the southern pine beetle, *Dendroctonus frontalis* (Zimmerman) (Coleoptera: Scolytidae), illustrates how climatic variables, along with other variables, are incorporated into a regionwide outbreak prediction model.

A. Spruce Budworm

Graham (1939) was apparently the first to suggest that outbreaks of the spruce budworm might be triggered by favorable climatic conditions. Wellington *et al.* (1950), Wellington, (1952, 1954a,b), Greenbank (1956, 1957, 1963), and Ives (1974) provided supporting evidence. Initially, the hypothesis was based on the observation that sunny, dry weather in the spring enhances continuous larval feeding, whereas cool, cloudy, and rainy conditions interrupt or prolong feeding. Greenbank (1956) showed that the correlations between the duration of larval stadia and weather conditions were high, such that larval development is completed during a shorter period of time with higher average temperatures and reduced cloudiness and rainfall. Since larval development time is positively correlated with pupal size and fecundity, these weather conditions should result in higher populations the next year. Evidence that higher temperatures increase survival was provided by Morris's (1963) "key-factor" analysis. He obtained a correlation of .69 between the mean daily maximum tem-

TABLE 1

Studies Attempting to Relate Forest Insect Outbreaks to Climate Conditions

Insect, location (reference)[a]	Index of population change	Outbreak criterion	Postulated duration of release period	Number of outbreaks in study
Choristoneura fumiferana, eastern Canada (1)	Tree growth rings	First year of reduced tree increment growth	3–4 years	8[c]
C. fumiferana, New Brunswick (2)	Defoliation level	First year of heavy defoliation	Not specified	2
C. fumiferana, central Canada (3)	CFIDS[d]	Year of maximum increase and maximum decrease in abundance	Not specified	57[c]
C. occidentalis, Washington (4)	Aerial survey of defoliation, beating, foliage samples	Relative intensity of defoliation	3 years	3[c]
C. occidentalis, Idaho, Montana (5)	Aerial survey of defoliation	Defoliation trend	Not specified	1
C. occidentalis, British Columbia (6)	CFIDS,[d] tree growth rings	First year of widespread defoliation	1 year	5[c]
Malacosoma disstria, northern Ontario	Not specified	Not specified	3–4 years	2
M. disstria, Minnesota (8)	Light trap data, 1928–1960	Relative abundance	2–3 years	2
M. disstria, Canadian prairie, Ontario (9)	CFIDS[d]	Year of maximum increase and maximum decrease in abundance	2–4 years	60[c]
Rhyaciona buoliana, Denmark (10)	No. of damage reports	First year of 4-fold increase in defoliation reports	2 years	7
Zeiraphera griseana, Europe (11)	Observed defoliation	Relative intensity of defoliation	4–10 years	3

Weather conditions considered favorable for population increase	Life stage and/or activity affected	Weather variables actually tested[b]	Tests
Warm dry spring, early and continuous snow Nov.–March, steady and continuous cold Nov.–March	Early larval feeding, overwintering hibernacula	Annu. no. cyclon. ctr., Mon. precip.	Graphic
Warm and dry June–July	Larval feeding and development	Mon. precip.	Graphic
Cool fall and early spring, warm winter, warm and dry June–Aug.	Diapausing small larval survival, large larval development and survival	Cum. heat units, Mon. min. temp., Mon. precip., Mon. days w/ precip.	Graphic students' t
Warm and dry spring and summer	Larval feeding	Mon. max. temp., Mon. days w/ precip.	Tabular (nonstatistical)
Warm dry summer	Larval feeding and survival	Mon. days w/ precip., Mon. max. temp.	Graphic, regression chi-square
Warm dry summer	Larval feeding and survival, phenological synchrony	23 weather "parameters"[e]	Graphic
Warm humid cloudy May–June	Larval feeding and development	Annu. no. cyclon. ctr.	Graphic
Continuous warm 3 weeks following eclosion	Early caterpillar feeding	Cum. heat units, Mean temp. after eclosion	Graphic
Continuous cold fall and early spring, continuous warm late spring	Overwintering pharate first instar, larval feeding	Cum. heat units	Graphic
Warm dry June–Sept., warm winter	Oviposition and egg development, early boring and diapausing larva	Mon. temp., Mon. precip.	Graphic
Cool June–Aug., cold winter, warm and humid Mar.–Apr.	Oviposition and egg development, diapausing egg survival, eclosion and early feeding	Mon. temp., Mon. precip.	Graphic, chi-square

(continued)

TABLE 1 (*Continued*)

Insect, location (reference)[a]	Index of population change	Outbreak criterion	Postulated duration of release period	Number of outbreaks in study
Hyphantria cunea, central eastern Canada (12)	CFIDS[d]	Relative abundance of nests	Not specified	4
Dendrolimus punctatus, southern China (13)	Foliage samples	Increase in density per tree	One generation	1
Recurvaria starki, western Canada (14)	Tree ring growth, defoliation	None indicated	6 years (3 generations)	1
Five species of forest Lepidoptera and Hymenoptera, central Europe (15)	Historical documents	Relative defoliation intensity	Not specified	27[f]
Twenty-one species of forest Lepidoptera and Hymenoptera, central Canada (16)	CFIDS[d]	Year of maximum increase and maximum decrease in abundance	Not specified	Not specified
Three species of phasmatids, southeastern Australia (17)	Historical documents, aerial surveys of defoliation	Not specified	2 years	3[g]
Several species of forest Lepidoptera (18)	Historical documents	Not specified	Not specified	Not specified

[a] References: (1) Wellington *et al.* (1950); (2) Greenbank (1956); (3) Ives (1974); (4) Twardus (1980); (5) Hard *et al.* (1980); (6) Thompson *et al.* (1984); (7) Wellington (1952); (8) Hodson (1977); (9) Ives (1973); (10) Bejer-Peterson (1972); (11) Baltensweiler (1966); (12) Morris (1964); (13) Hsiao and Yen (1964); (14) Stark (1959); (15) Cramer (1962); (16) Ives (1981); (17) Readshaw (1965); (18) White (1974).

[b] Annu. no. cyclon. ctr., annual number of cyclonic centers passing through the area of defoliation; Mon. precip., total monthly precipitation; Mon. temp., mean monthly temperature; Mon. days w/ precip., total number of days in a month with measurable precipitation; Mon. max. temp., maximum temperature each day averaged for all the days of the month; Mon. min. temp., minimum temperature each day averaged for all the days of the month; Cum. heat units, cumulative heating or cooling degree-days above or below a specified threshold.

[c] Not distinct outbreaks separated in time. Many may be disjunct infestations or "epicenters" of a large widespread outbreak. Distinction is not clearly made in the study.

[d] Canadian Forest Insect and Disease Survey, begun 1936.

[e] Twenty-three weather variables tested.

[f] Total of 27 outbreaks among five species.

[g] Intensive sampling done for most recent outbreaks.

Weather conditions considered favorable for population increase	Life stage and/or activity affected	Weather variables actually tested[b]	Tests
Warm and humid July–Sept.	Larval development, overwintering survival	Mon. min. temp.	Graphic
Cool humid spring, cool humid Aug., warm late winter	First-generation egg eclosion, second-generation egg eclosion, egg "hibernation"	Mon. temp., Mon. precip., Yearly temp. and precip.	Graphic
Warm winter	Larval "hibernation"	Mon. temp., yearly temp.	Graphic
Different for each species	Not specified	Mon. temp., Mon. precip.	Graphic
Different for each species	Not specified	Seasonal temp., Seasonal precip., Seasonal cum. heat units	Graphic
Cool summers (Sept.–Apr.)	Control of diapausing physiology in eggs	Mean summer temp., (Sept.–Apr.)	Graphic
Generally drought-stress conditions	Control of available nitrogen in host plant	Annual precip., drought-stress index	Graphic

perature for the average large-larval period (June 1 to July 31) and the residuals of a first-order autoregression in larval density.

Greenbank also postulated that during abnormally warm, sunny springs larvae may be developmentally more advanced relative to shoot growth. They would thus feed on tender, more palatable young foliage. In addition, a greater than normal crop of staminate flowers is produced during years of drought, providing larvae with protein-rich pollen. Ives (1974) determined correlations between nine weather variables and changes in spruce budworm abundance for 57 disjunct infestations. His results partially supported the conclusions of Greenbank's hypothesis. During the feeding period of large larvae (third to fifth instar) there was good correlation between temperature and changes in abundance, but not between precipitation and abundance.

Other studies have tried to relate graphically spruce budworm outbreaks to climatic anomalies. Wellington *et al.* (1950) showed that outbreaks were preceded by a reduction, during three or four consecutive years, in the annual number of cyclonic centers passing through areas of infestations, which, he argued, caused a change from abnormally wet, cloudy weather to relatively dry, clear weather. Drier weather also apparently preceded two outbreaks in New Brunswick (Greenbank, 1956). In addition, the weather conditions and the yearly rate of change in budworm populations exhibited coincidental trends. Greenbank concluded that weather controlled budworm populations through its effect on the survival of the large larvae.

Outbreaks of the western spruce budworm, *C. occidentalis* Freeman, whose biology and life history are similar to those of *C. fumiferana,* also appear to coincide with warm, dry summer weather. Hard *et al.* (1980), Twardus (1980), and Thompson *et al.* (1984) concluded that outbreaks of the western spruce budworm in British Columbia were related not only to warm, dry summers but also to an optimal synchrony between larval emergence and bud flush. Outbreak collapse appeared to be related to high autumn temperatures following moth flight.

B. Forest Tent Caterpillar

Unfavorable spring weather appears to be the most important natural control of the forest tent caterpillar. Warm and sunny weather is necessary for the early emergence and initiation of feeding of pharate first instars, but continuous dry, clear weather early in the spring permits larvae to feed only in the daytime (the nights being too cold). Later, in May and June, continuous clear and sunny conditions warm caterpillars too much, and they spend time moving to shaded areas rather than feeding. In contrast, continuous warm, humid, cloudy conditions permit night feeding as well as day feeding and reduce exposure to excessive radiant heating. Consequently, these conditions probably accelerate larval development (Wellington, 1952, 1954b).

Unusually warm weather early in the spring induces premature eclosion and budbreak. When this is followed by a prolonged period of cold, high mortality of young larvae and disruption of phenological synchrony result (Tothill, 1918; Tomlinson, 1938; Sweetman, 1940; Hodson, 1941; Ives, 1981). Blais *et al.* (1955) attributed the collapse of a 111,400-km^2 outbreak in central Canada to similar conditions.

Several studies suggest that forest tent caterpillar population buildup may be favored by opposite conditions. Hodson (1977) found that two outbreaks in Minnesota were preceded by 2 to 3 years of high minimum temperatures, and a low accumulation of degree-days with maximum temperatures below 15°C, for the 3-week period following eclosion in the spring. Wellington (1952) found that, for two outbreaks in northern Ontario, there appeared to be shifts in the general circulation just before outbreak, bringing about an increase in the annual number of warm, wet air masses and a decrease in cold, dry air masses. He believed that these conditions were favorable for forest tent caterpillar.

The overwintering pharate larva also appears to be especially susceptible to the adverse effects of weather. Eggs are laid in masses around twigs of the host tree in July. Embryonic development is completed in about 3 weeks and is followed by a period of obligatory diapause, which lasts about 3 months. Cold weather thereafter keeps the larvae dormant until the spring. Severe, prolonged winter cold can cause high mortality. However, an abnormally warm late fall can cause premature eclosion and larval activity. High mortality occurs when cold weather returns (Hodson and Weinman, 1945; Sippell *et al.*, 1964). A cool fall, cold winter, and warm spring are best for a high degree of survival. Ives (1973) found that years of increasing forest tent caterpillar populations (data from the Canadian Forest Insect and Disease Survey) in the Canadian prairie provinces and Ontario had cooler overwintering periods and warmer early feeding periods than those with decreasing populations. All known infestations in these areas were preceded (in the 4-year period before the infestation) by a single year with a relatively cool winter and an unusually warm spring. In addition, most population collapses were accompanied by cool springs and some by warm winters.

C. Southern Pine Beetle

The causes of rapid growth of pine beetle populations are not clearly understood, but climate apparently plays an important role. Site and stand conditions may predispose a stand to attack, and a climatic anomaly may trigger the outbreak, both by affecting the ability of the host to resist attack and by influencing the mortality and reproduction of beetle populations (Hicks, 1982).

Several studies have investigated the relationship between tree water balance and successful beetle attack. Lorio and Hodges (1968, 1977), for example, found that oleoresin exudation pressure of large loblolly pines was reduced during periods of drought-induced moisture stress, thus reducing the probability that pine beetles would be killed by exuding pitch. This, in turn, would increase the beetles' attack success. Artificially stressed trees succumbed to induced beetle attack more readily than nonstressed trees.

Initial attempts to relate rainfall and moisture balance to areawide fluctuations in pine beetle infestations yielded conflicting results. Kroll and Reeves (1978) found that in eastern Texas abundant rain during the summer was positively correlated whereas fall and spring precipitation was negatively correlated with infestation rate the following summer. King (1972) found low summer rainfall in Georgia, high winter rainfall in Texas, and high spring and low early summer rainfall in the Carolinas to be associated with epidemic years. Hansen *et al.* (1973) found that outbreaks in the Atlantic coastal plain of Delaware, Maryland, and Virginia were associated with extended drought periods. Conversely, Kalkstein (1974) found that southern pine beetle activity in Texas and Louisiana was associated with increased late winter moisture. Some results support the hypothesis that lowered host resistance due to water stress leads to an increase in infestations; other results do not. It seems that infestations in the Western Gulf coastal plain are associated with prior abundant moisture, whereas infestations along the Atlantic coastal plain are associated with prior drought.

Attempts to model the relationship between southern pine beetle outbreaks and climatic fluctuations for predictive purposes have had only limited success. The proportion of variation in infestation numbers explained by the model is often low. Nonclimatic independent variables (the number of infestations the previous season, stand and site characteristics, etc.) are often more significant than the climatic variables, and the capacity of the model to predict future infestations is often poor (Kroll and Reeves, 1978; Campbell and Smith, 1980). More recently Kalkstein (1981) and Michaels (1984) used principal component analysis to reduce the number of potential predictor variables, which, theoretically, should result in a more reliable predictor estimate. Kalkstein's (1981) analysis, based on 18 years of Texas data, reduced a large number of independent climatic variables to a three-component solution that explained 87% of the total variation in the climatic variables. A multiple linear regression based on the three components then predicted (with one exception) monthly variation in the number of southern pine beetle infestations for one county in Texas over a 5-year period. Michaels (1984) used similar methods to predict the yearly change in the number of infestations in Atlantic coastal and piedmont regions. Sixty-eight percent of the changes were correctly predicted. Michaels *et al.* (1986) improved the approach by increasing the area included in the model (the entire endemic range of the beetle) and by using a longer time series (53 years). The improved model is currently being implemented as a regionwide prediction tool in a southern pine beetle integrated pest management (IPM) program. However, the total explained variance due to climatic variables is small (33%). Michaels suggested that this is probably the largest amount that can be expected in such models.

V. METHODOLOGICAL PROBLEMS

Although the evidence suggests that fluctuations in climate can drive outbreak dynamics of some forest insects, serious weaknesses in methods and assumptions are often not adequately addressed, and these may cast doubt on conclusions. The procedural difficulties with bioclimatic studies should therefore be briefly reviewed (for more thorough discussions, see Wellington, 1957; Messenger, 1959; Ives, 1981).

A major source of error is the assumption that standard meteorological data are predictably related to conditions experienced by the insect. Often data are taken from weather stations a long distance away from the location of the study. Differences in topography, elevation, aspect, and so on may have an unknown influence on the actual conditions in the forest. Nor can we uncritically assume that weather data taken in the forest meaningfully reflect conditions in the insect microenvironment (Wellington, 1954a; Morris, 1963; Ferro *et al.*, 1979). Wellington, for example, showed that temperatures at a leaf surface in the summer may greatly exceed the air temperature during the day and drop far below the air temperature during the night. Likewise, Ives (1964) found that exposed tamarack bark temperature greatly exceeded air temperature. The microenvironment of a shelter-building or tissue-inhabiting insect will exhibit a greater difference from ambient conditions than will that of an insect that feeds exposed. Careful studies of the insect microenvironment are therefore needed to determine the exact effect of macroclimatic variables and where we might be misled by using such variables. Many of the data on the functional response of insects to varying physical conditions are derived from laboratory experiments. These data can also be inadequate or misleading when extrapolated to a field or forest situation. Compensating factors may be needed for better agreement between laboratory and field situations.

VI. PROBLEMS IN DRAWING INFERENCES

Implicit in all tests of climatic release is a statistical model: A group of independent variables (weather data) are tested for their effects on a dependent variable (change in insect abundance). In the studies reviewed here, there are two types of dependent variable. The first is an index of continuous population change, such as sweep net or light trap catches, number of "reports of occurrence," or number of infestation "spots." Do such simple indices truly reflect regional population trends? Few of the studies use data derived from sampling plans designed to account for sources of error in population estimates. The other type of dependent variable is derived from historical records or evidence of secondary effects of intense insect attack such as tree ring analysis or visible defoliation. From this information an outbreak history is inferred and the dependent variable becomes a binary, or two-state, variable (the presence or absence of outbreak). A criterion is then established for defining the point in time when an "outbreak" exists. The criterion is often more arbitrary than an index of population change: first year of noticeable or widespread defoliation, first year with reduced increment growth, first year with at least a fourfold jump in a population index, etc. (Table 1).

If we have no actual population data, we have no idea how long it takes the population to increase from the endemic to the outbreak level, or even if there is such a thing as an endemic level. In other words, we must guess the length of the release period. It may be different for each outbreak or different in each epicenter of the same outbreak.

There are other problems with an estimated release period when an outbreak history is inferred. The task becomes one of searching for the same climatic anomaly (selected a priori as "favorable") in the release period before each outbreak. An overestimated release period will increase the probability, and an underestimated period will decrease the probability, of finding the anomaly by chance alone. Therefore, there is the chance (the magnitude of which is unknown) that inferences may be based on a spurious correlation. This is even more likely when we draw inferences from only a few outbreaks.

Many of the studies rely on a graphic match between changes in insect abundance or the occurrence of outbreaks and the weather variables plotted as a time series as evidence for a relationship. If the match does not seem very strong at first, it would be too easy to use a different index of abundance or change the criterion for outbreak. With some studies, graphic methods are used to test hypotheses that are, by their nature, difficult to falsify. For example, White's (1974) hypothesis suggests that looper outbreaks should be associated with either abnormally dry or abnormally wet conditions. He used graphic methods to show that outbreaks of several species of loopers coincided with higher or lower than normal periods of precipitation. However, the possibility of such a coincidence occurring by chance alone would seem rather high.

What is the most meaningful way to present climatic or weather data? Monthly or yearly averages or summaries are fictional and overlook the possibility that insects are more sensitive to catastrophic weather events of short duration that may not be accounted for in long-term means. Many studies (e.g., Wellington, 1954b; Greenbank, 1956; Stark, 1959; Bejer-Peterson, 1972) use moving averages to reduce the noise and make trends more apparent in a weather-variable time series. This method can actually create an artificial trend or oscillation where none exists and is therefore questionable (Mitchell, 1964; Royama, 1984).

There are problems of interpretation in the case of outbreaks that occur over large, physiographically diverse areas. Should disjunct infestations that appear simultaneously be considered separate outbreaks, or should they be treated as part of a single widespread outbreak? With some forest insects the pattern of outbreak is very complex, infestations spreading and disappearing in an irregular manner. With others, there may be a fairly distinct physiographic trend. Wellington *et al.* (1950) plotted several isolated spruce budworm "outbreaks" with the annual number of cyclonic centers passing nearby, thus giving the impression of replication. However, several of these isolated "outbreaks" were very close together in space and time and may, in fact, have been part of a single regional population trend. Greenbank (1956) likewise treated the 1949 spruce budworm outbreak in New Brunswick as isolated, despite the fact that a large outbreak had been steadily building to the west (Ontario and Quebec) since about 1937. He recognized the dispersal potential of large spruce budworm populations (Greenbank, 1957) but still concluded that the New Brunswick outbreak was due to local climatic conditions. When many disjunct infestations occur simultaneously, it may be too easy to claim that they are isolated outbreaks and choose those with a climatic record that matches the hypothesis and overlook those that do not. To summarize, what constitutes independence in outbreak events is a serious question that has not been adequately addressed in most studies.

The main weakness of studies that attempt to relate long-term changes in insect abundance with climatic factors is that they are by nature correlative. They attempt to demonstrate a relationship between two sets of uncontrollable variables. As such, they are subject to all the possible errors of inference associated with any correlational study. Each life history scenario requires a unique set of assumptions. This confounds the risk of using a standard method of verification when more than one outbreak species is included in a study. Most studies base the choice of "favorable" weather conditions on previous life history studies or casual observation of the apparent effects of weather on abundance. Very few studies include life table analysis. Therefore, most of the evidence that weather conditions affect long-term fluctuations in forest insect abundance is circumstantial at best. Finally, there are the special statistical problems associated with time-series data. Yule (1926), for example, showed that the chance of getting a nonsense correlation between time-series data is higher than expected when inferences are made uncritically.

Morris (1963) based his research and conclusions concerning spruce budworm populations dynamics on 15 years of sampling data from the Green River project in New Brunswick. In a reanalysis of the Green River data, Royama (1978) argued that the coincidence in trends in larval survival claimed by Morris was probably spurious. Royama (1984) also reinterpreted the life table studies of Morris (1963), Miller (1963), and Greenbank (1956, 1963). He proposed a new hypothesis on the role of weather in spruce budworm population dynamics. He rejected the "double equilibrium" theory and the notion that favorable weather permits "release" from an endemic population level. Instead, he asserted that an oscillatory model with secondary fluctuations about the basic trend is more parsimonious with the data. The basic oscillation is determined by the combined action of density-dependent mortality factors during the large larval instars. These factors include parasitoids, disease, and a complex of unknown causes, which he called "the fifth agent." Weather is not the cause of the basic, universally occurring oscillation. Finally, Royama concluded that the data did not support the hypothesis that weather influences the survival of large larvae, as Morris thought.

VII. NATURE OF CLIMATIC VARIABILITY

A. Temporal Variation

The theory of climatic release as stated by Wellington and Greenbank assumes that a favorable climatic anomaly must be of sufficient duration to drive a forest insect population to the outbreak level. For example, Greenbank estimated that a spruce budworm population can rise from an endemic level in a period of 4 to 5 years and thus would require favorable climatic conditions of equal duration. How likely is it that the proper "favorable" climatic anomaly will be of sufficient duration? Would such an anomaly occur by chance as part of a random fluctuation in atmospheric events from year to year, or are climatic events more systematic or predictable? Both Wellington and Greenbank believed that climatic fluctuation is periodic or orderly on a time scale that fits the outbreak pattern of forest insects like the spruce budworm and forest tent caterpillar. For example, Greenbank (1956) stated,

> The jet stream is one important factor that determines if an area such as New Brunswick shall come predominantly under the influence of the favorable polar air or unfavorable tropical air. This . . . wind has a guiding influence on the pressure systems that develop at the earth's surface. Displacements in the mean course of the jet stream whereby it is farther south or north than normal in successive summers may result in a more or less orderly variation in climate over short periods, i.e., from dry, clear summers through a cloudier, moister period to a number of wet summers.

Wellington (1954b) attempted to demonstrate by graphic methods the existence of periodic (on a 3- to 10-year scale) latitudinal shifts in the paths of pressure centers over a region, thus changing the frequency with which air masses of certain types invade a region. To what extent are these assertions supported by the findings of modern meteorology and climatology? During the years since the work of Wellington and Greenbank, statistical procedures have been developed that can answer these questions.

The atmosphere is essentially a thermally active fluid that derives its kinetic energy from differential solar heating of the earth's surface. The resulting motion, modified by the rotation of the earth on its axis, is predominantly turbulent. The largest elements of turbulence are the so-called planetary waves that migrate erratically from west to east in middle latitudes and the cyclones and anticyclones we see on the daily weather map that are carried along in these planetary waves. Smaller elements of turbulence, hurricanes, thunderstorms, and shower clouds, are also thermally active. This picture of atmospheric behavior, of which the chaos typical of all turbulent systems is an outstanding feature, would seem to leave little room for cyclicity or persistence of events in climate or weather. The diurnal and annual cycles are due to astronomical "forcing functions" to which the atmosphere *must* respond (e.g., the differential radiant warming of the earth near the poles between winter and summer). Other cyclic or periodic events may exist, but they are difficult to verify statistically (Mitchell, 1964). Where they exist (e.g., the "biennial" cycle, the 27-year solar–lunar tidal influence, the 30-day lunar cycle, the 11-year sunspot cycle), a large amount of weather data is required to detect the signal (Mitchell *et al.*, 1979; Rasmussen *et al.*, 1981; Walsh *et al.*, 1982b), and the contribution to total climatic variation due to the cyclic model is usually swamped by other sources of variation (Gilman *et al.*, 1963; Mitchell, 1964). Their influence on surface weather is therefore usually statistically insignificant.

In North America, interannual and seasonal weather changes are intimately associated with the so-called polar front (Cole, 1980), an air mass discontinuity between cold polar air and warmer southern air. Along this front, frequent storms (low pressure areas) develop. The jet stream develops high above the polar front and is a consequence of the juxtaposition of warm and cold air at the surface. Pressure decreases with height faster in cold air than in warm air, which causes an intense pressure gradient in the upper air, the jet stream. Thus, the assertion of Greenbank is misleading in that it is the surface warm air and cold air that govern the system, not the jet stream, as he suggests (M. E. Landsburg, personal communication; E. R. Reiter, personal communication; Reiter and Westhoff, 1981). The path of the jet stream migrates farther south in the winter and farther north in the summer. Geographic features, such as mountain ranges and occasionally a persistent surface weather anomaly such as a stable depression or high, also influence the jet stream. Other than these influences, the pattern of the jet stream is essentially random. Consecutive-season persistence in weather variables (e.g., the tendency for consecutive summers to be similar), in the behavior of air masses, cyclonic paths, or jet stream paths over long time series, have never been statistically demonstrated. At best, there is an occasional 6-month persistence or anomaly in the frequency of cyclones passing through an area (Hayden, 1984).

Martinat (1984) used the Durbin–Watson test statistic (Draper and Smith, 1981) to test for autocorrelation in 35 weather variables near saddled prominent, *Heterocampa guttivitta* (Walker) (Lepidoptera: Notodontidae), outbreak areas in Vermont. (A weather variable was, e.g., total July rainfall at a given weather station for 80 consecutive years.) The null hypothesis (zero autocorrelation) was rejected only once, suggesting that anomalies in the variables were random in both occurrence and duration. Thus, climatic anomalies of the sort needed in a climatic release model are unlikely, at least in temperate climates near the polar front. If outbreaks exhibit periodicity, there is even less chance of finding an appropriate pattern of climatic anomalies.

There is, however, good evidence for the existence of quasi periodicity in the expression of drought in many regions of the world. Cook and Jacoby (1979) used tree ring analysis to reconstruct a drought time series back to 1694 in the Hudson Valley, New York. Using spectral analysis, they found statistically significant evidence for periodicity at 11 and 26 years. Karl and Koscielny (1982) found that droughts persist longer in the interior of the continent than in areas closer to the coasts; but, again, whether the cyclic drought signal is consistent and strong enough to affect the outbreak dynamics of a periodic forest insect defoliator should be carefully studied. Martinat and Allen (1987) found that several saddled prominent outbreaks occurred either during or immediately after drought periods in New England but concluded that the association was not consistent enough to claim a real correlation.

B. Spatial Variation

The spatial aspect of climatic variation must also be considered, since a single forest insect outbreak may occur over millions of square kilometers. Assuming that the role of migration and dispersal is minimal, one prediction of a climatic release hypothesis would be that anomalies in weather variables should be well correlated simultaneously at all outbreak epicenters. Martinat and Allen (1987) compared correlations of weather variables among several New England weather stations and more distant weather stations in Pennsylvania and Wisconsin, all located near epicenters of a single (1969–1972) saddled prominent outbreak. Temperature variables were the best-correlated factors among stations. Precipitation variables were very poorly correlated. This suggests that temperature anomalies occur on a larger spatial scale than precipitation anomalies. Most poorly correlated was "number of days with six inches of snow on ground," suggesting that snow cover in winter is spotty and unpredictable. These types of results might have implications if the objective were to base a prediction model on weather variables. Some variables would necessarily be ruled out as predictors of widespread outbreak merely by their poor correlation among stations, regardless of any effect they might have on the biology of the insect.

A weather "field" is a space–time matrix for a given weather variable (for example, July temperature averaged by station, for all the weather stations in a region, for a given number of years). Walsh *et al.* (1982a) used factor analysis to study precipitation fields in the United States using 78 years of data from 61 weather stations scattered throughout the country. A substantial proportion of the variation in the precipitation field was due to local "convective" scale precipitation. After this source of variation was statistically removed, the spatial pattern of precipitation variation (or the "spatial coherence" of precipitation) loaded onto eight or nine distinct regions, indicating that regions as far apart as the Midwest and New England, for example, are affected by different synoptic weather systems. Autocorrelations from year to year in each region for a given monthly total were very small, again suggesting that there is little or no persistence in the behavior of precipitation between years. However, with the exception of summer, seasonal continuity in the patterns is rather high. This suggests that precipitation is more strongly dominated by local convective circulation in the summer than in the other seasons. Analyses of this type also have implications for a climatic release hypothesis. There is probably little chance of finding a climatic signal behind an outbreak that occurs simultaneously across different synoptic weather regions.

VIII. CONCLUSIONS

Three characteristics of many forest insect outbreaks are simultaneous population buildup over large and sometimes disjunct areas, the subsequent appearance of widespread outbreaks, and the occurrence of new outbreaks in the same areas following outbreak collapse and a period of low abundance. These characteristics suggest the operation of a large-scale cyclic or periodic signal. It has been suggested that one source of such a large-scale driving force might be regional climatic variation. Thus, several authors have suggested a "climatic release" mechanism to explain widespread periodic outbreaks. However, at least in temperate North America, where the predominant influence on surface weather is the polar front, the pattern of climatic variation is unlikely to be of the sort assumed by the theory of climatic release. A climatic anomaly is unlikely to be of sufficient duration or spatial coherence to be the sole cause of a widespread and periodic forest insect outbreak. Quasi periodicity in the occurrence of drought is one possible exception.

Hypotheses of forest insect outbreaks that involve climatic anomalies are intrinsically difficult to test. Problems of methodology reduce the chance of finding a relationship between weather and a forest insect's outbreak dynamics. The fact that many studies claim to support a climatic release hypothesis despite their methodological problems is probably due to rather liberal "outbreak" criteria and a failure to consider carefully what constitutes independence in outbreak events. Many claims probably would not hold up under rigorous examination.

Studies of change in abundance over longer periods that use a per generation population index require fewer assumptions than studies that use only an outbreak history. Both types of studies, however, may require assumptions difficult to support or substantiate which reduce the efficiency of a dependent variable.

There is no doubt that weather does influence generation survival and that weather variables *can,* in some cases, improve predictability when they are included in prediction models of seasonal or generational population change. There is probably a rather low upper limit, however, on the amount of variability that can be explained by weather variables in such models. Studies of this type require fairly precise sampling data, with known error, over long periods and large areas. Data of this nature are, of course, unavailable for the majority of forest insects that exhibit outbreak behavior.

The concept of "release" from an assumed endemic level is questionable. More precise long-term population data before outbreak would be required to establish the existence of a release period for most forest insect outbreaks.

More attention should be paid to the possible effects of catastrophic weather events of short duration, such as late spring or early fall frosts, excessive cloudiness, rain, or cold of short duration during critical windows of vulnerability in the herbivore life cycle. Effects of this sort would not be accounted for in monthly weather summaries.

Clearly, any outbreak hypothesis must adequately explain all spatial and temporal aspects of outbreak behavior. In the development of future hypotheses, whether they involve climatic anomalies, an interaction between weather and a biotic factor, or a purely biotic factor, two outbreak features must be understood. First, what is the nature of outbreak foci, or epicenters? Is there a regionwide population buildup, with more rapid buildup in some spots (epicenters) due to local edaphic or weather conditions, or is population buildup strictly a local phenomenon with spread from the epicenters due to dispersal? Second, we must understand the factors controlling the time period between outbreaks. Some kind of "recovery" may be involved: a period in which host plants mature and return to health or normal nutritional quality or in which the herbivore population recovers vigor or increases in number to a critical level (following a catastrophic crash). An apparently periodic outbreak cycle may be due to the interaction of factors that control the duration of a recovery period and the probability that a set of favorable conditions, climatic or otherwise, will occur after the recovery period.

REFERENCES

Ahmed, M. S. H., Al-Saqur, A. M., and Al-Hakkak, F. S. (1982). Effect of different temperatures on some biological activities of the parasitic wasp *Bracon hebetor* (Say) (Hymenoptera). *Date Palm. J.* **1,** 239–247.

Allen, D. C. (1972). Insect parasites of the saddled prominent, *Heterocampa guttivitta* (Lepidoptera: Notodontidae) in northeastern United States. *Can. Entomol.* **104,** 1609–1622.

Anderson, R. M., and May, R. M. (1980). Infectious diseases and population cycles of forest insects. *Science* **210,** 658–661.

Andrewartha, H. G., and Birch, L. C. (1954). "The Distribution and Abundance of Animals." Univ. of Chicago Press, Chicago, Illinois.

Baltensweiler, W. (1966). Zur Erklarung der Massenvermehrung des Grauen Larchenwicklers (*Zeiraphera griseana* Hb. = Gn.). I. Die Massenvermehrungen in Mitteleurope. *J. For. Suisse* **117,** 466–491.

Barbosa, P., and Frongillo, E. A. (1979). Influence of light intensity and temperature on the locomotory and flight activity of *Brachymeria intermedia* (Hymenoptera: Chalcididae), a pupal parasitoid of the gypsy moth. *Entomophaga* **22**(4), 405–411.

Barney, R. J., Armburst, E. J., Pausch, R. D., and Roberts, S. J. (1979). Effect of constant versus fluctuating temperature regimes on *Bathyplectes curculionis* (Hymenoptera: Ichneumonidae) activity. *Great Lakes Entomol.* **12**(2), 67–71.

Bejer-Peterson, B. (1972). Relation of climate to the start of Danish outbreaks of the pine shoot moth (*Rhyaciona buoliana* Schiff.) *Forstl. Forsoegsvaes. Dan.* **33**(1), 41–50.

Berryman, A. A., and Stark, R. W. (1985). Assessing the risk of forest insect outbreaks. *Z. Angew Entomol.* **99,** 199–208.

Blais, J. R., Prentice, R. M., Sippell, W. L., and Wallace, D. R. (1955). Effects of weather on the forest tent caterpillar, *Malacosoma disstria* Hbn., in central Canada in the spring on 1953. *Can. Entomol.* **87,** 1–8.

Bursell, E. (1974a). Environmental aspects: Temperature. *In* "The Physiology of Insecta" (M. Rockstein, ed.), 2nd ed., Vol. 2, pp. 1–41. Academic Press, New York.

Bursell, E. (1974b). Environmental aspects: Humidity. *In* "The Physiology of Insecta" (M. Rockstein, ed.), 2nd ed., Vol. 2, pp. 43–84. Academic Press, New York.

Campbell, R. W., and Sloan, R. J. (1978). Numerical biomodality among North American gypsy moth populations. *Environ. Entomol.* **7,** 641–646.

Campbell, J. B., and Smith, K. E. (1980). Climatological forecasts of southern pine beetle infestations. *Southeast. Geogr.* **20**(1), 16–30.

Cole, F. W. (1980). "Introduction to Meteorology," 3rd ed. Wiley, New York.

Cook, E. R., and Jacoby, G. C. (1979). Evidence for quasi-periodic July drought in the Hudson Valley, New York. *Nature* (*London*) **282,** 390–392.

Cramer, H. H. (1962). Moeglichkeiten der Forstschaedlingsprongnose mit Hilfe meteorologischer Daten. *Schriftenr. Forstl. Abt. Albert-Ludwigs Univ. Freiburg im Breisgau* **1,** 238–245.

Draper, N. R., and Smith, H. (1981). "Applied Regression Analysis," 2nd ed. Wiley, New York.

Ehrlich, P. R., and Birch, L. C. (1967). The "balance of nature" and "population control." *Am. Nat.* **101,** 97–107.

Elton, C. (1949). Population interspersion: An essay on animal community patterns. *J. Ecol.* **37,** 1–23.

Ferro, D. N., Chapman, R. B., and Penman, D. R. (1979). Observations on insect microclimate and insect pest management. *Environ. Entomol.* **8,** 1000–1003.

Gilman, D. L., Fuglister, F. J., and Mitchell, J. M., Jr. (1963). On the power spectrum of "red noise." *J. Atmos. Sci.* **20,** 182–184.

Graham, S. A. (1939). Forest insect populations. *Ecol. Monogr.* **9,** 301–310.

Green, T. R., and Ryan, C. A. (1972). Wound-induced proteinase inhibitor in plant leaves: A possible defense mechanism against insects. *Science* **175,** 776–777.

Greenbank, D. O. (1956). The role of climate and dispersal in the initiation of outbreaks of the spruce budworm in New Brunswick. I. The role of climate. *Can. J. Zool.* **34,** 453–476.

Greenbank, D. O. (1957). The role of climate and dispersal in the initiation of spruce budworm outbreaks in New Brunswick. II. The role of dispersal. *Can. J. Zool.* **35,** 385–403.

Greenbank, D. O. (1963). Climate and the spruce budworm. *Mem. Entomol. Soc. Can.* **31,** 174–180.

Hansen, J. B., Baker, B. H., and Barry, P. J. (1973). Southern pine beetles on the Delaware Peninsula in 1971. *J. Ga. Entomol. Soc.* **8**(3), 157–164.

Hard, J., Tunnock, S., and Eder, R. (1980). Western spruce budworm defoliation trend relative to weather in the northern region, 1969–1979. *USDA For. Serv. Northeast. Reg. State Priv. For.*

Haukioja, E. (1980). On the role of plant defences in the fluctuation of herbivore populations. *Oikos* **35,** 202–213.

Haukioja, E., and Niemela, P. (1977). Retarded growth of a geometrid larva after mechanical damage to leaves of its host tree. *Ann. Zool. Fenn.* **14,** 48–52.

Hayden, B. P. (1984). Climate prediction. I. Cyclone frequency. *In* "Classification of Coastal Environments," Tech. Rep. No. 30, For. Serv. North Reg. State Priv. For. Rep. No. 80–4. Dep. Environ. Sci., University of Virginia, Charlottesville.

Hicks, R. R. (1982). Climatic, site, and stand factors. *USDA For. Serv. Sci. Educ. Admin., Tech. Bull.* **1631.** 55–68.

Hodson, A. C. (1941). An ecological study of the forest tent caterpillar, *Malacosoma disstria* Hbn., in northern Minnesota. *Minn., Agric. Exp. Stn., Tech. Bull.* **148,** 1–55.

Hodson, A. C. (1977). Some aspects of forest tent caterpillar population dynamics. *Minn., Agric. Exp. Stn., Tech. Bull.* **310,** 5–16.

Hodson, A. C., and Weinman, C. J. (1945). Factors affecting recovery from diapause and hatching of eggs of the forest tent caterpillar, *Malacosoma disstria* Hbn. *Minn., Agric. Exp. Stn., Tech. Bull.* **170,** 1–31.

Hsiao, K. J., and Yen, C. C. (1964). Studies on the population dynamics of the pine caterpillar (*Dendrolimus punctatus* Walker) in China. *Sci. Silvae (Peking)* **9,** 201–220 (translation from Chinese by Can. Dep. For., Ottawa, Ontario: Transl. No. 55).

Ives, W. G. H. (1964). "Temperatures On or Near an Exposed Tamarack Tree," Interim Res. Rep. Can. Dep. For., For. Entomol. Pathol. Branch, Winnipeg, Manitoba.

Ives, W. G. H. (1973). Heat units and outbreaks of the forest tent caterpillar. *Can. Entomol.* **105**(4), 529–543.

Ives, W. G. H. (1974). Weather and outbreaks of the spruce budworm. *Can. For. Serv. North. For. Res. Cent.* Information Rept. **NOR-X-118,** 1–28.

Ives, W. G. H. (1981). Environmental factors affecting 21 forest insect defoliators in Manitoba and Saskatchewan, 1945–69. *Can. For. Serv. North. For. Res. Cent.* Information Rept. **NOR-X-233,** 1–142.

Jackson, C. G., and Butler, G. D. (1984). Development time of three species of *Bracon* (Hymenoptera: Braconidae) on the pink bollworm (Lepidoptera: Gelechiidae) in relation to temperature. *Ann. Entomol. Soc. Am.* **77,** 539–542.

Juillet, J. A. (1964). Influence of weather on flight activity of parasitic Hymenoptera. *Can. J. Zool.* **42,** 1133–1141.

Kalkstein, L. S. (1974). The effect of climate upon outbreaks of the southern pine beetle. *Publ. Climatol.* **27,** 1–65.

Kalkstein, L. S. (1981). An improved technique to evaluate climate–southern pine beetle relationships. *For. Sci.* **27**(3), 579–589.

Karl, T. R., and Koscielny, A. J. (1982). Drought in the United States: 1895–1981. *J. Climatol.* **2,** 313–329.

Kaya, H. K., and Anderson, J. F. (1972). Parasitism of elm spanworm eggs of *Ooencyrtus clisiocampae* in Connecticut. *Environ. Entomol.* **1,** 523–524.

Kaya, H. K., and Anderson, J. F. (1974). Collapse of the elm spanworm outbreak: Role of *Ooencyrtus* sp. *Environ. Entomol.* **3,** 359–363.

King, E. (1972). Rainfall and epidemics of the southern pine beetle. *Environ. Entomol.* **1,** 279–285.

Klomp, H. (1962). The influence of climate and weather on the mean density level, the fluctuations and the regulation of animal populations. *Arch. Neerl. Zool.* **15,** 68–109.

Knight, F. B., and Heikkenen, H. J. (1980). "Principles of Forest Entomology, 5th Edition." McGraw-Hill, New York.

Kroll, J. C., and Reeves, H. C. (1978). A simple model for predicting annual numbers of southern pine beetle infestations in East Texas. *South. J. Appl. For.* **2,** 62–64.

Lorio, P. L., Jr., and Hodges, J. D. (1968). Oleoresin exudation pressure and relative water content of inner bark as indicators of moisture stress in loblolly pines. *For. Sci.* **14,** 392–398.

Lorio, P. L., Jr., and Hodges, J. D. (1977). Tree water status affects induced southern pine beetle attack and brood production. *U.S., For. Serv., Res. Pap.* **SO-135.**

Magnoler, A. (1974). Effect of cytoplasmic polyhedrosis on larval and post larval stages of the gypsy moth. *J. Invertebr. Pathol.* **23,** 263–274.

Martinat, P. J. (1984). The affect of abiotic factors on saddled prominent, *Heterocampa guttivitta* (Walker) (Lepidoptera: Notodontidae), population biology. Ph.D. Thesis, State University of New York, Coll. Environ. Sci. For., Syracuse, New York.

Martinat, P. J., and Allen, D. C. (1987). Relationship between saddled prominent, *Heterocampa guttivitta* (Lepidoptera: Notodontidae), and drought. *Environ. Entomol.* **16,** 246–249.

Mason, R. R., and Luck, R. F. (1987). Population growth and regulation. *In* "The Douglas-Fir Tussock Moth: A Synthesis" (M. H. Brookes, R. W. Campbell, and R. W. Stark, eds.). USDA For. Serv. Tech. Bull. 1585, pp 41–47.

Messenger, P. S. (1959). Bioclimatic studies with insects. *Annu. Rev. Entomol.* **4,** 183–206.

Michaels, P. J. (1984). Climate and the southern pine beetle in Atlantic coastal and piedmont regions. *For. Sci.* **30,** 143–156.

Michaels, P. J., Sappington, D. E., Spengler, P. J., and Philip, J. (1986). SPBCMP—A program to assess the likelihood of major changes in the distribution of southern pine beetle infestations. *S. J. Appl. For.* **10,** 158–161.

Miller, C. A. (1963). The analysis of pupal survival, female proportion, fecundity proportion in the unsprayed area. *Mem. Entomol. Soc. Can.* **31,** 63–86.

Mitchell, J. M., Jr. (1964). A critical appraisal of periodicities in climate. *In* "Weather and Our Food Supply," Proceedings of a Conference, May 3-6, 1964, pp. 189–227. Cent. Agric. Econ. Dev., Rept. No. 20, Iowa State University, Ames.

Mitchell, J. M., Jr., Stockton, C. W., and Meko, D. M. (1979). Evidence of a 22-year rhythm of drought in the western United States related to the Hale solar cycle since the 17th century. *In* "Solar–Terrestrial Influences on Weather and Climate" (B. M. McCormac and T. A. Soliga, eds.), pp. 125–144. Reidel Publ., Dordrecht, Netherlands.

Morris, R. F. (1963). The development of prediction equations for the spruce budworm based on key-factor analysis. *Mem. Entomol. Soc. Can.* **31,** 116–121.

Morris, R. F. (1964). The value of historical data in population research, with particular reference to *Hyphantria cunea* Drury. *Can. Entomol.* **96,** 356–368.

Nicholson, A. J. (1933). The balance of animal populations. *J. Anim. Ecol.* **2,** 132–178.

Nicholson, A. J. (1947). Fluctuation of animal populations. *Rep. Meet. Aust. N. Z. Assoc. Adv. Sci.* **26,** 134–148.

Rasmussen, E. M., Arkin, P. A., Chen, W. Y., and Jalickee, J. B. (1981). Biennial variations in surface temperature over the United States as revealed by singular decomposition. *Mon. Weather Rev.* **109,** 587–598.

Readshaw, J. L. (1965). A theory of phasmatid outbreak release. *Aust. J. Zool.* **13,** 475–490.

Reiter, E. R., and Westhoff, D. (1981). A planetary-wave climatology. *J. Atmos. Sci.* **38,** 732–750.

Rhoades, D. F. (1979). Evolution of plant chemical defense against herbivores. *In* "Herbivores: Their Interaction with Secondary Plant Metabolites" (G. A. Rosenthal and D. H. Janzen, eds.), pp. 3–54. Academic Press, New York.

Rhoades, D. F. (1983). Herbivore population dynamics and plant chemistry. In "Variable Plants in Natural and Managed Systems" (R. F. Denno and M. F. McClure eds.) pp. 155–220. Academic Press, New York.

Richards, O. W. (1961). The theoretical and practical study of natural insect populations. *Annu. Rev. Entomol.* **6,** 147–162.

Royama, T. (1978). Do weather factors influence the dynamics of spruce budworm populations? *Can. For. Serv. Bi-Mon. Res. Notes* **34**(2), 9–10.

Royama, T. (1984). Population dynamics of the spruce budworm, *Choristoneura fumiferana*. *Ecol. Monogr.* **54**(4), 429–462.

Schultz, J. C., and Baldwin, I. T. (1982). Oak leaf quality declines in response to defoliation by gypsy moth larvae. *Science* **217,** 149–151.

Sippell, W. L., MacDonald, J. E., and Rose, A. H. (1964). Forest insect conditions (in Ontario). *In* "Annual Report of the Forest Insect and Disease Survey, 1963." Can. Dep. For., Ottawa.

Smith, H. S. (1935). The role of biotic factors in the determination of population densities. *J. Econ. Entomol.* **28,** 873–898.

Solomon, M. E. (1949). The natural control of animal populations. *J. Anim. Ecol.* **18,** 1–35.

Stairs, G. R. (1972). Pathogenic microorganisms in the regulation of forest insect populations. *Annu. Rev. Entomol.* **17,** 355–372.

Stark, R. W. (1959). Population dynamics of the lodgepole needle miner in the Canadian Rocky Mountain Parks. *Can. J. Zool.* **37,** 917–943.

Surgeoner, G. A. (1976). Life history and population dynamics of the variable oak leaf caterpillar, *Heterocampa manteo* (Dbldy), in Michigan. Ph.D. Dissertation, Dep. Entomol., Michigan State University, East Lansing.

Sweetman, H. I. (1940). The value of hand control for the tent caterpillar *Malacosoma americana* Fabr. and *Malacosoma disstria* Hbn. *Can. Entomol.* **72,** 245–250.

Tanada, Y. (1959). Microbial control of insect pests. *Annu. Rev. Entomol.* **4,** 277–302.

Tauber, M. J., and Tauber C. A. (1978). Evolution of phenological strategies in insects: A comparative approach with eco-physiological and genetic considerations. In "Evolution of Insect Migration and Diapause." (H. Dingle, ed.) pp. 53–71. Springer-Verlag, Berlin and New York.

Thompson, A. J., Shepherd, R. F., Harris, J. W. E., and Silversides, R. H. (1984). Relating weather to outbreaks of western spruce budworm. *Can. Entomol.* **116,** 375–381.

Thompson, H. M. (1960). A list and brief description of the microsporidia infecting insects. *J. Insect Pathol.* **2,** 346–385.

Thompson, W. R. (1956). The fundamental theory of natural and biological control. *Annu. Rev. Entomol.* **1,** 379–402.

Ticehurst, M., and Allen, D. C. (1973). Notes on the biology of *Telonomus coelodasidis* (Hymenoptera: Scelionidae) and its relationship to the saddled prominent, *Heterocampa guttivitta* (Lepidoptera: Notodontidae). *Can. Entomol.* **105,** 1133–1143.

Tomlinson, W. E. (1938). Fluctuations in tent caterpillar abundance and some of the factors influencing it. Ph.D. Thesis, Massachusetts State College, Amherst.

Tothill, J. D. (1918). The meaning of natural control. *Proc. Entomol. Soc. Nova Scotia* **4,** 10–14.

Twardus, D. B. (1980). Evaluation of weather patterns in relation to western spruce budworm outbreak in northcentral Washington, 1970–1980. Forest Insect and Disease Management. *U.S., For. Serv., Pac. Northwest Reg., Mimeo. Rep.*

Uvarov, B. P. (1931). Insects and climate. *Trans. Entomol. Soc. London* **79,** 1–247.

Uvarov, B. P. (1957). The aridity factor in the ecology of locusts and grasshoppers of the Old World. *Arid. Zone Res.* **8,** 164–198.

Walker-Simmons, M., and Ryan, C. A. (1977). Wound-induced accumulation of trypsin inhibitor activities in plant leaves. *Plant Physiol.* **59,** 437–439.

Wallner, W. E., and Walton, G. S. (1979). Host defoliation is a possible determinant of gypsy moth population quality. *Ann. Entomol. Soc. Am.* **72,** 62–67.

Walsh, J. E., Richman, M. B., and Allen, D. W. (1982a). Spatial coherence of monthly precipitation in the United States. *Mon. Weather Rev.* **110,** 272–286.

Walsh, J. E., Tucek, D. R., and Peterson, M. R. (1982b). Seasonal snow cover and short-term climatic fluctuations over the United States. *Mon. Weather Rev.* **110,** 1474–1485.

Wellington, W. G. (1952). Air-mass climatology in Ontario north of Lake Huron and Lake Superior before outbreaks of the spruce budworm, *Choristoneura fumiferana* (Clem.), and the forest tent caterpillar, *Malacosoma disstria* Hbn. *Can. J. Zool.* **30,** 114–127.

Wellington, W. G. (1954a). Weather and climate in forest entomology. *Meteorol. Monogr.* **2,** 11–18.

Wellington, W. G. (1954b). Atmospheric circulation processes and insect ecology. *Can. Entomol.* **86,** 312–333.

Wellington, W. G. (1957). The synoptic approach to studies of insects and climate. *Annu. Rev. Entomol.* **2,** 143–162.

Wellington, W. G., Fettes, J. J., Turner, K. B., and Belyea, R. M. (1950). Physical and biological indicators of the development of outbreaks of the spruce budworm, *Choristoneura fumiferana* (Clem.) (Lepidoptera: Tortricidae). *Can. J. Res., Sect. D* **28,** 308–331.

White, T. C. R. (1969). An index to measure weather-induced stress on trees associated with outbreaks of psyllids in Australia. *Ecology* **50**(5), 905–909.

White, T. C. R. (1974). A hypothesis to explain outbreaks of looper caterpillars, with special reference to populations of *Selidosema suavis* in a plantation of *Pinus radiata* in New Zealand. *Oecologia* **16,** 279–301.

White, T. C. R. (1976). Weather, food, and plagues of locusts. *Oecologia* **22,** 119–134.

Yule, U. (1926). Why do we sometimes get nonsense correlations between time series? *J. R. Stat. Soc.* **89,** 1–64.

Chapter 11

Pathogen-Induced Cycling of Outbreak Insect Populations

PAUL W. EWALD

Department of Biology
Amherst College
Amherst, Massachusetts 01002
and
Department of Pure and Applied Biology
Imperial College
London SW7 2BB, England

I. HISTORICAL PERSPECTIVE

Before the 1970s if one opened an ecology text to the section on population dynamics, one encountered analyses of the roles of predation, the physical environment, and perhaps multicellular parasites (e.g., Ricklefs, 1979; Krebs, 1978; Hutchinson, 1978) but little if any reference to the role of pathogens (i.e., microbial parasites). It is not clear why the potential importance of pathogens was overlooked. Perhaps the most important culprit was a misconception about the outcomes of coevolution between pathogens and their hosts. For most of the twentieth century, writers on this subject have presumed that parasites and their hosts should, as a rule, evolve toward benign or commensal relationships (Smith, 1934; Swellengrebel, 1940; Simon, 1960; Dubos, 1965; Thomas, 1972; Hoeprich, 1977). Although such arguments generally invoke untenable mechanisms of group- and species-level selection, they are still echoed in more recent literature (Alexander, 1981; Palmieri, 1982; Balashov, 1984), and even in some of the best ecological texts (e.g., Ricklefs, 1979).

Recent recognition of this problem has led to analyses of situations in which host–parasite coevolution might have severe effects on hosts in terms of extreme tissue damage, death, and reduction of population. The evolutionary stability of such severe host–parasite systems has been demonstrated theoretically (Levin, 1983; Anderson and May, 1982; May and Anderson, 1983), as have processes of group selection that could drive evolution toward benignness. In addition, characteristics of hosts, pathogens, and their environments that should favor severe evolutionary outcomes have been identified (Anderson and May, 1981; Bremerman and Pickering, 1983; Ewald, 1983). The recognition that severe disease is a feasible evolutionary outcome bears directly on our understanding of the relationships between populations of insects and their pathogens. If pathogens always evolved toward benignness, they would have at most transient relevance to the population dynamics of their hosts; for pathogen populations to regulate host populations, substantial reductions in host fitness are necessary (Anderson and May, 1981). As described in this chapter, for pathogens to be responsible for the dramatic fluctuations in host populations manifested as outbreaks, negative effects of pathogens on host fitness generally must be even more extreme.

Whatever the reasons for the past neglect of this area, relationships between pathogens and fluctuations in insect populations have now become the focus of intense interest, partly because they seem integral to a general understanding of population dynamics and partly because of their relevance to biological control programs. By determining the causes of fluctuations, we should be able to identify (1) parameters that must be manipulated and (2) pathogens that are most suitable for regulating insect populations at suboutbreak levels.

When insect outbreaks are viewed over long periods, they often comprise the peaks of regular cyclic changes (e.g., Varley, 1949; Baltensweiler, 1964; Neilson and Morris, 1964). The regularity of these fluctuations suggests that biological processes are responsible, but the kind of biological process has been a source of controversy. Interactions between insects and their predators, parasitoids, or food supplies have been suggested as causes of such cycles (Varley, 1949; Klomp, 1966; May, 1974; Ludwig *et al.*, 1978).

The high prevalence of pathogens during outbreaks has led others to suggest that pathogens may play a role in the downfall of insect populations at the end of outbreaks (Stairs, 1972; Thompson and Scott, 1979). Because of this prevalence and the cyclic nature of population fluctuations, the primary theoretical effort has been aimed at determining whether changes in populations of hosts could result solely from the negative effects of pathogens on the hosts. The approach has been to modify traditional epidemiological models by treating the population size of hosts as a variable rather than a constant and then analyzing the conditions that yield stable and cylic regulation of the insect populations. More generally, the task involves unification of interactions at the individual level with those at the population level.

The first steps in this direction were taken starting in the mid-1970s. They involved theoretical analyses of population-level effects of changes in population parameters and in characteristics associated with hosts and pathogens. Thus, characteristics of host–pathogen relationships at the organismal level, such as pathogenicity, are used to estimate population parameters such as mortality, which, in turn, are used to predict how populations change over time. These theoretical steps have paved the way for more rigorous experimental tests in two ways: They have revealed the parameters that should have major effects on the tendency for cyclic change in host populations, and they have generated precise predictions about the values of these parameters that should lead to cyclic as opposed to stable equilibria (*sensu* May and Anderson, 1983).

II. CYCLING VERSUS STABLE REGULATION

A. General Model

Anderson and May (1981) studied the effects of several modifications of a model described by differential equations that compartmentalize the host population into infected individuals Y and uninfected susceptibles X. In the general model they described the rates of change of these two subpopulations as

$$dX/dt = a(X + Y) + \gamma Y - bX - \beta XY \tag{1}$$

$$dY/dt = \beta XY - (\alpha + b + \gamma)Y \tag{2}$$

The entrance of susceptible individuals into the population is assumed to depend on a per capita birth rate a, which is the same for infected and uninfected individuals and the per capita rate γ at which infected individuals recover from infections. Susceptible individuals are eliminated from the population based on a per capita rate of death due to causes other than infection b and the rate at which an infected individual will transmit its infection per susceptible individual (i.e., the transmission coefficient β) times both the number of infected and the number of susceptible individuals. The rate at which the infected subpopulation changes equals the rate at which susceptible individuals are infected (βXY) diminished by the rate at which infected individuals are lost as defined by per capita rates of pathogen-induced mortality (α), death from all other causes (b), and recovery (γ). Addition of Eqs. (1) and (2) yields the rate at which the total population of hosts N changes,

$$dN/dt = rN - \alpha Y \tag{3}$$

where r, which equals $a - b$, is the intrinsic growth rate of the population. In this model pathogens will regulate the host population if $\alpha > 4$, but this regulation will always be a globally stable equilibrium (Anderson and May, 1981); thus, according to this model pathogens will not induce cycles in host populations.

This model, however, is unrealistic in several respects; for example, it does not incorporate an incubation period. Anderson and May (1981) incorporated an incubation period into the model by defining a third subgroup of the population, M, which represents the number of hosts infected but not yet infectious. In this case,

$$dX/dt = a(X + M + Y) - bX - \beta XY + \gamma Y \tag{4}$$

$$dM/dt = \beta XY - (b + \nu)M \tag{5}$$

$$dY/dt = \nu M - (\alpha + b + \gamma)Y \tag{6}$$

$$dN/dt = rN - \alpha Y \tag{7}$$

where ν is the rate at which an infected but not yet infectious individual becomes infectious. With this modification, stable limit cycles can occur if α is similar in magnitude to ν and large relative to a, b, and r (Anderson and May, 1981).

This model provides a useful starting point for understanding population fluctuations: It illustrates that, in a simple model, the tendency for a particular parameter to induce oscillations depends on the values of the other parameters.

B. Pathogens, Multicellular Consumers, and Environmental Heterogeneity

Among multicellular parasites, parasitoids, herbivores, and predators, substantial lags generally occur between the time when an individual first exists (i.e., in the form of a fertilized egg) and when it makes extensive use of resources (i.e., usually in the adult or late juvenile stages). Such lags are destabilizing factors. In multicellular predators, parasitoids, and herbivores, lags tend to favor oscillations by causing the creation of greater numbers of consumers than can be supported on available resources (Nicholson, 1958; May *et al.,* 1974). When a time delay between the production of transmission stages and their availability for infection is incorporated into a model generally relevant to multicellular parasites, host–parasite relationships also tend to be destabilized (May and Anderson, 1978). Using a model generally relevant to infectious diseases of vertebrates, Grossman (1980) showed that lags tend to destabilize host–pathogen situations. Similarly, the incorporation of an incubation period into Eqs. (4)–(7) introduces a lag between the production of pathogens within a host and the time at which they may infect new hosts. The consequence is a slight destabilization of the system (see Section II,A). However, in contrast to multicellular consumers, lags in resource use due to developmental events are often absent or greatly reduced among pathogens. In the absence of such lags we expect a closer tracking of host resources and, as a consequence, a relative dampening of oscillations; thus, in this regard consumer-induced cycles may tend to be less prevalent among pathogens than among multicellular consumers.

Some experimental data are consistent with this contention; for example, in the laboratory, populations of the flour beetle, *Tribolium casteneum* Herbst, infected with the protozoan *Adelina triboli* Bhatia were only about one-third the size of uninfected populations. Dramatic oscillations were apparent in the uninfected populations but not in the infected population, which tended to remain below the minimum levels of the uninfected population (Park, 1948). In this case, infection seemed to stabilize the beetle population by preventing overexploitation of the food sources, thereby dampening rather than inducing cycles in the host population.

The potential limitations of laboratory experiments must be borne in mind when such results are interpreted. Increased environmental heterogeneity (i.e., discontinuously distributed resources) should affect the trackability of insects by pathogens in a manner analogous to the development lags described above. Consider a "sit-and-wait" pathogen, that is, a pathogen whose transmission depends on its remaining viable at or near the point of release from the infectious host until a susceptible host contacts it. Because insect hosts tend to be far more mobile than such pathogens, subpopulations of the hosts should temporarily escape the pathogens: Hosts should reach pathogen-free portions of the habitat and remain pathogen-free until the pathogen arrives or until the host density increases sufficiently to permit contact with the infected areas. The lag in this case occurs between the time when host resources are produced and the time when they are used, rather than a lag between when the consumer is produced and when the use of resources per consumer is greatest. The similarity is that in both cases the host population is able to escape temporarily from the pathogens, but in the latter case it is a spatiotemporal escape whereas in the former it is only temporal. As the mobility of the pathogen increases relative to that of the host (e.g., through transmission by wind or hosts that are infected but not killed as larvae), influences of pathogen-free space become less clear, because the potential for a temporary escape from infection by a portion of the host population is reduced.

In these respects such lethal sit-and-wait pathogens (e.g., the nuclear polyhedrosis virus of the Douglas fir tussock moth, *Orygia pseudotsuga* McDun.; see Thompson and Scott, 1979) differ substantially from parasitoids and predators, for whom environmental heterogeneity should often stabilize interactions. The differences result in part from the greater relative mobility, lethality, and the developmental time lags associated with these multicellular consumers. For example, in the absence of refuges created by environmental heterogeneity, predators can readily cause extinction of the prey. In comparable situations, pathogens may themselves become extinct, but extinction of their hosts will be unlikely because of the failure of pathogens to increase in prevalence as the host population falls below a threshold termed N_t. This threshold is the population size at which an infected individual introduced into a susceptible population infects only one susceptible individual throughout the entire period of its infectiousness; for example, for the relationships specified by Eqs. (4)–(7)

$$N_t = (\alpha + b + \gamma)(b + \nu)\beta b \quad (8)$$

[Anderson and May (1981) provide a derivation of N_t.]

In the absence of heterogeneity, pathogens may regulate a host population to a low, stable equilibrium, termed N^*_{nh} (as apparently was the case in the experiments using *A. triboli*). In this case, heterogeneity may destabilize the system by permitting areas in which the host population can expand to high levels, resulting in rises in the average host density above N^*_{nh}, greater eventual production of pathogens per unit area, and hence a decline in the host population perhaps to a level below N^*_{nh}. This decline might occur through increased frequency of infection or the positive effect of dose on parasite-induced mortality, as described in Section II,C.

If, in the absence of heterogeneity, pathogens regulate hosts in a cyclic equilibrium, the effect of heterogeneity on stability will be more complex. Destabilization should tend to occur when the pulses of hosts from pathogen-free areas coincide with the outbreaks of hosts in infected areas (a situation akin to resonance). A temporal staggering of these two events, however, should tend to stabilize the system.

Heterogeneity, therefore, tends to stabilize predator–prey systems by reducing the probabilities of extinction of the consumed species (e.g., Huffaker, 1958), but it may destabilize interactions between lethal, sit-and-wait pathogens and their hosts by permitting periods of unregulated expansion of the consumed species. Formal analyses are needed to elucidate the specific conditions under which heterogeneity contributes to cycles in host–pathogen systems.

C. Sit-and-Wait Pathogens and Cycles in Homogeneous Environments

The duration of time between the release of a pathogen from an infected host into the environment and the subsequent infection of a susceptible host represents one kind of lag between pathogen reproduction and use of host resources. The effects of such lags have been analyzed by Anderson and May (1980, 1981) through modification of Eqs. (1)–(3) to incorporate the number of free-living stages,

$$dX/dt = a(X + Y) + \gamma Y - bX - \nu WX \tag{9}$$

$$dY/dt = \nu WX - (\alpha + b + \gamma)Y \tag{10}$$

$$dN/dt = rN - \alpha Y \tag{11}$$

where ν represents that rate at which a susceptible host will be infected per free-living infective stage; thus, ν is a transmission coefficient analogous to β, and νWX represents the rate at which susceptible individuals in the population become infected. To complete this model the rate at which the population of free-living infective stages changes must be described. Additions to W depend on the average rate at which an infected individual sheds pathogens (λ). Deletions from W depend on the rate at which a free-living pathogen dies or otherwise become noninfective (μ) and the rate at which a free-living pathogen is picked up by the host population (νN); thus,

$$dW/dt = \lambda Y - (\mu + \nu N)W \tag{12}$$

where λY represents the rate at which pathogens are shed from the infected subpopulation and $(\mu + \nu N)W$ represents the rate at which pathogens leave the population of free-living infective pathogens. In this model, low values of α yield persistence of both pathogens and hosts without regulation of the host population (Fig. 1). Increases in α result in linear

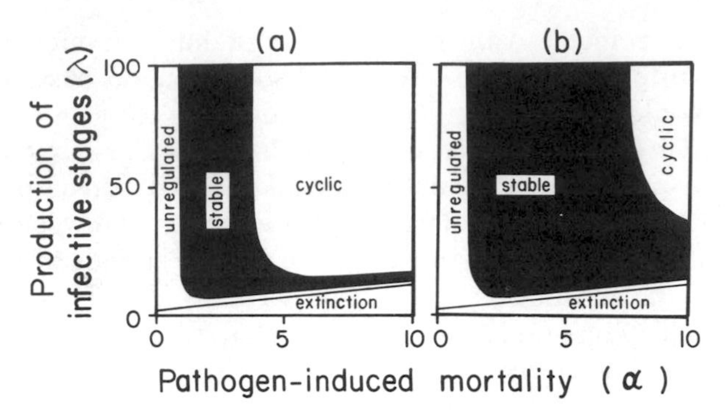

Fig. 1. Effects of the rate of production of infective stages (λ) and rate of pathogen-induced mortality (α) on the regulation of hosts by pathogens, according to Eqs. (9)–(12). In each plot $r = 1.0$ week^{-1}, $b = 1.0$ week^{-1}, and $\gamma = 0$ week^{-1}. The mortality of a pathogen in its free-living infective state (μ) is set equal to 0.02 week^{-1} for plot (a) and 14 for plot (b). Each plot demonstrates that increases in α tend to increase the pathogen-induced cycling of host populations. Comparison of plot (a) with plot (b) shows the reduced tendency for cycling resulting from increased mortality of free-living pathogens. Redrawn from Anderson and May (1981).

increases in the probabilities of extinction of the pathogen, which tend to occur only when λ is extremely small (Fig. 1). In situations in which extinction does not occur, increases in α first lead to stable regulation of the host population and then, generally, to limit cycles, which are increasingly prevalent as λ increases (Fig. 1).

The effects of increases in the longevity of the pathogen outside of the host (i.e., decreases in μ) are analogous to the effects of increases in α. The negative effect on the host population per free-living pathogen is increased and cyclic equilibria become more prevalent (compare Fig. 1a with 1b). Modification of this model to incorporate reductions in fecundity due to infection expands the range of conditions that result in cycles (Anderson and May, 1981).

Anderson and May (1980, 1981) tested predictions of this model against data from one of the most extensively studied species of forest insects exhibiting outbreaks, the larch budmoth, *Zeiraphera diniana* Guenee, which is subject to infection with granulosis virus. The sizes of host populations, amplitudes and periods of the cycles, and the timing of pathogen maxima relative to host maxima derived from the model agreed closely with the corresponding empirical observations. In the model, however, the cause of these changes was a pathogen prevalence that approached 80% during the peak of each cycle. In contrast, the maximum prevalence per cycle in the natural systems was generally between 25 and 50%. The roughness of parameter estimates in the simulation necessitates a cautious interpretation of the close fit between the data and the model; for example, the intrinsic growth rate of the host population (r) and the death rate of free-living pathogens (μ) could be estimated only roughly. According to this model, the period of the cycles is approximately $2\pi(r\mu)^{-1/2}$ (Anderson and May, 1981), and the predicted periods are subject to substantial uncertainty.

The key role of pathogen-induced mortality in determining whether regulation of host populations will be cyclic draws attention to an assumption in great need of empirical evaluation. Models that demonstrate pathogen-induced cycling [e.g., Eqs. (8)–(11)] assume that pathogens can induce high death rates. For example, the numerical analyses of Eqs. (9)–(12) by Anderson and May (1980, 1981) assumed that infected hosts die either from their infections or from other causes, but never recover; stable oscillations could occur only when α was relatively large (e.g., greater than about five times b; see Fig. 1). Such high mortality is consistent with estimates derived from laboratory reports (Anderson and May, 1981), but the high doses typically used in the laboratory (see Table 1) probably inflate mortality above that occurring in nature. Many studies have demonstrated a positive correlation between dose and the mortality of all individuals exposed to the dosage (e.g., Whitcomb *et al.*, 1966; Doane, 1967; Hunter *et al.*, 1973; Henry and Oma, 1974; Lacey and Mulla, 1977; see Table 1), but data generally are not reported in such a way that one can calculate the percentage of death among infected individuals over a relevant range of doses. The few available data suggest that increases in mortality take place over doses spanning greater than one order of magnitude; for example, when the grasshopper *Melanoplus bivittatus* Say was infected with spores of the microsporidian *Nosema cuneatum* Henry and maintained in isolation (which prevented cannibalism), mortality measured relative to number of infected individuals increased from 33 to 94% (see the first three figures of the last column of Table 2; $p < .01$; G test with the Williams correction). Unfortunately, doses in nature are difficult to estimate. To help resolve this issue, naturally occurring distributions of free-living pathogens and susceptible hosts must be simulated in the laboratory in order to quantify pathogen-induced mortality associated with populations on these distributions.

The relevance of mortality in the laboratory to that in nature seems especially tenuous when experiments involve purified pathogens and artificial inoculation or saturation of the environment or when the pathogen does not naturally infect the insect but was being investigated for the purpose of biological control. Although the adaptation of pathogens to hosts may lead to increased pathogenicity (Ewald, 1983; see also data reported by Hunter *et al.*, 1973), investigating new host–pathogen associ-

TABLE 1

Mortality and Characteristics of Infection of Natural Pathogens of Insects That Are Known or Likely to Undergo Outbreaks

Pathogen[a]	Host	b^b	α^b	α/b	Dose	Route of infection[a]	Reference
npv	*Lymantria dispar* (L.) (gypsy moth)	0.08[c]	1.09	13.3	7000[d]	F	Doane (1967)
npv	*L. dispar* (L.) (gypsy moth)	0.08[c]	0.32	3.9	700[d]	F	Doane (1967)
npv	*L. dispar* (L.) (gypsy moth)	0.08[c]	0.04	0.5	70[d]	F	Doane (1967)
npv	*L. dispar* (L.) (gypsy moth)	0.08[c]	0[c]	0	7[d]	F	Doane (1967)
npv	*Malacosoma americanum* Fabr. (eastern tent caterpillar)	0.06	0.68	11.6	5×10^6[e]	F	Smirnoff (1967)
npv	*Cadra cautella* Walker (almond moth)	0.03	0.58	17.0	400[d]	F	Hunter *et al.* (1973)
npv	*Hyphantria cunea* Drury (fall webworm)	0–0.02[f]	0.59	≥29.5	10^3[g]	F	Nordin and Maddox (1972)
Rhabdion virus	*Oryctes rhinocerus* L.	0.14	0.24	1.7	2×10^{-4}[h]	I	Zelazny (1973)
Erwinia sp.?	*Colladonus montanus* Van Duzee	0.06	0.15	2.7	NG[i]	F	Whitcomb *et al.* (1966)
Pleistophora shubergi	*Hyphantria cunea* Drury (fall webworm)	0–0.02[f]	0.02	≥1.0	10^5[g]	F	Nordin and Maddox (1972)
Mattesia dispora	*Laemophloeus minutus* Oliv.	0.025	0.06	2.4	NG[i]	F	Finlayson (1950)

[a] npv, Nuclear polyhedrosis virus; F, feeding on contaminated food or drink; I, injection.

[b] Units of α and b are $week^{-1}$. The values differ slightly from those of previous summaries (Anderson and May, 1980, 1981) because of (1) different groupings of data and (2) a slightly different method of calculation, which involved a weighted average over all intervals of observation.

[c] Mortality data were not given as a function of time for uninfected individuals; however, because the percentage of mortality of controls in a related set of data was slightly (but not significantly) greater than that of the 7-cm^{-2} dosage (4 deaths out of 166 larvae versus 0 deaths out of 30), the mortality associated with this dose ($0.082\ week^{-1}$) was assumed to represent the mortality not due to infection. All data from Doane (1967).

[d] Units are number of polyhedrus inclusion bodies (PIBs) per square centimeter of food. At a dose of 7000 $PIBs/cm^2$, a larvae would consume approximately 4×10^4 during its first three instars (Doane, 1967). Assuming that each inclusion body contains between 10 and 100 viruses, the dosage for this treatment would be between 4×10^5 and 4×10^6 viruses.

[e] Units are presumably number of polyhedra per dose, but Smirnoff's description is unclear.

[f] No deaths occurred in the control group during the 12-day study period, but approximately 0.02 deaths per week in the infected groups were due to causes other than infection.

[g] Units are PIBs per larvae, incorporating Nordin and Maddox's assumption that only 10% of the available PIBs were consumed.

[h] This dose represents milliliters of undiluted virus-containing hemolymph extracted from infected individuals.

[i] NG, No value given in report.

TABLE 2

Mortality of *Melanoplus bivittatus* with Different Doses of *Nosema cuneatum*[a]

Dose (number of spores)	Number of grasshoppers: Total tested	Examined for infection	Infected	Total dead	No. dead / no. infected	No. dead / no. tested
5.5×10^3	20	20	3	1	0.33	0.05
5.5×10^4	20	19	14	8	0.57	0.40
5.5×10^5	20	18	16	15	0.94	0.75
Control	46	—	—	3	—	0.07

[a] Data from Henry (1971).

ations for purposes of biological control will probably create a bias in the literature toward reporting highly pathogenic associations (i.e., those presumed to be good agents for biological control) through emphasis on associations whose pathogenicity is above that which can be maintained in nature.

Even when large doses are not intentionally used, laboratory procedures may tend to yield doses far greater than those in nature. For example, Finlayson (1950) maintained 100 cucujid beetles, *Laemophloeus minutus,* on flour and brewer's yeast that had been used previously by beetles infected with the protozoan *Mattesia dispora* Naville. Forty *L. minutus* maintained on uncontaminated flour were used as a control. The pathogen-induced mortality was over twice the mortality of the uninfected population (Table 1). In nature, however, food is rarely if ever so spatially concentrated and temporally stable; the flour probably yielded unnaturally high doses of *Mattesia*.

Although the value of applying such mortality data to population dynamics in nature is dubious, some systems probably do yield high doses of pathogens; for example, the density of the tent caterpillar, *Malacosoma disstria* Hubn., can be as great as 10,000–20,000 per tree, and calculations suggest that the density of nuclear polyhedrosis virus in such environments may be comparable to that used in laboratory situations (Stairs, 1972).

In sum, at first glance values of pathogen-induced mortality from the literature lead one to believe that the high mortality required for pathogen-induced cycling is ubiquitous. Recognizing the biases inherent in obtaining these values, however, forces a more cautious conclusion: The levels of pathogen-induced mortality needed to generate cycling might be widespread in nature, especially in agricultural systems and when hosts are gregarious, but it will not be possible to draw definite conclusions until natural doses of natural pathogens are quantified and used in experiments designed to provide estimates of pathogen-induced mortality.

The preceding model treats mortality as a parameter, independent of infective doses; thus, the model implicitly assumes that mortality is a step function of the number of pathogens encountered. However, as the density of infectious hosts increases, the number of pathogens encountered per susceptible host will tend to increase. If the magnitude of the infective dose is positively correlated with pathogen-induced mortality (e.g., see Table 2), increased pathogen density should result in increased pathogen-induced mortality. Inapparent, latent, and lethal infections all can occur with a single kind of pathogen in a single host species (Bailey *et al.*, 1963; Tanada and Tanabe, 1964; Smith, 1967). Yet the relative frequencies of such infections in nature are not well understood. Nor is it clear whether the frequency distributions of pathogenicity within host populations tend to be narrow or broad. Although the functional relationship between α and host density cannot be deduced from available data, the rather gradual increases in mortality with increases in dosage (e.g., see Table 2) suggest that step functions, or even steeply rising sigmoidal functions, are unlikely.

In the absence of mathematical analyses it is not clear whether treating pathogen-induced mortality as a variable dependent on W would tend to have a stabilizing or destabilizing effect. For example, if we represent such pathogen-induced mortalities by the term $\hat{\alpha}W$, where $\hat{\alpha}$ is the pathogen-induced mortality per pathogen, $\hat{\alpha}W$ will generally be greater than α at high host densities but less than α at low host densities (assuming that the value used for α is an average of measured values over a range of host densities). The difference between $\hat{\alpha}W$ and α will be most dramatic when N is large because a greater density of pathogens can be produced. According to one line of reasoning, when the population is regulated at a large size by the pathogen, $\hat{\alpha}W$ will be greater than α, and parameters resulting in a stable equilibrium when used in Eqs. (9)–(12) may instead yield limit cycles. Because $\hat{\alpha}W$ will be less than α at low densities, extinction of the pathogens will be less feasible. The net result of the substitution might therefore be an increase in the range of conditions over which limit cycles will occur.

Alternatively, by reducing the negative effects of pathogens at low host densities and increasing the negative effects of pathogens as host densities are displaced above their equilibrium values, the substitution might damp out oscillations, especially if the changes in pathogenicity around the equilibrium values are great. Clearly, to resolve this issue, the functional relationships between pathogen-induced mortality and dose must be determined experimentally and then incorporated into the mathematical models.

III. EVOLUTIONARY CONSIDERATIONS

Unlike predators and parasitoids, pathogens often have generation times several orders of magnitude shorter than those of the organisms on which they feed. As a result of this asymmetry, changes in gene frequencies of pathogens can occur within a few host generations or even a single generation. The parameters of the models describing the population dynamics of host–microparasite systems, therefore, may change on time scales that are similar to those over which the sizes of host populations are changing. When pathogen-induced mortality and/or pathogen-reduced fecundity of hosts are high, both host and pathogen may undergo substantial changes in gene frequency during such short time scales (May and Anderson, 1979). When such negative effects of pathogens are low, the changes in pathogen characteristics represent the primary complication. These complex interactions are to be understood thoroughly only through experiments in which both the key parameters and variables (e.g., gene frequency) are held constant and experimentally manipulated.

Some insight into the interplays between evolution and population dynamics, however, can be obtained by analyzing characteristics of hosts and parasites in terms of fitness costs and benefits. As already described, pathogens with long-lived, free-living stages, high pathogenicity, and a large number of propagules produced per infection are especially likely to induce cycling in their host populations. Analysis of these characteristics from an evolutionary perspective underscores the significance of this conclusion. Specifically, if the situation is analyzed in terms of the fitness costs and benefits accrued by the pathogens, this combination of characteristics seems especially likely to occur. To begin with, pathogens transmitted by a sit-and-wait mechanism (e.g., as opposed to transovarial or vector-borne transmission) should benefit from characteristics that increase survival outside the host; hence, characteristics resulting in prolonged survival outside the host should be especially well developed for such pathogens. Through increased reproduction within hosts, pathogens incur both costs and benefit. The benefit is that a greater numer of pathogens are shed into the environment. One important cost is decreased mobility or death of the host, which generally will result in a decrease in the number of potential contacts with new hosts. High levels of pathogenicity should evolve when such benefits are exceptionally great and/or costs exceptionally low. A long-lived free-living stage is one characteristic that should lower the rate at which the above-mentioned cost of immobilization is incurred as a function of the level of pathogen reproduction. Essentially this characteristic permits an increased use of the movements of susceptible hosts for transmission, thereby reducing dependence on the mobility of the infected host. The evolutionary consequence of these relationships is that high levels of reproduction within a host and hence high pathogenicity should tend to occur among pathogens with long-lived free-living stages; thus, evolution should tend to favor at least one combination of characteristics that leads to pathogen-induced cycling of insect hosts.

IV. CONCLUSIONS

Subjective evaluations of the agreement between predictions of host–pathogen models and the empirical observations [e.g., as demonstrated by Anderson and May (1980, 1981)] have focused on the potential importance of pathogens as a cause of fluctuations in insect populations. Further advances in this area require the development of alternative predictions from models based on interactions with pathogens as well as from models based on interactions with predators, other parasites, and food sources of the host. Only then can the relative importance of these different classes of interactions be distinguished. For example, in most models the peak of pathogen prevalence would tend to occur slightly after the peak in host population regardless of whether pathogen populations were causing the decline in the host population or merely responding to changes in host populations due to other factors. Alternative models might, however, differ in their predictions about the magnitude of the prevalence peaks or the degree to which the prevalence peaks lag behind the peaks in host abundance. Such differences between the prediction of alternative models should be determined and then compared with the actual data by means of statistical tests.

The most immediate obstacle to the use of such tests is the need to assign values to the components of the models. Not only do errors associated with the rough estimates of parameters such as r, μ, and α have a strong potential impact on the resulting population dynamics, but there is also uncertainty as to whether quantities such as pathogen-induced mortality are best treated as parameters or as variables. Data such as those from *N. cuneatum* and *M. bivittatus* (see Section II,C) suggest that treating these quantities as variables is appropriate. The models suggest that relatively benign classes of pathogens such as cytoplasmic polyhedrosis viruses will tend not to cause cycles. Relatively severe classes such as nuclear polyhedrosis viruses may cause cycles if pathogen densities are sufficiently high to yield heavy doses per susceptible individuals. The most pressing need is for experiments designed to obtain accurate values for the components of the models. Coupling such estimates with strong inference tests using both observational data and experimental manipulations of the key parameters and variables (e.g., pathogen-induced mortality through the use of different strains; the presence or absence of pathogens) should greatly clarify the role of pathogens in the population cycles of insects.

ACKNOWLEDGMENTS

This work was supported by a NATO–National Science Foundation Postdoctoral Fellowship and by a National Institutes of Health, Biomedical Research Support Grant awarded through Amherst College. I thank R. M. May, R. M. Anderson, P. Barbosa, and J. C. Schultz for comments on the manuscript and S. Bittenbender for assistance in preparing the manuscript.

REFERENCES

Alexander, M. (1981). Why microbial predators and parasites do not eliminate their prey and hosts. *Annu. Rev. Microbiol.* **35,** 113–133.

Anderson, R. M., and May, R. M. (1980). Infectious diseases and population cycles of forest insects. *Science* **210,** 658–661.

Anderson, R. M., and May, R. M. (1981). The population dynamics of microparasites and their invertebrate hosts. *Philos. Trans. R. Soc. London, Ser. B* **291,** 451–524.

Anderson, R. M., and May, R. M. (1982). Coevolution of hosts and parasites. *Parasitology* **85,** 411–426.

Bailey, L., Gibbs, A. J., and Woods, R. D. (1963). Two viruses from the adult honey bee (*Apis mellifera* Linn.). *Virology* **21,** 390–395.

Balashov, Y. S. (1984). Interaction between blood-sucking arthropods and their hosts, and its influence on vector potential. *Annu. Rev. Entomol.* **29,** 137–156.

Baltensweiler, W. (1964). *Zeiraphera griseana* Hubner (Lepidoptera: Tortricidae) in the European Alps: A contribution to the problem of cycles. *Can. Entomol.* **96,** 792–800.

Bremerman, N., and Pickering, H. J. (1983). A game-theoretical model of parasite virulence. *J. Theor. Biol.* **100,** 411–426.

Doane, C. C. (1967). Bioassay of nuclear-polyhedrosis virus against larval instars of the gypsy moth. *J. Invertebr. Pathol.* **9,** 376–386.

Dubos, P. W. (1965). "Man Adapting." Yale Univ. Press, New Haven, Connecticut.

Ewald, P. W. (1983). Host–parasite relations, vectors, and the evolution of disease severity. *Annu. Rev. Ecol. Syst.* **14,** 465–485.

Finlayson, L. H. (1950). Mortality of *Laemophloeus* (Coleoptera: Cucujidae) infected with *Mattesia dispora* Naville (Protozoa, Schizogregarina). *Parasitology* **40,** 261–264.

Grossman, Z. (1980). Oscillatory phenomena in a model of infectious diseases. *Theor. Popul. Biol.* **18,** 204–243.

Henry, J. E. (1971). *Nosema cuneatum* sp. n. (Microsporida: Nosematidae) in grasshoppers (Orthoptera: Acrididae). *J. Invertebr. Pathol.* **17,** 164–174.

Henry, J. E., and Oma, E. A. (1974). Effects of infections by *Nosema locustae* Canning, *Nosema acridrophagus* Henry, and *Nosema cuneatum* Henry (Microsporidia: Nosematidae) in *Melanoplus bivittatus* (Say) Orthoptera: Acrididae. *Acrida* **3,** 223–231.

Hoeprich, P. D. (1977). Host–parasite relationships and the pathogenesis of infectious disease. *In* "Infectious Disease" (P. D. Hoeprich, ed.), pp. 34–45. Harper & Row, New York.

Huffaker, C. B. (1958). Experimental studies on predation: Dispersal factors and predator–prey oscillations. *Hilgardia* **27,** 343–383.

Hunter, D. K., Hoffman, D. F., and Collier, S. J. (1973). Pathogenicity of a nuclear polyhedrosis virus of the almong moth *Cadra cautella. J. Invertebr. Pathol.* **21,** 282–286.

Hutchinson, H. (1978). "An Introduction to Population Ecology." Yale Univ. Press, New Haven, Connecticut.

Klomp, H. (1966). The dynamics of a field population of the pine looper (Lep., Geom.). *Adv. Ecol. Res.* **3,** 207–303.

Krebs, C. J. (1978). "Ecology: The Experimental Analysis of Distribution and Abundance," 2nd ed. Harper & Row, New York.

Lacey, L. A., and Mulla, M. S. (1977). Evaluation of *Bacillus thuringiensis* as a biocide of blackfly larvae (Diptera: Simulidae). *J. Invertebr. Pathol.* **30,** 46–49.

Levin, S. (1983). Some approaches to modelling of coevolutionary interactions. *In* "Coevolution" (M. Nitecki, ed.), pp. 21–65. Univ. of Chicago Press, Chicago, Illinois.

Ludwig, D., Jones, D. D., and Holling, C. S. (1978). Qualitative analysis of insect outbreak systems: The spruce budworm and forest. *J. Anim. Ecol.* **47,** 315–332.

May, R. M. (1974). "Stability and Complexity in Model Ecosystems." Princeton Univ. Press, Princeton, New Jersey.

May, R. M., and Anderson, R. M. (1978). Regulation and stability of host–parasite population interactions. II. Destabilizing processes. *J. Anim. Ecol.* **47,** 249–267.

May, R. M., and Anderson, R. M. (1979). Population biology of infectious diseases. Part II. *Nature (London)* **280,** 455–461.

May, R. M., and Anderson, R. M. (1983). Epidemiology and genetics in the coevolution of parasites and hosts. *Proc. R. Soc. London, Ser. B* **219,** 281–313.

May, R. M., Conway, G. R., Hassell, M. P., and Southwood, T. R. E. (1974). Time delays, density dependence and single species oscillations. *J. Anim. Ecol.* **43,** 907–915.

Neilson, M. M., and Morris, R. F. (1964). The regulation of European sawfly numbers in the maritime provinces of Canada from 1937–1963. *Can. Entomol.* **96,** 773–784.

Nicholson, A. J. (1958). The self-adjustment of populations to change. *Cold Spring Harbor Symp. Quant. Biol.* **22,** 153–173.

Nordin, G. L., and Maddox, J. V. (1972). Effects of simultaneous virus and microscopic infection of larvae of *Hyphantria cunea. J. Invertebr. Pathol.* **20,** 66–69.

Palmieri, J. R. (1982). Be fair to parasites. *Nature (London)* **298,** 220.

Park, T. (1948). Experimental studies of interspecies competition. I. Competition between populations of the flour beetles *Tribolium confusum* (Duval) and *Tribolium casteneum* (Herbst). *Ecol. Monogr.* **18,** 207–307.

Ricklefs, R. E. (1979). "Ecology," 2nd ed. Chiron Press, New York.

Simon, H. J. (1960). "Attenuated Infection." Lippincott, Philadelphia, Pennsylvania.

Smirnoff, W. A. (1967). Effect of gamma radiation on the larvae and the nuclear-polyhedrosis virus of the Eastern tent caterpillar, *Malacosoma americanum. J. Invertebr. Pathol.* **9,** 264–266.

Smith, K. M. (1967). "Insect Virology." Academic Press, New York.

Smith, T. (1934). "Parasitism and Disease." Princeton Univ. Press, Princeton, New Jersey.

Stairs G. (1972). Pathogenic microorganisms in the regulation of forest insect populations. *Annu. Rev. Entomol.* **17,** 355–373.

Swellengrebel, N. H. (1940). The efficient parasite. *Proc. Int. Congr. Microbiol., 3rd,* pp. 119–127.

Tanada, Y., and Tanabe, A. M. (1964). Response of the adult of the armyworm, *Pseudaletia unipuncta* (Haworth), to cytoplasmic-polyhedrosis-virus infection in the larval stage. *J. Insect Pathol.* **6,** 486–490.

Thomas, L. (1972). Germs. *N. Engl. J. Med.* **287,** 553–555.

Thompson, C. G., and Scott, D. W. (1979). Production and persistence of nuclear-polyhedrosis virus of the Douglas-fir tussock moth, *Orygia pseudotsugata* (Lepidoptera: Lymantriidae), in the forest ecosystem. *J. Invertebr. Pathol.* **33,** 57–65.

Varley, G. C. (1949). Population changes in German forest pests. *J. Anim. Ecol.* **18,** 117–122.

Whitcomb, R. F., Shapiro, M., and Richardson, J. (1966). An Erwinia-like bacterium pathogenic to leafhoppers. *J. Invertebr. Pathol.* **8,** 299–307.

Zelazny, B. (1973). Studies on *Rhabdionvirus oryctes*. III. Incidence in the *Oryctes rhinocerus* population of western samoa. *J. Invertebr. Pathol.* **22,** 359–363.

Chapter 12

The Role of Natural Enemies in Insect Populations

PETER W. PRICE

Department of Biological Sciences
Northern Arizona Univeristy
Flagstaff, Arizona 86011
and
Museum of Northern Arizona
Flagstaff, Arizona 86001

I. INTRODUCTION

Our views on the role of natural enemies in insect population outbreaks are colored by the history of our science of ecology. Therefore, a historial perspective is important in evaluating how well our conventional wisdom matches the evidence for and against the proposition that natural enemies are a significant part of the regulation of insect populations. Areas that have provided information on the role of natural enemies are treated roughly in historical order. This chapter deals with vertebrate and invertebrate predation and parasitoids as natural enemies of insects; pathogens are discussed in Chapter 11.

"Regulation" in this chapter means the maintenance of a population at an equilibrium level defined by "finite average populations around which the animal numbers fluctuate with steady average variances" (May, 1973, p. 17). The implication is that regulation occurs before resources are depleted, so that intraspecific competition for food and depletion of food resources and economic damage to crops are not evident. Ultimately, if all other regulating factors fail, regulation by intraspecific competition and resource depletion may come into play, suggesting that populations regulated in this way are unstable, characterized by relatively large fluctuations in population numbers (May, 1973). In this chapter I examine the evidence that natural enemies regulate insect populations. "Control" means applied regulation.

II. ENEMIES AS AGENTS IN NATURAL SELECTION

Early naturalists were impressed by the array of insect defenses against visually hunting predators. Modern science has strengthened the view that predators act as potent selective agents on the evolution of insect populations and species. Crypsis, catalepsis, aposematic coloration, Müllerian and Batesian mimicry, chemical defense, polymorphism, protean displays, and cellular defense against internal parasitoids illustrate the diverse evolutionary responses to enemies. Their commonness in nature demonstrates the pervasive effects that enemies have had, and still have, on insect populations.

Indeed, responses of insects to hunting predators have provided evolutionary biologists with some of the best evidence available on the force of selection and the precision of the evolutionary process. Commonly, a defensive mechanism confers a 30% advantage or more over individuals without the defense. This has been demonstrated for flash coloration (32–63% advantage), eyespots (76% advantage), industrial melanism in polluted environments (52% advantage), nonmelanic moths in unpolluted environments (66% advantage), and protean displays to escape hunting bats (40% advantage) (examples reviewed in Price, 1984). Such strong selective pressure is also illustrated by the precision with which models are copied by mimetic insects. These models include backgrounds against which cryptic species are exposed during the day and aposematic species for Batesian mimics.

Probably every insect species can be classified into one form of defense against enemies or another. No naturalist today is likely to dispute the importance of natural enemies in the evolutionary biology of any insect species. There is good evidence that natural selection continues to "tune" insect populations to changing environments, as in studies on *Biston betularia* (L.) in relation to industrial mechanism, and that enemies play a significant role (e.g., Kettlewell, 1973; Bishop and Cook, 1979, 1980). That parasitoids act as strongly as vertebrate predators in natural selection for defense is not well documented, although the evidence compiled by Paul Gross (review in preparation) suggests that they are equally potent. Salt (1970) reviewed internal insect defense against parasitoids and parasites.

Such consensus on the role of natural enemies in population regulation of insects is not likely to be reached. The kinds of evidence are extremely diverse, the difficulties with analysis and detection of effects are significant, and insect taxa are so numerous and diverse in their ecology that it would be remarkable if simple generalizations became apparent. Nevertheless, through much of the history of insect population ecology, natural enemies have been regarded as an essential part of the regulating factors in insect populations. A brief historical view of the various kinds of evidence will show the reasons for opinion becoming polarized in this direction.

III. EVIDENCE FROM ATTEMPTS AT BIOLOGICAL CONTROL

The history of applied biological control has been covered effectively by Doutt (1964), DeBach (1974), and Van den Bosch and Messenger (1973). The spectacular early successes of biological control helped to establish the important role of natural enemies in the population regulation of epidemic insect and plant populations: cottony-cushion scale in California, 1888–1889; sugarcane leafhopper in Hawaii, 1904–1920; sugarcane beetle borer in Hawaii, 1907–1910; citrus whitefly in Florida, 1910–1911; prickly pear cactus in Australia, 1920–1925; coconut moth in Fiji, 1925 (examples and dates from DeBach, 1974). More recent examples have reinforced the view that single species of enemy can be of paramount importance in regulating insect pests (Huffaker, 1971; DeBach, 1974; Van den Bosch and Messenger, 1973): citriculus mealybug in Israel, 1939–1940; winter moth in Canada, 1955–1960; California red scale in California, 1957.

These studies on biological control under essentially natural conditions can be likened to experiments using prerelease populations as controls and postrelease populations as treatments. However, many factors change with time, so alternative hypotheses can be invoked to account for population regulation of insect pests from epidemic to endemic levels.

However, the most parsimonious hypothesis is usually that natural enemies were of prime importance.

Several important lessons have been learned from these experiments:

1. Populations of insects can erupt when introduced into new geographic areas without their natural enemies, the strong implication being that indigenous populations are regulated by enemies.

2. Populations of weeds can become exceedingly large when the weeds are released in new geographic areas without their naturally occurring herbivorous insects (see Chapter 7). Since such population eruption of weeds does not occur in native regions, the implication is that herbivores may be important in regulating populations and that their natural enemies do not suppress this ability. When herbivores are introduced for the biological control of weeds without their natural enemies, however, they can become spectacularly abundant and devastating to plant populations.

3. When herbivores are introduced to regulate weedy plant populations where herbivore outbreaks are desirable, failure of biological control frequently results from natural-enemy regulation of the herbivore population. Goeden and Louda (1976) estimated that about half of the 23 weed control attempts they reviewed were unsuccessful for reasons ranging from prevention of herbivore establishment to reduced impact of herbivores.

4. Many attempts at biological control have failed. Even when enemies have become established they have remained at low levels, implying that, at least in exotic situations, the regulatory role of natural enemies is unlikely to be ubiquitous [e.g., *Cryptochaetum iceryae* (Will.) introduced for the control of cottony-cushion scale]. In addition, biological control is not attempted on outbreak species of no medical, agricultural, or forest importance or on species that typically remain at low population levels—the vast majority of species. Also, little work involving predators and parasitoids has been done on some outbreak species such as grasshoppers and flies of medical importance such as mosquitoes.

5. When released in exotic locations, natural enemies can have a much greater impact than when released in indigenous areas. Coquillett received only 129 individuals of the vedalia beetle from Koebele during the Australian summer of 1888–1889, because they were so rare (Doutt, 1964). In contrast Koebele sent about 12,000 individuals of the parasitic fly *Cryptochaetum iceryae*, also an enemy of cottony-cushion scale, but it was largely the vedalia beetle that became important as the biological control agent on the cottony-cushion scale. The tachinid fly *Cyzenis albicans* (Fall.), is much more important in regulating winter moth larvae in Nova Scotia than in its indigenous locations in Europe (Hassell, 1980).

The reasons for such differences have not been adequately studied. One population dynamics model for *Cyzenis albicans* works both for a population in Wytham Wood in England and for populations in Nova Scotia, Canada (Hassell, 1980). The higher densities of hosts in Nova Scotia resulted in stronger regulation by *Cyzenis*. Such an explanation does not apply to *Cryptochaetum,* which usually became less important on the high-density populations of cottony-cushion scale in California than in native populations in Australia. Autecological factors such as adaptation to new environments were probably important. Such differences in exotic and indigenous populations are in need of more study.

Studies of biological control have provided important insights into the role of natural enemies in regulating insect population outbreaks. Many examples show that regulation is achieved largely by the action of natural enemies. The many failed attempts at regulation, even though establishment of enemies has been achieved, demonstrate equally well that natural enemies are unlikely to be important in the dynamics of every outbreak insect species and in every location where these species occur.

IV. PESTICIDES AND NATURAL ENEMIES

It has been shown repeatedly that after the use of pesticides target pests may increase and new pests may materialize (Ripper, 1956; Rudd, 1964; Varley *et al.,* 1973). These responses are commonly so rapid that evolutionary adaptation to toxins can be excluded as a short-term mechanism. One frequently recognized effect of pesticide application is that natural enemies suffer greater impact than the pest species. Some stages of enemies are more mobile than their prey, and their exposure to insecticide is greater. Predators eat many prey and concentrate toxins, and poisoned cadavers of this pest may be eaten by predators. In addition, natural enemies are frequently more innately susceptible to pesticides than the herbivores on which they feed (Mullin and Croft, 1985). Ripper (1956) cited 50 cases of resurgence of pest populations after insecticide treatment. He evaluated three hypotheses accounting for such resurgence: (1) reduction of natural enemies, (2) favorable influences of pesticides such as stimulation of natality or improvements in plant quality, and (3) removal of competing species. He found the cases for hypotheses (2) and (3) unsound, but on the role of natural enemies he concluded, "There is a great deal of evidence that many pesticides much reduce the natural enemies of the pests or of potential pests, and there is proof that the elimination of the natural enemies is, at least in some cases, the sole cause of the resurgence of pests'' (p. 416). Detailed studies have shown that natural enemies are frequently the major regulating factor before insecticide treatment. Moreton (1969) reviewed several cases in detail, providing much evidence that pest species become more abundant after insecticide treatment because their enemies have been suppressed. He concentrated on aphids, caterpillars of butterflies and moths on crucifer crops, cabbage root fly, fruit tree red spider mite, and other spider mites. Moreton (1969, p. 82) called the fruit tree red spider mite a ''man-made pest'' because it was unknown as a pest before 1923, when weaker pesticides were used. Red spider mites became common because lichens on tree trunks and branches were killed, after tar-oil washes were introduced to kill aphid eggs and other pests. Enemy populations were reduced catastrophically through direct death, loss of protection from lichens, and increased predation.

Cases of insecticide treatment can be regarded as experiments on the role of natural enemies. The experimental approach is often better than in biological control, because treatments and controls can be run simultaneously and on a relatively small scale. The difficulty of interpreting results is greater, however, because the application of toxins has multiple and complex effects. Thus, unraveling the real mechanisms of increased pest abundance or emergence of new pests is difficult. However, many studies have been careful enough to make a convincing case that in the absence of natural enemies many species reach outbreak proportions (e.g., Ripper, 1956; Moreton, 1969). The judicious use of selective and biodegradable insecticides may offer an experimental tool for longer-term studies of insect population dynamics in the absence of enemies.

V. POPULATION MODELS

Models of predator–prey and parasitoid–host population interactions have been influenced from their inception by early concepts of the balance of nature, a state of equilibrium between interacting species, and the nature of insect and enemy population cycles. Lotka (1925) was clearly impressed by the empirical observations of early entomologists such as L. O. Howard and W. R. Thompson. For example, he quoted Howard (1897, p. 48) as follows: ''With all very injurious lepidopterous larvae we constantly see a great fluctuation in numbers, the parasite rapidly increasing immediately after the increase of the host species, overtaking it numerically, and reducing it to the bottom of another ascending period of development.'' This pattern of change was effectively mimicked by Lotka's (1925) equations, which were actually based on insect host–parasitoid interactions.

The Lotka (1925) and Volterra (1926) equations stimulated the formulation of a new model by Nicholson and Bailey (1935). Further modeling aimed to produce stability in the host–enemy interaction based on nonlinear functional responses of enemies to prey populations (e.g., Holling, 1959, 1965; Tinbergen and Klomp, 1960; Holling and Ewing, 1971) and interference between enemies as host populations increase (Hassell and Varley, 1969). Later, evidence accumulated that many invertebrate predators and parasitoids show sigmoid (type III) functional responses in experimental arenas, which may play a role in stabilizing prey populations at relatively low levels (Hassell *et al.*, 1977; Hassell, 1978). Sophisticated models explaining many dynamic aspects of predator–prey and parasitoid–host relationships have resulted. Much of the development of this field is reviewed by Hassell (1978, 1981) and May (1981). Recent developments in modeling suggest that the patchiness of the environment is an important factor in host–enemy interactions (e.g., May, 1978; Hassell, 1982, 1984), but we know very little about its importance in nature.

Modeling of ecological interactions has been held in high esteem, partly because it is one of the most rigorous aspects of ecology in the sense that assumptions are clearly stated and deductions are reached using defined logic. Therefore, this field has a strong impact on the way we perceive nature. Models repeatedly suggest that natural enemies have the potential to regulate insect populations. However, models do not describe nature; they tell us only what is possible or likely. The sophistication of the models is far beyond our current understanding of nature, so greater emphasis on testing models in natural systems is desirable.

VI. LIFE TABLE ANALYSIS

Life tables of human populations used by actuaries for estimating insurance rates per age class were the basis for developing similar tables for animals, an undertaking pioneered by Pearl and Parker (1921), Pearl and Miner (1935), and Deevey (1947). The major impetus for the analysis of natural insect populations came from Morris and Miller (1954), who added a "factor responsible for death" (d_xF) column and quantified these factors for each age interval of the spruce budworm life history. Their approach offered a clear method for organizing sampling programs, tabulating data, and analyzing results. Soon the methods of single-factor analysis (Morris, 1959, 1963) and key-factor analysis (Varley and Gradwell, 1960, 1968) were developed. Life tables and these methods for analysis have been used extensively since that time to study the population dynamics of insects, interpret the causes of population change, and define the factors responsible for population regulation (e.g., Harcourt, 1969; Southwood,

1975; Podoler and Rogers, 1975; Caughley and Lawton, 1981; Dempster, 1983; Price, 1984).

However, the analytical tools for life table data do not provide an unambiguous picture of cause-and-effect relationships. The methods and strengths of making deductions from such data have been widely debated (e.g., Hassell and Huffaker, 1969; Morris and Royama, 1969; Maelzer, 1970; St. Amant, 1970; Luck, 1971; Podoler and Rogers, 1975; Royama, 1977, 1981a,b; Strong, 1984; Hassell, 1985). Of central concern is that correlation does not imply causation, and causation can normally be established by what Morris (1969) called "process studies" involving detailed experiments on mechanisms causing population change (see also Wellington, 1957, 1977; Eberhardt, 1970; Morris, 1971; Price, 1971; McNeill, 1973). Royama (1977) made the distinction between type A density dependence, which is *causally related* to the rate of change of population density, and type B density dependence, which is *statistically related,* not causally related, to population change. He then pointed out that "the concept of density-dependent regulation in terms of the Type B notion is simply an uninspiring tautology, from which we get little insight into the type of unknown generating mechanisms" (p. 32). Royama also explained that to analyze for a type A effect it is incorrect to compare population change directly with the type A factor: "To do this properly, the effect of the Type A density-dependent factors must be removed" (p. 33). Then the dynamic properties of the population without this factor can be explored. Such removal can be achieved by experimental means or perhaps with modeling, but it is seldom practiced in the analysis of life table data.

Experimental "removal" of a suspected causative (type A) factor occurs in biological control attempts when natural enemies are removed by transporting an insect to a new location and its population dynamics are observed largely in the absence of natural enemies. Natural enemies are then introduced so that this treatment effect can be observed. Where life table data are available before and after the introduction of natural enemies, the evidence for their effect as type A density-dependent regulators of insect populations is particularly strong. An excellent example was reported by Embree (1966, 1971), who documented the effects of the parasitoids *Cyzenis albicans* and *Agrypon flaveolatum* (Grav.) on winter moth populations in Nova Scotia (Fig. 1). The combined mortality caused by these introduced enemies reached almost 80% of the larval host population in the seventh year of a winter moth outbreak, and by the eighth year the population had crashed to an endemic level. Unfortunately, such good data on the indigenous populations of enemies and before and after their introduction are almost unique.

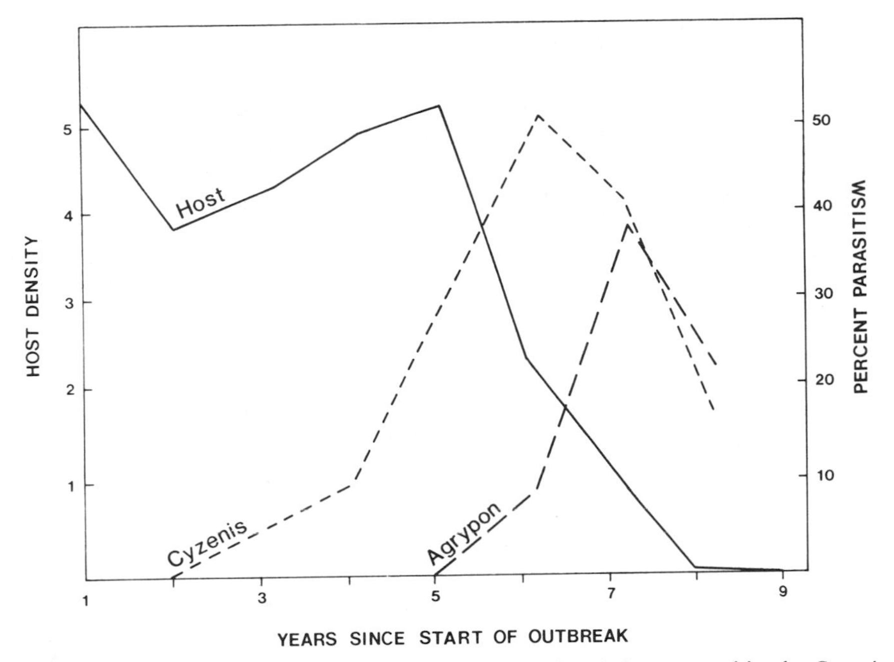

Fig. 1. Change in host density of winter moth larvae in relation to parasitism by *Cyzenis albicans* and *Agrypon flaveolatum*. After Embree (1971).

In spite of the problems of analyzing life table data and the shortage of good experimental work on process studies, life table analysis has provided us with the largest body of data for evaluating the role of natural enemies in relatively natural insect populations. One simple approach to evaluating the role of natural enemies is to develop survivorship curves for cohorts of insects and to subtract from these the effects of natural enemies. Then a comparison can be made of generation survival with enemies present and with no enemies, and the population-depressing effect of enemies can be estimated. This can be done, for example, using two populations of spruce budworm, one epidemic and one relatively low, with life tables for each provided by Morris and Miller (1954). At both low and high budworm densities this analysis reveals a strong impact by natural enemies (Fig. 2). At the low population level, enemies caused an eventual 17.5% difference (18% survival without enemies versus 0.5% with enemies) in the number of adults per unit of branch surface. The population would be 36 times higher in the absence of enemies. At the high population level, enemies caused a 7.95% difference (8% survival without enemies versus 0.05% with enemies) and the population would

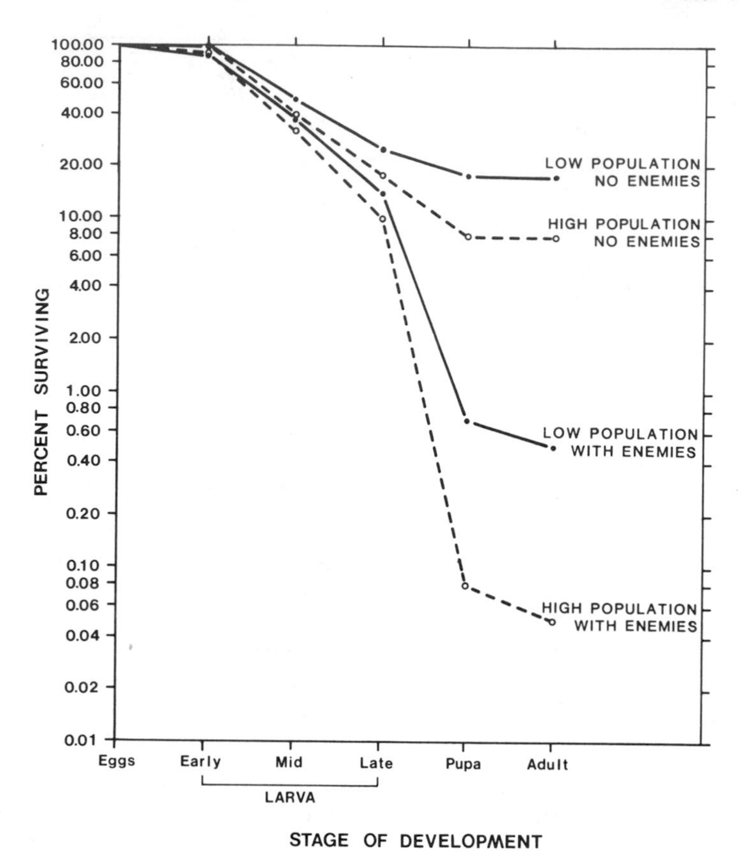

Fig. 2. Survivorship curves for spruce budworm, *Choristoneura fumiferana* (Clem), derived from life tables developed by Morris and Miller (1954). The complete survivorship curves (with enemies) are for the 1952–1953 generation in the Green River watershed, New Brunswick, for a large population (dashed lines, open circles) and a relatively small population (solid lines, closed circles). Survivorship curves are also given in which the effects of predators and parasitoids have been removed (no enemies). Percentage of survival is on a logarithmic scale.

have been 160 times higher without natural enemies. Many life tables on other species would reveal a similar impact.

These figures are impressive but probably very misleading. They assume that no compensation exists in the system. For example, in the large population, if no parasitoids or predators had killed eggs, when they had their strongest effect, subsequent losses in dispersal, winter mortality, and starvation would probably have been greater. Such compensating factors cannot be evaluated by simple removal from a life table of one cause of death or by the modeling approach, unless it is based on a detailed understanding of the basic natural history of the system. Removal of the factor under study must occur in the field, in as natural a setting as possible, so that compensating factors can play their role, if present. Experiments are essential.

Another approach to estimating the role of natural enemies is to search for consensus among those who have summarized extensive life table data. I have taken seven such summaries and 10 species that are treated by at least four of them and recorded the major factors identified in the population dynamics of each species (Table 1). Few other species could be added to this list of species that have been so broadly reviewed. Conclusions in reviews are not independent of others because the original life table data were the same for each of the seven summaries. Principal analyses were performed by Varley *et al.* (1973), Podoler and Rogers (1975), and Dempster (1983), and the other reviews depend heavily on conclusions therein.

For several species, emphasis differed substantially among authors because their methods of analysis differed. Some authors relied mainly on Varley and Gradwell's (1960) method of key-factor analysis, in which the killing power of a factor, k, is correlated with insect density (Varley *et al.*, 1973; Stubbs, 1977; Caughley and Lawton, 1981; Dempster, 1983; Strong *et al.*, 1984). Other authors relied on Podoler and Rogers's (1975) method of k-factor analysis, in which the killing power of a factor, k, is correlated with total killing power per generation, K (Southwood, 1975; Podoler and Rogers, 1975). Key-factor analysis is intended to identify factors responsible for population change and density-dependent factors that are likely to regulate and stabilize population fluctuations, whereas k-factor analysis identifies those k factors that correlate most with total population mortality, whether they are density dependent or not. Hence, for winter moth in England, winter disappearance is the key factor in population change, and pupal predation is the density-dependent factor most important in population regulation. Winter disappearance, or death among life stages in the tree, from egg to late larvae, is recognized by k-factor analysis. Further differences in emphasis arise among authors using key-factor analysis, because some concentrate on key factors and others on density-dependent factors (noted in Table 1).

Several points are worthy of note in Table 1. Natural enemies are identified repeatedly as a major factor in population regulation (7 of 10 species). Of the examples in which a consensus is achieved among authors (4 of 10 species) three concern natural enemies (i.e., parasites are the major factor in population change and the major density-dependent factor). The one example that received identical emphasis among authors

TABLE 1

Summary of Life Table Analysis and Major Factors in Population Dynamics[a]

Order and species	Location	Varley *et al.* (1973) key factor (key)	Southwood (1975) *k* factor	Podoler and Rogers (1975) *k* factor	Stubbs (1977) key factor (d.d.)	Dempster (1983) key factor (key)	Strong *et al.* (1984)[b] key factor (d.d.)
Lepidoptera							
Winter moth, *Operophtera brumata* (L.)	England	Winter disappearance	Winter disappearance	Winter disappearance	Pupal predation	Dispersal and mortality of young larvae	Pupal predation and disease
Winter moth,[c] *Operophtera brumata* (L.)	Canada (after 1959)	Parasitism of larvae	Parasitism of larvae	Parasitism of larvae	—	Parasitism of larvae	Parasitism of larvae
Pine looper, *Bupalus piniarius* L.	Holland	Parasitism	Parasitism, predation, and disease of larvae	Total larval mortality, mainly parasitism	Parasitism, predation, and disease of larvae	Egg and early larval mortality and reduced fecundity	Larval parasitism and disease
Gray larch moth, *Zeiraphera diniana* Hübner	Switzerland	Fecundity	Fecundity, egg, pupal, and adult survival	Fecundity, parasitism, and predation	—	Starvation, dispersal, and fecundity	Predation and parasitism of eggs and pupae, and disease
Black-headed budworm,[d] *Acleris variana* (Fern.)	Canada	Parasitism and "residual mortality"	Parasitism	Parasitism	—	—	Predation and parasitism of eggs and pupae, and disease

Cinnabar moth, *Tyria jacobaeae* L.	England	—	Larval starvation	—	—	Starvation, dispersal, and adult fecundity	Starvation, dispersal, and adult fecundity
Coleoptera							
Colorado potato beetle, *Leptinotarsa decemlineata* (Say)	Canada	—	Migration	Emigration of adults	Parasitism	—	Larval starvation and pupal parasitism
Diptera							
Cabbage root fly, *Erioischia brassicae* (Bouché)	Canada	—	Adult mortality, migration, and fecundity	Fecundity, pupal parasitism, and predation	Pupal parasitism and predation	—	Pupal parasitism and predation
Frit fly, *Oscinella frit* L.	England	—	Adult mortality, migration, and fecundity	Adult mortality, migration, and fecundity	—	—	Adult mortality, migration, and fecundity
Hemiptera							
Grass mirid, *Leptopterna dolabrata* (L.)	England	—	Fecundity, larval predation, and accident	Fecundity, loss of late nymphs, potential fecundity	Natality competition, weakening, and weather susceptibility	—	Adult and nymphal competition

[a] The method of data analysis is given below each author citation. Where key-factor analysis was used, a distinction is made between authors who listed key factors (key) and those who listed density-dependent factors (d.d.).
[b] The conclusions of Caughley and Lawton (1981) are almost identical to those of Strong *et al.* (1984), so one column is used for both studies.
[c] An example of biological control in which parasites were not functioning in a typical community.
[d] Morris (1959) made no claim that parasites were regulating host population.

is winter moth in Canada after 1959, involving an experiment that removed natural enemies and then reintroduced them (Fig. 1). Apart from this investigation, in no other study showing the importance of natural enemies were experiments performed to exclude their effect. In all such studies a cause-and-effect relationship between population size and the role of natural enemies was not established. In one case when consensus was reached, on the black-headed budworm, parasitism was recognized as the most important regulatory factor. Thus, the seven reviews extrapolated well beyond Morris's (1959) original conclusion. He wrote, "It is concluded that parasitism is a key factor in the two populations, in the sense that it has useful predictive value" (p. 588), and "a key factor was defined as any mortality factor that has useful predictive value and no attempt has been made to establish cause and effect, because single-factor data are scarcely suitable for this purpose" (p. 587).

The review by Dempster (1983) on the natural regulation of butterfly and moth populations will be heavily debated because he concluded that natural enemies were important in a small minority of cases. In 24 populations of 21 species he found parasitoids acting in a density-dependent way in only one population and predation in another, and questioned their ability to regulate. He concluded that in most cases food resource availability was the most important regulating factor in lepidopteran populations. The differences between Dempster's conclusions and those of other authors are based on his detailed reanalyses, and so the original paper should be consulted. Of course, our whole perception of this field is colored by an emphasis on species that do conspicuous damage to crops and trees, meaning pest species that commonly deplete their food resources. Dempster's review identified 16 species some populations of which regularly overexploit food, causing defoliation and subsequent starvation. Most insect species are not pests, and we know very little about their population dynamics. An important aspect of Dempster's review is his demonstration that rational scientists can reach very different conclusions about population regulation on the basis of life table analysis. This emphasizes the inadequacies of the approach and the need for a more rigorous scientific method in the study of insect population dynamics. Hassell (1985) criticized Dempster's conclusions, arguing that key-factor analysis often fails to detect the regulatory role of natural enemies. He argues that much more detailed information is necessary for each generation, emphasizing both temporal and spatial aspects of sampling.

In general, one feels rather impotent in evaluating this body of work because the basic population data, the analytical tools, the experimental work, and, as a result, any real understanding of these systems are not well developed. This statement, made after more than 40 years of intensive study, is a rather damning reflection of the rigor of this science.

VII. EXPERIMENTAL STUDIES

A brighter side to the gloomy conclusions in the preceding section is that experimental work on the role of natural enemies is gathering momentum. In early experiments with hole-nesting birds, nest boxes were used to increase the number of breeding birds per unit area for the control of forest pests like the pine looper and the green oak leaf roller. For example, Herberg (1960) summarized more than 30 years of study showing that supplemental bird boxes increased the number of breeding birds, and no outbreaks of pine looper occurred between 1926 and 1958. In a control area the abundance of pine looper was much higher, reaching outbreak numbers several times in the 32 years reported (more than 5 pupae to more than 13 pupae per square meter) (Fig. 3). These results suggest that hole-nesting birds can be effective in regulating populations but that they are usually not effective in these carefully managed European forests because of limited nesting sites. Dahlsten and Copper (1979) report equivalent studies carried out over 10 years in northeastern California and experiments indicating that the predation by mountain chickadee and other birds on eggs and pupae of tussock moth was significant.

More recently, experiments have excluded the effects of birds. Using

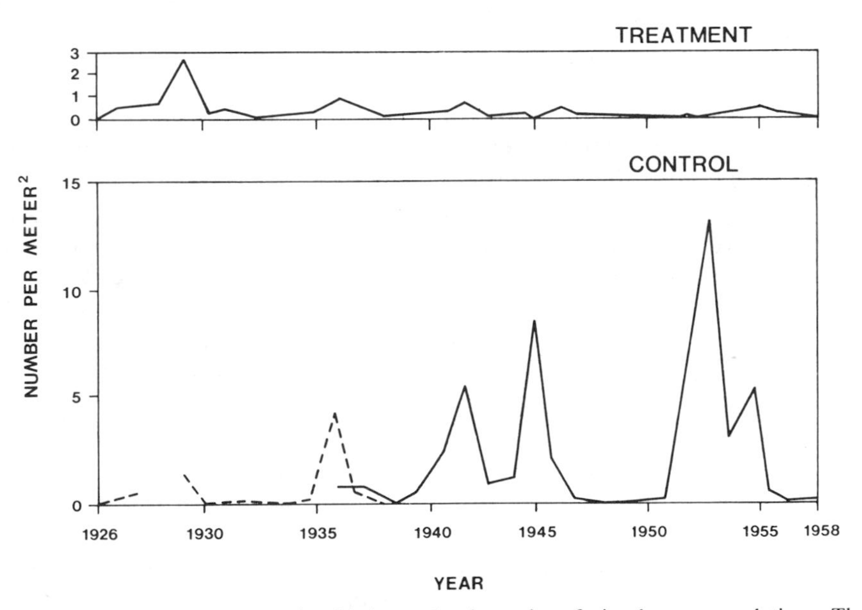

Fig. 3. Effect of hole-nesting birds on the dynamics of pine looper populations. The number of pupae per meter square from 1926 to 1958 is illustrated for a treatment area with an increased number of birds because of supplemental nesting boxes and for a control area with a natural number of birds present. After Herberg (1960).

exclosures effective against birds but not insects, Holmes *et al.* (1979) demonstrated a consistently significant effect of birds in lowering the number of leaf-feeding caterpillars over one season in New Hampshire. Screen exclosures effective against woodpeckers showed that populations of southern pine beetle, *Dendroctronus frontalis* (Zimm.), in the absence of woodpeckers could be from 7 to 25 times higher than when not excluded (Kroll and Fleet, 1979). Campbell and associates (1981, 1983; Campbell and Torgersen, 1983a,b) used whole-tree exclosures effective against birds and ants on 9-m trees. On western spruce budworm, birds and ants together contributed about 29% to mortality (estimated as total killing power) from egg to adult and 86% to mortality from fourth instar to adult, with usually strong compensation by one taxon in the absence of the other. In similar experiments birds and small mammals were excluded from gypsy moth populations. Together they contributed about 27% to the total killing power during the whole generation and 47% to mortality from fifth instar to adult (Campbell and Torgersen, 1983a). The strong influences of these predators on budworm and gypsy moth are apparent from these experiments. However, such short-term experiments do not inform us about predators' effects on the dynamics of insect populations through time. They leave two questions unanswered: Are predators essential in determining the dynamic properties of insect populations? How would dynamics differ in the absence of predators?

Experiments on other invertebrate enemies are also becoming more common. The influence of the predaceous carabid beetle *Bembidion lampros* (Herbst.) on cabbage root fly populations was estimated using exclosures of straw rope soaked in DDT (Wright *et al.*, 1960). At some stages in crop development damage by root fly larvae was twice as high in the control as in the treatment. When Lawton (1984) excluded ants from bracken fern by using tree-banding grease on fronds, he concluded that ants did not have an impact on the insect fauna. In 2 years of study, predators and parasitoids were excluded from an oak tree infested with leaf-mining moths in the genus *Cameraria* (Faeth and Simberloff, 1981). In the cage, the emergence of adults was very high (>80% each year), but in uncaged controls emergence was low (<10% each year). Mortality in the uncaged population was due largely to parasitoids and predators that caused about 60 and 70% mortality in 1978 and 1979, respectively. Parasitoid exclosures resulted in populations of the leaf miner, *Hydrellia valida* Loew, on *Spartina alterniflora* Loisel that were more than 10 times higher than in control areas with access to parasitoids (Stiling *et al.*, 1982). The authors concluded that gradients in density of the leaf miner were imposed by differences in parasitoid abundance. The delphacid *Prokelisia marginata* Van Duzee, in the same *Spartina* community, increased in egg density in caged populations to seven times the size of that of control populations (Stiling and Strong, 1982). The egg parasitoid *Anagrus delicatus* Dozier did not respond in a density-dependent way to increased delphacid density but tended to show inverse density dependence and therefore a lack of regulatory influence. However, the screen cages with 2-mm mesh used in this experiment did not preclude movement of the small mymarid parasitoid across the barrier, with the possibility of greater emigration than immigration.

Thus, experimental approaches have usually provided very convincing evidence for and against the view that natural enemies play an important part in the regulation of insect numbers. Much more experimental work would be desirable. The challenge is to convert these short-term experiments in such a way that longer-range study can offer a better explanation of the role of enemies in population dynamics and insect outbreaks. In addition, to complement more detailed sampling programs advocated by Hassell (1985), experiments on spatial variation and patch dynamics in the herbivore–enemy interaction are needed.

VIII. CONCEPTS OF MORE THAN ONE POPULATION EQUILIBRIUM

Concepts of two or more levels of equilibrium in insect populations have been inherent in many views on population regulation (see Chapter 1). For example, according to the concept of climatic release advocated by Greenbank (1956), Stark (1959), Morris (1964), and others, populations usually at low levels are regulated by poor weather and perhaps enemies but suddenly erupt when the climate becomes unusually favorable (see Chapter 10). Under such conditions natural enemies become ineffective and food shortage may become the ultimate regulating mechanism. Holling (1965) has three equilibria in his model of population regulation by natural enemies with a type 3 sigmoid functional response: an unstable threshold density for extinction, a stable equilibrium density, and an unstable threshold for population escape. This view was extended in the synoptic model by Southwood and Comins (1976). In this model a "natural enemy ravine" imposes a stable equilibrium, but if the ravine is habitually weak, as in disturbed habitats, or temporarily disrupted, the population escapes to a very high equilibrium density.

The stick insect, *Didymuria violescens* (Leach), in New South Wales shows two major phases in population density. One, at low density, is regulated by egg parasitoids and/or birds, and the other, at high density, is limited by food shortage (Readshaw, 1965). *Cardiaspina albitextura* Taylor (Psyllidae) on *Eucalyptus* in eastern Australia also showed two radically different levels of density (summary in Clark *et al.*, 1967). At low density, natural enemies played an important role in population regulation. Unusually low temperatures coupled with above-normal rainfall disrupted parasitoid searching and synchrony with the host and resulted in psyllid outbreaks. At outbreak densities, competition for food results in reduced fecundity and ultimately starvation of nymphs, and after a few generations the population may crash to endemic levels again.

Other concepts of population regulation do not invoke a role for natural enemies but emphasize the plant–herbivore interaction. Change in plant quality causes the release of populations to outbreak levels or suppresses populations to endemic levels. White (e.g., 1974, 1976) emphasized the role of plant stress in increasing the nutritional quality of food for herbivores. Rhoades (1979) noted that plant stress also changes plant defense, and defenses induced by herbivore attack are probably important in regulation (e.g., Rhoades, 1983a; Baldwin and Schultz, 1983). Berryman (1982) concluded that the key to understanding bark beetle outbreak was knowledge of the interaction between beetle populations and tree defense systems: "Other factors, such as parasitism, predation, and competing species may modify this behavior, largely through their effect on beetle population size, but they will not change the general qualitative properties of the system" (p. 313).

The regulatory role of the plant–herbivore interaction in the absence of natural enemies is without doubt very important, at least in some systems, and perhaps in many. It is too early, however, to recognize broad patterns in which the plant–herbivore interaction becomes more important than the herbivore–enemy interaction. We must recognize the probability that natural enemies play a rather passive role in some insect population dynamics. As Varley *et al.* (1973) noted about the role of parasitoids in winter moth population dynamics in England, "they respond to the changes in their host's densities rather than cause them" (p. 126).

Concepts of population regulation at low and outbreak densities have both emphasized and deemphasized the importance of natural enemies. A real dichotomy in nature may exist, although the linkage between three trophic levels may be much tighter than generally recognized by students of insect population dynamics.

IX. THREE-TROPHIC-LEVEL INTERACTIONS

Feeny (1976) argued that any form of digestibility reduction through plant defenses would cause an insect herbivore to feed longer and therefore become more vulnerable to enemies. Part of plant defense involves the third trophic level. Indeed, it has been argued that defense through digestibility reduction could not have evolved in the absence of natural enemies, because the original plants with decreased digestibility would have suffered more damage since insects would have eaten more to compensate for the low quality of food (Price *et al.,* 1980). An experimental test of the value of digestibility reducers demonstrated the importance of natural enemies in the system (Price *et al.,* 1980). Lawton and McNeill (1979) recognized that herbivores are "between the devil and the deep blue sea," the sea being one of low-quality plant food, and the devil taking the form of natural enemies. Only rarely was plant food adequate and were natural enemies weak enough to permit an insect outbreak (see also Rhoades, 1983b). Schultz (1983) argued that the major role of plant defensive chemicals was to increase the impact of herbivore diseases, parasitoids, and predators. Population variation in such defenses posed serious problems in adaptation by herbivores, making difficult an evolutionary response to plant defenses. Campbell and Duffey (1979) have demonstrated that increased plant toxic defenses may well be more deleterious to enemies than to herbivores. Thus, moderate levels of innate plant defense coupled with effective extrinsic defense from natural enemies may prove to be the best evolutionary solution for some plant species.

As our understanding of plant–herbivore interactions becomes more sophisticated, it will be important to recognize that the survival of predators and parasitoids depends on sophisticated mechanisms of host discovery and exploitation. Natural enemies are very sensitive to plant and herbivore semiochemicals (Price, 1981), so if plants are "talking to each other" (e.g., Rhoades, 1983a; Baldwin and Schultz, 1983) enemies may intercept the signals. The possibility exists that induced plant defenses and volatile signals between plants are also adaptive through attraction to natural enemies and such chemicals have evolved to maximize reception at the third trophic level. Tests are needed. Rhoades (1983b, 1985) suggests ever more complex and subtle interactions between plants and herbivores, such as suppression by herbivores of communication between plants and the emission of countersignals by herbivores. Such chemical skulduggery is likely to be used by enemies since we know they can learn to employ novel odors as kairomones, and they evolve to use almost any form of B.O. from plants or herbivores (Price, 1981).

Probably, understanding the dynamics of herbivorous insect populations will necessitate a close integration of three-trophic-level interactions.

X. CONCLUSIONS

Excellent evidence exists that natural enemies are important agents of natural selection in the evolution of insect defenses. Evidence from biological control is conclusive that natural enemies can regulate insect populations. In the absence of enemies, outbreaks occur; a cause-and-effect relationship is often established. These large-scale experiments have contributed significantly to our understanding of regulation, especially in simplified systems where the insect pest is an exotic. However, many attempts at biological control have failed, although the interpretation of such failures is difficult.

The effects of pesticides on natural enemies have clearly demonstrated the normal regulatory role of these enemies, for when their activity is disrupted, pest outbreaks occur. Cause and effect are frequently clear. A strong experimental approach is available, and judicious application of insecticide treatments may still serve as an incisive approach to the study of insect outbreaks.

Population models of predator–prey and parasitoid–host interactions indicate that under many hypothetical conditions enemies can regulate prey–host populations.

Life table analysis using single-factor analysis, key-factor analysis, or k-factor analysis has only rarely established a convincing link between cause and effect, rendering such analysis a very weak tool for investigating insect outbreaks. Correlation is totally inadequate for demonstrating mechanisms of population regulation. Seldom have process studies involving experimentation been coupled with these correlations to establish a convincing scenario in which natural enemies play an important role in producing endemic or outbreak insect levels. Clearly, much more detailed experimental work with treatments that exclude the effects of enemies are required. Experiments that last several insect generations are necessary, for as with biological control and insecticide treatments, results on enemies are not necessarily apparent immediately.

Experimental studies have frequently demonstrated a strong impact by enemies on insect herbivore populations, enemy exclosure treatments being much more convincing than treatments that augment natural-enemy populations. Other experiments have demonstrated a lack of any regulatory role for enemies. The results are clear, cause and effect are evident, and continued experimentation will ultimately reveal where and when and on which taxa enemies play a significant role in population regulation. Some concepts of two equilibrium levels in insect population dynamics have stressed the role of natural enemies in establishing a stable equilibrium density at low levels. Others emphasize the role of climate or plant physiology and defense in regulation at low levels. Much evidence supporting these concepts is based on life table studies, so the concepts remain as tenuous as the data on which they are based. However, the developing synthesis of interactive mechanisms should yield a more realistic perception of nature and, therefore, better field studies and experiments on mechanisms.

The new synthesis of animal population dynamics also includes an integrated picture of three-trophic-level interactions, with plants, herbivores, and enemies as essential components of a dynamic system. Concepts are in formative stages, but they open many areas of research on the role of natural enemies in insect outbreaks, and they are mechanistic in their formulation.

In the past we have been satisfied with, and willing to accept, any evidence that natural enemies play a role in the dynamics of insect populations because this is what our expectation has been—from Darwin (1859) to today. As a result of this uncritical atmosphere most effort has been expended on gathering the weakest kind of evidence—using life table analysis in the absence of detailed studies on mechanisms.

We have available potent methods for investigating the role of natural enemies in insect populations: the mechanisms by which enemies change insect population performance or their absence. The areas of biological control and insecticide treatment suggest many smaller-scale experiments that could be extended over several insect generations. The increasing emphasis on experimental exclosure studies will also result in long-term studies that evaluate not only the short-term impact of enemies, but their role in the longer-term dynamic properties of populations. Although much evidence *suggests* that natural enemies are important in insect population dynamics, these experimental approaches will *demonstrate* quantitatively their real effects.

ACKNOWLEDGMENTS

I am grateful to Dr. M. J. Crawley, Dr. M. P. Hassell, and Dr. Jeffrey K. Waage and the editors for their critical comments on this chapter. Financial support was provided by National Science Foundation Grant BSR-8314594.

REFERENCES

Baldwin, I. T., and Schultz, J. C. (1983). Rapid changes in tree leaf chemistry induced by damage: Evidence for communication between plants. *Science* **221,** 277–279.

Berryman, A. A. (1982). Population dynamics of bark beetles. *In* ''Bark Beetles in North

American Conifers'' (J. B. Mitton and K. B. Sturgeon, eds.), pp. 264–314. Univ. of Texas Press, Austin.

Bishop, J. A., and Cook, L. M. (1979). A century of industrial melanism. *Antenna 3,* 125–128.

Bishop, J. A., and Cook, L. M. (1980). Industrial melanism and the urban environment. *Adv. Ecol. Res.* **11,** 373–404.

Campbell, B. C., and Duffey, S. S. (1979). Tomatine and parasitic wasps: Potential incompatibility of plant antibiosis with biological control. *Science* **205,** 700–702.

Campbell, R. W., and Torgersen, T. R. (1983a). Compensatory mortality in defoliator population dynamics. *Environ. Entomol.* **12,** 630–632.

Campbell, R. W., and Torgersen, T. R. (1983b). Effect of branch height on predation of western spruce budworm (Lepidoptera: Tortricidae) pupae by birds and ants. *Environ. Entomol.* **12,** 697–699.

Campbell, R. W., Torgersen, T. R., Forest, S. C., and Youngs, L. C. (1981). Bird exclosures for branches and whole trees. *USDA For. Serv. Gen. Tech. Rep. PNW* **PNW-125.**

Campbell, R. W., Torgersen, T. R., and Srivastava, N. (1983). A suggested role for predaceous birds and ants in the population dynamics of the western spruce budworm. *For. Sci.* **29,** 779–790.

Caughley, G., and Lawton, J. H. (1981). Plant–herbivore systems. *In* ''Theoretical Ecology'' (R. M. May, ed.), 2nd ed. pp. 132–166. Sinauer Assoc., Sunderland, Massachusetts.

Clark, L. P., Geier, P. W., Hughes, R. D., and Morris, R. F. (1967). ''The Ecology of Insect Populations in Theory and Practice.'' Methuen, London.

Dahlsten, D. L., and Copper, W. A. (1979). The use of nesting boxes to study the biology of the mountain chickadee (*Parus gambeli*) and its impact on selected forest insects. *In* ''The Role of Insectivorous Birds in Forest Ecosystems'' (J. G. Dickson, R. N. Conner, J. C. Kroll, R. R. Fleet, and J. A. Jackson, eds.), pp. 217–260. Academic Press, New York.

Darwin, C. (1859). ''The Origin of Species by Means of Natural Selection.'' Murray, London.

DeBach, P. (1974). ''Biological Control by Natural Enemies.'' Cambridge Univ. Press, London and New York.

Deevey, E. S. (1947). Life tables for natural populations of animals. *Q. Rev. Biol.* **22,** 283–314.

Dempster, J. P. (1983). The natural control of populations of butterflies and moths. *Biol. Rev. Cambridge Philos. Soc.* **58,** 461–481.

Doutt, R. L. (1964). The historical development of biological control. *In* ''Biological Control of Insect Pests and Weeds'' (P. DeBach, ed.), pp. 21–42. Van Nostrand-Reinhold, Princeton, New Jersey.

Eberhardt, L. L. (1970). Correlation, regression, and density dependence. *Ecology* **51,** 306–310.

Embree, D. G. (1966). The role of introduced parasites in the control of winter moth in Nova Scotia. *Can. Entomol.* **98,** 1159–1168.

Embree, D. G. (1971). The biological control of the winter moth in eastern Canada by introduced parasites. *In* ''Biological Control'' (C. B. Huffaker, ed.), pp. 217–226. Plenum, New York.

Faeth, S. H., and Simberloff, D. (1981). Population regulation of a leaf-mining insect, *Cameraria* sp. nov., at increased field densities. *Ecology* **62,** 620–624.

Feeny, P. (1976). Plant apparency and chemical defense. *In* ''Biochemical Interaction Between Plants and Insects'' (J. W. Wallace and R. L. Mansell, eds.), pp. 1–40. Plenum, New York.

Goeden, R. D., and Louda, S. M. (1976). Biotic interference with insects imported for weed control. *Annu. Rev. Entomol.* **21,** 325–342.

Greenbank, D. O. (1956). The role of climate and dispersal in the initiation of outbreaks of the spruce budworm in New Brunswick. I. The role of climate. *Can. J. Zool.* **34,** 453–476.

Harcourt, D. G. (1969). The development and use of life tables in the study of natural insect populations. *Annu. Rev. Entomol.* **14,** 175–196.

Hassell, M. P. (1978). "The Dynamics of Arthropod Predator–Prey Systems." Princeton Univ. Press, Princeton, New Jersey.

Hassell, M. P. (1980). Foraging strategies, population models and biological control: A case study. *J. Anim. Ecol.* **49,** 603–628.

Hassell, M. P. (1981). Arthropod predator–prey systems. *In* "Theoretical Ecology" (R. M. May, ed.), 2nd ed., pp. 105–131. Sinauer Assoc., Sunderland, Massachusetts.

Hassell, M. P. (1982). Patterns of parasitism by insect parasitoids in patchy environments. *Ecol. Entomol.* **7,** 365–377.

Hassell, M. P. (1984). Parasitism in patchy environments: Inverse density dependence can be stabilizing. *I.M.A.J. Math. Appl. Med. Biol.* **1,** 123–133.

Hassell, M. P. (1985). Insect natural enemies as regulating factors. *J. Anim. Ecol.* **54,** 323–334.

Hassell, M. P., and Huffaker, C. B. (1969). The appraisal of delayed and direct density-dependence. *Can. Entomol.* **101,** 353–351.

Hassell, M. P., and Varley, G. C. (1969). New inductive population model for insect parasites and its bearing on biological control. *Nature (London)* **223,** 1133–1137.

Hassell, M. P., Lawton, J. H., and Beddington, J. R. (1977). Sigmoid functional responses by invertebrate predators and parasitoids. *J. Anim. Ecol.* **46,** 249–262.

Herberg, M. (1960). Drei Jahrzehnte Vogelhege zur Niederhaltung waldschadlicher Insekten durch die Ansiedlung von Hohlenbrotern. *Arch. Forstwes.* **9,** 1015–1048.

Holling, C. S. (1959). The components of predation as revealed by a study of small mammal predation of the European pine sawfly. *Can. Entomol.* **91,** 293–320.

Holling, C. S. (1965). The functional response of predators to prey density and its role in mimicry and population regulation. *Mem. Entomol. Soc. Can.* **45,** 1–60.

Holling, C. S., and Ewing, S. (1971). Blind man's buff: Exploring the response space generated by realistic ecological simulation models. *In* "Statistical Ecology" (G. P. Patil, E. C. Pielou, and W. E. Waters, eds.), Vol. 2, pp. 207–223. Pennsylvania State Univ. Press, University Park.

Holmes, R. T., Schultz, J. C., and Nothnagle, P. (1979). Bird predation on forest insects: An exclosure experiment. *Science* **206,** 462–463.

Howard, L. O. (1897). A study in insect parasitism: A consideration of the parasites of the white-marked tussock moth, with an account of their habits and interrelations and with descriptions of new species. *USDA Tech. Ser.* **5,** 1–157.

Huffaker, C. B., ed. (1971). "Biological Control." Plenum, New York.

Kettlewell, H. B. D. (1973). "The Evolution of Melanism: The Study of a Recurring Necessity with Special Reference to Industrial Melanism in the Lepidoptera." Oxford Univ. Press, London and New York.

Kroll, J. C., and Fleet, R. R. (1979). Impact of woodpecker predation on overwintering within-tree populations of the southern pine beetle (*Dendroctonus frontalis*). *In* "The Role of Insectivorous Birds in Forest Ecosystems" (J. G. Dickson, R. N. Conner, J. C. Kroll, R. R. Fleet, and J. A. Jackson, eds.), pp. 269–281. Academic Press, New York.

Lawton, J. H. (1984). Herbivore community organization: General models and specific tests with phytophagous insects. *In* "A New Ecology: Novel Approaches to Interactive

Systems'' (P. W. Price, C. N. Slobodchikoff, and W. S. Gaud, eds.), pp. 329–352. Wiley, New York.

Lawton, J. H., and McNeill, S. (1979). Between the devil and the deep blue sea: On the problem of being a herbivore. *Symp. Br. Ecol. Soc.* **20,** 223–244.

Lotka, A. J. (1925). ''Elements of Physical Biology.'' Williams & Wilkins, Baltimore, Maryland.

Luck, R. F. (1971). An appraisal of two methods of analyzing life tables. *Can. Entomol.* **103,** 1261–1271.

McNeill, S. (1973). The dynamics of a population of *Leptopterna dolabrata* (Heteroptera: Miridae) in relation to its food resources. *J. Anim. Ecol.* **42,** 495–507.

Maelzer, D. A. (1970). The regression of log N_{n+1} on log N_n as a test of density dependence: An exercise with computer-constructed density-dependent populations. *Ecology* **51,** 810–822.

May, R. M. (1973). ''Stability and Complexity in Model Ecosystems.'' Princeton Univ. Press, Princeton, New Jersey.

May, R. M. (1978). Host–parasitoid systems in patchy environments: A phenomenological model. *J. Anim. Ecol.* **47,** 833–843.

May, R. M. (1981). Models for two interacting populations. *In* ''Theoretical Ecology'' (R. M. May, ed.), 2nd ed., pp. 78–104. Sinauer Assoc., Sunderland, Massachusetts.

Moreton, B. D. (1969). Beneficial insects and mites. 6th ed. *Bull.—Minist. Agric., Fish. Food (G.B.)* **20,** 1–118.

Morris, R. F. (1959). Single-factor analysis in population dynamics. *Ecology* **40,** 580–588.

Morris, R. F. (1963). Predictive population equations based on key factors. *Mem. Entomol. Soc. Can.* **32,** 16–21.

Morris, R. F. (1964). The value of historical data in population research, with particular reference to *Hyphantria cunea* Drury. *Can. Entomol.* **96,** 356–368.

Morris, R. F. (1969). Approaches to the study of population dynamics. *USDA For. Serv. Res. Pap. NE* **NE-125,** 9–28.

Morris, R. F. (1971). Observed and simulated changes in genetic quality in natural populations of *Hyphantria cunea. Can. Entomol.* **103,** 893–906.

Morris, R. F., and Miller, C. A. (1954). The development of life tables for the spruce budworm. *Can. J. Zool.* **32,** 283–301.

Morris, R. F., and Royama, T. (1969). Logarithmic regression as an index of responses to population density. *Can. Entomol.* **101,** 361–364.

Mullin, C. A., and Croft, B. A. (1985). An update on development of selective pesticides favoring arthropod natural enemies. *In* ''Biological Control in Agricultural IPM Systems'' (M. A. Hoy and D. C. Herzog, eds.), pp. 123–150. Academic Press, New York.

Nicholson, A. J., and Bailey, V. A. (1935). The balance of animal populations. Part I. *Proc. Zool. Soc. London* pp. 551–598.

Pearl, R., and Miner, J. R. (1935). Experimental studies on the duration of life. XIV. The comparative mortality of certain lower organisms. *Q. Rev. Biol.* **10,** 60–79.

Pearl, R., and Parker, S. L. (1921). Experimental studies on the duration of life. I. Introductory discussion of the duration of life in *Drosophila. Am. Nat.* **55,** 481–509.

Podoler, H., and Rogers, D. (1975). A new method for the identification of key factors from life-table data. *J. Anim. Ecol.* **44,** 85–114.

Price, P. W. (1971). Toward a holistic approach to insect population studies. *Ann. Entomol. Soc. Am.* **64,** 1399–1406.

Price, P. W. (1981). Semiochemicals in evolutionary time. *In* ''Semiochemicals: Their Role in Pest Control'' (D. A. Nordlund, R. L. Jones, and W. J. Lewis, eds.), pp. 251–279. Wiley, New York.

Price, P. W. (1984). "Insect Ecology," 2nd ed. Wiley, New York.
Price, P. W., Bouton, C. E., Gross, P., McPheron, B. A., Thompson, J. N., and Weis, A. E. (1980). Interactions among three trophic levels: Influence of plants on interactions between insect herbivores and natural enemies. *Annu. Rev. Ecol. Syst.* **11,** 41–65.
Readshaw, J. L. (1965). A theory of phasmatid outbreak release. *Aust. J. Zool.* **13,** 475–490.
Rhoades, D. F. (1979). Evolution of plant chemical defense against herbivores. *In* "Herbivores: Their Interaction with Secondary Plant Metabolites" (G. A. Rosenthal and D. H. Janzen, eds.), pp. 3–54. Academic Press, New York.
Rhoades, D. F. (1983a). Responses of alder and willow to attack by tent caterpillars and webworms: Evidence for pheromonal sensitivity of willows. *In* "Plant Resistance to Insects" (P. A. Hedin, ed.), pp. 55–68. Am. Chem. Soc., Washington, D. C.
Rhoades, D. F. (1983b). Herbivore population dynamics and plant chemistry. *In* "Variable Plants and Herbivores in Natural and Managed Systems" (R. F. Denno and M. S. McClure, eds.), pp. 155–220. Academic Press, New York.
Rhoades, D. F. (1985). Offensive–defensive interactions between herbivores and plants: Their relevance in herbivore population dynamics and ecological theory. *Am. Nat.* **125,** 205–238.
Ripper, W. E. (1956). Effect of pesticides on balance of arthropod populations. *Annu. Rev. Entomol.* **1,** 403–438.
Royama, T. (1977). Population persistence and density dependence. *Ecol. Monogr.* **47,** 1–35.
Royama, T. (1981a). Fundamental concepts and methodology for the analysis of animal population dynamics, with particular reference to univoltine species. *Ecol. Monogr.* **51,** 473–493.
Royama, T. (1981b). Evaluation of mortality factors in insect life table analysis. *Ecol. Monogr.* **51,** 495–505.
Rudd, R. L. (1964). "Pesticides and the Living Landscape." Univ. of Wisconsin Press, Madison.
St. Amant, J. L. S. (1970). The detection of regulation in animal populations. *Ecology* **51,** 949–965.
Salt, G. (1970). "The Cellular Defence Reactions of Insects." Cambridge Univ. Press, London and New York.
Schultz, J. C. (1983). Impact of variable plant defensive chemistry on susceptibility of insects to natural enemies. *In* "Plant Resistance to Insects" (P. A. Hedin, ed.), pp. 37–54. Am. Chem. Soc., Washington, D. C.
Southwood, T. R. E. (1975). The dynamics of insect populations. *In* "Insects, Science and Society" (D. Pimentel, ed.), pp. 151–199. Academic Press, New York.
Southwood, T. R. E., and Comins, H. N. (1976). A synoptic population model. *J. Anim. Ecol.* **45,** 949–965.
Stark, R. W. (1959). Population dynamics of the lodgepole needle miner, *Recurvaria starki* Freeman, in Canadian Rocky Mountain parks. *Can. J. Zool.* **37,** 917–943.
Stiling, P. D., and Strong, D. R. (1982). Egg density and the intensity of parasitism in *Prokelisia marginata* (Homoptera: Delphacidae). *Ecology* **63,** 1630–1635.
Stiling, P. D., Brodbeck, B. V., and Strong, D. R. (1982). Foliar nitrogen and larval parasitism as determinants of leafminer distribution patterns on *Spartina alterniflora*. *Ecol. Entomol.* **7,** 447–452.
Strong, D. R. (1984). Density-vague ecology and liberal population regulation in insects. *In* "A New Ecology: Novel Approaches to Interactive Systems" (P. W. Price, C. N. Slobodchikoff, and W. S. Gaud, eds.), pp. 313–327. Wiley, New York.
Strong, D. R., Lawton, J. H., and Southwood, T. R. E. (1984). "Insects on Plants: Community Patterns and Mechanisms." Harvard Univ. Press, Cambridge, Massachusetts.

Stubbs, M. (1977). Density dependence in the life-cycles of animals and its importance in *K*- and *r*-strategies. *J. Anim. Ecol.* **46,** 677–688.

Tinbergen, L., and Klomp, H. (1960). The natural control of insects in pinewoods. II. Conditions for damping of Nicholson oscillations in parasite–host systems. *Arch. Neerl. Zool.* **13,** 344–379.

Van den Bosch, R., and Messenger, P. S. (1973). "Biological Control." Intext, New York.

Varley, G. C., and Gradwell, G. R. (1960). Key factors in population studies. *J. Anim. Ecol.* **29,** 399–401.

Varley, G. C., and Gradwell, G. R. (1968). Population models for the winter moth. *Symp. R. Entomol. Soc. London* **4,** 132–142.

Varley, G. C., Gradwell, G. R., and Hassell, M. P. (1973). "Insect Population Ecology: An Analytical Approach." Blackwell, Oxford.

Volterra, V. (1926). Variazioni e fluttuazioni del numero d'individui in specie animali conviventi. *Atti R. Accad. Naz. Lincei, Mem. Cl. Sci. Fis., Mat. Nat.* **2,** 31–113.

Wellington, W. G. (1957). Individual differences as a factor in population dynamics: The development of a problem. *Can. J. Zool.* **35,** 293–323.

Wellington, W. G. (1977). Returning the insect to ecology: Some consequences for pest management. *Environ. Entomol.* **6,** 1–8.

White, T. C. R. (1974). A hypothesis to explain outbreaks of looper caterpillars, with special reference to populations of *Selidosema suavis* in a plantation of *Pinus radiata* in New Zealand. *Oecologia* **16,** 279–301.

White, T. C. R. (1976). Weather, food and plagues of locusts. *Oecologia* **22,** 119–134.

Wright, D. W., Hughes, R. D., and Worrall, J. (1960). The effect of certain predators on the numbers of cabbage root fly [*Erioischia brassicae* (Bouche)] and on the subsequent damage caused by the pest. *Ann. Appl. Biol.* **48,** 756–763.

Chapter 13

The Role of Weather and Natural Enemies in Determining Aphid Outbreaks

PAUL W. WELLINGS[1]

CSIRO Division of Entomology Research Station
Warrawee, NSW 2074, Australia

A. F. G. DIXON

School of Biological Sciences
University of East Anglia
Norwich NR4 7TJ, England

[1] Present address: CSIRO Division of Entomology, Canberra, ACT 2601, Australia.

I. INTRODUCTION

The populations of many insects may fluctuate over as many as five orders of magnitude, and in some cases, periods of scarcity regularly follow periods of abundance (Itô, 1980). These fluctuations cannot be considered to be truly cyclical, because they vary in relation to both the interval between peak periods of abundance and the amplitude of the peaks. Changes in the population densities of aphids are broadly similar to this general pattern. The abundance of aphids is highly seasonal and may vary considerably from one year to another. In addition, the patterns of fluctuations of populations may differ in different geographic regions (Bejer-Peterson, 1962) and among populations developing in the same region over a number of years (Wellings *et al.*, 1985). On a finer scale, there may be considerable variation in the abundance of neighboring populations at the same time.

Severe crop losses as a result of aphid outbreaks have been documented since early in the nineteenth century (e.g., Kirby and Spence, 1858). The specific mechanisms leading to such outbreaks are poorly described. However, some understanding of aphid outbreaks can be gained by considering the ecological processes operating on aphid populations in agroecosystems. We follow Joyce's (1983) description of a pest outbreak and define an outbreak as an occasion when a large number of insects on host plants causes a loss of productivity. Thus, an outbreak may be due to (1) the multiplication of a small number of initial immigrants; (2) invasion by insects whose numbers exceed the injury level; (3) continuous immigration, resulting in temporary or permanent accumulation on host plants; (4) redistribution of populations among host plants (Joyce, 1983).

The factors causing fluctuations in the population density of aphids may include (1) climatic conditions, (2) the action of natural enemies, (3) the action of pathogens, (4) the host plant condition, (5) the presence of biotic factors that enhance aphid multiplication (e.g., ants), (6) cultural measures, and (7) the action of chemicals (de Fluiter, 1966). In this chapter we consider the importance of the first two factors in the population dynamics of aphids and illustrate how they might affect the likelihood of aphid outbreaks.

II. MODELS INCORPORATING DENSITY INDEPENDENCE

The role of abiotic factors in simple analytical models has been largely ignored, primarily because their action is density independent and, therefore, unable to provide the negative feedback necessary for regulation. However, some efforts have been made to incorporate the influences of density-independent factors into strategic models, and in this section we investigate some of their effects and consider their significance in the development of insect outbreaks. These models do not provide detailed descriptions of particular systems; indeed, they have never been intended to do so. As May (1973) points out, "They [strategic models] are at best caricatures of reality, and thus, have both the truth and falsity of caricatures." Strategic models have provided general insights into the dynamics of populations (May, 1974; Hassell *et al.*, 1976) and, at the same time, have accelerated the development of descriptions of density-dependent processes (Bellows, 1981). Such models have also highlighted the fact that even the simplest descriptions of populations may have complex properties (May, 1976; May and Oster, 1976).

Single-species models describe changes in the per capita replacement rate of a population as a function of population density. Bellows (1981) has catalogued a variety of forms of these functions and provides a general relationship that incorporates density-independent effects into a single-species population model,

$$N_{t+1} = dN_t f(N_t) \tag{1}$$

where N_{t+1} and N_t represent the number of individuals at times $t+1$ and t, respectively, $f(N_t)$ describes the density-dependent term, and d describes the density-independent term and is restricted such that $0 < d < 1$. This model assumes that the density-independent mortality acts on the survivors of the density-dependent term. An obvious alternative is the case in which density-independent mortality precedes the density-dependent term:

$$N_{t+1} = N_t f(dN_t) \tag{2}$$

The form of the function describing density dependence that we have considered here is

$$f(N_t) = a \exp(-bN_t) \tag{3}$$

where a represents the net rate of increase and b the constant describing the density-dependent term. This model is similar to that described by Moran (1950) and Ricker (1954) and is analogous to the continuous time form of the logistic equation (Charlesworth, 1980). It has been extensively used; some authors have investigated its dynamic properties, while others have used it to describe density dependence (e.g., Cook, 1965; May, 1974, 1976; May *et al.*, 1974; May and Oster, 1976; Hassell *et al.*, 1976; Berryman, 1978; Southwood, 1981). The model has a single equilibrium point at ln a/b, which is locally stable when ln $a < 2$ and unstable if ln $a > 2$. If $0 <$ ln $a < 1$ the populations exhibit monotonic damping, and when $1 <$ ln $a < 2$ the model populations exhibit oscillations, converging back to the equilibrium point. Once ln $a > 2$, the model populations generate oscillations of increasing amplitude (May, 1974).

This model underpins the analysis of tsetse fly (*Glossina morsitans submorsitans*) population dynamics, presented by Rogers (1979). Previous studies of the population dynamics of tsetse had shown that survival was strongly influenced by abiotic factors and had suggested that fluctuations in numbers were purely a consequence of prevailing weather conditions. However, Rogers (1975) has shown that tsetse are subject to density-dependent mortality. Three components are central to Rogers's (1979) analysis: (1) Female tsetse flies have relatively low fecundities and give birth to well-developed larvae; (2) the density-independent mortality incurred takes the form of a sine curve, covering the whole year (Fig. 1); and (3) density-dependent mortality above a particular lower threshold was assumed to be direct and acted throughout the year. By incorporating these factors, Rogers demonstrated that the monthly catches of non-teneral tsetse flies were approximately bounded by a Moran curve (Fig. 2). The deviations to the left of the Moran curve in Fig. 2 suggest either that Rogers's estimate of a is too low or that (seasonal?) immigration may

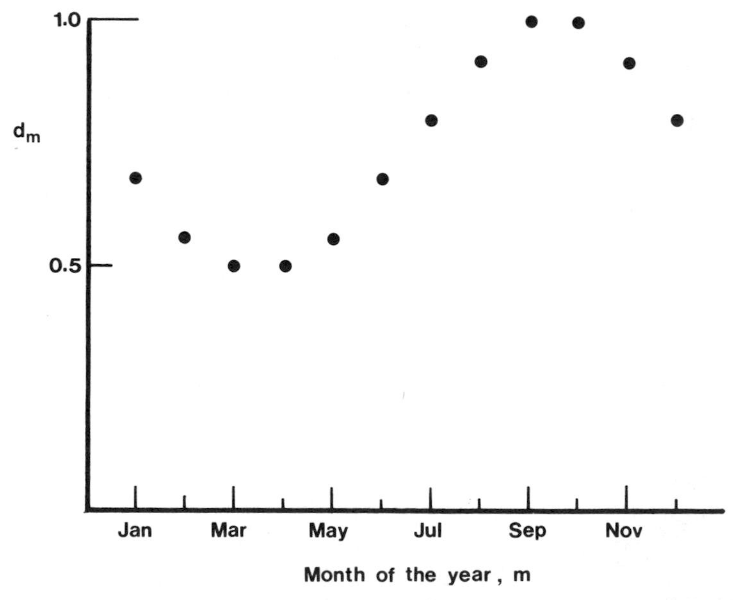

Fig. 1. Sinusoidal density-independent survivorship d_m. After Rogers (1979).

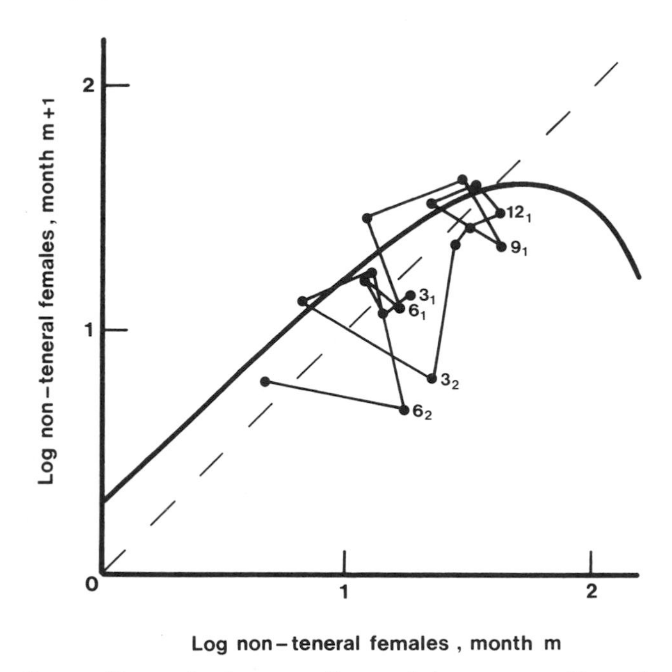

Fig. 2. Rogers–Moran plot for tsetse fly populations over an 18-month period. The numbers represent the month and the subscripts the year. After Rogers (1979).

be important in the tsetse population. His approach, however, highlights the potential influence of seasonal density-independent mortality in insect population dynamics (Southwood, 1981).

The general model underlying this analysis is of the form

$$N_{m+1} = d_m N_m a \exp(-bN_m) \tag{4}$$

where N_m and N_{m+1} are the number of females in month m and $m+1$, respectively, and d_m is the proportion surviving density-independent mortality in month m. In Fig. 3 we present Rogers–Moran diagrams, plotting population densities at equilibrium. We use the density-independent survivorship schedule from Fig. 1 and a constant value for b, and investigate the effects of changing a. For low values of a (Fig. 3a,b,c) the model populations are "driven" around the illustrated ellipses (obviously, the exact form depends on the seasonal changes in density-independent mortality), and as the value of a increases, the model populations exhibit disturbed limit cycles but return to the same point at the end of each year (Fig. 3d,e). Only when the net rate of increase a is very high (Fig. 3f) does the population fail to return to its starting point: The population is then exhibiting "chaotic" behavior. However, in each case the model popula-

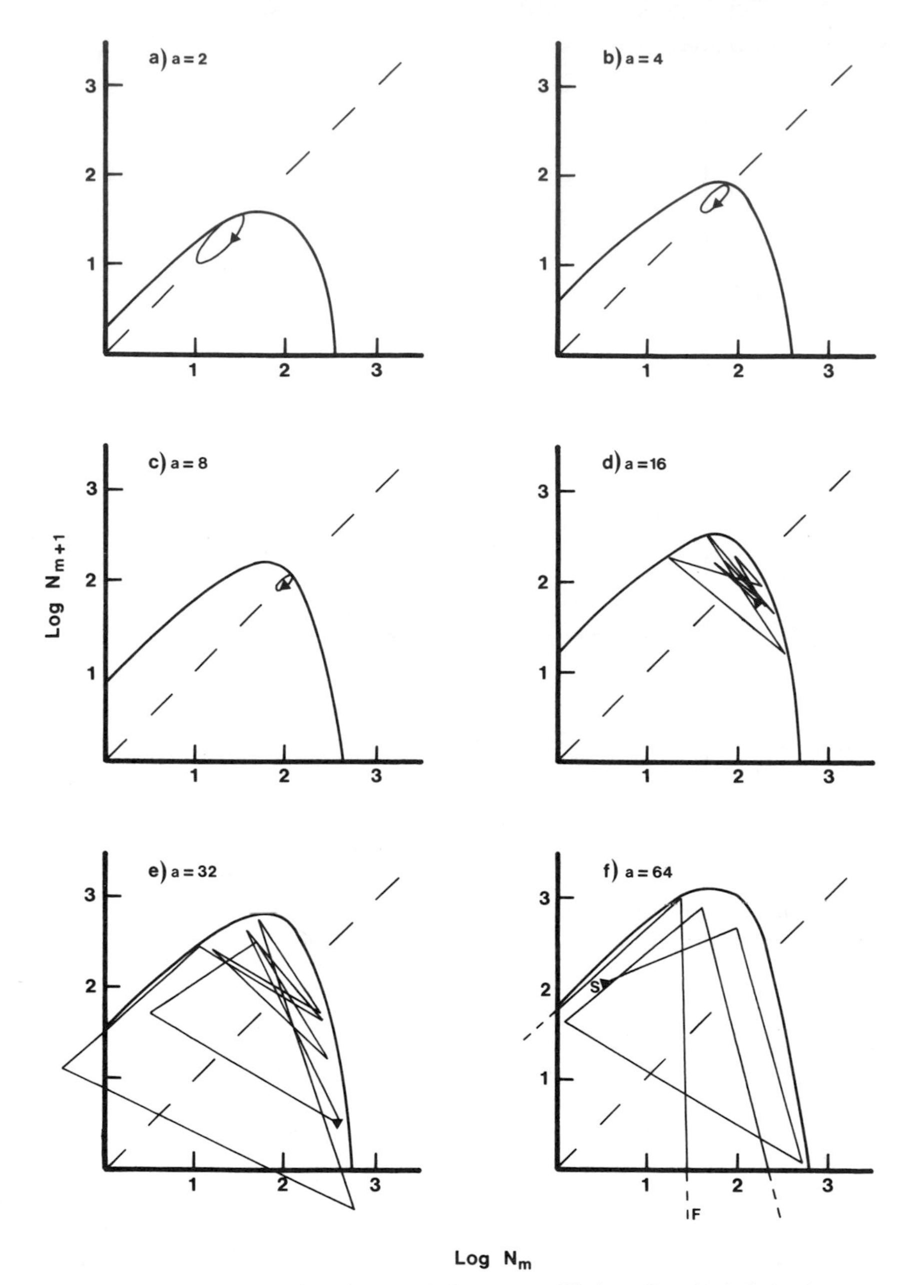

Fig. 3. Rogers–Moran plots for populations at equilibrium. Density-independent mortality acts on the survivors of the density-dependent effect [Eq. (4)]. Here, $b = 0.018$ and d_m varies as in Fig. 1. The unbroken curves illustrate the Moran plot in the absence of density-independent mortality, and the single equilibrium point lies at the intercept of this curve and the broken 45° line. The arrowheads indicate January. The start and finish of plots of populations displaying chaotic behavior are marked S and F, respectively.

tion remains bounded by a Moran curve. Two conditions should be noted. First, the value of $\bar{d}_m a$ must be greater than 1 if the populations are to persist without immigration. Second, the boundaries for the various forms of dynamic behavior now depend on $\bar{d}_m a$, where $\bar{d}_m$ is the mean value of the density-independent survivorship during the year.

The model population patterns are very different for cases in which density-independent mortality precedes the action of density dependence. Equation (4) becomes

$$N_{m+1} = N_m a \exp(-bd_m N_m) \quad (5)$$

Rogers–Moran plots for Eq. (5) are presented in Fig. 4. In this model the populations are "driven" around the illustrated ellipses at the lowest values of a (Fig. 4a,b) and exhibit disturbed limit cycles as a increases (Fig. 4c). For higher values of a the populations show chaotic patterns (Fig. 4d,e,f). The boundaries of the various forms of dynamic behavior are those described by May (1974), and the Moran curves now no longer act as boundaries to the model populations. The order in which density dependence and density independence act in an insect's life cycle may have a profound influence on its population dynamics and contribute to the likelihood of an outbreak. If density-independent effects precede the action of density dependence, then population eruptions above an equilibrium point are possible, even for those species with low rates of increase.

Berryman (1978) outlined a slightly different form of analysis, which used Eq. (3) as a starting point. He suggested that the equilibrium point of this model may not be static in time and space and formulated the hypothesis that epidemics may be caused by a sudden increase in the favorability of a population's environment. Inspection of Eqs. (1) and (2) indicates that such sudden changes can be brought about by changes in a (e.g., by qualitative changes in individual reproductive statistics), b (by the failure of density-dependent influences), and d (by changes in climatic conditions). Once such changes revert to the initial conditions, populations then return to their original positions.

The predictions made by such simple models incorporating seasonal and/or stochastic variations in environment outline the underlying causes of epidemics in many systems subject to high levels of disturbance (Berryman, 1978). However, understanding the fluctuations in pest numbers and, more important, predicting pest abundance, require detailed knowledge of the ecological processes involved and the management practices in particular agricultural systems. In the following sections we examine a number of studies that give some insight into the causes of fluctuations in numbers and the factors influencing the outbreaks of a number of species of aphids.

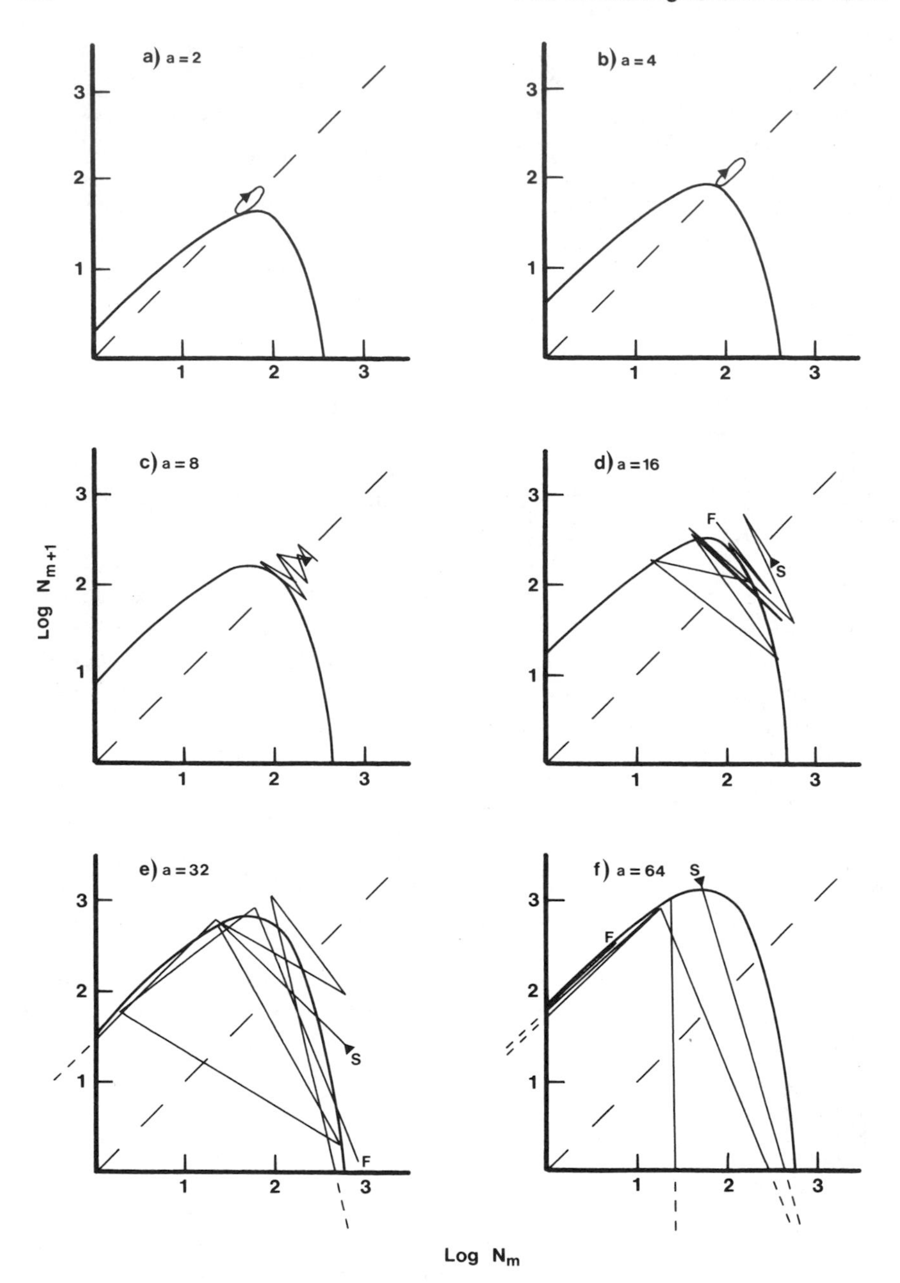

Fig. 4. Rogers–Moran plots as in Fig. 3 but now density-dependent mortality acts on the survivors of the density-independent effect [Eq. (5)].

III. APHID LIFE CYCLES

Most species of aphid are cyclically parthenogenetic; thus, there may be several parthenogenetic generations between a single, usually annual sexual generation. This mode of reproduction is important in determining population structure, because it has led to the evolution of polymorphism and pedogenesis (Dixon, 1985a) and also enables aphids to achieve prodigious rates of increase (Dixon, 1987). About 10% of aphid species exhibit the phenomenon of host alternation (heteroecy), utilizing a woody winter (or primary) host, on which eggs are laid, and herbaceous summer (or secondary) hosts. This behavior is associated with predictable seasonal changes in host plant suitability, and these species of aphid are usually highly polymorphic. Some morphs are unwinged (apterous) and others are winged (alate). In addition, the alate morphs differ in migratory function. Thus, those that fly between primary and secondary hosts in spring and autumn are obligate migrants, whereas those produced on secondary hosts in the summer are facultative migrants. Autoecious species are not so polymorphic and do not show a seasonal shift in host plant. They colonize and lay their eggs on only one or a few closely related species of woody or herbaceous hosts. The life cycles of some species are complicated by the fact that some or all clones in some geographic areas do not produce sexual forms. These lineages are anholocyclic, as opposed to the holocyclic forms, which produce sexuales and overwinter or oversummer as eggs.

The evolution of alary dimorphism in some species allows aphids to track environmental conditions very closely. This enables them to produce the more fecund and faster-developing apterae when environmental conditions and habitat quality are suitable and to switch to producing alatae, which disperse to other plants as conditions deteriorate (Dixon, 1985a). Aphid morphs differ in their life history strategies, and some differences appear to be adaptations that spread risks in highly heterogeneous habitats (Ward and Dixon, 1984). There does not appear to be a direct association between outbreaks and life cycle complexity. For example, populations of the spotted alfalfa aphid, *Therioaphis trifolii* f. *maculata* (usually an anholocyclic species with two morphs), the sycamore aphid, *Drepanosiphum platanoidis* (an autoecious species with three morphs), and the bird cherry-oat aphid, *Rhopalosiphum padi* (usually an heteroecious species with eight morphs), can all reach outbreak levels.

IV. APHID QUALITY AND PHENOTYPIC PLASTICITY

Individual aphids may differ from one another in a variety of ways (e.g., age, morph, sex, reproductive condition, size), and such differences may have important consequences at a population level. Polymorphism is the most commonly observed type of qualitative variation among individuals of the same species. Aphid polymorphism may be controlled by various extrinsic factors such as photoperiod, temperature, and competitor density; in addition, the sequence of some morphs may have a preprogrammed, intrinsic element.

Variation in weight among individual aphids may be as much as an order of magnitude (Way and Banks, 1967), and these differences may have a marked influence on the rate of increase of aphid populations. An understanding of the factors determining weight is important; age at first reproduction, age-specific fecundity patterns, and progeny size have all been correlated with adult weight (e.g., Murdie, 1969; Dixon and Wratten, 1971). These differences in adult weights arise through variations in environmental conditions experienced during development. Dixon *et al.* (1982) showed that temperature, food quality, and intraspecific competition influence the size of aphids because of their effects on growth and development rates (Fig. 5). Developmental experience is also important because it may act as a constraint on the life history traits of adults subjected to sudden changes in environmental conditions (Wellings and Dixon, 1983).

As with other insects, the rate of development of aphids is temperature dependent and approximately linear (Campbell *et al.*, 1974). Many laboratory studies have investigated the effects of temperature on the intrinsic rate of increase r_m of aphid populations. The general form of these relationships is broadly similar, although the lower and upper thermal thresholds and the location and amplitude of the maximum r_m values differ among species (Fig. 6). As Williamson (1972) points out, r_m is a density-free parameter that defines the maximum rate of increase under specific conditions. There is no reason to believe that abundance and r_m are related. Temperature is only one factor influencing r_m, and other investigations have shown that it is also affected by light intensity (Wyatt and Brown, 1977) and host plant (Komazaki, 1982; Berg, 1984).

In highly polymorphic species developmental experience can also influence the production of winged individuals within a population. At high population densities a greater proportion of individuals develop with wings than at low population densities (Hughes, 1963; Shaw, 1970). Differences among winged individuals may also be important. Cockbain (1961) demonstrated that small individuals of *Aphis fabae* have smaller fat reserves than large individuals. The availability of fat reserves is an index of the resources available for active flight and may be important in dispersal. Environmental conditions experienced by winged individuals have a direct influence on flight activity (Taylor, 1963; Walters and Dixon, 1984).

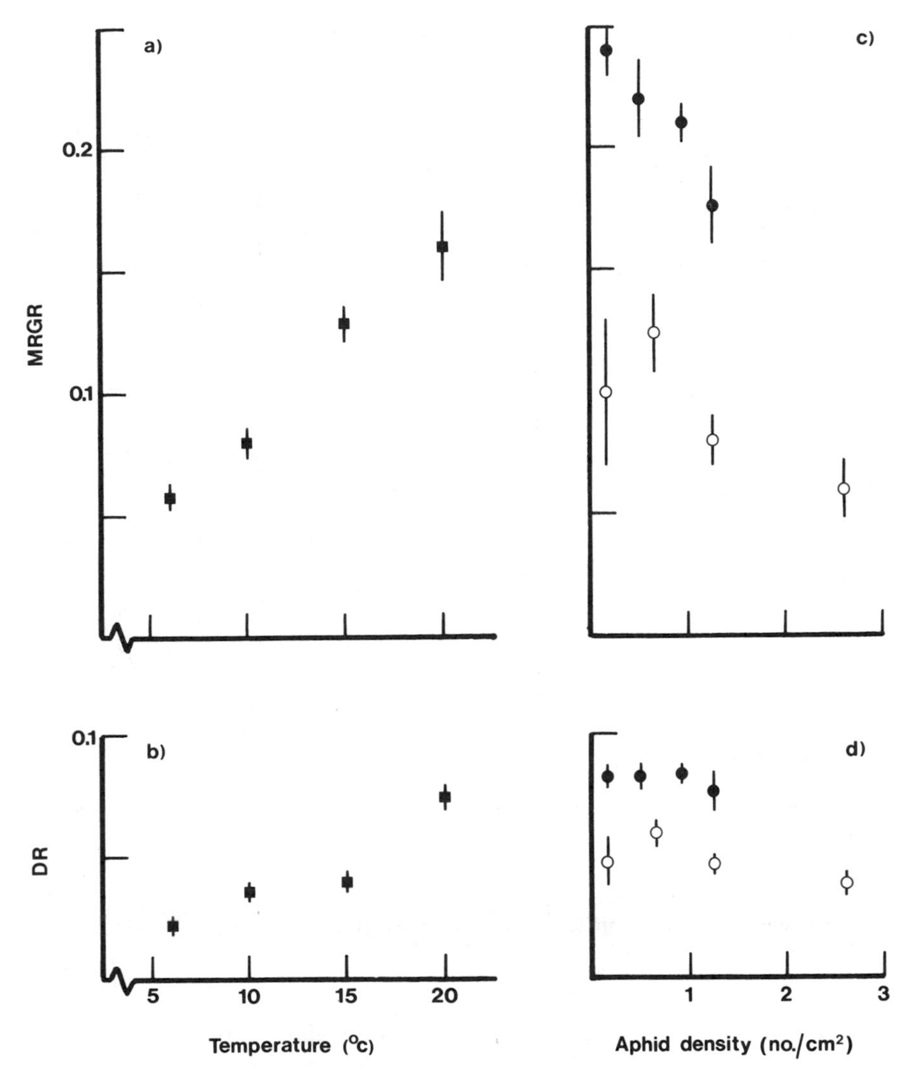

Fig. 5. Growth and development of the sycamore aphid. The relationships between (a) mean relative growth rate (MRGR, $\mu g\ \mu g^{-1}\ day^{-1}$) and temperature (°C), and between (b) development rate (DR, $days^{-1}$) and temperature are shown, as are the relationships between (c) MRGR on unfurling leaves ("high-quality nutrition," filled circles) and on mature leaves ("poor nutrition," empty circles) and aphid density (number cm^{-2}) at 15°C and between (d) DR on unfurling leaves (filled circles) and mature leaves (empty circles) and aphid density at 15°C. After Wellings (1981) and Chambers *et al.* (1985).

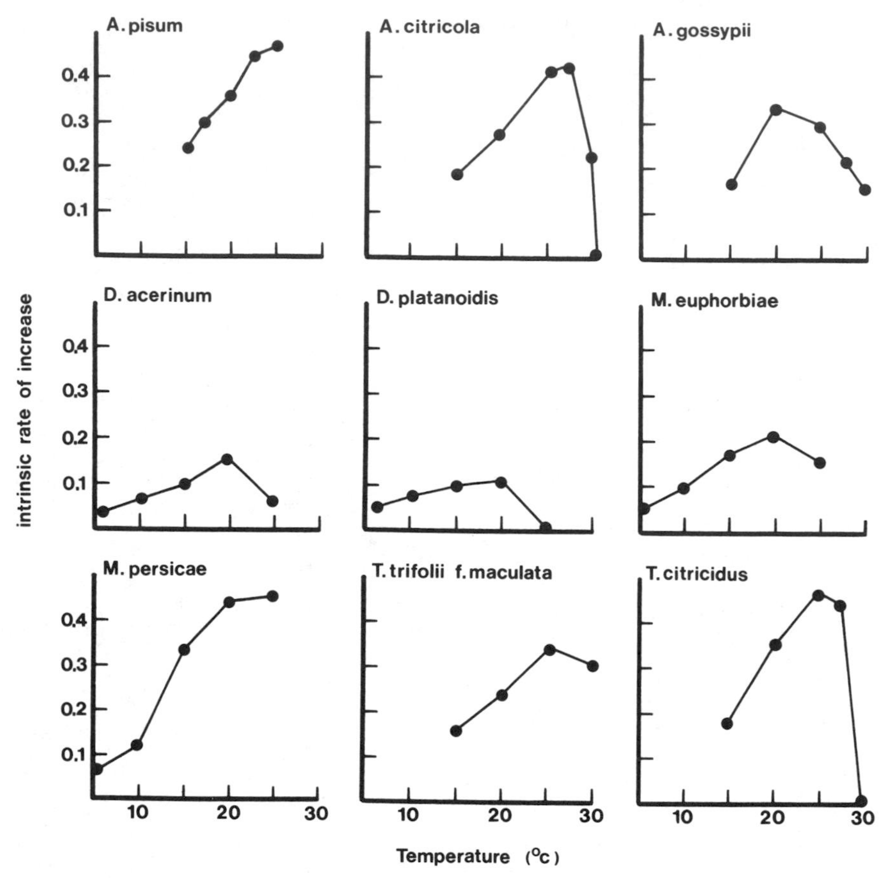

Fig. 6. Relationship between the intrinsic rate of increase and temperature in nine species of aphid. Based on data given by Graham (1959), Barlow (1962), and Siddiqui *et al.* (1973), Wellings (1981), and Komazaki (1982).

It seems likely that adverse weather conditions may delay but not prevent migration.

Phenotypic plasticity is a characteristic feature of aphid biology and arises mainly through the effects of current developmental experience. This means that aphids are very sensitive to changes in environmental conditions, and because their development times are short, these responses are rapidly translated into marked changes in population growth rates. For example, the rate of increase of populations is positively correlated with the mean relative growth rate of individual aphids (Leather and Dixon, 1984). Since individual growth rates are determined by developmental conditions (Fig. 5) and may be coupled with preadult survival

(Dixon, 1985b), relatively small shifts in environmental conditions and habitat quality may result in marked changes in aphid abundance and in the structure of their populations.

V. WEATHER AND APHID NUMBERS

The number of aphids both on host plants and in the aerial plankton has frequently been shown to be correlated with several meteorological factors. Such correlations do not necessarily imply that there are underlying causal relationships, but they are useful in that they indicate associations between the number of aphids or particular life history events and climatic conditions. They therefore provide a starting point for understanding aphid outbreaks.

The overall climatic conditions impose constraints on aphid abundance in that climate largely determines the availability of host plants. Maelzer (1981), working in Adelaide, Australia, noted that seasonal changes in weather largely determined the quality and quantity of aphid food supply and also had a direct effect on the rate of increase of aphids and their dispersal. In Adelaide, the number of aphids appears to be dependent on the length of time that the host plants remain favorable. This period is curtailed by a severe moisture deficit during the summer, and the agricultural season is approximately delineated by the time during which the rainfall–evaporation ratio is greater than 0.3. This constrains many aphid species, which appear to be adapted to take advantage of flushes of new growth on their host plants. In temperate regions the climatic conditions in spring may influence the hatching rates of overwintering eggs. For example, egg hatch in the sycamore aphid, *D. platanoidis,* is influenced by temperature (Dixon, 1976). Similarly, the time of budbreak of sycamore is correlated with spring temperatures (Wellings *et al.,* 1985). The timing of egg hatch relative to the time of budbreak is important because it is critical to the survival of young aphids and synchronization allows the aphid to exploit the highly nutritious flush of new growth (Dixon, 1976).

There is a considerable amount of data that provide correlations between number of aphids on crops and climatic conditions. Jones (1979) gives information on the number of aphids on winter and spring cereals at Rothamsted, United Kingdom. She monitored the abundance of certain species of aphids over a 5-year period (1973–1977) and also recorded meteorological conditions. (Figure 7 illustrates four of these years.) The number of aphids was correlated with accumulated temperatures above 5°C, and temperature accounted for between 53 and 99% of the variation in abundance. Jones (1979) also noted the adverse effects of heavy rainfall

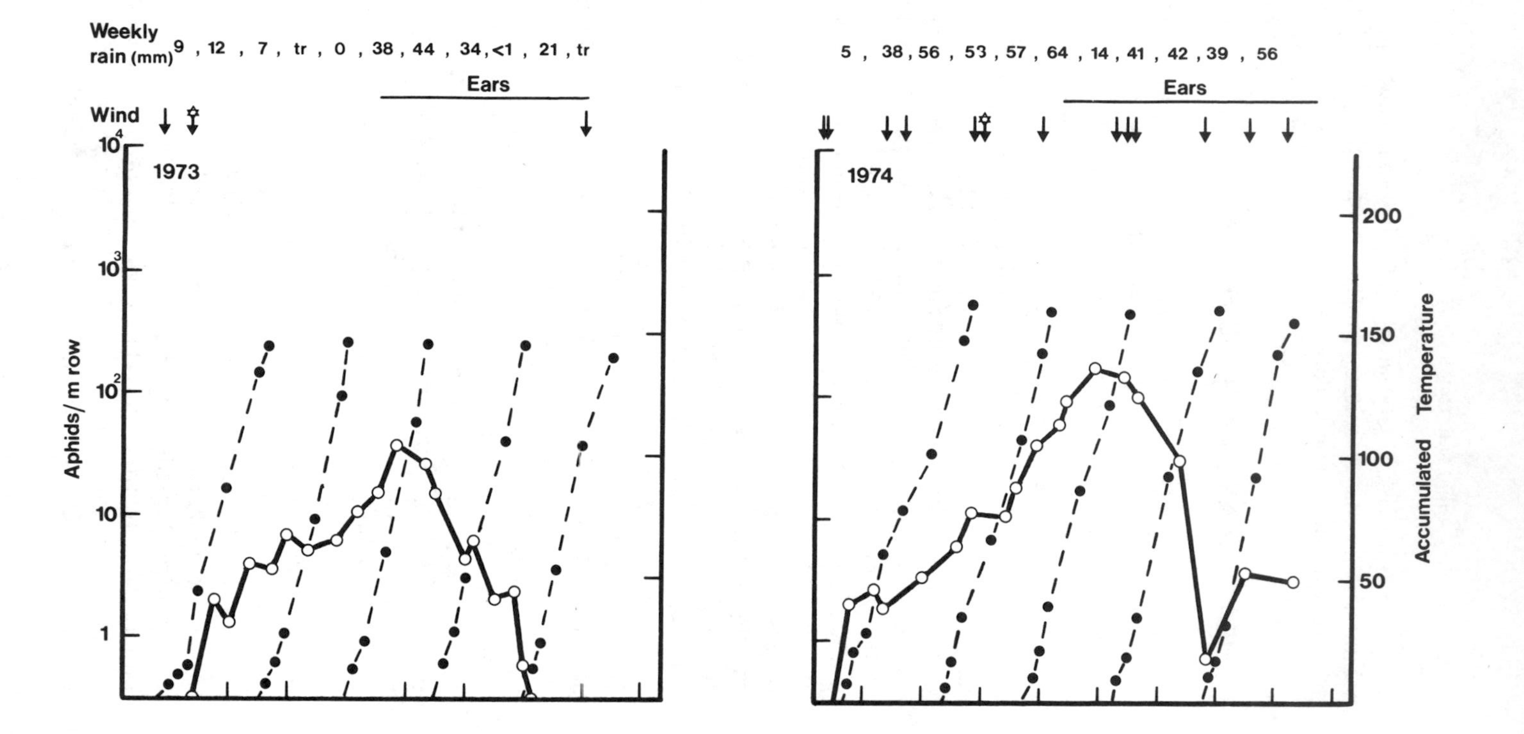
Weekly rain (mm)
9 , 12 , 7 , tr , 0 , 38 , 44 , 34 , <1 , 21 , tr
5 , 38 , 56 , 53 , 57 , 64 , 14 , 41 , 42 , 39 , 56
Ears
Ears
Wind
1973
1974
Aphids/ m row
10^4
10^3
10^2
10
1
200
150
100
50
Accumulated Temperature

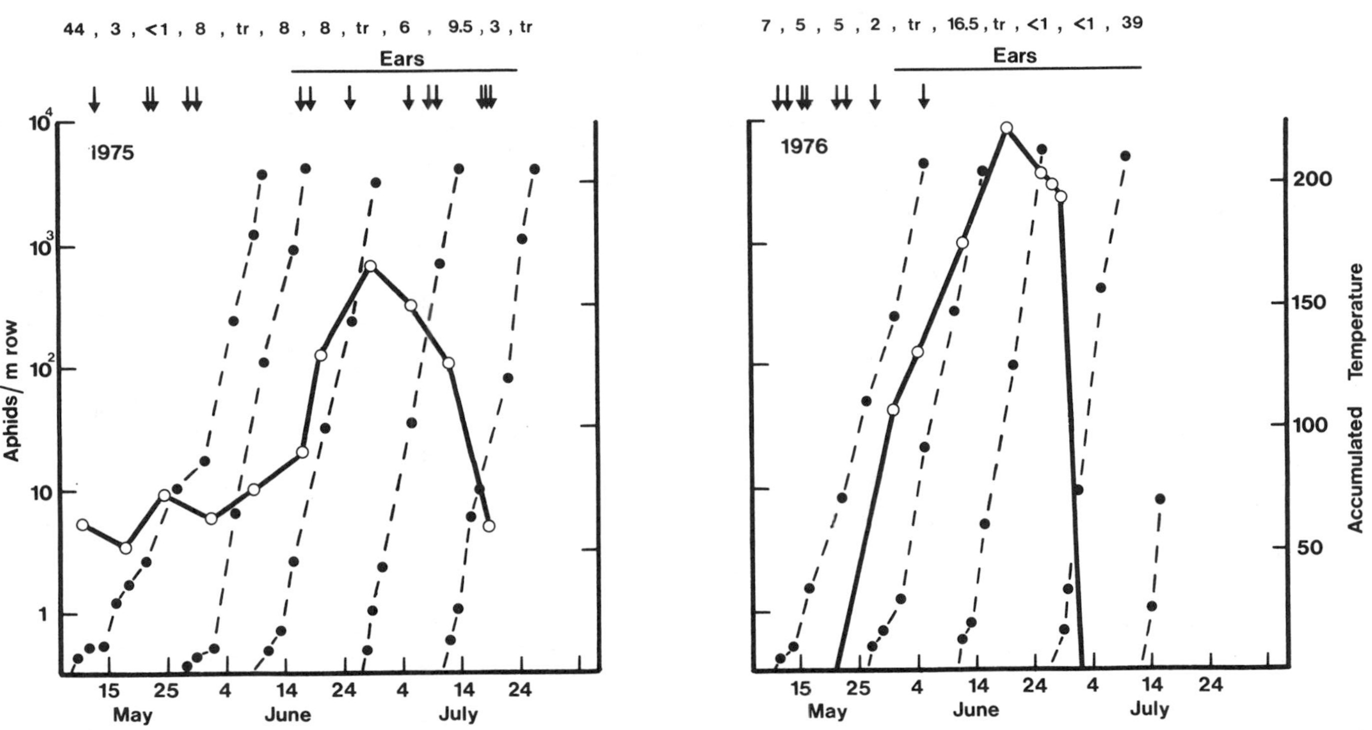

Fig. 7. Log number of cereal aphids per meter row of winter and spring cereals (empty circles) during May–July 1973–1976. Accumulated temperatures in day-degrees above 5°C (filled circles), weekly rainfall (mm), winds stronger than 14.42 m/sec (arrows), and winds stronger than 18.03 m/sec (arrows with star) are shown. The time when ears were present is also indicated. After Jones (1979).

and high winds on aphid abundance. In 1973 and 1974, heavy rain in June washed many aphids off the plants, whereas in 1975, 1976, and 1977 little rain fell in June and aphid populations increased rapidly. Strong winds may also dislodge aphids, but some species are protected in their preferred feeding sites. Thus, *Metopolophium dirhodum,* which colonizes the leaves of grasses, may be adversely influenced by high winds, whereas *Sitobion avenae* feeds in the developing ears and is less easily dislodged. Jones's study emphasizes the importance of favorable weather for the buildup of aphid populations.

Individual weather events may have an influence on aphid numbers. For example, Dunn and Wright (1955) showed that the number of pea aphids, *Acrythosiphum pisum,* feeding on lucerne was greatly reduced by heavy showers of rain. Dixon (1976) noted that there is evidence implicating rain as a sporadic factor in the disappearance of a large number of first-instar sycamore aphids before budbreak. Once budbreak has begun and new leaves are unfurling, the newly hatched aphids are sheltered within the expanding buds. The distribution of aphids within a crop may also be associated with prevailing weather conditions. Way and Cammell (1973) mapped the distribution of the black bean aphid, *A. fabae,* in a single field of beans in Sussex, United Kingdom. The distribution appeared to be highly dependent on the direction of the prevailing wind. Wind gusts may cause individual aphids to fall off plants, causing redistribution when they resettle. In addition, local topography (e.g., the presence of hedgerows) may have a marked influence on the settling patterns of winged individuals (Lewis, 1965).

There are also associations between the number of winged aphids trapped from the aerial plankton and weather conditions. The peak number of cereal aphids flying during the summer and autumn migrations is correlated with the accumulated rainfall and accumulated day-degree temperatures (Robert and Rouze-Jouan, 1976). Similarly, A'Brook (1983) demonstrates correlations between monthly trap catches of aphids and various meteorological statistics (e.g., accumulated rainfall, rain days, air and ground frosts, soil and air temperatures). The number of *Rhopalosiphum padi* and *Rhopalosiphum insertum* trapped in the autumn is associated with high summer rainfall and a high level of soil moisture throughout August (A'Brook, 1981). These aphids originate from grasses in late summer, and grass growth is related to air temperature and soil moisture. A'Brook (1983) suggests that there is a causal relationship between prior weather conditions and aphid abundance through the productivity of their grass hosts.

A long-term field study of eight populations of *D. platanoidis* indicates that the yearly changes in population growth tend to be synchronous

(Wellings *et al.*, 1985). These patterns arise because weather conditions are the major variables governing aphid populations (e.g., Barlow and Dixon, 1980). The main components of climate influencing aphid numbers are temperature, rainfall, evaporation, and wind. These factors may influence abundance, either through effects mediated via the host plants or through direct effects on the aphids themselves.

VI. ROLE OF NATURAL ENEMIES IN LIMITING APHID POPULATIONS

A wide variety of predators (e.g., birds, coccinellids, syrphid larvae, cecidomyid larvae, chryopsids) and parasites (e.g., aphidids, aphelinids) have been recorded as attacking aphids. The main groups have been the subjects of major reviews (Hagen and van den Bosch, 1968; Starý, 1970; Hodek, 1973). Vickerman and Wratten (1979) note that natural enemies could be important in two respects. First, they may reducc aphid numbers within crops and so prevent outbreaks. Second, they may regulate aphid numbers from year to year.

A. Within Crops

Hodek (1973) reported a number of factors thought to be important in the buildup of coccinellid populations: (1) the distribution and direction of movement of migratory adult coccinellids from hibernation sites, (2) the qualities of habitat that attract immigrants, (3) the influence of meteorological conditions, and (4) the population density of aphids necessary first to retain the predators and second to allow egg production. A similar set of factors may be significant for the development of parasitoid populations.

The proximity of hibernation sites to the location of the aphid populations, especially in the case of annual crop plants (where predator complexes are re-formed each year), is important in determining the role of predator populations. The abundance of predators may differ widely in different fields, being higher in fields close to wooded areas and lower in unwooded areas. In fields near woods the activity of predators may begin earlier and abundance may be higher during the buildup of aphid populations (Galecka, 1966). The fate of many aphid populations is significantly influenced by the number of immigrant aphids, so that the synchronization of natural enemies and aphids in time and space may be critical in preventing outbreaks.

Temperature has the most important climatic influence on natural-enemy populations. Development rates are influenced much as are those of aphids (Campbell *et al.*, 1974; Scopes and Biggerstaff, 1977), and in general the lower thermal threshold is higher in natural enemies than in the hosts. Laboratory studies of the intrinsic rate of increase of the spotted alfalfa aphid and various parasitoids have demonstrated that the growth rates of parasitoid populations are limited at low temperatures (e.g., Force and Messenger, 1964). At low temperatures aphid populations are able to continue growing, whereas at higher temperatures aphid populations may be driven to extinction by the action of predators and parasitoids (Messenger and Force, 1963; van Emden, 1966). Temperature has a direct influence on various essential components of natural enemy biology. For example, feeding rate and egg production are temperature dependent (Ives, 1981; Mills, 1981), as are functional responses (Messenger, 1968), searching rates (Mack and Smilowitz, 1982), and handling time (Mills, 1981; Mack and Smilowitz, 1982). In haplodiploid parasitoids sex ratio may also be a function of temperature, with a trend of male-biased sex ratios at low temperatures (Messenger and Force, 1963; van den Bosch *et al.*, 1966). In addition, the incidence of diapause in parasitoids is temperature-related (Starý, 1970). The net effect of low temperatures on all these factors is to reduce the effectiveness of natural enemies. In these natural-enemy–prey systems there appears to be a transient enemy-free space lying within a temperature band in which aphid populations are able to increase in the absence of actively searching natural enemies.

The density of aphid populations is also critical to the effectiveness of natural enemies. There appears to be a lower density threshold below which coccinellids tend to emigrate (e.g., Sluss and Hagen, 1966; Honěk, 1980) and a higher threshold for egg production (Honěk, 1980). In addition, it has been amply demonstrated that many facets of foraging and oviposition behavior are density-related (see Hassell, 1978, for a general review).

B. Year to Year

A feature commonly observed in long-term studies on aphid populations is a 2-year cycle in abundance, with years of aphid scarcity following years of aphid outbreaks. Blattný (1925) hypothesized that a large number of coccinellids leave crops at the end of outbreaks and, given favorable weather conditions, successfully overwinter to colonize crops in the following spring. These immigrants arrive in sufficient number to have a marked impact on aphid populations, and as a result, relatively few larval coccinellids have sufficient food to complete their development later in the season. This results in few adults overwintering. Thus, in the following spring there is low predation pressure, and aphid populations are able to increase to outbreak levels. Similar observations have been made by various authors (e.g., Behrendt, 1966; Müller, 1966; Way, 1967). This hypothesis may be especially relevant to studies of aphids in field crops, whereas in perennial crop systems, spring weather conditions appear to be important in determining the level of predator activity, and different population patterns have been observed in different geographic regions (e.g., Smith and Hagen, 1966). Long-term studies of aphids on woody hosts indicate that, at best, predators are of marginal importance (e.g., Sluss and Hagen, 1966).

C. Impact of Natural Enemies on Aphid Numbers

The assessment of the impact of natural enemies is technically difficult, and no one method of studying the effectiveness of natural enemies is free of limiting assumptions (see Chapter 12, this volume). As a result, the impact of natural enemies on aphid numbers is unclear. Some authors (e.g., Bodenheimer and Swirski, 1957) suggest that natural enemies become important only when the numbers of aphids is at a peak or already decreasing; others suggest that aphid populations may be regulated by predator complexes (e.g., Neuenschwander *et al.*, 1975). In the latter study, however, the authors pooled data on a regional basis. Within alfalfa fields over the same period, marked fluctuations in aphid abundance were recorded. Similar observations were made by Sluss and Hagen (1966), who monitored walnut aphid populations in eight orchards. In three of these, coccinellid numbers were always low; in three other predator numbers trailed aphid numbers; and predation was considered to be the key regulatory factor at the remaining two sites. Studies on the impact of natural enemies on cereal aphid populations indicate that parasitoids may play a minor role in limiting aphid numbers (Powell, 1983), whereas a more detailed study indicates that the natural-enemy complex may substantially limit aphid populations under some circumstances (Chambers and Sunderland, 1983). Decreases in *Schizaphis graminum* densities also appear to be correlated with the number of natural enemies (Hamilton *et al.*, 1982).

The most comprehensive study of the effect of natural enemies on aphid populations is presented in a series of papers by Baumgaertner *et al.* (1981). These authors studied the influence of coccinellids on pea aphid populations at Vancouver, Canada. Predators appeared to have a continuous and marked effect on aphid number. The main findings of this study were that the immigration rates of the coccinellids were temperature dependent and their emigration rates were dependent on both temperature and aphid density. The key component of almost every facet of the study was the influence of temperature. This observation helps explain the results of studies in which natural enemies have been mass-released into aphid populations (e.g., Tamaki and Weeks, 1973; Kieckhefer and Olson, 1974). Both these investigations recorded a low level of recapture of predators, and the impact of natural enemies on aphid populations was marginal. The factors influencing natural-enemy movement may be the key to understanding their role in preventing aphid outbreaks.

Frazer and Gilbert (1976) suggest that the coccinellid and pea aphid interaction is at marked variance with the existing theories of predator–prey dynamics (e.g., Hassell, 1978). The relationship appears to be completely unstable but at the same time extremely resilient. Other evidence indicates that natural mortality declines as the exploitation of aphid populations increases; this compensatory response is rapid and almost complete (Charnov *et al.,* 1976).

VII. POPULATION DYNAMICS

The multiplication of a small number of initial colonists is undoubtedly the most significant factor in determining the probability of an aphid outbreak. The key factors in the buildup of aphid populations are (1) overwintering and oversummering on crop and noncrop host plants, (2) the factors influencing immigration, (3) the factors influencing multiplication in crops, and (4) the factors causing emigration.

A. Overwintering and Oversummering

Aphids may overwinter or oversummer as a mixture of nymphs and adults or as eggs. The former feature is characteristic of anholocyclic life cycles, whereas the production of eggs is characteristic of autoecious and heteroecious holocyclic life cycles. In Mediterranean and subtropical environments oversummering appears to be a response to hot, dry weather conditions. In contrast, overwintering is a feature of aphid biology in temperate regions. The severity of the climatic conditions during winter has an important influence on the proportion of nymphs and adults surviving the winter among those species that overwinter viviparously. For example, temperatures lower than −12.2°C for one night and lower than 0°C the following day caused slight reductions in the size of *S. graminum* populations. In contrast, temperatures lower than −12.2°C. for three successive nights and lower than 0°C in the intervening days caused a reduction of 75% in population size (Arnold, 1981). The aphids *Aulacorthum solani, Macrosiphum euphorbiae,* and *Myzus persicae* can overwinter anholocyclically (Heathcote *et al.*, 1965; Turl, 1983), and their overwintering success also depends on the severity of the winter. However, this relationship may depend as much on the relative cold-hardiness of the preferred host plants as on the cold-hardiness of the aphids themselves (Turl, 1983).

The cereal aphids *S. avenae, M. dirhodum,* and *R. padi* can overwinter as adults and nymphs in southern England. Griffiths and Wratten (1979) investigated the responses of these aphids to subzero temperatures and showed significant differences in mortality among clones of a species but not among species. They demonstrated a significant relationship between the mortality of clones and the winter temperatures of the place of origin of the clones. These differences appear to be more important than acclimation and indicate genetic differences among clones associated with geographic distribution.

The survival of eggs is less likely to be influenced by winter temperature (e.g., the eggs of *R. padi* can survive temperatures of −37°C; Sömme, 1969). Nevertheless, the egg stage may suffer high mortality (Leather, 1983), and the mortality rate of eggs is greatest just before egg hatch (Way and Banks, 1964). Some species of aphid may overwinter as nymphs, adults, and eggs in certain parts of their geographic range (e.g., *M. persicae, R. padi*). The winter temperatures have their greatest influence on the survivorship of the free-living stages, and this may have a marked influence on the probability of outbreaks, because the relative effects of winter conditions on eggs and adults directly influence aphid numbers in spring before the colonization of summer host plants.

B. Colonization and Dispersal

Aphids are exceptionally good colonizers, and the extent of their potential for increase has been observed on occasions when nonnative aphids have arrived in uncolonized areas. Smith (1959) detailed the spread of the spotted alfalfa aphid, *T. trifolii* f. *maculata,* in California, and more recently, these events have been documented for the blue-green lucerne aphid, *Acyrthosiphon kondoi,* in Oklahoma (Berberet *et al.*, 1983). The advent of the spotted aphid in Australia in the late 1970s saw a similar spectacular colonization of lucerne and medic growing regions (Fig. 8). Prevailing wind conditions are important determinants of the dispersal patterns of aphids in these circumstances.

Such colonization events are relatively rare; more frequently aphid outbreaks are the product of recurrent local movements of endemic species. Aphids are poor fliers but can actively leave the aerial plankton and settle on plants. The interaction between settling behavior and flight be-

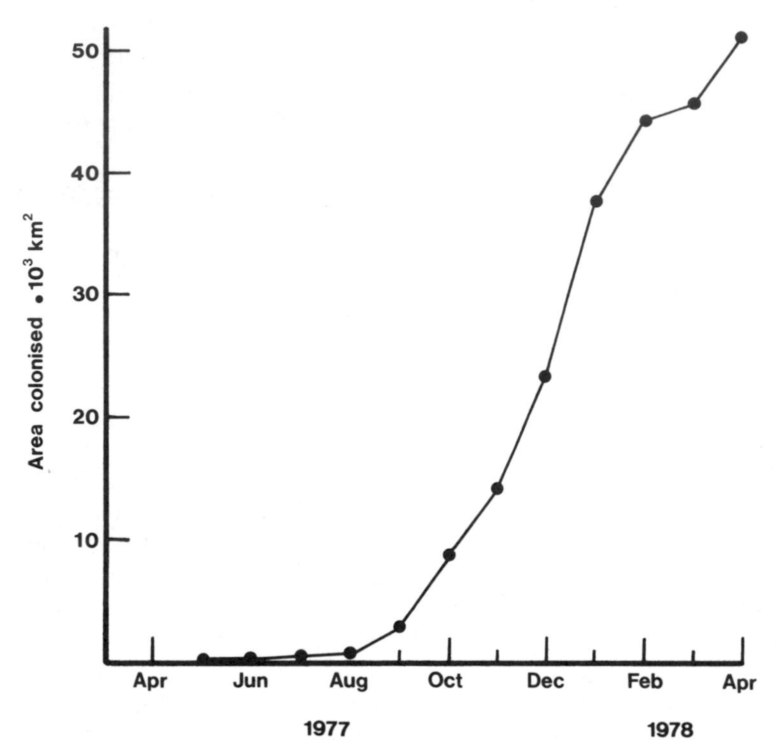

Fig. 8. Spread of the spotted alfalfa aphid following initial colonization of South Australia. Based on data in Wilson *et al.* (1981).

havior influences host finding, and aphids settle more readily after long flights than after short flights (Johnson, 1958; Kennedy and Booth, 1963). In addition, the food plant influences settling behavior, and within host plants the physiological condition of the plant may be important (Walters and Dixon, 1982). The peak number of aphids on a crop is related to the number of immigrants arriving earlier in the season, and the timing of colonization may be important (Walters and Carter, 1983). In addition, the pattern of colonization may also influence the timing and amplitude of the peak, although this point does not seem to have been investigated.

Many species exhibit alary polymorphism, and photoperiod, host plant conditions, and population density have all been shown to influence the production of winged individuals within a population (Schaefers, 1972; Yagamuchi, 1976). Winged individuals are able to fly soon after the adult molt, and their flight behavior appears to be a graded response to environmental circumstances (Dixon *et al.*, 1968). In many species an increasing proportion of alate individuals are found as population density increases (e.g., Hughes, 1963; Shaw, 1970). Density-dependent alate production may have a marked influence on population dynamics, with the overall effect of shaping population trajectories (Fig. 9). If the density dependence of alate production is sufficiently strong, populations may "crash" even in the absence of adverse weather conditions, deteriorating food quality, or natural enemies.

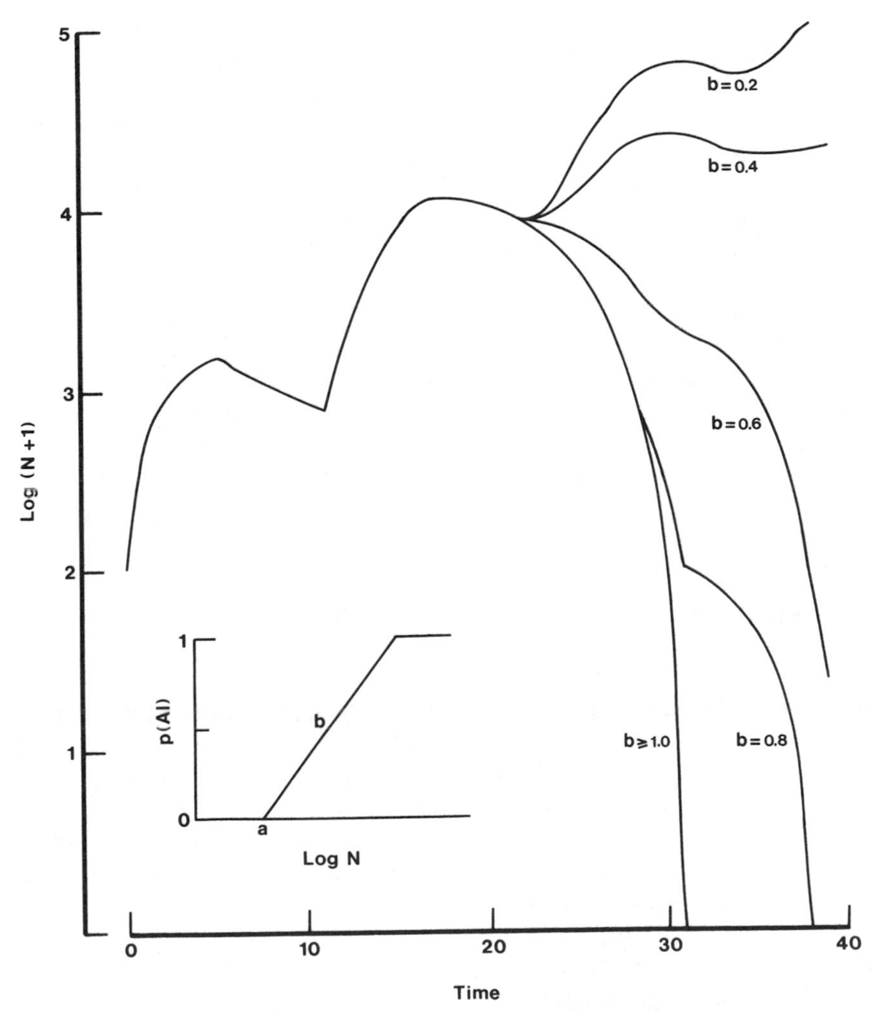

Fig. 9. Population trajectories of a hypothetical polymorphic aphid with constant development times, reproductive rates, and daily survivorship. The proportion of offspring that become alate $p(Al)$ and fly without reproducing is density dependent (see inset), with a lower threshold a, which is held constant. Variation in the slope of the density-dependent relationship b *shapes* the overall population trajectory.

C. Population Buildup

Parthenogenesis, pedogenesis, high reproductive rates, and short generation time are features of aphid biology. All contribute to the high rates of increase of population that may occur on host plants following colonization, if environmental circumstances are favorable. Intrinsic rates of increase are dependent on temperature, food quality, and a number of intraspecific processes. The relative changes in growth and development rates influence adult weight, and fecundity has been shown to be a function of weight in a large number of species (e.g., Murdie, 1969; Taylor, 1975; Watt, 1979). Intraspecific processes affect the size of aggregations of *Brevicoryne brassicae* (Way, 1968; Way and Cammell, 1970), and similar effects are significant in the population dynamics of *Eucallipterus tiliae* and *D. platanoidis* (Barlow and Dixon, 1980; Chambers *et al.*, 1985). In addition, seasonal changes in the quality of the host plant influence population dynamics (Watt, 1979).

Such features as polymorphism and overlapping generations mean that conventional approaches to understanding the major factors regulating populations are inappropriate for aphids. Gilbert (1982) suggests that aphid population dynamics cannot adequately be described by one or two parameters. Other studies indicate that the effects of these processes are not additive; rather, aphid populations are governed by a system of hierarchical regulation (Barlow and Dixon, 1980).

VIII. FORECASTING APHID OUTBREAKS

Most of our knowledge of the dynamics of aphid populations has been derived from detailed simulation studies (e.g., Hughes and Gilbert, 1968; Gilbert and Gutierrez, 1973; Barlow and Dixon, 1980; Carter *et al.*, 1982). Such studies are important, because they provide a functional basis for events observed in the field and illustrate which processes are significant in the overall dynamics (Barlow and Dixon, 1980). The use of physiological time scales in some of these studies has led to the development of temperature-dependent models for predicting the size of aphid populations in the field (e.g., Whalon and Smilowitz, 1979). At the same time other workers have attempted to devise forecasting systems that give long-term warnings of outbreaks.

In the case of heteroecious species there may be distinct movements between the primary and secondary hosts (e.g., Taylor *et al.*, 1979), and the sizes of the populations overwintering in the egg stage can be assessed with a high degree of accuracy. Area forecasting of *A. fabae* relies heavily on this feature of the aphid's life cycle. Hodek *et al.* (1966) noted that the size of *A. fabae* populations is dependent on the influx of the aphids moving from primary to secondary hosts. In their study about 30% of cases were clearly destined to be either outbreak or nonoutbreak areas. The remaining cases were intermediate: It is these instances that forecasting schemes should address if they are to be useful.

In the United Kingdom, crop losses of spring beans are related to the proportion of plants colonized by immigrant *A. fabae* in June, and area forecasts based on the sizes of egg populations in the previous winter have proved effective in providing early warning of both outbreaks and the need for chemical control (Way and Cammell, 1973, 1982; Way *et al.*, 1977). This forecasting system also uses estimates of the density of the migrating aphid populations, based on catches of aphids in suction traps. Way *et al.* (1981) have demonstrated that the proportion of bean crops infested with *A. fabae* is related to (1) the size of the autumn migration in the previous year, (2) the winter egg count, (3) the spring population size on the primary host, and (4) the size of the spring migration. Forecasts based on a knowledge of these relationships are accurate on about 90% of occasions.

A similar forecasting scheme based on egg counts of *R. padi* has been suggested by Leather (1983). However, he notes that such systems may be feasible only when aphids are totally dependent on the egg stage in winter. Thus, such systems may be practicable for *R. padi* in countries like Finland, where the species is heteroecious, but not in the United Kingdom, where an unknown proportion of the aphids successfully overwinter as parthenogenetic individuals. Other investigators have attempted to forecast the likelihood of cereal aphid outbreaks in the United Kingdom and Europe. Some studies have demonstrated relationships between the abundance of cereal aphids in trap catches and the subsequent maximum density of the aphid in cereal crops (Wiktelius, 1982), whereas others have found no relationship (Dean, 1974). The start of the spring migration of cereal aphids is correlated with winter and spring temperatures in France (Robert and Rouze-Jouan, 1976); in this region some aphids overwinter as viviparae. In contrast, no correlation has been found in Sweden, where the aphids overwinter as eggs (Wiktelius, 1982). It appears that colonies of aphids are more directly influenced by temperature than are eggs in diapuase. The variability in life history strategies clearly complicates the development of forecasting systems for cereal aphids.

Vickerman (1977) proposed a hypothesis aimed at explaining cereal aphid outbreaks. He suggested that cold springs precede outbreaks. In mild springs natural enemies are active and reach higher densities in cereal crops early; this reduces the likelihood of an aphid outbreak. In contrast, natural enemies are not abundant in cold springs, and they are unable to prevent the rapid buildup of aphids within the crop. An alternative view is that the size and timing of the spring migration of cereal aphids have an important effect on the likelihood of cereal aphid outbreaks (Watson and Carter, 1983; Walters *et al.,* 1983). In the United Kingdom the size of the spring migration of *S. avenae* is inversely related to winter temperatures below 0°C (Watson and Carter, 1983), and simulation studies suggest that peak number on the crop is directly related to the number of immigrants (Carter *et al.,* 1982). In addition, the growth stage of the cereal plants at the time of initial spring migration may also have a marked influence on peak number (Henderson and Perry, 1978; Carter and Rabbinge, 1980). However, the relationship between input and peak aphid number is not simple (Watson and Carter, 1983), and prevailing meteorological conditions in the postmigratory buildup phase may have marked effects on peak number in some years (e.g., Jones, 1979).

General forecasting schemes are adequate on a regional basis, but they often perform poorly on a field-to field basis. Rautapää (1976) attempted to address this problem and devised a simple method for predicting aphid outbreaks in individual fields based on extrapolating maximum densities from successive estimates of field populations. The problem of field-to-field variation is also dealt with by the impressive EPIPRE forecasting system, which now operates in Europe. Participating farmers monitor their own crops and send data to a centralized data bank. Simulation models produce estimates of damage caused by fungal diseases and aphids, and appropriate decision strategies are returned to the farmers (Rabbinge and Rijdijk, 1983). This sytems shows the value of clearly defined objectives and is an outstanding example of the potential of forecasting systems.

IX. CLASSIFYING APHID OUTBREAKS

As a result of their phenotypic plasticity and polymorphism, aphids are able to track environmental conditions very closely. This suggests that aphids might most generally be classified as exhibiting gradient outbreaks (see Chapter 1). Wellings *et al.* (1985) present graphs of k values plotted against population density for the yearly changes in eight *D. platanoidis* populations. Since population growth rate r is equal to $-k$, these plots are analogous to Berryman's ''phase portraits'' (see Chapter 1), and in each case the phase portraits are very narrow and fulfill the criterion of a gradient outbreak. However, these data are for the yearly changes in population density and do not provide any insight into the within-year dynamics of populations. The degree of synchrony among these sycamore aphid populations suggests that they are driven by weather and calm wind conditions in autumn, factors that appear to be important in generating high population densities (Chambers *et al.*, 1985). In contrast, *A. fabae* populations exhibit attributes of cyclical outbreaks (see Chapter 1) as a consequence of the delayed effects of natural enemies (Way, 1967). However, this species is heteroecious and its dynamics could also be viewed in terms of a pulse outbreak (Chapter 1) driven by immigrants arriving from the primary host (Way *et al.*, 1981). This view is also supported by the fact that *A. fabae* outbreaks occur most frequently on secondary hosts in regions where the primary host is most abundant (Way and Cammell, 1982). The primary and secondary hosts of some heteroecious species have contiguous distributions, whereas the hosts of other species have disjunct distributions (Taylor *et al.*, 1979). The overwintering or oversummering sites of anholocyclic species have similar spatial properties. For example, the overwintering success of *S. avenae* depends on the severity of the winter (Watson and Carter, 1983) and overwintering sites are contiguous with the distribution of summer hosts. In contrast, the oversummering sites of other anholocyclic species may be confined, in some years, to regions with mild summers. However, outbreaks in disjunct regions with harsh summers may arise if a large number of aphids are able to migrate into these regions. It is clear that migrating aphids are able to explore large regions for potential hosts and, if successful, may generate outbreaks despite, in some cases, arising from localized sources (Taylor *et al.*, 1979).

Aphid outbreaks are seasonal and rarely occur over successive years; thus, they should not be classified as eruptive outbreaks (*sensu* Berryman, Chapter 1, this volume). Most outbreaks appear to be noneruptive and are a function of environmental conditions. Some outbreaks arise through conditions permitting prolonged periods of positive population growth; they may be gradient, cyclic, or some combination of both. However, others, especially in heteroecious and anholocyclic species, depend in part on high levels of immigration, and these might best be classified as pulse noneruptive outbreaks.

X. CONCLUSIONS

There does not appear to be an association between life cycle complexity and the incidence of outbreaks in aphids. Climatic conditions, especially temperature and wind, are critical to outbreaks. Prolonged favorable weather conditions allow aphids to achieve their rapid population increase to epidemic levels. In host-alternating and anholocyclic species, climatic conditions may also influence the number of immigrants settling on crop plants, and this has a direct effect on the peak number of aphids. In addition, prevailing weather conditions influence host plant quality, and this also determines the potential for increase in aphid populations. In many pest species immigration and emigration processes are of fundamental importance in determining the level of abundance of transient populations. The impact of natural enemies on these populations is variable. In some cases they may contribute to the reduction in aphid number around the peak in abundance, but in many crop systems losses in yield will have already occurred. Natural-enemy/prey ratios have to be relatively high soon after the aphids colonize their host plants if outbreaks are to be prevented by the action of predators and parasitoids. The effect of climatic conditions on natural enemies frequently prevents such ratios from being achieved.

REFERENCES

A'Brook, J. (1981). Some observations in West Wales on the relationship between number of alate aphids and weather. *Ann. Appl. Biol.* **97,** 11–15.

A'Brook, J. (1983). Forecasting the incidence of aphids using weather data. *Bull. OEPP* **13,** 229–233.

Arnold, D. C. (1981). Effects of cold temperatures and grazing on greenbug populations in wheat in Noble County, Oklahoma, 1975–1976. *J. Kans. Entomol. Soc.* **54,** 571–577.

Barlow, C. A. (1962). The influence of temperature on the growth of experimental populations of *Myzus persicae* (Sulzer) and *Macrosiphum euphorbiae* (Thomas). *Can. J. Zool.* **40,** 146–156.

Barlow, N. D., and Dixon, A. F. G. (1980). "The Simulation of Lime Aphid Population Dynamics." Pudoc, Wageningen.

Baumgaertner, J. U., Frazer, B. D., Gilbert, N., Gill, B., Gutierrez, A. P., Ives, P. M., Nealis, V., Raworth, D. A., and Summers, C. G. (1981). Coccinellids (Coleoptera) and aphids (Homoptera) [a series of 9 papers]. *Can. Entomol.* **113,** 975–1048.

Behrendt, K. (1966). Über längjahrige massenwechselbeobachtungen an der Schwarzen Bohnenblattlaus, *Aphis fabae* Scopoli (Homoptera: Aphididae). *Wanderversamml. Dt. Entomol.* **10.**

Bejer-Peterson, B. (1962). Peak years and regulation of numbers in the aphid *Neomyzaphis abietina* Walker. *Oikos* **13,** 155–168.

Bellows, T. S., Jr. (1981). The descriptive properties of some models for density-dependence. *J. Anim. Ecol.* **50,** 139–156.

Berberet, R. C., Arnold, D. C., and Soteres, K. M. (1983). Geographical occurrence of *Acyrthosiphon kondoi* Shinji in Oklahoma and its seasonal incidence in relation to *Acrythosiphon pisum* (Harris) and *Therioaphis maculata* (Buckton) (Homoptera: Aphididae). *J. Econ. Entomol.* **76,** 1064–1068.

Berg, G. N. (1984). The effect of temperature and host species on the population growth potential of the cowpea aphid, *Aphis craccivora* (Homoptera: Aphididae). *Aust. J. Zool.* **32,** 345–352.

Berryman, A. A. (1978). Towards a theory of insect epidemiology. *Res. Popul. Ecol.* **19,** 181–196.

Blattný, C. (1925). Het voorspellen van het massaal optreden van schadelikjke insekten. *Tijdschr. Plantenziekten* **31,** 139–144.

Bodenheimer, F. S., and Swirski, E. (1957). "The Aphidoidea of the Middle East." Weizmann Sci. Press, Jerusalem.

Campbell, A., Frazer, B. D., Gilbert, N., Gutierrez, A. P., and Mackauer, M. (1974). Temperature requirements of some aphids and their parasites. *J. Appl. Ecol.* **11,** 431–438.

Carter, N., and Rabbinge, R. (1980). Simulation models of population development of *Sitobion avenae*. *I.O.B.C. Bull.* (*W.P.R.S.*) **3,** 93–98.

Carter, N., Dixon, A. F. G., and Rabbinge, R. (1982). "Cereal Aphid Populations: Biology, Simulation and Prediction." Pudoc, Wageningen.

Chambers, R. J., and Sunderland, K. D. (1983). The abundance and effectiveness of natural enemies of cereal aphids on two farms in southern England. *In* "Aphid Antagonists" (R. Cavallaro, ed.), pp. 83–87. A. A. Balkema, Rotterdam.

Chambers, R. J., Wellings, P. W., and Dixon, A. F. G. (1985). Sycamore aphid numbers and population density. II. Some processes. *J. Anim. Ecol.* **54,** 425–442.

Charlesworth, B. (1980). "Evolution in Age-Structured Populations." Cambridge Univ. Press, London and New York.

Charnov, E., Frazer, B. D., Gilbert, N., and Raworth, D. (1976). Fishing for aphids: The exploitation of a natural population. *J. Appl. Ecol.* **13,** 379–389.

Cockbain, A. J. (1961). Low temperature thresholds for flight in *Aphis fabae* Scop. *Entomol. Exp. Appl.* **4,** 211–219.

Cook, L. M. (1965). Oscillation in the simple logistic models. *Nature* (*London*) **207,** 316.

Dean, G. J. (1974). The overwintering and abundance of cereal aphids. *Ann. Appl. Biol.* **76,** 1–7.

de Fluiter, J. (1966). The aspects of integrated control with reference to aphids and scale insects. *In* "Ecology of Aphidophagous Insects" (I. Hodek, ed.), pp. 291–295. Junk, The Hague.

Dixon, A. F. G. (1976). Timing of egg hatch and viability of the sycamore aphid, *Drepanosiphum platanoidis* (Schr.), at bud burst of sycamore, *Acer pseudoplatanus* L. *J. Anim. Ecol.* **45,** 593–603.

Dixon, A. F. G. (1985a). Structure of aphid populations. *Annu. Rev. Entomol.* **30,** 155–174.

Dixon, A. F. G. (1985b). "Aphid Ecology." Blackie, Glasgow.

Dixon, A. F. G. (1987). Parthenogenetic reproduction and the intrinsic rate of increase in aphids. *In* "Aphids: Their Biology, Natural Enemies and Control" (P. Harrewijn and A. Minks, eds.), pp. 269–287. Elsevier, Amsterdam.

Dixon, A. F. G., and Wratten, S. D. (1971). Laboratory studies on aggregation, size and fecundity in the black bean aphid, *Aphis fabae* Scop. *Bull. Entomol. Res.* **61,** 97–111.

Dixon, A. F. G., Burns, M. D., and Wangboonkong, S. (1968). Migration in aphids: Response to current adversity. *Nature* (*London*) **220,** 1337–1338.

Dixon, A. F. G., Chambers, R. J., and Dharma, T. R. (1982). Factors affecting size in aphids with particular reference to the black bean aphid *Aphis fabae*. *Entomol. Exp. Appl.* **31,** 123–128.

Dunn, J. A., and Wright, D. W. (1955). Population studies of the pea aphid in East Anglia. *Bull. Entomol. Res.* **46,** 369–387.

Force, D. C., and Messenger, P. S. (1964). Fecundity, reproductive rates, and innate capacity for increase of three parasites of *Therioaphis maculata* Buckton). *Ecology* **45,** 706–715.

Frazer, B. D., and Gilbert, N. (1976). Coccinellids and aphids: A quantitative study of the impact of adult ladybirds (Coleoptera: Coccinellidae) preying on field populations of pea aphids (Homoptera: Aphididae). *J. Entomol. Soc. B. C.* **73,** 33–56.

Galecka, B. (1966). The effectiveness of predators in control of *Aphis nasturtii* Kalt. and *Aphis frangulae* Kalt. on potatoes. *In* "Ecology of Aphidophagous Insects" (I. Hodek, ed.), pp. 255–258. Junk, The Hague.

Gilbert, N. (1982). Comparative dynamics of a single-host aphid, III. Movement and population structure. *J. Anim. Ecol.* **51,** 469–480.

Gilbert, N., and Gutierrez, A. P. (1973). A plant–aphid–parasite relationship. *J. Anim. Ecol.* **42,** 323–340.

Graham, H. F. (1959). Effects of temperature and humidity on the biology of *Therioaphis maculata* (Buckton). *Univ. Calif. Publ. Entomol.* **16,** 47–80.

Griffiths, E., and Wratten, S. D. (1979). Intra- and inter-specific differences in cereal aphid low-temperature tolerance. *Entomol. Exp. Appl.* **26,** 161–167.

Hagen, K. S., and van den Bosch, R. (1968). Impact of pathogens, parasites, and predators on aphids. *Annu. Rev. Entomol.* **13,** 325–384.

Hamilton, G. C., Kirkland, R. L., and Peries, I. D. R. (1982). Population ecology of *Schizaphis graminum* (Rondani) (Homoptera: Aphididae) on grain sorghum in central Missouri. *Environ. Entomol.* **11,** 618–628.

Hassell, M. P. (1978). "The Dynamics of Arthropod Predator-Prey Systems." Princeton Univ. Press, Princeton, New Jersey.

Hassell, M. P., Lawton, J. H., and May, R. M. (1976). Patterns of dynamical behaviour in single-species populations. *J. Anim. Ecol.* **45,** 471–486.

Heathcote, G. D., Dunning, R. A., and Wolfe, M. D. (1965). Aphids on sugar beet and some weeds in England and notes on weeds as a source of beet viruses. *Plant Pathol.* **14,** 1–10.

Henderson, I. F., and Perry, J. N. (1978). Some factors affecting the buildup of cereal aphid infestations in winter wheat. *Ann. Appl. Biol.* **89,** 177–183.

Hodek, I. (1973). "Biology of Coccinellidae." Academia, Prague.

Hodek, I., Holman, J., Novák, K., Skuhravý, V., Starý, P., Weismann, L., and Zelený, J. (1966). The present possibilities and prospects of integrated control of *Aphis fabae* Scop. *In* "Ecology of Aphidophagous Insects" (I. Hodek, ed.), pp. 331–335. Junk, The Hague.

Honěk, A. (1980). Population density of aphids at the time of settling and ovariole maturation in *Coccinella septempunctata* [Col: Coccinellidae]. *Entomophaga* **25,** 427–430.

Hughes, R. D. (1963). Population dynamics of the cabbage aphid, *Brevicoryne brassicae* (L.). *J. Anim. Ecol.* **32,** 393–424.

Hughes, R. D., and Gilbert, N. (1968). A model of an aphid population—a general statement. *J. Anim. Ecol.* **37,** 553–563.

Itô, Y. (1980). "Comparative Ecology" (J. Kikkawa, ed. tranl.). Cambridge Univ. Press, London and New York.

Ives, P. M. (1981). Feeding and egg production of two species of coccinellids in the laboratory. *Can. Entomol.* **113,** 999–1005.

Johnson, B. (1958). Factors affecting locomotor and settling responses of alate aphids. *Anim. Behav.* **6,** 9–26.

Jones, M. G. (1979). Abundance of aphids on cereals from before 1973 to 1977. *J. Appl. Ecol.* **16,** 1–22.

Joyce, R. J. V. (1983). Aerial transport of pests and pest outbreaks. *Bull. OEPP* **13,** 111–119.

Kennedy, J. S., and Booth, C. O. (1963). Coordination of successive activities in an aphid. The effect of flight on the settling responses. *J. Exp. Biol.* **40,** 351–369.

Kieckhefer, R. W., and Olson, G. A. (1974). Dispersal of marked coccinellids from crops in South Dakota. *J. Econ. Entomol.* **67,** 52–54.

Kirby, W., and Spence, W. (1858). "An Introduction to Entomology" 7th Ed. Longman, Brown, Green, Longmans & Roberts, London.

Komazaki, S. (1982). Effects of constant temperatures on population growth of three aphid species, *Toxoptera citricidus* (Kirkaldy), *Aphis citricola* van der Goot, and *Aphis gossypii* Glover (Homoptera: Aphididae) on citrus. *Appl. Entomol. Zool.* **17,** 75–81.

Leather, S. R. (1983). Forecasting aphid outbreaks using winter egg counts: An assessment of its feasibility and an example of its application in Finland. *Z. Angew. Entomol.* **96,** 282–287.

Leather, S. R., and Dixon, A. F. G. (1984). Aphid growth and reproductive rates. *Entomol. Exp. Appl.* **35,** 137–140.

Lewis, T. (1965). The effects of shelter on the distribution of insect pests. *Sci. Hortic. (Canterbury, Engl.)* **17,** 74–84.

Mack, T. P., and Smilowitz, Z. (1982). Using temperature-mediated functional response models to predict the impact of *Coleomegilla maculata* (De Geer) adults and 3rd-instar larvae on green peach aphids. *Environ. Entomol.* **11,** 46–52.

Maelzer, D. A. (1981). Aphids—introduced pests of man's crops. *In* "The Ecology of Pests: Some Australian Case Histories" (R. L. Kitching and R. E. Jones, eds.), pp. 89–106. CSIRO, Melbourne.

May, R. M. (1973). "Stability and Complexity in Model Ecosystems." Princeton Univ. Press, Princeton, New Jersey.

May, R. M. (1974). Biological populations with non-overlapping generations: Stable points, stable cycles and chaos. *Science* **186,** 645–647.

May, R. M. (1976). Simple mathematical models with very complicated dynamics. *Nature (London)* **261,** 459–467.

May, R. M., and Oster, G. F. (1976). Bifurcations and dynamic complexity in simple ecological models. *Am. Nat.* **110,** 573–600.

May, R. M., Conway, G. R., Hassell, M. P., and Southwood, T. R. E. (1974). Time delays, density-dependence and single species oscillations. *J. Anim. Ecol.* **43,** 747–770.

Messenger, P. S. (1968). Bioclimatic studies of the aphid parasite *Praon exsoletum*. I. Effects of temperature on the functional response of females to varying host densities. *Can. Entomol.* **100,** 728–741.

Messenger, P. S., and Force, D. C. (1963). An experimental host–parasite system: *Therioaphis maculata* (Buckton)–*Praon palitans* Muesebeck (Homoptera: Aphididae–Hymenoptera: Braconidae). *Ecology* **44,** 532–540.

Mills, N. J. (1981). Some aspects of the rate of increase of a coccinellid. *Ecol. Entomol.* **6,** 293–299.

Moran, P. A. P. (1950). Some remarks on animal population dynamics. *Biometrics* **6,** 250–258.

Müller, H. J. (1966). Über mehrjährige Coccinelliden–Fänge auf ackerhohnen mit hohem *Aphis fabae*–Besatz. *Z. Morphol. Öekol. Tiere* **58,** 144–161.

Murdie, G. (1969). The biological consequences of decreased size caused by crowding or rearing temperatures in apterae of the pea aphid *Acyrthosiphon pisum* Harris. *Trans. R. Entomol. Soc. London* **121,** 443–455.

Neuenschwander, P., Hagen, K. S., and Smith, R. F. (1975). Predation on aphids in California's alfalfa fields. *Hilgardia* **43,** 53–78.

Powell, W. (1983). The role of parasitoids in limiting cereal aphid populations. *In* "Aphid Antagonists" (R. Cavallaro, ed.), pp. 50–56. A. A. Balkema, Rotterdam.

Rabbinge, R., and Rijdijk, F. H. (1983). EPIPRE: A disease and pest management system for winter wheat, taking account of micrometeorological factors. *Bull. OEPP* **13,** 297–305.

Rautapää, J. (1976). Population dynamics of cereal aphids and method of predicting population trends. *Ann. Agric. Fenn.* **15,** 272–293.

Ricker, W. E. (1954). Stock and recruitment. *J. Fish. Res. Board Can.* **11,** 559–623.

Robert, Y., and Rouze-Jouan, J. (1976). Neuf ans de piège de pucerons des céréales: *Acyrthosiphon (Metopolophium) dirhodum* Wlk., *A. (M.) festucase* Wlk., *Macrosiphum (Sitobion) avenae* F., *M. (S.) fragariae* Wlk. et *Rhopalosiphum padi* L. en Bretagne. *Rev. Zool. Agric. Pathol. Vég.* **75,** 67–80.

Rogers, D. J. (1975). Ecology of *Glossinia.* Natural regulation and movement of tsetse fly populations. Revue d'élévage et de médecine vétérinaire des pays tropicaux. Supplement. Les moyens de lutte contre trypanosomes et leur vecteurs. *Actes Colloq., Paris, 1974,* pp. 35–38.

Rogers, D. J. (1979). Tsetse population dynamics and distribution: A new analytical approach. *J. Anim. Ecol.* **48,** 825–849.

Schaefers, G. A. (1972). The role of nutrition in alary polymorphism among the Aphididae—an overview. *Search: Agric.* **2,** 1–8.

Scopes, N. E. A., and Biggerstaff, S. B. (1977). The use of a temperature integrator to predict the development period of the parasite *Aphidius matricariae. J. Appl. Ecol.* **14,** 799–802.

Shaw, M. J. P. (1970). Effect of population density on alienicolae of *Aphis fabae* Scop. II. The effects of crowding on the expression of migratory urge among alatae in the laboratory. *Ann. Appl. Biol.* **65,** 197–203.

Siddiqui, W. H., Barlow, G. A., and Randolph, P. A. (1973). Effects of some constant and alternating temperatures on population growth of the pea aphid, *Acyrthosiphon pisum* (Homoptera: Aphididae). *Can. Entomol.* **105,** 145–156.

Sluss, R. R., and Hagen, K. S. (1966). Factors influencing the dynamics of walnut aphid populations in northern California. *In* "Ecology of Aphidophagous Insects" (I. Hodek, ed.), pp. 243–248. Junk, The Hague.

Smith, R. F. (1959). The spread of the spotted alfalfa aphid, *Therioaphis maculata* (Buckton), in California. *Hilgardia* **28,** 647–682.

Smith, R. F., and Hagen, K. S. (1966). Natural regulation of alfalfa aphids in California. *In* "Ecology of Aphidophagous Insects" (I. Hodek, ed.), pp. 297–315. Junk, The Hague.

Sömme, L. (1969). Mannitol and glycerol in overwintering aphid eggs. *Nor. Entomol. Tidsskr.* **16,** 107–111.

Southwood, T. R. E. (1981). Stability in field populations of insects. *In* "The Mathematical Theory of the Dynamics of Biological Populations II" (R. W. Hiorns and D. Cooke, eds.), pp. 31–45. Academic Press, New York.

Starý, P. (1970). "Biology of Aphid Parasites (Hymenoptera: Aphididae) with Respect to Integrated Control." Junk, The Hague.

Tamaki, G., and Weeks, R. E. (1973). The impact of predators on populations of greenpeach aphids on field-grown sugerbeets. *Environ. Entomol.* **2,** 345–349.

Taylor, L. R. (1963). Analysis of the effect of temperature on insects in flight. *J. Anim. Ecol.* **32,** 99–117.

Taylor, L. R. (1975). Longevity, fecundity and size; control of reproductive potential in a polymorphic migrant, *Aphis fabae* Scop. *J. Anim. Ecol.* **44,** 135–163.

Taylor, L. R., Woiwod, I. P., and Taylor, R. A. J. (1979). The migratory ambit of the hop aphid and its significance in aphid population dynamics. *J. Anim. Ecol.* **48,** 955–972.
Turl, L. A. D. (1983). The effect of winter weather on the survival of aphid populations on weeds in Scotland. *Bull. OEPP* **13,** 139–143.
van den Bosch, R., Schlinger, E. I., Lagace, C. F., and Hall, J. C. (1966). Parasitization of *Acyrthoiphon pisum* by *Aphidius smithii,* a density-dependent process in nature (Homoptera: Aphididae) (Hymenoptera: Aphididae). *Ecology* **47,** 1049–1055.
van Emden, H. F. (1966). The effectiveness of aphidophagous insects in reducing aphid populations. *In* "Ecology of Aphidophagous Insects" (I. Hodek, ed.), pp. 227–235. Junk, The Hague.
Vickerman, G. P. (1977). Monitoring and forecasting insect pests of cereals. *Proc. Br. Crop Prot. Conf.—Pests Dis.,* pp. 227–234.
Vickerman, G. P., and Wratten, S. D. (1979). The biology and pest status of cereal aphids (Hemiptera: Aphididae) in Europe: A review. *Bull. Entomol. Res.* **69,** 1–32.
Walters, K. F. A., and Carter, N. (1983). Settling behaviour of cereal aphids and forecasting outbreaks. *Proc. Br. Crop. Prot. Conf.—Pests Dis.,* pp. 207–215.
Walters, K. F. A., and Dixon, A. F. G. (1982). The effect of host quality and crowding on the settling and take-off of cereal aphids. *Ann. Appl. Biol.* **101,** 211–218.
Walters, K. F. A., and Dixon, A. F. G. (1984). The effect of temperature and wind on the flight activity of cereal aphids. *Ann. Appl. Biol.* **104,** 17–26.
Walters, K. F. A., Watson, S. J., and Dixon, A. F. G. (1983). Forecasting outbreaks of the grain aphid *Sitobion avenae* in East Anglia. *Proc. Int. Congr. Plant Prot., 10th, 1983,* Vol. 1, p. 168.
Ward, S. A., and Dixon, A. F. G. (1984). Spreading the risk, and the evolution of mixed strategies: Seasonal variation in aphid reproductive biology. *Adv. Invertebr. Reprod.* **3,** 367–386.
Watson, S. J., and Carter, N. (1983). Weather and modelling cereal aphid populations in Norfolk (U.K.). *Bull. OEPP* **13,** 223–227.
Watt, A. D. (1979). The effect of cereal growth stages on the reproductive activity of *Sitobion avenae* and *Metopolophium dirhodum. Ann. Appl. Biol.* **91,** 147–157.
Way, M. J. (1967). The nature and causes of annual fluctuations in numbers of *Aphis fabae* Scop. on field beans (*Vicia faba*). *Ann. Appl. Biol.* **59,** 175–188.
Way, M. J. (1968). Intra-specific mechanisms with special reference to aphid populations. *In* "Insect Abundance" (T. R. E. Southwood, ed.), pp. 18–36. Blackwell, Oxford.
Way, M. J., and Banks, C. J. (1964). Natural mortality of eggs of the black bean aphid *Aphis fabae* on the spindle tree *Euonymus europaeus. Ann. Appl. Biol.* **54,** 255–267.
Way, M. J., and Banks, C. J. (1967). Intra-specific mechanisms in relation to the natural regulation of numbers of *Aphis fabae* Scop. *Ann. Appl. Biol.* **59,** 189–205.
Way, M. J., and Cammell, M. E. (1970). Aggregation behaviour in relation to food utilization by aphids. *In* "Animal Populations in Relation to Their Food Resources" (A. Watson, ed.) pp. 229–247. Blackwell, Oxford.
Way, M. J., and Cammell, M. E. (1973). The problem of pest and disease forecasting—possibilities and limitations as exemplified by work on the bean aphid, *Aphis fabae. Proc. Br. Insectic. Fungic. Conf.,* Vol. 7, pp. 933–954.
Way, M. J., and Cammell, M. E. (1982). The distribution and abundance of the spindle tree, *Euonymus europaeus,* in southern England, with particular reference to forecasting infestations of the black bean aphid, *Aphis fabae. J. Appl. Ecol.* **19,** 929–940.
Way, M. J., Cammell, M. E., Alford, D. V., Gould, H. J., Graham, C. W., Lane, A., Light, W. I. St.G., Rayner, J. M., Heathcote, G. D., Fletcher, K. E., and Seal, K. (1977).

Use of forecasting in chemical control of the black bean aphid, *Aphis fabae* Scop., on spring-down field beans, *Vicia faba* L. *Plant Pathol.* **26,** 1–7.

Way, M. J., Cammell, M. E., Taylor, L. R., and Woiwood, I. P. (1981). The use of egg counts and suction trap samples to forecast the infestation of spring-sown field beans, *Vicia faba,* by the black bean aphid, *Aphis fabae*. *Ann. Appl. Biol.* **98,** 21–34.

Wellings, P. W. (1981). The effect of temperature on the growth and reproduction of two closely related aphid species on sycamore. *Ecol. Entomol.* **6,** 209–214.

Wellings, P. W., and Dixon, A. F. G. (1983). Physiological constraints on the reproductive activity of the sycamore aphid: The effect of developmental experience. *Entomol. Exp. Appl.* **34,** 227–232.

Wellings, P. W., Chambers, R. J., Dixon, A. F. G., and Aikman, D. P. (1985). Sycamore aphid numbers and population density. 1. Some patterns. *J. Anim. Ecol.* **54,** 411–424.

Whalon, M. E., and Smilowitz, Z. (1979). GPA-CAST, a computer forecasting system for predicting populations and implementing control of the green peach aphid on potatoes. *Environ. Entomol.* **8,** 908–913.

Wiktelius, S. (1982). Flight phenology of cereal aphids and possibilities of using suction trap catches as an aid in forecasting outbreaks. *Swed. J. Agric. Res.* **12,** 9–16.

Williamson, M. (1972). "The Analysis of Biological Populations." Arnold, London.

Wilson, C. G., Swincer, D. E., and Walden, K. J. (1981). The origins, distribution and host range of the spotted alfalfa aphid, *Therioaphis trifolii* (Monell) f. *maculata,* with a description of its spread in South Australia. *J. Entomol. Soc. South. Afr.* **44,** 331–341.

Wyatt, I. J., and Brown, S. J. (1977). The influence of light intensity, daylength and temperature on increase rates of four glasshouse aphids. *J. Appl. Ecol.* **14,** 391–399.

Yagamuchi, H. (1976). Biological studies on the todo-fir aphid *Cinara todicola* Inouye with special reference to its population dynamics and morph determination. *Bull. Gov. For. Exp. Stn.* (*Jpn.*) **238,** 1–102.

Chapter 14

Amino Acid Nutrition of Herbivorous Insects and Stress to Host Plants

BRENT BRODBECK and DONALD STRONG

Department of Biological Science
Florida State University
Tallahassee, Florida 32306

I. INTRODUCTION

Outbreaks are short-lived increases in population density that are often spectacular among insects (Strong, 1984). Outbreaks are part and parcel of loose and liberal population dynamics, which involve only slight, subtle, and delayed regulation (if any regulation at all!) within a broad range of medial densities. Above this medial range, some sort of density-dependent decrease in population, caused by emigration, resource depletion, natural enemies, or a shortage of safe sites that provide refuge from inclement weather, is increasingly likely. However, characteristic of the liberal nature of insect population dynamics, not all decreases from numerical peaks are caused by density-dependent mechanisms. The change to poor environmental conditions at the end of a growing season readily causes a correlation between density and population change that only mimics regulation. Below the medial range at extremely low densities, some of the possibilities include the inverse density dependence of the "Allee effect" or, conversely, density dependence caused by low levels of background immigration, which might rescue a local population from extinction. However, little is known empirically about dynamics of very low densities, and we only guess about feedback at low extremes of population density.

Elements of this sort of "density vagueness" have long been a big part of theory about insect populations (Strong, 1985, 1986). Density vagueness might be viewed as "broad-banded" rather than point equilibria. Theory concerned with organisms other than insects has also addressed dynamics that are density vague with broad-banded rather than point equilibria (e.g., Caswell, 1982; Chesson, 1987; Connell and Sousa, 1983; Hubbell, 1979).

Understanding population dynamics means understanding the variation in the rate of population change with changing densities. Much of insect population change is governed by autecological and synecological factors that are only vaguely related to population density; this is obvious from the topics in this volume. This chapter addresses one of these synecological factors that is important among herbivorous insects: the amino acid nutrition afforded by host plants. We are particularly concerned with how this nutrition is influenced by factors external to the insect population, especially by stresses to the plant. Outbreaks of herbivorous insects can, theoretically, be caused by stresses to host plants that cause increases in the concentration of certain amino acids crucial to the fecundity or survival of insects, as suggested by White (1978). We are also concerned with how amino acid nutrition should vary among different types of insects. Because the composition and concentration of the amino acid pool vary greatly among plant tissues, especially in response to stress, insects feeding in different ways and on different tissues should respond differently to the same plant and to the same stress.

Our focus on the amino acid nutrition of herbivorous insects is an extension of an earlier productive focus on the total nitrogen content of host tissue (Southwood, 1972; McNeil and Southwood, 1978; White, 1978; Mattson, 1980; Scriber and Slansky, 1981). Because plants usually contain less nitrogen than is optimal for insect diets, ecological entomologists have learned a tremendous amount by studying insect performance in relation to the total nitrogen available in plant tissue. One of the most reliable generalizations in all of ecology is that differences in insect performance (i.e., in growth, survivorship, and sometimes even fecundity) depend on differences in the nitrogen and water content of host plant tissue (Scriber and Slansky, 1981; Mattson, 1980). There are exceptions, of course (Faeth *et al.*, 1981; Vince *et al.*, 1981).

Food quality is an obvious consideration in population dynamics of phytophagous insects, yet the value of nutritive dietary components like amino acids has received little recent attention from ecologists, and ecological studies have emphasized noxious phytochemistry more than insect nutrition (e.g., Rosenthal and Janzen, 1979). Perhaps the apparent abundance of food, as formalized by the "earth is green" philosophy (Hairston *et al.,* 1960), discouraged consideration of the possible effects of food quality. In a parallel fashion Fraenkel's (1953) statement that nutritional requirements for herbivorous insects are "essentially similar" may have also reduced interest in these studies; Fraenkel (1969) later softened this position. The fact that many plants are poor food for insects strictly in terms of basic nutrition—protein, and amino acid availability—indicates that knowledge of these nutritive components is as necessary for an understanding of the ecology of insects on plants as is knowledge of the noxious components of plant tissue.

II. AMINO ACID NUTRITION OF INSECTS

Insect physiologists have shown that individual amino acids differ greatly in their effect on insect performance (Rock and King, 1967; Dadd and Krieger, 1968; Auclair, 1963); subsets of amino acids are the mechanistic basis of the correlations between total plant nitrogen and insect performance. As well, correlations between total nitrogen and insect performance mean that a large component of total nitrogen in plants is often in a form usable by insects. We infer that much of the nitrogen usable by herbivorous insects is a component of photosynthetic enzymes, which account for roughly half of the nitrogen in many plants (Bjorkman, 1973). Few ecological studies with insects have approached these subjects in terms of specific amino acids, so we rely mainly on literature from laboratory studies on insect and plant physiology, on the value to insects of specific amino acids in artificial diets, and the changes in amino acid concentrations under plant stress conditions.

The amino acid nutrition of insects is similar among diverse feeding modes and is not unlike that of mammals. Dietary deficiencies of insects result from insufficient concentrations of essential amino acids rather than from a lack of total nitrogen. Most require the same 10 essential amino acids needed for mammalian growth (isoleucine, leucine, lysine, methionine, phenylalanine, threonine, tryptophan, valine, arginine, histidine; Taylor and Medici, 1966). If any essential amino acid is too rare in the diet, insect development does not occur regardless of the amount of total nitrogen. If concentrations of essential amino acids are adequate, dietary supplementation even of essential amino acids improves insect performance only slightly, if at all (Tsiropoulos, 1978; Vanderzant, 1958).

For any amino acid, an "adequate" threshold exists above which further dietary supplementation leads to toxicity as ammonia accumulates when specific metabolic pathways are overloaded. Amino acid profiles of plant tissues suggest that dietary concentrations of essential amino acids rarely exceed and are often below the intermediate "adequate" concentrations determined by laboratory studies, especially in vascular fluids, xylem, and phloem.

Different amino acids are usable (essential), nonusable (nonessential), or toxic to herbivorous insects. Nitrogen also occurs in such forms as alkaloids, cyanogenic glycosides, nitrates, or "nonprotein" amino acids that are not usable by insects. Different plants allocate nitrogen differently among amino acids, proteins, and other molecules. For example, most dicots convert nitrates to the amine form shortly after absorption, whereas many grasses maintain high levels of nitrogen in nitrate, which is unusable by most herbivorous insects (Meyer *et al.,* 1973). For this reason, one might expect ecological interpretations of nutritional adequacy for insects based on total plant nitrogen concentrations to be less accurate for grasses than for dicots, but this question has not been well investigated.

Another facet of the amino acid nutrition of insects that has received little attention is the distribution of these molecules among plant tissues; in their chemical analyses, ecologists often treat plants as homogeneous tissue. Amino acid profiles can vary greatly among plant tissues and to a lesser extent spatially within the same type of tissue (Mattson, 1980). Amino acid concentrations vary systematically among categories of plant tissues fed on by insects (Fig. 1), but little information relevant to the performance of insects in nature is available regarding the variability of amino acid profiles within plant tissues.

Total nitrogen concentration in plant tissues is likely to be grossly correlated with the nutritional adequacy of amino acid composition, yet the basis of the gross nitrogen correlations in terms of individual amino acids is necessary for an understanding of underlying mechanisms. Even fairly recently, chemical analysis of individual amino acids was too cumbersome for general ecological studies (McNeill and Southwood, 1978). However, new automatic amino acid analyzers and high-performance liquid chromatographs, as well as new filtration and sequestering techniques that isolate amino acids from interfering compounds in plant tissue, have much improved the analytical potential available to ecologists. We cannot stress how important these new tools and machines are for gaining a mechanistic understanding of processes occurring in natural populations. The great amount of data on different amino acids in one sample will generate new statistical problems; 10 variables (individual essential amino

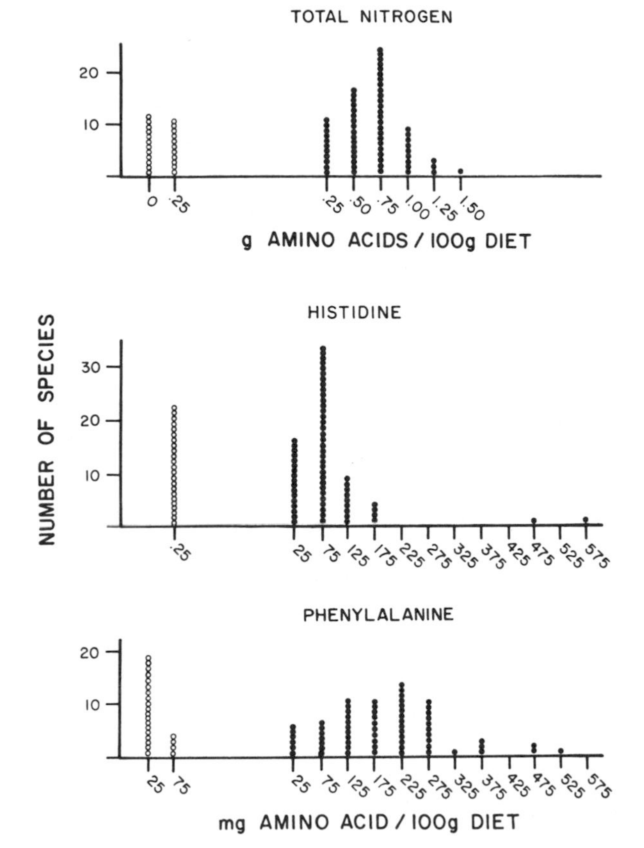

Fig. 1. Concentrations of two essential amino acids and total nitrogen in phloem (○) and leaves (●). Points represent values for different plant species [Food and Agriculture Organization (FAO), 1970].

acids) replace 1 (total nitrogen). The problem may not be too great, however, because some essential amino acids ("target" amino acids) are more likely to affect insect performance than others.

III. TARGET AMINO ACIDS

A preliminary discrimination of what we call "target" amino acids comes from comparisons of insect requirements in artificial diets with concentrations of amino acids in plants (Fig. 2). The plant data are from two phytochemical catalogues [Food and Agriculture Organization (FAO), 1970; Ziegler, 1975]. These data have an obvious agricultural bias,

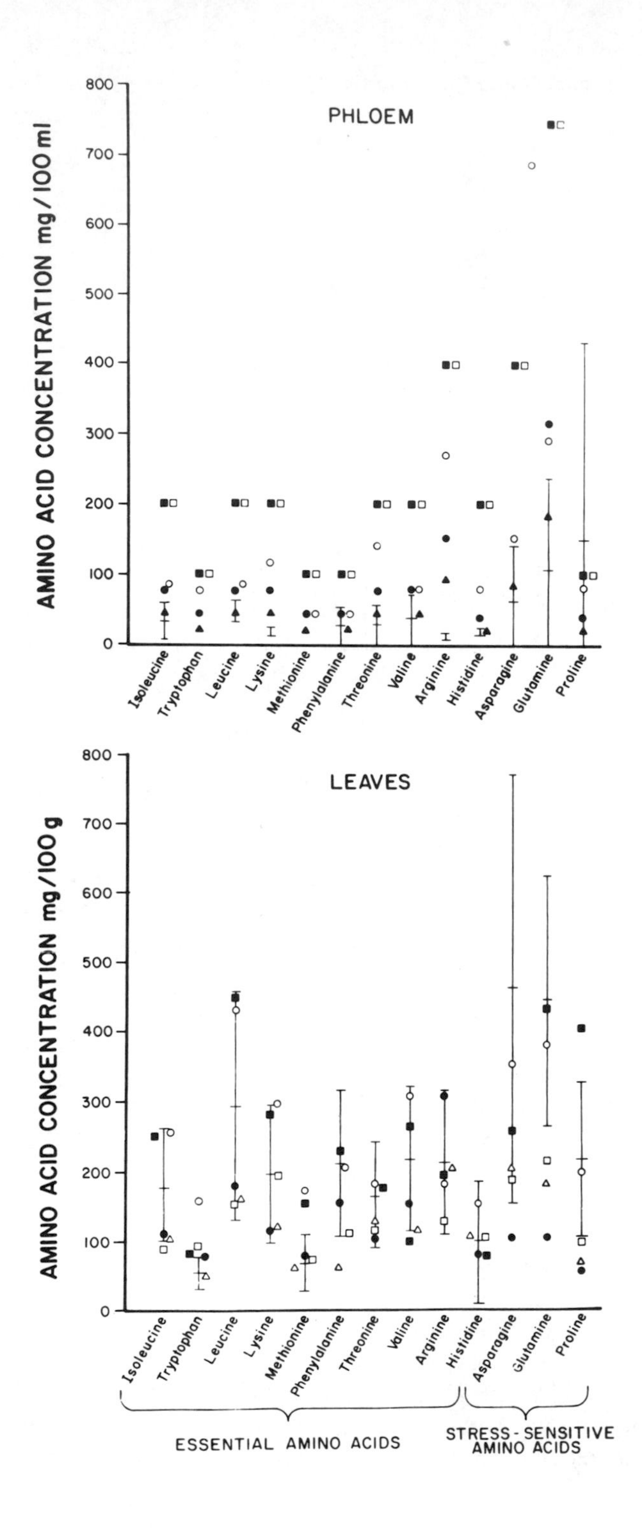
PHLOEM
AMINO ACID CONCENTRATION mg/100 ml
0
100
200
300
400
500
600
700
800
Isoleucine
Tryptophan
Leucine
Lysine
Methionine
Phenylalanine
Threonine
Valine
Arginine
Histidine
Asparagine
Glutamine
Proline
LEAVES
AMINO ACID CONCENTRATION mg/100 g
ESSENTIAL AMINO ACIDS
STRESS-SENSITIVE AMINO ACIDS

and although we do not think artificial-diet values can be extrapolated directly to the nutritional quality of plants in nature, the comparison is a place to begin the search for limiting "target" amino acids. Similar comparisons use a "protein score" based on relative concentrations of essential amino acids to determine the nutritional value of plants in human diets (FAO, 1970). This approach was also suggested for entomological use by insect physiologists (Rock and King, 1967).

What is an adequate level of an amino acid in an artificial diet? Artificial diets for insects have several complications: different food form and texture, interaction of dietary components, and pH control (Mittler, 1972); these can depress insect performance to levels below those of insects on natural foods even with the "current best" artificial diet. Conversely, some synthetic diets increase performance to levels above those of insects on natural diets (e.g., Hou and Brooks, 1975). For Fig. 2, we selected diets from studies of different insects that varied amino acid composition of food in the laboratory. We do not assume that any diet is optimal. We interpret these laboratory diets to be adequate.

One interesting feature of Fig. 2 is the similarity of adequate diets among insect feeding modes. For instance, phloem feeders' natural food is often roughly an order of magnitude lower in concentration of component amino acids than that of leaf feeders (Figs. 1 and 2), yet there is no commensurate difference in adequate levels of particular amino acids for the artificial diets. *Ceteris paribus,* this finding would suggest that nutritional hindrances to insect performance from low amino acid concentration in food are more likely for phloem feeders, mainly homopterans. Obversely, increases in phloem concentration of amino acids would have a greater effect on the performance of these insects according to this "other things being equal" logic.

However, additional considerations (Raven, 1983) can greatly complicate inferences from our simple comparison of amino acid needs and plant concentrations in Fig. 2. First, although the concentration of nitrogen in phloem is quite low relative to that in whole leaves, the net flux in phloem can be very high, sometimes approaching 100% of the plant nitrogen

Fig. 2. Amino acid concentrations in adequate artificial diets of insects (symbols) compared with concentrations of amino acids in plants (vertical lines). Vertical marks indicate means and standard deviations of plant values. Insect and mite data from the following studies: *Heliothis zea* (Rock and Hodgson, 1971), *Bombyx mori* (Ito and Narihiko, 1967), *Tetrachynus urticae* (Rodriquez and Hampton, 1966), *Argyrotaenia veluntinana* (Rock and King, 1967), *Locusta* (Dadd, 1960), *Macrosteles fascifrons* (Hou and Brooks, 1975), *Myzus persicae* (Dadd and Krieger, 1968), *Aphis fabae* (Dadd and Krieger, 1967), *Acyrthosiphon pisum* (Akey and Beck, 1971), and *Nephotettix cincticeps* (Hou and Lin, 1979).

concentration; "all of the organic carbon and most of the organic nitrogen involved in plant growth pass through the phloem" (Raven 1983, p. 212). Second, nitrogen in phloem is incorporated in small molecules, which are more readily used by insects than large protein molecules; free amino acids and amides can be immediately assimilated by insects. Phloem is also a source of abundant water, which is often dear to herbivorous insects (Scriber and Slansky, 1981). In addition, phloem contains neither the variety nor the concentration of potentially noxious phytochemicals found in leaf tissue (Raven, 1983). Finally, because phloem is under pressure, a large volume of its contents can be processed by insects at low metabolic expense. In sum, although low in amino acids, plant phloem is a potentially rich food for insects. That high net flux rates can allow insects to be nourished adequately by fluids low in nitrogen nutrients may also be the reason that xylem, which is the most dilute food of phytophagous insects, supports so many species of insects (e.g., most Cercopidae, Cicadellidae, and Cicadidae; Raven, 1983).

For leaf chewers, two amino acids have conspicuously low plant concentrations relative to dietary "adequacy" for insects: methionine and tryptophan (Fig. 2a). Human diets are also deficient in methionine (FAO, 1970). Methionine may be the source of the majority of sulfur in some plant species, and it ranges down to less than 25% in others (Bigwood, 1972). Low methionine levels in food may even cause an overall sulfur deficiency. If so, the amino acid cysteine, which contains sulfur, can be an alternative sulfur source in some cases.

Tryptophan levels may also be less than adequate in insect diets (Fig. 2a). Pirie (1978) warned that inaccurately low concentrations of tryptophan may be reported. However, FAO techniques have been carefully reviewed, and we view tryptophan as a target amino acid for nutritional studies of many herbivorous insects. For phloem feeders, all the plant tryptophan concentrations are low relative to insect requirements, and our data do not include methionine or tryptophan in phloem fluid (Fig. 2b). Insect dietary requirements for histidine may be greater than plant concentrations for both leaf and phloem feeders.

Other essential amino acids are seldom dilute enough to be inadequate for insect diets as long as methionine, tryptophan, and histidine levels are adequate, according to our perusal of the literature.

IV. PLANT STRESS

Correlations between climatic extremes and phytophagous insect outbreaks underly White's hypothesis, especially with regard to drought.

Desiccating leaves can be poisoned by ammonia accumulation. Tolerable degrees of stress allow plants to detoxify ammonia by converting it to free amino acids (Saunier *et al.*, 1968). Free amino acid levels also increase following salinity exposure in some plants (Jeffries *et al.*, 1979), ectoparasitic infection (Epstein and Cohn, 1971), nutrient depletion (Court *et al.*, 1972), and/or exposure to pollutants (Malhotra and Sarkar, 1979). The concentrations of single amino acids probably never change much without changes in the concentrations of other amino acids in the plant. A variety of amino acid profiles can be created by the application of different stress conditions to a single plant. Variations in fertilizer form (e.g., urea versus nitrates) also affect plant amino acid composition. Similarly, fluctuations in soil pH alter amino acid profiles. Plants in acidic soil absorb nitrogen primarily as nitrates; those in soils of higher pH tend to absorb more ammonium ions (Meyer *et al.*, 1973). Changes in these abiotic conditions yield different concentrations of specific amino acids in plants, which can then be correlated with phytophage success or abundance.

Increases in free amino acids consistently follow stress, but the total nitrogen concentration of plant tissue may not vary with stress. Stress-induced increases in free amino acids are often accompanied by a slight decrease in protein amino acid concentration, but the increase in free amino acids is not solely a result of protein degradation. Stress to plants can also cause decreased protein synthesis and increased free amino acid production (Cooke *et al.*, 1979). It is important to note that both the composition and concentration of total amino acids (free plus those in protein form) may change with stress. Figure 3 contrasts the great variability of free amino acid concentrations caused by plant stress with the great consistency of amino acids in plant proteins.

White's hypothesis hinges on whether such amino acid changes can be nutritionally advantageous to phytophagous insects. He argues that insects absorb nitrogen through the gut primarily in the form of free amino acids or very small peptides. Thus, the initial cost of proteolysis is saved if amino acids are ingested in this form, but at least one study of dietary efficiency revealed no significant nutritional differences for insects between amino acids in free and protein form (Vanderzant, 1963). The metabolic costs of protein and amino acid turnover are much greater than those of proteolysis, but one assumption of White's hypothesis, and indeed most views of herbivore nutrition based nitrogen limitation, is that dietary energy is far less critical to insect performance than are amino acids, amides, or usable proteins. Without further evidence we hesitate to assume that proteolysis is a rate-limiting step in insect nutrition. The distinction between free amino acids and those in proteins may be important for insects that are physiologically incapable of ingesting large pep-

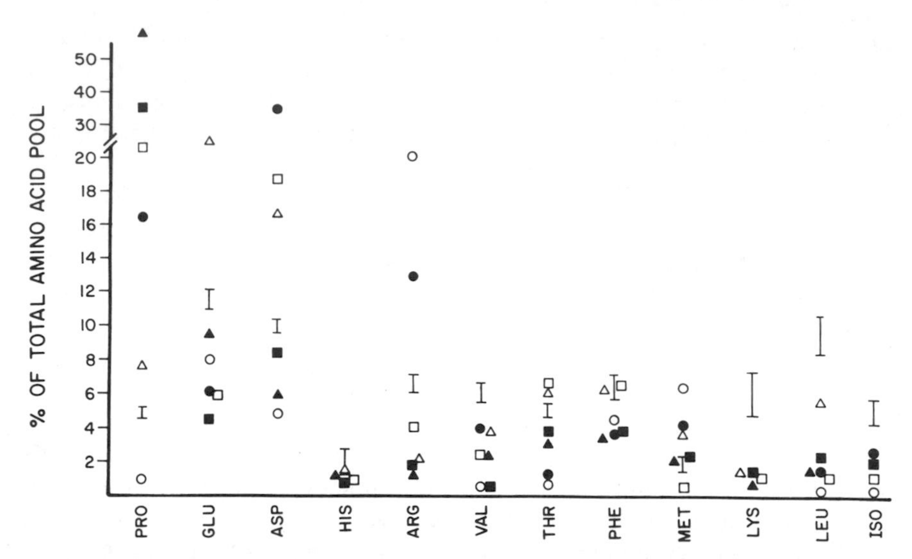

Fig. 3. Concentration of individual amino acids expressed as a percentage of amino acid pool. Closed symbols represent free amino acids in drought-stressed plants; open represent control plants (Baskin and Baskin, 1974; Singh *et al.*, 1973; Protsenko *et al.*, 1968). Vertical lines represent the range of percent composition of amino acids in protein form for nine plant proteins (Byers, 1971).

tides (e.g., aphids; Auclair, 1963), but this issue is separate from the energy costs of proteolysis.

Proteolysis, the process by which protein amino acids are broken down to free amino acids, may be far less important to insects than other sources of change in the plant's amino acid pool, and there are several different ways that amino acid levels can increase following stress to host plants. Much of the increase is likely to involve nonessential amino acids: proline, glutamine, or asparagine (Fig. 3). Amide levels are often higher in plants after stress and amides are a common vehicle for nitrogen transport. Proline and a few other small organic molecules such as betaines have osmoregulatory or "dehydration" resistance functions at a cellular level in the cytoplasm among a wide variety of organisms from bacteria through green plants and invertebrates (Le Rudlier *et al.,* 1984).

The nutritional value to insects of nonessential amino acids that are increased by stress is unclear; there may be indirect advantages, but these have not been well established. Studies suggest that proline can be used as an energy source for flight (Weeda and deKort, 1979), yet since insects are able to synthesize this amino acid the relative advantages of ingesting it intact are difficult to assess. Studies show that proline is not an essential source of dietary nitrogen (Tsiropoulos, 1978; Vanderzant, 1963). Amides occur in high concentrations in many successful artificial diets, which may indicate a nutritive function for insects. Some arginine, lysine, and methionine may be found in small "polyamines," levels of which often increase under plant stress (Galston, 1983); they may be particularly concentrated in phloem, because most plants transport nitrogen through phloem in amide or polyamine form. Increases in essential amino acids may also occur with plant stress. Although minor compared with increases in amides and proline, a stress-induced increase in target amino acids could affect insects suffering from deficiencies of these compounds. This problem requires study.

Total concentration is not well correlated with the mix of different amino acids. Total protein concentrations decrease during host plant stress, but the amino acid mix (profile) in leaf protein is remarkably consistent among tissue types, plant species, and environmental conditions (Byers, 1971). Stress changes the mix of free amino acids in plants (Fig. 3). The composition of free amino acids in plants under stress varies greatly, whereas that of amino acids in protein stays relatively constant. Phloem concentrations of amino acids often increase as nutrients are translocated away from foliage about to be abscised (Peel, 1974). This finding reinforces the expectation that homopterans that drink phloem fluid are the best candidates for outbreaks due to White's hypothesized mechanisms.

Amino acid concentration can affect fecundity in phytophagous species (Vanderzant, 1958). More subtle, but potentially important, nutritional effects may be felt by polymorphic species. Both the amide glutamine and proline have been shown to increase the ratio of long-winged to short-winged individuals among various homopterans (Dadd, 1968; Mittler and Kleinjan, 1970).

Stress-induced phytochemical changes can affect insect ingestion by means of phagostimulants. Phagostimulants increase consumption by stimulating feeding among searching insects or by detaining potential dispersers. Some amino acids are either phagostimulants themselves or precursors of phagostimulants (Schoonhoven, 1968). It is often difficult to determine whether a certain chemical affects phytophages as a nutrient, a phagostimulant, or both. Similarly, plant amino acids have been shown to affect ovipositional preferences of the rice planthopper, a phloem feeder (Sogawa, 1982).

Of course, stress to plants affects herbivorous insects in ways other than altered amino acid concentration; one example is changing water concentration in host plant tissues. Some trees maintain foliar water concentrations during drought by shedding some leaves (Kozlowski, 1974); herbs that do not drop leaves facultatively are more susceptible to desiccation. Drought can also affect carbohydrate metabolism, with the reduction of complex carbohydrates to simple sugars (Vaadia *et al.*, 1961). Plant phospholipids can undergo changes in form during drought that may have long-range effects: Unlike changes in nitrogen metabolism, lipid alterations are often irreversible upon removal of the stress condition (Chetal *et al.*, 1980). Stress-induced changes in carbohydrate metabolism may affect some insects. Pollination rates by honeybees change with sugar metabolism during drought (Goldman and Dovrat, 1980), and herbivorous insects sensitive to plant sugars may be similarly affected (e.g., aphids).

Potentially noxious phytochemicals, "secondary products," are changed by plant stress (Gershenzon, 1984), and these surely can affect insects, sometimes in a manner counter to White's hypothesis. Gould (1978) found that drought increased the resistance of cucumbers to spider mites by increasing concentrations of cucurbitacin, and Schultz and Baldwin (1983) surmised that physical damage could have similar effects. Variation of phytochemical production during stress and different insect response to noxious plant compounds allows for a broad spectrum of net insect response to plant stress. We expect that amino acids and some noxious phytochemicals can covary during and after stress, since many amino acids are precursors of these natural products (Bell and Charlwood, 1980). Again, these effects may be minimized for phloem feeders because fewer and smaller noxious compounds are found in vascular tissue than in whole-leaf concentrations (Raven, 1983).

A major problem with insect nutritional studies and our preliminary assessment is that of applying laboratory results to field situations. Stress may affect the nutrient status of a host plant in laboratory environments, but in the field these effects can readily be swamped by biotic interactions or direct effects of stress on insects themselves. For example, cold-stressed plants, as food, increase survivorship for the grasshopper *Aulocara elliotti,* yet this increase is insignificant compared with the direct effects of the cold on the grasshopper (Visscher *et al.*, 1979). Similarly, survival of some aphids is higher on food plants that are stressed by water-logged soils than on unstressed plants on drier soils (Wearing and van Emden, 1967), but the structural frailty of aphids can result in extremely high mortality from rain (up to 99%; Maezler, 1977). Nutritional changes in plants caused by environmental stress affect herbivory, yet the magnitude of these changes may not propagate proportionately through the life history to affect the population dynamics of insects. Field experiments are needed to assess the net effect on insect populations of amino acid variability due to host plant stress, and laboratory experiments are needed to elucidate the physiological mechanisms.

We are experimenting with amino acid changes in stressed marsh grass,

Spartina alterniflora, and resultant insect herbivore performance. The stresses are high salinity and rhizosphere deoxygenation (due to poor drainage), which induce high levels of proline and other amino acids and amides in the plant. The influence on insects is complicated, however. Although proline is not an essential amino acid for insects, changes in other plant constituents can boost insect performance. However, the effects on natural enemies of insects, the weather, and plant phenology, in light of the seasonal evolution of morphology and chemistry through the progression from sprouting in the spring, herbaceous growth in the summer, to flowering in the fall, are so great as to make stress-induced changes difficult to detect. An additional factor is that the carbon metabolism of this plant is drastically altered by stress from poor drainage (Mendelssohn *et al.,* 1981). The result is that the system appears to have very high contingency; the net improvement in diet from increased amino acids and amides in stressed *S. alterniflora* may be felt in the population dynamics of insects that eat the plant only when other variables do not overwhelm the effect.

V. SUMMARY

Environmental stress changes plant amino acid concentrations, and this may lead to insect outbreaks in some circumstances. These changes can increase the performance of insects feeding on stressed plant tissue. As insect food, phloem fluids, which have much lower concentrations of amino acids than whole-leaf tissue, appear most likely to be improved by stress. Phloem conducts free amino acids and small polyamines that are mobilized by plant stress. However, the low absolute concentrations of amino acids in phloem may not limit insects to a great extent, because they can process large quantities of this liquid food, which requires little energy to extract. Phloem liquid is desirable as insect food because it is provided under pressure, contains much water, and contains relatively few large noxious compounds compared with leaf tissue.

Although low in nutrients, phloem has a very high flux rate, and the phloem vessels normally conduct most of the total carbon and nitrogen in the development of the plant. It is quite plausible for stress conditions to increase short-term concentrations of one or a few limiting free amino acids sufficiently for outbreaks of phloem-feeding insect populations to occur. As yet, no insect outbreak has been well documented to have occurred by means of this sequence of events, but new tools and techniques have greatly facilitated detailed investigations of this sort, which only recently were virtually beyond the scope and extensiveness required of high-quality field ecology. The wide availability of rapid and inexpensive high-performance liquid chromatography, automatic amino acid analyzers, and associated filtration techniques have greatly increased the resolution available to ecologists on the details of herbivorous insect nutrition in field situations.

Three essential amino acids are "targets" for stress-induced effects on insects: methionine, tryptophan, and sometimes histidine. The concentrations of these amino acids are frequently lower in plants than in the "adequate" laboratory diets of insects. Cysteine, if it can be substituted as a sulfur source for methionine, might also affect insect performance when its concentration in plant tissue is increased. Stress can also produce accumulations in plant tissue of molecules with less, if any, nutritional value to insects, such as amides, proline, or ammonia. Phloem feeders in particular may benefit from increases in amides, but this is not yet well established.

Plants that maintain high water concentrations during drought conditions may be more prone to herbivore outbreaks; trees rather than herbs may better maintain normal osmotic pressures because they can drop some leaves in response to drought. Herbivores very sensitive to host plant water concentration may be less likely to benefit from drought stress to plants. Homopterans that feed on phloem fluids of plants appear particularly likely to benefit nutritionally from plant stress. Nitrogen nutrition for many species of phloem feeders comes solely from free amino acids, amides, and small polypeptides. These insects may benefit from stress-induced increases in these compounds, yet will not be affected by the corresponding loss of large proteins in the plant. Insects with unusual or high requirements of amino acids are more likely to benefit during drought, as are those that perceive certain amino acids as phagostimulants.

Much current research highlights effects on insects of "secondary compounds," that is, noxious phytochemicals. We believe that the complement, primary nutrition of insects, has been underemphasized. There is ample evidence that amino acids and protein are often deficient in plants on which insects feed and that these deficiencies greatly affect insect performance. Changes in amino acid composition underlie a great deal of variability in plant quality as food for insects. Only careful field experiments, based on a good understanding of both natural history and nutritional physiology, will adequately address these possibilities.

REFERENCES

Akey, D. H., and Beck, S. D. (1971). Continuous rearing of the pea aphid, *Acrythosiphon pisum,* on a holidic diet. *Ann. Entomol. Soc. Am.* **64,** 353–356.

Auclair, J. L. (1963). Aphid feeding and nutrition. *Annu. Rev. Entomol.* **8,** 439–490.

Baskin, C. C., and Baskin, J. M. (1974). Responses of *Astragalus tennesseensis* to drought. *Oecologia* **17,** 11–16.

Bell, A., and Charlwood, B. V., eds. (1980). "Secondary Plant Products." Springer-Verlag, Berlin and New York.

Bigwood, E. J. (1972). Amino acid patterns of animal and vegetable proteins—common features and diversities. *In* "Protein and Amino Acid Functions" (E. J. Bigwood, ed.), pp. 215–258. Pergamon, Oxford.

Bjorkman, O. (1973). Comparative studies on photosynthesis in higher plants. *In* "Photophysiology" (A. C. Geise, ed.), Vol. 8, pp. 1–63. Academic Press, New York.

Byers, J. M. (1971). The amino acid composition of some leaf protein preparations. *In* "Leaf Protein: Its Agronomy, Preparation, Quality and Use" (N. W. Pirie, ed.), p. 95. Blackwell, Oxford.

Caswell, H. (1982). Life history and the equilibrium status of populations. *Am. Nat.* **120,** 317–339.

Chesson, P. L. (1986). Environmental variation and the coexistence of species. *In* "Community Ecology" (J. Diamond and T. Case, eds.), pp. 240–256. Harper & Row, New York.

Chetal, S., Wagle, D. S., and Nainawatee, H. S. (1980). Phospholipid changes in wheat and barley leaves under water stress. *Phytochemistry* **19,** 1393–1395.

Connell, J. H., and Sousa, W. P. (1983). On the evidence needed to judge ecological stability or persistence. *Am. Nat.* **121,** 789–824.

Cooke, R. J., Oliver, J., and Davies, D. D. (1979). Stress and protein turnover in *Lemna minor. Plant Physiol.* **64,** 1109–1113.

Court, R. D., Williams, W. T., and Megarty, M. P. (1972). The effect of mineral nutrient deficiency on the content of free amino acids in *Setaria sphacelata. Aust. J. Biol. Sci.* **25,** 77–87.

Dadd, R. H. (1960). The nutritional requirements of locusts. V. *J. Insect Physiol.* **6,** 134–139.

Dadd, R. H. (1968). Dietary amino acids and wing determination in the aphid, *Myzus persicae. Ann. Entomol. Soc. Am.* **61,** 1201–1210.

Dadd, R. H., and Krieger, D. L. (1967). Continuous rearing of the *Aphis fabae* complex on sterile synthetic diet. *J. Econ. Entomol.* **60,** 1512–1514.

Dadd, R. H., and Krieger, D. L. (1968). Dietary amino acid requirements of the aphid *Myzus persicae. J. Insect Physiol.* **14,** 741–764.

Epstein, E., and Cohn, E. (1971). Biochemical changes in terminal root galls caused by an ectoparasitic nematode, *Longidorus africanus:* Amino acids. *J. Nematol.* **3,** 334–340.

Faeth, S. H., Mopper, S., and Simberloff, D. (1981). Abundances and diversity of leaf mining insects on three oak host species: Effects of host-plant phenology, nutrition and geographic changes. *Oikos* **37,** 238–251.

Food and Agriculture Organization (FAO) (1970). "Amino Acid Concent of Food and Biological Data on Proteins." Food Policy and Food Sci. Serv., Nutr. Div., FAO, Rome.

Fraenkel, G. (1953). The nutritional value of green plants for insects. *Trans. Int. Congr. Entomol., 9th, 1951,* Vol. 2, pp. 90–100.

Fraenkel, G. (1969). Evaluation of our thoughts on secondary plant substances. *Entomol. Exp. Appl.* **12,** 473–486.

Galston, A. W. (1983). Polyamines as modulators of plant development. *BioScience* **33,** 382–388.

Gershenzon, J. (1984). Plant secondary metabolite production under stress. *Recent Adv. Phytochem.* **18,** 273–320.

Goldman, A., and Dovrat, A. (1980). Irrigation regime and honeybee activity as related to seed yield in alfalfa. *Agron. J.* **72,** 961–965.

Gould, F. (1978). Resistance of cucumber varieties to *Tetrachynus urticae:* Genetic and environmental determinants. *J. Econ. Entomol.* **71,** 680–683.

Hairston, N. G., Smith, F. E., and Slobodkin, L. B. (1960). Community structure, population control, and competition. *Am. Nat.* **94,** 421–425.

Hou, R. F., and Brooks, M. A. (1975). Continuous rearing of the aster leafhopper, *Macrosteles fascifrons,* on a chemically defined diet. *J. Insect Physiol.* **21,** 1481–1483.

Hou, R. F., and Lin, L. (1979). Artificial rearing of the rice green leafhopper, *Nephotettix cincticeps,* on a holidic diet. *Entomol. Exp. Appl.* **25,** 158–164.

Hubbell, S. P. (1979). Tree dispersion, abundance, and diversity in a tropical dry forest. *Science* **203,** 1299–1309.

Ito, T., and Narihiko, A. (1967). Nutritive effects of alanine, cystine, glycine, serine, and tyrosine on the silkworm, *Bombyx mori. J. Insect Physiol.* **13,** 1813–1824.

Jeffries, R. L., Rudmik, T., and Dillon, E. M. (1979). Responses of halophytes to high salinities and low water potentials. *Plant Physiol.* **64,** 989–994.

Kozlowski, T. T. (1974). Extent and significance of shedding of plant parts. *In* "Shedding of Plant Parts" (T. T. Kozlowski, ed.), pp. 1–45. Academic Press, New York.

Le Rudlier, D., Stom, A. A., Dandekar, A. M., Smith, L. T., and Valentine, R. C. (1984). Molecular biology of osmoregulation. *Science* **224,** 1064–1068.

McNeil, S., and Southwood, T. R. E. (1978). The role of nitrogen in the development of insect/plant relationships. *In* "Biochemical Aspects of Insect/Plant Interactions" (J. B. Harborne and H. F. van Emden, eds.), pp. 77–98. Academic Press, New York.

Maezler, D. A. (1977). The biology and main causes of changes in numbers of the rose aphid, *Macrosiphum rosae* (L.), on cultivated roses in South Australia. *Aust. J. Zool.* **25,** 269–284.

Malhotra, S. S., and Sarkar, S. K. (1979). Effects of sulphur dioxide on sugar and free amino acid content of pine seedlings. *Physiol. Plant.* **47,** 223–228.

Mattson, W. (1980). Herbivory in relation to plant nitrogen content. *Annu. Rev. Ecol. Syst.* **11,** 119–161.

Mendelssohn, I. A., McKee, K. L., and Patrick, W. H. (1981). Oxygen deficiency in *Spartina alterniflora* roots: Metabolic adapatation to anoxia. *Science* **214,** 439–441.

Meyer, B. S., Anderson, D. B., Bohning, R. H., and Fratianne, D. G. (1973). "Introduction to Plant Physiology." Van Nostrand-Reinhold, Princeton, New Jersey.

Mittler, T. E. (1972). Interactions between dietary components. *In* "Insect and Mite Nutrition" (J. G. Rodriguez, ed.), pp. 211–223. North-Holland Publ., Amsterdam.

Mittler, T. E., and Kleinjan, J. E. (1970). Effect of artificial diet composition on wing-production by the aphid *Myzus persicae. J. Insect Physiol.* **13,** 289–318.

Peel, A. J. (1974). *In* "Transport of Nutrients in Plants" (A. J. Peel, ed.), pp. 49–50. Butterworth, London.

Pirie, N. W. (1978). "Leaf Protein and Other Aspects of Fodder Fractionation." Cambridge Univ. Press, London and New York.

Protsenko, D. F., Shmatko, I. G., and Rubanyuk, E. A. R. (1968). Drought resistance of winter wheats in relation to their amino acid content. *Aziol. Rast.* **15,** 680–687.

Raven, J. A. (1983). Phytophages of xylem and phloem: A comparison of animal and plant sap-feeders. *Annu. Rev. Ecol. Syst.* **13,** 135–234.

Rock, G. C., and Hodgson, E. (1971). Dietary amino acid requirements for *Heliothis zea* determined by dietary deletion and radiometric deletion techniques. *J. Insect Physiol.* **17,** 1087–1097.

Rock, G. C., and King, K. W. (1967). Quantitative amino acid requirements of the red-

banded leaf roller, *Argyrotaenia velutinana* (Lepidoptera: Torticidae). *J. Insect Physiol.* **13,** 59–68.
Rodriguez, J. G., and Hampton, R. E. (1966). Essential amino acids determined in two spotted spider mite, *Tetrachynus urticae* Koch (Acarina, Tetranychidae) with glucose U–C. *J. Insect Physiol.* **12,** 1209–1216.
Rosenthal, G. A., and Janzen, D. H., eds. (1979). "Herbivores: Their Interaction with Secondary Plant Metabolites." Academic Press, New York.
Saunier, R. E., Hull, H. M., and Ehrenreich, J. H. (1968). Aspects of the drought tolerance in creosotebush. *Plant Physiol.* **43,** 401–404.
Schoonhoven, L. M. (1968). Chemosensory bases of host plant selection. *Annu. Rev. Entomol.* **13,** 115–136.
Schultz, J. C., and Baldwin, I. T. (1983). Oak leaf quality declines in response to defoliation by gypsy moth larvae. *Science* **217,** 149–150.
Scriber, J. M., and Slanksy, F., Jr. (1981). The nutritional ecology of immature insects. *Annu. Rev. Entomol.* **26,** 183–211.
Singh, T. N., Paleg, L. G., and Aspinall, D. (1973). Stress metabolism. I. Nitrogen metabolism and growth in the barley plant during water stress. *Aust. J. Biol. Sci.* **26,** 45–56.
Sogawa, K. (1982). The rice brown planthopper: Feeding physiology and host plant interactions. *Annu. Rev. Entomol.* **27,** 49–73.
Southwood, T. R. E. (1972). The insect/plant relationship—an evolutionary perspective. *Symp. R. Entomol. Soc. London* **6,** 3–30.
Strong, D. R. (1984). Density-vague ecology and liberal population regulation in insects. *In* "A New Ecology: Novel Approaches to Interactive Systems" (P. W. Price, C. N. Slobodchikoff, and W. S. Gaud, eds.), pp. 313–327. Wiley, New York.
Strong, D. R. (1986). Density vagueness: Abiding the variance in the demography of real populations. *In* "Community Ecology" (J. Diamond and T. Case, eds.), pp. 257–268. Harper & Row, New York.
Strong, D. R. (1986). Population theory and understanding pest outbreaks. *In* "Ecological Theory and IPM Practice" (M. Kogan, ed.). Wiley, New York.
Strong, D. R., Lawton, J. H., and Southwood, T. R. E. (1984). "Insects on Plants: Community Patterns and Mechanisms." Blackwell, Oxford.
Taylor, M. W., and Medici, J. C. (1966). Amino acid requirements of grain beetles. *J. Nutr.* **88,** 176–180.
Tsiropoulos, G. J. (1978). Holidic diets and nutritional requirements for survival and reproduction of the adult walnut husk fly. *J. Insect Physiol.* **24,** 239–242.
Vaadia, Y., Roney, F. C., and Hagen, R. M. (1961). Plant water deficits and physiological processes. *Annu. Rev. Plant Physiol.* **12,** 293–326.
Vanderzant, E. S. (1958). The amino acid requirements of the pink bollworm. *J. Econ. Entomol.* **51,** 309–311.
Vanderzant, E. S. (1963). Nutrition of the adult boll weevil: Oviposition on defined diets and amino acid requirements. *J. Insect Physiol.* **9,** 683–691.
Vince, S. W., Valiela, I., and Teal, J. M. (1981). An experimental study of the structure of herbivorous insect communities in a salt marsh. *Ecology* **62,** 1662–1678.
Visscher, S. N., Lund, R., and Whitmore, W. (1979). Host plant growth temperatures and insect rearing temperatures influence reproduction and longevity in the grasshopper, *Aulocara elliotti* (Orthoptera: Acrididae). *Environ. Entomol.* **8,** 253–258.
Wearing, C. H., and van Emden, H. F. (1967). Effects of water stress in host plants on infestation by *Aphis fabae* Scop., *Myzus persicae* (Sulz.), and *Brevicoryne brassicae* (L.) *Nature (London)* **213,** 1051–1052.

Weeda, E., and deKort, C. A. D. (1979). Fuels for energy metabolism in the Colorado potato beetle, *Leptinotarsa decemlineata* Say. *J. Insect Physiol.* **25,** 951–955.
White, T. C. R. (1978). The importance of a relative shortage of food in animal ecology. *Oecologia* **33,** 71–86.
Ziegler, D. (1975). The nature of transported substances. *In* "Transport in Plants I" (M. H. Zimmerman and J. A. Milburn, eds.), pp. 59–136. Springer-Verlag, Berlin and New York.

Chapter 15

The Role of Drought Stress in Provoking Outbreaks of Phytophagous Insects

WILLIAM J. MATTSON and ROBERT A. HAACK

U.S. Department of Agriculture, Forest Service
North Central Forest Experiment Station
East Lansing, Michigan 48823

I. INSECT OUTBREAKS IN RELATION TO WATER DEFICITS

The hypothesis of plant stress participating somehow in the precipitation of phytophagous insect outbreaks presumes that in its absence the host plant is usually capable (in concert with natural enemies) of regulating the population densities of its herbivores. Whether this is true is, of course, pivotal in the decision to focus attention on plant-stressing factors. If plants are incapable of regulating their herbivores, why concern oneself with plant stress? After all, if the plant's condition has no bearing on the dynamics of the herbivore, it matters not whether plants are vigorous, avigorous, stressed, or unstressed. If, however, plants indeed participate in the regulation of their herbivores, the question of plant stress becomes much more interesting. Stress becomes a factor that could upset the basic regulatory mechanisms.

We believe that water stress (to be considered synonymous with water deficits in this chapter) alters the plant and its thermal environment so that stressed plants become progressively more susceptible (through erosion of their defense systems) and suitable (through enhancement of certain constitutive traits) to their adapted consumers, allowing them to achieve faster growth, higher survival, and more realized reproduction. In this chapter, we identify those morphological, spectral, thermal, physiological, and biochemical changes that may be of potential relevance to herbivores and discuss how these changes might provoke outbreaks of phytophagous insects (for additional discussion see Mattson and Haack, 1987).

Evidence to support the hypothesis that plant water deficits promote insect outbreaks is laregly circumstantial. This is not unexpected, of course, because the detailed experimental unraveling of all the factors involved in the population dynamics of herbivores in their natural ecosystems is nearly impossible. Most of the evidence, then, consists of numerous observations that outbreaks are often preceded by periods of unusually warm, dry weather (Table 1). For example, because rainfall and temperature have been so frequently correlated with outbreaks of the southern pine beetle, *Dendroctonus frontalis,* these variables have been incorporated into outbreak models for this insect (Michaels *et al.*, 1985). Other supporting evidence is the frequent occurrence of outbreaks on poorer sites having soils of low moisture-holding capacity (Grimalsky, 1961; Mason and Tigner, 1972; Otto, 1970; Mattson and Addy, 1975; Kolomiets *et al.,* 1979; Stoszek *et al.,* 1981; Kemp and Moody, 1984; Larsson and Tenow, 1984).

These outbreak insects seem to fall mainly into three classes: phloem feeders (Coleoptera), leaf feeders (Lepidoptera, Hymenoptera, and Orthoptera), and sap suckers (Homoptera). Our purpose here is not to discuss the individual cases and critique the evidence, but only to demonstrate that it exists for the reader's reference.

II. EFFECTS OF MOISTURE STRESS ON PLANT TRAITS

We propose that plant–insect relationships must certainly be influenced by water stress, because virtually every plant process is affected during drought (Table 2). The degree to which any single plant process changes will depend on the severity and duration of water stress as well as the stage of plant development when the stress occurs (Kramer, 1983). In this section we discuss some of the morphological, physiological, and biochemical changes that occur in water-stressed plants (primarily mesophytes) that are of potential relevance to insect performance.

A. Plant Growth

Significantly less tissue will be produced and thus made available to insect herbivores if water stress occurs during the active growth phase of a plant. A reduction in plant size is the most common result of water stress, because cell division, enlargement, and differentiation are all reduced during drought (Barlow *et al.*, 1980). Cell growth is apparently the first plant process affected by water deficits, cell division and cell wall synthesis being the next most sensitive plant processes (Table 2; Hsiao, 1973; Hsiao *et al.*, 1976). Small water deficits inhibit the growth of leaves, cambial tissue, roots, flowers, buds, and reproductive organs (Kozlowski, 1982). Zahner (1968) estimated that up to 90% of the annual variation in the xylem increment of forest trees in arid zones, and up to 80% in humid regions, could be attributed to water deficits.

B. Leaf Texture

Insect performance may be affected by textural changes in leaves of certain water-stressed plants. In addition to smaller leaf size, water stress induces increased (1) thickness, (2) waxiness, and (3) density and size of trichomes in the leaves that develop during or after drought (Table 3; Turner, 1979; Ehleringer, 1980; Kramer, 1983). Such morphological changes would probably aid water-stressed plants by increasing the reflection of incident radiation and decreasing the transpirational water loss (Begg and Turner, 1976; Ehleringer, 1980; Parsons, 1982). However, the leaves of several herbaceous and woody species were reported to be thinner during a year of severe drought (Turrell and Turrell, 1943). If certain insects are restricted to feeding only on new foliage, textural changes as mentioned above could lower their performance.

TABLE 1

Outbreaks of Forest and Range Insects Associated with Drought

Species	Family	Genus of host	Location	Reference
Coleoptera				
Agrilus bilineatus	Buprestidae	*Quercus*	United States	Haack and Benjamin (1982)
Corthylus colambianus	Scolytidae	*Acer*	United States	McManus and Giese (1968)
Dendroctonus brevicomis	Scolytidae	*Pinus*	United States	Vite (1961)
D. frontalis	Scolytidae	*Pinus*	United States	Craighead (1925); St. George (1930)
D. ponderosae	Scolytidae	*Pinus*	Canada	Thomson and Shrimpton (1984)
Ips calligraphus	Scolytidae	*Pinus*	United States	St. George (1930)
I. grandicollis	Scolytidae	*Pinus*	United States	St. George (1930)
	Scolytidae	*Pinus*	Australia	Witanachchi and Morgan (1981)
I. paraconfusus	Scolytidae	*Pinus*	United States	Vite (1961)
Ips spp.	Scolytidae	*Pinus, Picea*	Europe, Africa	Chararas (1979)
Scolytus quadrispinosa	Scolytidae	*Carya*	United States	Blackman (1924); St. George (1929, 1930)

S. ventralis	Scolytidae	*Abies*	United States	Ferrell and Hall (1975); Berryman (1973)
Tetropium abietis	Cerambycidae	*Abies*	United States	Ferrell and Hall (1975)
Homoptera				
Aphis pomi	Aphididae	*Crataegus*	Switzerland	Braun and Fluckiger (1984)
Cardiaspina densitexta	Psyllidae	*Eucalyptus*	Australia	White (1969)
Hymenoptera				
Neodiprion sertifer	Diprionidae	*Pinus*	Sweden	Larsson and Tenow (1984)
Lepidoptera				
Bupalus piniarius	Geometridae	*Pinus*	Europe	Schwenke (1968)
Choristoneura fumiferana	Tortricidae	*Abies, Picea*	Canada	Wellington (1950); Ives (1974)
Lambdina fiscellaria	Geometridae	*Abies, Picea*	Canada	Carroll (1956)
Lymantria dispar	Lymantriidae	*Picea*	Denmark	Bejer-Peterson (1972)
Selidosema suavis	Geometridae	*Pinus*		White (1974)
Orthoptera				
Several species	Acrididae	Grasses	Worldwide	White (1976)

TABLE 2

Effects of Water Stress on Certain Plant Processes in Decreasing Order of Sensitivity[a]

Process	Trend[b]	Water potential when process is first affected[c] (bars)	Genera of plants commonly studied
Cell wall synthesis	−	−1 to −5	*Avena, Helianthus, Triticum, Zea*
Protein synthesis	−	−1 to −5	*Helianthus, Nicotiana, Triticum, Zea*
Chlorophyll formation	−	−2 to −6	*Triticum*
Nitrate reductase level	−	−3 to −7	*Hordeum, Gossypium, Sorghum, Triticum, Zea*
Abscisic acid synthesis	+	−3 to −9	*Gossypium, Nicotiana, Phaseolus, Triticum*
Stomatal opening	−	−5 to −10	*Glycine, Gossypium, Pisum, Sorghum, Vicia*
CO_2 Assimilation	−	−5 to −10	*Glycine, Gossypium, Hordeum, Oryza, Vicia*
Respiration	−	−6 to −16	*Glycine, Helianthus, Lycopersicon, Malus, Zea*
Proline accumulation	+	−8 to −16	*Festuca, Glycine, Gossypium, Helianthus, Hordeum, Lycopersicon, Phaseolus, Sorghum, Zea*
Sugar accumulation	+	−11 to −16	*Acer, Glycine, Gossypium, Helianthus, Hordeum, Malus, Phaseolus, Pisum, Quercus, Sorghum, Zea*

[a] Adapted from Hsiao (1973), Hsiao *et al.* (1976), and Bradford and Hsiao (1982).
[b] Trend: +, increase; −, decrease.
[c] Values are for mesophytes.

C. Leaf Color and Spectral Qualities

Color and spectral changes in the leaves of water-stressed plants may influence the ability of insects to locate hosts. Healthy, green leaves reflect and transmit little radiation at ultraviolet and visible (except in the green) wavelengths, whereas reflection and transmission at the infrared wavelengths are generally high (Gates, 1980). As water stress becomes progressively more severe, leaf reflectance increases in both the visible and infrared wavelengths (Drake, 1976). In severely water stressed

TABLE 3

Water-Stress-Induced Changes in Leaf Morphology and Orientation

Species	Induced change[a]	Reference
Brassica oleracea	Cuticular waxes +	Baker (1974)
Encella farinosa	Trichome density +	Ehleringer (1980)
Glycine max	Cuticular waxes +	Clark and Levitt (1956)
Gossypium hirsutum	Cuticular waxes +	Weete *et al.* (1978)
Helianthus argophyllus	Trichome size +	Begg and Turner (1976)
Lupine arizonicus	Cupping occurs	Wainwright (1977)
Nicotiana glauca	Cuticle thickness +	Skoss (1955)
Pelargonium zonale	Layers of palisade cells +	Amer and Williams (1958)
Phaseolus vulgaris	Orient parallel to sunlight	Dubetz (1969)
Populus clones	Stomatal waxes +, cuticle thickness +	Pallardy and Kozlowski (1980)
Sorghum bicolor	Cuticular waxes +	Blum (1975)
Stylosanthes humilis	Leaf cupping +	Begg and Torssell (1974)
Triticum aestivum	Trichome density +	Quarrie and Jones (1977)

[a] Trend: +, increase.

plants, inhibition of chlorophyll production (Table 3; Hsiao, 1973) and nutrient uptake (Viets, 1972; Begg and Turner, 1976) often results in leaf yellowing (Kozlowski, 1976). If insects utilize spectral cues in host location, the changes noted here may allow them to locate water-stressed plants more efficiently.

D. Plant Temperature

Because water-stressed plants are warmer than well-watered plants due to reduced transpirational cooling (Begg, 1980), insects feeding on stressed plants will probably maintain higher body temperatures. Temperature differences between well-watered and water-stressed plants are commonly 2–4°C, but they can be as great as 10–15°C (Table 4; Drake 1976; Bucks *et al.*, 1984). If the elevated temperatures of water-stressed plants allow insects to live closer to their thermal optima, their developmental and reproductive rates may approach the maximal values attainable for a given species.

Some plants reduce the energy load on their leaves by orienting them parallel to incoming radiation or by means of leaf flagging, rolling, cupping, or wilting (Table 3; Turner, 1979; Begg, 1980). Insects feeding on

TABLE 4

Maximal Leaf or Canopy Temperature Differences between Well-Watered and Water-Stressed Plants

Species	Temp (°C)	Plant part measured	Reference
Brassica oleracea	2–3	Leaf	Clum (1926)
Gossypium hirsutum	3.6	Leaf	Wiegand and Namken (1966)
	6	Canopy	Bartholic *et al.* (1972)
Oryza sativa	4	Canopy	O'Toole *et al.* (1984)
Parthenium argentatum	7–15	Canopy	Bucks *et al.* (1984)
Solanum tuberosum	3–8	Canopy	Tanner (1963); Stark and Wright (1985)
Sorghum vulgare	3	Leaf	Ehrler and van Bavel (1967)
Triticum aestivum	4.8	Canopy	Ehrler *et al.* (1978)
Xanthium strumarium	6–10	Leaf	Drake and Salisbury (1972)

water-stressed plants whose leaves are oriented away from incoming radiation may be less capable of elevating their body temperature even though leaf temperatures are still higher than those of well-watered plants.

E. Leaf Longevity

The amount of leaf biomass available to consumers is regulated, in part, by the severity of the drought. Mild to moderate water stress may actually suspend leaf aging (Begg and Turner, 1976), and thus consumers may encounter physiologically younger tissue for longer periods of time. However, during periods of severe stress, insect herbivores may actually encounter less food because premature senescense and shedding of fully expanded leaves is often induced. This is a mechanism by which herbaceous and woody plants reduce water loss during dry summers (Kozlowski, 1976). Leaf shedding usually begins with the lower, physiologically older leaves (Parsons, 1982). Premature leaf shedding could result in greater mortality for some insect species, especially those with a sessile feeding stage, such as scale insects. However, because many insects are early-season defoliators and typically feed on younger foliage (Kramer and Kozlowski, 1979), premature shedding of older leaves would probably be detrimental to few defoliators.

F. Plant Metabolism

As water stress becomes progressively more severe, a series of physiological and biochemical changes systematically occur within the plant

(Table 2; Hsiao, 1973; Hsiao *et al.*, 1976). As a result, water-stressed plants may at first become more capable, but then less and less capable of defending themselves against herbivory as stress intensifies. We now discuss how water deficits affect some of the major physiological processes in plants.

1. Stomatal Behavior

The principal role of stomata is to regulate water vapor loss and CO_2 uptake and thereby control the balance between transpiration and assimilation (Ludlow, 1980). By means of stomatal closure (Table 2), water-stressed plants reduce transpiration, restore turgor and growth, and protect leaf organelles that are sensitive to desiccation (Mansfield and Davies, 1981). Stomatal opening and closing are governed largely by the uptake or loss of K^+ ions as well as complex metabolic changes involving organic acids and growth regulators such as abscisic acid (Milborrow, 1981; Kramer, 1983; Zeiger, 1983). After rewatering, stomatal opening recovers slowly, usually over a period of days (Kozlowski, 1982; Mansfield and Davies, 1985).

2. Photosynthesis

The photosynthetic ability of a plant is determined largely by its total leaf area and the level of photosynthetic activity in its leaves (Parsons, 1982). Not only does stomatal closure bring about reduced transpiration and increased leaf temperature; it also reduces photosynthesis by inhibiting CO_2 conductance (Hsiao, 1973). Besides stomatal effects, CO_2 assimilation (Table 2) is also reduced in water-stressed plants by nonstomatal factors such as increased mesophyll resistance to CO_2 diffusion and depression of chloroplast and enzyme activity (Begg and Turner, 1976; Kriedemann and Downton, 1981; Hanson and Hitz, 1982; Kozlowski, 1982). The rate at which photosynthesis recovers after rewatering is influenced by the severity and duration of the water deficits as well as the extent to which leaf shedding and injury to stomata, chloroplasts, and roots occur (Kramer and Kozlowski, 1979).

3. Respiration

Respiration is generally depressed whenever water stress is sufficiently severe to cause stomatal closure and inhibit photosynthesis (Table 2; Begg and Turner, 1976). Reduced energy demand as a result of inhibited cell division and elongation is probably the primary cause of the initial reduction in respiration (Bradford and Hsiao, 1982). Under water stress, the carbon budget of a plant is generally altered so that proportionately more photosynthate is used for maintenance than for continued growth

(Hanson and Hitz, 1982). Because water stress depresses photosynthesis more than respiration, carbohydrate reserves usually decrease during drought (Kramer, 1983).

4. *Translocation*

Water stress usually reduces the long-distance transport of photosynthate as a result of its effect on the rates of one or more of the following processes: (a) product assimilation, (b) product utilization, (c) phloem loading and unloading, and (d) product movement in the sieve tubes (Wardlaw, 1968; Begg and Turner, 1976). However, the translocation system itself is highly resistant to desiccation and continues to operate at water potentials that severely inhibit photosynthesis (Kramer, 1983).

5. *Osmotic Adjustment*

Osmotic adjustment is a mechanism by which water-stressed plants lower their osmotic potential by accumulating solutes in cells and thereby maintain turgor, cell enlargement and growth, stomatal opening, and photosynthesis to lower significantly water potentials that would otherwise be inhibitory (Kramer, 1983; Morgan, 1984). Osmotic adjustment enables plants not only to reduce water loss to a dry environment, but also to take up water from the drying soil (Turner and Jones, 1980; Tyree and Jarvis, 1982). Some of the solutes responsible for osmotic adjustment are soluble carbohydrates, sugar alcohols, amino acids, organic acids, and inorganic ions (Tables 5–7; Cram, 1976; Morgan, 1984; Wyn Jones, 1984). Salts in general are not ideal osmolytes because their abundance affects catalytic rates and the Michaelis constant K_m of enzymes (Yancey *et al.*, 1982). Hence, their use under water stress may be less than that for other, more biologically safe solutes such as proline, polyols, and sugars.

Osmotic adjustment is apparently not as well developed in woody shrubs and trees as in herbaceous plants (Tyree and Jarvis, 1982). The net metabolic cost of osmotic adjustment is probably not excessive, because the solutes accumulated during water stress generally become available for other processes within days after rewatering (Kramer, 1983). Because many of the solutes that increase in concentration in response to water stress serve as feeding stimulants and primary nutrients of insects, water-stressed plants may be more attractive and nutritious to insects than are nonstressed plants.

6. *Nutrient Uptake and Mineral Content*

The absorption of inorganic ions from soil is limited during drought because ion movement is slow in drying soil, root growth is decreased, and increased root suberization decreases root permeability (Pitman,

TABLE 5

Changes in the Nitrogen Fraction of Plants in Response to Water Deficits

Species	Organ or tissue	Fraction and trend[a]	Reference
Woody plants			
Artemisia tridentata	Leaves	Total N −	Dina and Klikoff (1973)
	Stem	Total N +	
Citrus (2 spp.)	Leaves, stem	Total N +	Chen *et al.* (1964)
	Roots	Total N −	
Pinus taeda	Inner bark	Amino N =	Hodges and Lorio (1969)
Quercus velutina	Roots	Pro +, Asp +, Thr +	Parker (1979)
Lycopersicon esculentum	Leaves, stem	Total N +, Sol. N +	Walgenbach *et al.* (1981)
Herbaceous plants			
Brassica rapa	Leaves	Pro +	Stewart *et al.* (1966)
Cynodon dactylon	Leaves	Pro +, Asp +	Barnett and Naylor (1966)
Festuca arundinacea	Leaves	Sol. N +	Belesky *et al.* (1982)
Glycine max	Leaves	Pro +, Asp +	Fukutoku and Yamada (1984)
Gossypium hirsutum	Leaves	Pro +	McMichael and Elmore (1977)
Helianthus annuus	All parts	Pro +	Wample and Bewley (1975)
Hordeum vulgare	Leaves	Pro +	Hanson *et al.* (1977)
	Leaves, roots	Pro +	Singh *et al.* (1973a)
	Leaves	Betaine + Pro +	Hanson and Nelson (1978)
Lolium perenne	Leaves	Pro +	Kemble and MacPherson (1954)
Lycopersicon esculentum	Leaves	Pro +	Aloni and Rosenshtein (1984)
Medicago sativa	Leaves, stems	Total N +, Sol. N +	Walgenbach *et al.* (1981)
Phaseolus vulgaris	Leaves	Free amino acids +	Jager and Meyer (1977)
Trifolium repens	Leaves	Pro +, Sol. N =	Routley (1966)
Triticum aestivum	Leaves, roots	Pro +	Singh *et al.* (1973b)
	Leaf apex	Pro +, Asp +	Munns *et al.* (1979)
Zea mays	Mesocotyl	Polyribosomes −	Bewley and Larsen (1982)
	Roots	Pro +	Pahlich and Grieb (1983)
Desert annuals	Leaves	Pro +	Treichel *et al.* (1984)
Forage crops (8 spp.)	Leaves	Total N +	Abdel Rahman *et al.* (1971)
Tropical legumes (9 spp.)	Leaves	Pro +	Ford (1984)

[a] Pro, Proline; Asp, asparagine; Thr, threonine, Sol. N, soluble nitrogen; total N, total nitrogen; +, increase; −, decrease; =, no change.

TABLE 6

Changes in the Carbohydrate Fraction of Plants in Response to Water Deficits

Species	Organ or tissue	Fraction and trend[a]	Reference
Woody plants			
Acer saccharum	Root bark	Sucrose −, starch −	Wargo (1984)
	Root sapwood	Glucose +, fructose +	
Arbutus (2 spp.)	Leaves	Sugars +	Diamantoglou and Kull (1984)
	Branch bark	Sugars +, starch −	
Artemisia tridentata	Leaves, root	Sugars +	Dina and Klikoff (1973)
Liriodendron tulipifera	Stem phloem	Glucose +, fructose +, sucrose −	Roberts (1963)
Malus pumila	Leaves	Sugars +, starch −	Magness *et al.* (1932)
	Bark and wood	Sugars +, starch −	
Pinus halepensis	Leaves	Sugars +, starch −	Diamantoglou and Kull (1982)
	Branch bark	Sugars +, starch +	
P. silvestris	Leaves	Glucose +	Otto (1970)
P. taeda	Inner bark	Sugars +, starch −, TC +	Hodges and Lorio (1969)
Pistacia lentiscus	Leaves	Sugars +, starch −	Diamantoglou and Meletiou-Christou (1980)
	Branch bark	Sugars +, starch +	
Quercus coccifera	Leaves, bark	Sugars +, starch −	Diamantoglou and Kull (1982)
Q. rubra	Roots	Starch −, sucrose −	Parker (1979)
		Glucose +, fructose +	
Q. velutina	Root bark	Sucrose −, starch −	Wargo (1984)
	Root sapwood	Glucose +, fructose +	
Schinus molle	Leaves	Sugars +, starch −	Diamantoglou and Kull (1982)
	Branch bark	Sugars +, starch +	
Vitis viniera	Leaves	Starch −, sugars +	Ahrns (1924)
Many temperate trees	Leaves	Soluble sugars +	Schwenke (1968)
Many Rhizophoraceae	Leaves	Sugar alcohols +	Popp (1984)

Herbaceous plants			
Gossypium hirsutum	Leaves	Glucose +, fructose +, sucrose +	Cutler *et al.* (1977)
	Leaves	Glucose +, starch +	Cutler and Rains (1978)
	Stems, roots	Sugars +, starch +	
	Leaves	Starch −, TC −	Eaton and Ergle (1948)
Helianthus annuus	Leaves	Starch −, sugars +	Ahrns (1924)
Phaseolus vulgaris	Leaves	Sucrose +, starch −	Stewart (1971); Ahrns (1924)
	All parts	Sucrose −, starch − (SWS)	Woodhams and Kozlowski (1954)
Pisum sativum	Leaves	Starch −, sugars +	Ahrns (1924)
Plantago maritima	Leaves	Sorbitol +	Ahmad *et al.* (1979)
Saacharum officinarum	Stem	Sucrose +	Hartt (1967)
Sorghum bicolor	Leaves	Sugars +	Acevedo *et al.* (1979)
	Leaves	Glucose +, sucrose +	Jones *et al.* (1980)
Trifolium aestivum	Leaves	Sugars +, sucrose + (MWS)	Vassiliev and Vassiliev (1936)
		Sugars +, sucrose − (SWS)	
	Leaves	Glucose +, sucrose +	Munns and Weir (1981)
Tropaeolum majus	Leaves	Starch −, sugars +	Ahrns (1924)
Zea mays	Leaves	Soluble sugars +	Barlow *et al.* (1976)
Forage crops (8 spp.)	Leaves	TC +	Abdel Rahman *et al.* (1971)
Pasture grasses (8 spp.)	Roots	Sucrose +, sugars =	Julander (1945)
Tropical grasses (4 spp.)	Leaves	Sugars +	Ford and Wilson (1981)
Tropical legumes (14 spp.)	Leaves	Inositol +	Ford (1984)

[a] TC, total carbohydrates, +, increase; −, decrease; =, no change; MWS, mild water stress; SWS, severe water stress.

TABLE 7

Changes in Levels of Organic Acids and Inorganic Ions in Response to Water Stress

Species	Constituent and trend[a]	Reference
Cenchrus ciliaris	Chlorine +, potassium +	Ford and Wilson (1981)
Gossypium hirsutum	Malate +	Cutler and Rains (1978)
	Citrate +, malate +, potassium +	Cutler *et al.* (1977)
Helianthus annuus	Calcium +, potassium +, magnesium +	Jones *et al.* (1980)
Heteropogon contortus	Chlorine +, potassium +	Ford and Wilson (1981)
Macroptilium atropurpureum	Chlorine +	Ford and Wilson (1981)
Panicum maximum	Chlorine +, sodium	Ford and Wilson (1981)
Ricinus communis	Potassium +	Smith and Milburn (1980)
Sorghum bicolor	Chlorine +, potassium +	Jones *et al.* (1980)

[a] Trend: +, increase.

1981; Kramer, 1983). In addition, a slowing of the transpiration stream and inhibition of nutrient loading into the xylem result in reduced nutrient uptake in water-stressed plants (Bradford and Hsiao, 1982). Reduced uptake of nitrogen, phosphorus, and several other elements as a result of water deficits is well documented (Viets, 1972; Begg and Turner, 1976). Nevertheless, the total mineral ash content is generally greater in water-stressed plants, probably because less "dilution" has occurred due to restricted growth.

In several plant species, foliar levels of potassium, sodium, calcium, magnesium, and chlorine increase, whereas phosphorus (Kilmer *et al.*, 1960; Bates, 1971; Nuttall, 1976) and iron levels decrease during water stress (Abdel Rahman *et al.*, 1971). In contrast, Bates (1971), Naylor (1972), and Viets (1972) concluded that potassium concentration usually declined but that the levels of other minerals varied inconsistently. Our own research on balsam fir also indicates that levels of potassium declines, whereas those of phosphorus, iron, and zinc increase.

Viets (1972) emphasized that, depending on plant rooting architecture, increasing water stress can have either a diluting or a concentrating effect on mineral nutrients. If, under water stress, more and more of the water budget is absorbed from deeper and deeper roots, nutrients may become diluted in plant tissue, because although deep roots are taking up water, they are taking up very few nutrients since the latter occur mainly in the rich, upper soil horizons, where soil moisture stress typically becomes highest. If, however, both water and nutrient uptake by roots are limited primarily to upper soil horizons, many nutrients may become concentrated, because water becomes limiting for growth long before nutrients do.

7. Nitrogen Metabolism

In general, water stress disturbs nitrogen metabolism so that protein decreases while amino acids (especially proline) increase in concentration (Table 5; Kramer, 1983). The observed increases in amino acid content result in part from protein hydrolysis, as noted by increased enzymatic activity (Todd, 1972), as well as from reduced plant growth. Proline synthesis appears to be supported ultimately by the oxidation of stored sugars and starch (Stewart *et al.*, 1966).

Nitrogen levels, especially soluble nitrogen levels, increase under water stress; however, the effects on total nitrogen are not as clear-cut (Table 5). For example, old leaves may senesce during drought and their nitrogen transported out to more juvenile leaves and other critical growing points of the plant (Viets, 1972; Naylor, 1972). Hence, total nitrogen may decline in older tissues (except for their phloem, which acts as a pipeline for the transport of hydrolyzed protein, etc.), whereas total nitrogen may actually increase in young tissues because of (1) new import from senescing plant parts and (2) exclusive import from the whole root system, which used to service both older and juvenile tissues. Many authors have concluded that water stress usually results in an increase in aboveground tissue concentrations of nitrogen (Chen *et al.*, 1964; Abdel Rahman *et al.*, 1971; Bates, 1971; Viets, 1972; Naylor, 1972; Nutall, 1976; Walgenbach *et al.*, 1981). However, the increase occurs only until severe tissue wilting, after which a decline usually takes place (Fukutoku and Yamada, 1984).

8. Carbon Metabolism

Overall, water stress disturbs carbon metabolism so that starch levels generally decrease while sugar levels increase (Table 6; Abdel Rahman *et al.*, 1971; Dina and Klikoff, 1973; Kramer, 1933). These trends result in part from starch hydrolysis (Todd, 1972), as well as from reduced plant growth. Our work on water-stressed balsam fir showed a 2.5-fold increase in foliar sugars owing to water stress. However, during very prolonged water stress, both starches and sugars can be depleted (Eaton and Ergle, 1948; Woodhams and Kozlowski, 1954).

G. Allelochemical Content

Although few detailed studies have been conducted, foliar concentrations of secondary compounds such as cyanogenic glycosides, glucosinolates, and other sulfur compounds, alkaloids, and terpenoids tend to increase in water-stressed plants (Table 8). However, no clear relationship between foliar phenolic levels and water stress has been established

TABLE 8

Changes in the Allelochemical Fraction of Plants in Response to Water Deficits

Species	Organ or tissue	Fraction and trend[a]	Reference
Amalanchier ainifolia	Leaves	Cyanogens +	Majak *et al.* (1980b)
Cynodon dactylon	Leaves	Cyanogens +	Blohm (1962)
Manihot esculenta	Leaves	Cyanogens +	Bruijn (1973)
Sorghum bicolor	Leaves	Cyanogens +	Blohm (1962)
	Leaves	Cyanogens +	Nelson (1953)
S. sudanense	Leaves	Cyanogens +	Boyd *et al.* (1938)
Trifolium repens	Leaves	Cyanogens +	Rogers and Frykolm (1937)
Triglochin maritima	Leaves	Cyanogens +	Majak *et al.* (1980a)
Braccica oleracea	Leaves	Glucosinolates +	Bible *et al.* (1980)
Rorippa nasturtium	Leaves	Alkyl sufides +	Freeman and Mossadeghi (1971)
Cinchona ledgeriana	Leaves	Quinine +	Waller and Nowacki (1978)
Conium maculatum	Leaves	Alkaloids +	Blohm (1962)
Datura innoxia	Leaves	Alkaloids +	Sokolov (1959)
Festuca arundinacea	Leaves	Alkaloids +	Kennedy and Bush (1983)
Hyoscyamus muticus	Leaves	Alkaloids +	Ahmed and Fahmy (1949)
Lobelia sessifolia	Leaves	Alkaloids +	Sokolov (1959)
Lupinus spp.	Leaves	Alkaloids +	Waller and Nowacki (1978)
Nicotiana tabacum	Leaves	Nicotine +	van Bavel (1953)
	Leaves	Nicotine +	Waller and Nowacki (1978)
Papaver somniferum	Fruit capsule	Alkaloids +	Bunting (1963)
Phalaris aquatica	Leaves	Alkaloids +	Ball and Hoveland (1978)
	Leaves	Alkaloids +	Majak *et al.* (1979)
Phalaris tuberosa	Leaves	Alkaloids +	Williams (1972)
Quercus velutina	Seedling	Alkaloids +	Parker and Patton (1975)
Senecio longilobus	All parts	Alkaloids +	Briske and Camp (1982)
Marjorana hortensis	Leaves	Essential oils +	Fluck (1963)
Satureja douglasii	Leaves	Monoterpenes +	Gershenzon *et al.* (1978)

[a] Trend: +, increase.

(Gershenzon, 1984). Secondary metabolites may accumulate in foliage during periods of water stress because (1) plant growth is reduced more than is synthesis of allelochemicals and thus there is less plant tissue to "dilute" these compounds, and/or (2) synthesis of allelochemicals is enhanced because greater quantities of carbon and nitrogen become available as a result of both slower growth and hydrolyzing of starch and protein. In contrast, production of the terpenoid-rich oleoresin of conifers (see Section II,H) and latex of *Hevea* trees is generally reduced (Buttery and Boatman, 1976) during severe drought.

H. Oleoresin Exudation Pressure and Composition

The initial pressure at which xylem oleoresin is forced from a wound is termed the oleoresin exudation pressure (OEP). Relatively large values of OEP are detrimental to many bark beetles (Vite, 1961; Vite and Wood, 1961; Wood, 1962) and needle-feeding Lepidoptera (Grimalsky, 1961; 1966; Otto and Geyer, 1970) that attack living conifers. Measured along the trunk, OEP is highly correlated with the relative water content of needles and the xylem water potential of twigs (Hodges and Lorio, 1971). Water stress lowers the exudation pressure, rate of flow, and total flow of oleoresin in conifers (Vite, 1961; Lorio and Hodges, 1968, 1974; Mason, 1971; Blanche *et al.,* 1985).

Changes in the composition of xylem oleoresin occur during water stress. Overall, the resin acid/monoterpene ratio decreases, and this probably causes reductions in oleoresin viscosity and rate of crystallization (Runckel and Knapp, 1946; Hodges and Lorio, 1975). Considering only the monoterpene fraction, water stress usually brings about an increase in α-pinene and a reduction in myrcene and limonene content (Hodges and Lorio, 1975; Gilmore, 1977; Blanche *et al.,* 1985). Considering that α-pinene is a common bark beetle attractant (Rudinsky, 1966) and that limonene and myrcene are deterrents and toxicants (Smith, 1963, 1975; Coyne and Lott, 1976; Gollob, 1980), such changes in the relative amounts of these monoterpenes could influence host finding, host acceptance, and colonization success by bark beetles.

I. Wound Healing

Water-stressed plants may be more susceptible to insects because the process of wound healing is slowed. In *Abies* species, water stress reduces the rate of wound healing in stem bark and is correlated with susceptibility to the balsam woolly adelgid, *Adelges piceae* (Puritch and Mullick, 1975). Similarly, first periderm formation in trees is delayed by water stress (Borger and Kozlowski, 1972). To our knowledge, no studies have been conducted on the influence of water stress on the hypersensitive (induced) response of conifers.

III. IMPLICATIONS FOR INSECT EPIDEMIOLOGY

How are the effects of moisture stress on plants translated into more insect biomass? We recognize that insects respond on both a behavioral and physiological level to certain classes of plant traits (Mattson *et al.,* 1982; Ahmad, 1983; Visser, 1983; Haack and Slansky, 1987; Mattson *et al.*, 1987). This then causes one to consider how moisture stress might change these critical traits to enhance or inhibit either the behavioral or physiological processes of the insects.

Although we are focusing on morphological and biochemical changes in plants, we fully recognize that usually plant and ambient air temperatures are also significantly elevated during moisture deficits. Because insects are ectotherms, their basic behavior and physiology will therefore be directly affected. If the concomitant rise in temperatures and solar insolation permit the insects to grow in a more nearly optimal thermal regime, theoretically they could without any change in food quality grow faster and larger and with lower mortality (Morris and Fulton, 1970; Begon, 1983; Scriber and Lederhouse, 1983; Haack *et al.*, 1984, 1985; Reichenbach and Stairs, 1984). We believe that changes in the insect's thermal environment and changes in host plant quality interact synergistically to allow the development of insect outbreaks during periods of droughty weather. Mattson and Scriber (1987) concluded that grass and arboreal folivores, in particular, may be critically dependent on a proper thermal regime for maximal performance owing to the unique enzyme and protein systems of folivores adapted to such nutrient-impoverished foods. We will not address this aspect of problem any further because our main focus is on purely plant mediated effects of moisture deficits on insects.

A. Changes Affecting Insect Behavior

1. Finding and Accepting Hosts

A thorough knowledge of all the plant traits that have an important bearing on the finding and accepting of hosts by insects is still lacking (Miller and Strickler, 1984). To this point most research has focused on plant chemical properties that may affect this behavior, but there is a growing body of evidence about the significance of visual and textural plant properties (Ahmad, 1983; Prokopy and Owens, 1983; Bell and Carde, 1984).

a. Chemical Cues In the case of chemical properties, the general consensus is that insects respond not so much to single compounds as to blends of compounds in the process of host finding and accepting (Miller and Strickler, 1984). These blends contain both positive and negative molecular species whose critical balance determines the exact course of insect behavior—assuming, of course, that one holds other significant environmental factors (visual and textural plant cues, temperature, humidity, etc.) constant. We propose that water deficits could affect the qualitative as well as the quantitative compositions of these blends confronting an insect. In fact, we hypothesize that the insect's sensory systems, particularly the chemoreceptors, are especially appropriate for detecting the changes in plant chemical composition that typically result from water deficits. For example, phytophagous insects commonly have special contact chemoreceptors, which are sensitive to (1) sugars, (2) salt/water, (3) amino acids, and (4) secondary plant substances (Stadler, 1984; Visser, 1983). It is peculiar that amino acid receptors have been found only in larvae of phytophagous insects, not in other trophic categories of insects (Stadler, 1984). In contrast, salt receptors seem to be ubiquitous among insects and other classes of invertebrates, even mammals, birds, and fishes (Dethier, 1977). The significance of these two receptors for phytophages seems obvious because their diet predisposes them to suffer possible shortages of these nutrients (Mattson, 1980; Mattson and Scriber, 1987). Stadler (1984), however, reported that there also appears to be an inositol receptor in many lepidopteran larvae even though most insects have no dietary need for this compound. Even more puzzling is that fact that for some it also functions as a phagostimulant. One can reconcile this apparent anomaly when one realizes that inositol is one of many common plant osmolytes [solutes that contribute to cell osmoregulation: sugars, polyols (e.g., inositol), amino acids (especially proline), organic acids, and salts (see Section II,F,5)].

Hence, phytophagous insects have apparently evolved the unique sensory apparatus that enables them to detect the presence and abundance of these well-known osmolytes. More than that, most of these compounds are also feeding stimulants or feeding cofactors (Table 9; Stadler, 1984). Therefore, when these osmolytes increase as water deficits intensify, they may simultaneously stimulate insect feeding behavior. This assumes, of course, that under normal water conditions the concentrations of these stimulatory compounds in the plant are below the level at which the insect's optimal responses occur. This is probably true for late-instar spruce budworm (*Choristoneura fumiferana*) larvae, which are very sensitive to sucrose levels in their food. In foliage of unstressed balsam fir, *Abies balsamea,* sucrose levels are usually close to 0.004 *M,* whereas in moisture-stressed fir sucrose concentrations are nearly threefold higher (0.011 *M*) (B. M. McLaughlin, unpublished). According to Albert *et al.* (1982), the peak feeding response to sucrose by sixth-instar budworm larvae occurs between 0.01 and 0.05 *M*. Hence, in the case of this single stimulant, plant water stress would increase sucrose levels and that should increase budworm feeding responses (Albert and Jerrett, 1981). Apparently, the same would be true for *Locusta migratoria,* the geometrid *Lambdina fiscellaria,* and *Pieris brassicae,* because their optimal

TABLE 9

Selected Insect Feeding Stimulants That May Be Altered by Plant Moisture Stress

Species	Family	Principal stimulants	Reference
Nitrogen-containing compounds			
Coleoptera			
Leptinotarsa decemlineata	Chrysomelidae	Three amino acids	Hsiao and Fraenkel (1968)
Sericesthis geminata	Scarabaeidae	Six amino acids	Wensler and Dudzinski (1972)
Lepidoptera			
Choristoneura fumiferana	Tortricidae	Proline	Heron (1965)
Orthoptera			
Locusta migratoria	Acrididae	Proline, serine	Cook (1977)
Carbohydrates			
Hypera postica	Curculionidae	Sucrose, fructose	Hsiao (1969)
Leptinotarsa decemlineata	Chrysomelidae	Sucrose	Hsiao and Fraenkel (1968)
Sericesthis geminata	Scarabaeidae	Sucrose	Wensler and Dudzinski (1972)
Sitona cylindricollis	Curculionidae	Sucrose	Åkeson *et al.* (1970)
Homoptera			
Dysdercus koenigii	Pyrrhocoridae	Sucrose, fructose	Saxena (1965)
Myzus persicae	Aphididae	Glucose, sucrose	Mittler and Dadd (1964)
Lepidoptera			
Choristoneura fumiferana	Tortricidae	Sucrose, fructose	Heron (1965)
Lambdina fiscellaria	Geometridae	Glucose, sucrose	Ouellet *et al.* (1983)
Sphinx pinastris	Sphingidae	Glucose	Otto (1970)
Spodoptera exempta	Noctuidae	Sucrose	Ma (1977)
Uraba lugens	Nolidae	Sucrose, fructose	Cobbinah *et al.* (1982)

Orthoptera			
Locusta migratoria	Acrididae	Eight sugars	Cook (1977)
Schistocerca gregaria	Acrididae	Sucrose, glucose	Dadd (1960)
Allelochemicals			
Coleoptera			
Dendroctonus pseudotsugae	Scolytidae	α-Pinene, camphor	Rudinsky (1966)
Diabrotica undecimpuctata howardi	Chrysomelidae	Cucurbitacins	Chambliss and Jones (1966)
Gnathotrichus sulcatus	Scolytidae	α-Pinene	Rudinsky (1966)
Hylastes ater	Scolytidae	α-Pinene	Perttunen (1957)
Phyllotreta spp.	Chrysomelidae	Six glucosinolates	Nielsen (1978)
Phyllotreta cruciferae	Chrysomelidae	Five glucosides	Hicks (1974)
Pissodes strobi	Curculionidae	Limonene, camphor	Alfaro *et al.* (1980)
Homoptera			
Brevicoryne brassicae	Aphididae	Glucoside	Wensler (1962)
Lepidoptera			
Choristoneura fumiferana	Tortricidae	Glucose	Heron (1965)
		Sucrose	Albert *et al.* (1982)
Papillio spp.	Papilionidae	Essential oils	Dethier (1941)
Plutella maculipennis	Plutelidae	Nine glycosides	Nayar and Thorsteinson (1963)
Spodoptera eridania	Noctuidae	Cyanide	Brattsten *et al.* (1983)
Orthoptera			
Manduca sexta	Sphingidae	Glycoside	Yamamoto and Fraenkel (1960)

responses to sucrose are so high (>0.10 *M*) that any increase of it in a host plant would probably always be below that optimal level and would therefore always elicit positive feeding responses (Cook, 1977; Ouellet *et al.*, 1983; Visser, 1983).

Undoubtedly, the various concentrations of osmolytes interact with one another to promote further host acceptance and feeding. For example, Ma (1972) concluded that L-proline and sucrose interacted synergistically to stimulate feeding by *P. brassicae,* as did Bentley *et al.* (1982) for *C. fumiferana*. The fact that proline is a feeding stimulant for many phytophages and usually only at high levels (Cook, 1977) is interesting because proline is not an essential amino acid and it is usually not abundant except under conditions of moisture stress (Table 5). Hence, in this respect it is like inositol, a compound that "signals" plant stress.

Not only primary nutrients and osmolytes change under water stress; so do the levels and perhaps kinds of secondary metabolites. In general, there is a tendency for many allelochemical concentrations to increase during moisture stress (Table 8). In fact, if water-stress-tolerant plants (xerophytes) represent the ultimate morphological and physiological response to water stress, it is clear that such chronic stress results in the allocation of greater than normal amounts of energy and carbon to the production of secondary metabolites (Rodriquez, 1983; Hoffman *et al.*, 1984a,b; DiCosmo and Towers, 1984). From the consumer's point of view, the plant's increase in secondary metabolites under stress would seem to be inhibiting, and for marginally adapted insect species it probably is. However, for the well-adapted species the increases may not be substantial enough to exceed their tolerance; in the case of some compounds, the increase may actually be behaviorally enhancing. For example, the glucosinolate sinalbin interacts synergistically with sucrose to promote feeding by *Pieris brassicae* (Visser, 1983). Lechowicz (1983) hypothesized that the magnitude of the sugar/tannin (S/T) ratio in host foliage is positively linked to host plant selection or preference by the very polyphagous gypsy moth, *Lymantria dispar*. Our research on balsam fir shows that water stress increases the size of the S/T ratio by causing sugar concentrations to increase more than those of tannin (3 : 1.3). We found the same to be true for water-stressed red pine, *Pinus resinosa*. If most phytophagous insects employ across-fiber patterns in sensory coding of information from receptors, as do Lepidoptera larvae, then the absolute concentrations of the critical compounds in the plant tissue may be less important than their relative concentrations (to one another) because identical ratios (e.g., S/T) at both high and low concentrations evoke identical across-fiber patterns (Stadler, 1984). Mustaparta (1984), however, also argued that the sensitivity of insects to precise ratios of compounds in pheromone plumes is linked to their specialist-type receptors for these compounds and their labeled lines to the central nervous system (CNS). Hence, it appears that the precise ratios of plant and pheromone compounds are an important piece of information to insects regardless of the kind of neurological network connecting the receptor cells to the CNS.

There are, of course, many adapted species of insects that apparently employ one or more of the host plant allelochemicals as part of their host-finding and/or host-accepting cues (Renwick, 1983; Visser, 1983). However, most attempts to single out the "few" key guiding compounds have been unsuccessful. In fact, the consensus emerging is that full behavioral responses probably require a very complicated profile of components along with the appropriate visual and tactile host properties (Miller and Strickler, 1984; Feeny *et al.*, 1983). Therefore, the entire question about the real nature of plant primary attractants is largely unresolved, with only weak evidence to support the general hypothesis of key compounds (Miller and Strickler, 1984). It is obviously difficult, then, to address the effects of moisture stress on allelochemicals and the resulting consequences for insect host finding or accepting. In the case of conifers (*Pinus, Picea, Abies*), water stress may cause a reduction in the emission of most terpene molecules into the atmosphere owing to lower actual production of foliar terpenes and perhaps less gaseous exchange due to closed stomata (Chararas, 1979; Cates and Alexander, 1982). Furthermore, the levels of all compounds apparently are not equally reduced and some, such as α-pinene, actually increase (Cates and Alexander, 1982; Section II,H), resulting in increased dominance of such compounds in the stressed plant's gaseous atmosphere. It may be more than coincidence that α and β-pinenes are common attractants for scolytids attacking living conifers and at the same time are among the least toxic of all the terpenoids (Chararas, 1979; Cates and Alexander, 1982; Brattsten, 1983). According to Chararas (1979), the reduction in terpene emissions by water-stressed conifers moves the concentrations of most of these compounds from a repellent to an attractant level for many bark beetles.

When more is understood about the exact role of allelochemicals in insect host plant finding and accepting, it will be possible to decide whether stress changes those important plant properties in such a way that it either usually enhances or discourages the insect.

b. Electromagnetic Cues Electromagnetic cues include those that are normally perceived with vision and those that are perceived with special infrared (IR) receptors (Callahan, 1975; Evans and Kuster, 1980). As a result of water stress, vegetative growth slows or ceases, as do transpiration and protein and leaf chlorophyll synthesis (Table 2). Leaves may begin to curl or cup and may form thicker layers of waxes, pubescence, and so on (Table 3), causing changes in leaf spectral properties in both the visual part of the spectrum (including ultraviolet in the case of insects) and the IR wavelengths. As a result of these changes plants may be more "attractive" to insects. Prokopy and Owens (1983) suggest that particularly the hue and intensity of spectral quality may be the principal inducements to insects to alight on plants. Apparently intensity changes more than hue with plant stress. Nevertheless, many herbivorous insects are attracted to yellow hues. In particular, many aphids are attracted to unsaturated yellow hues as are common in newly developing leaves (Prokopy and Owens, 1983) or in stressed plants where chlorophyll synthesis has stopped or been reduced.

In view of the fact that leaf temperature changes under water stress (Table 4) one might expect some insects to have "heat" receptors sensitive to the changes in IR radiation that results (Altner and Loftus, 1985). The buprestid beetle *Melanophila acuminata* apparently has such receptors adjacent to its mesocoxal cavities, which assist it in finding fire-scorched, weakened conifers for breeding and reproduction (Evans and Kuster, 1980). Evidence for IR perception by other insects is weak, but the hypothesis is nevertheless intriguing and should be addressed (Callahan, 1975).

B. Changes Affecting Insect Physiological Processes

The changes in plant chemistry and temperature that result from water stress may have a substantial impact on the physiological processes of insects. These processes can be placed in two broad categories: (1) primary metabolism (basic growth and maintenance) and (2) secondary metabolism (the insect's detoxification system and its immune system).

1. Enhanced Growth, Developmental Rates, and Fecundity

We hypothesize that, as a result of water stress, many plant phagostimulants and nutrients (amino acids, sugars, minerals, etc.) are either better balanced relative to one another or more concentrated in the plant, thereby stimulating consumption rates and also allowing improved growth and development of herbivores. It is widely suspected, for example, that many herbivores—particularly those of woody plants—are nutrient-limited. Nitrogen is probably the key nutrient (White, 1984; Scriber, 1984; Mattson, 1980), but there may also be shortages of available carbohydrates and some mineral elements (Mattson and Scriber, 1987). There can be little doubt that an increase in nitrogen, sugar, and perhaps minerals generally elicits better growth and development of most folivorous insects. The ratios of these nutrients may be just as important as their absolute concentration (House, 1966; Tsiropoulos, 1981; Waldbauer *et al.*, 1984). Cates *et al.* (1983) and Mattson *et al.* (1983) reported that low to moderate moisture stress increased the growth of western and eastern spruce budworm, respectively. However, neither group had a mechanistic explanation for such an effect. It is probably due to higher sugar levels, because Harvey (1974) demonstrated that the body size of female spruce budworms increased until total dietary sugar levels attained at least 4% wet weight, a level seldom attained even when their normal host plants undergo water stress. There is little other direct experimental evidence linking water-stressed plants to better insect performance. Most is circumstantial and has been thoroughly addressed by White (1974, 1976, 1984) and Rhoades (1983, 1985).

2. *Detoxification Ability*

The frequent correlation of insect outbreaks with hot, dry weather has caused many to conclude that moisture stress must somehow lower plant defensive properties (Mattson and Addy, 1975; Rhoades, 1983, 1985). Indeed, in the case of some, this is true. For example, oleoresin exudation pressure clearly declines with moisture stress (Berryman, 1972; Cates and Alexander, 1982; see Section II,H).

However, the evidence in Table 8 suggests the contrary. That is, many plant allelochemicals increase under moisture stress. How can one reconcile this apparent contradiction? One possibility is that these traits are not a part of the plant's system of defense against adapted herbivores. Another possibility is that the adapted herbivores, for some peculiar reason, are able to overcome these normally deterring, debilitating levels of allelochemicals. In other words, it is not that the plant's defense system fails but that the herbivore's detoxification system is somehow so enhanced that these normally effective defenses are rendered ineffective. This possibility has not been considered in full, but it has been addressed primarily from the point of view of the proliferation of a more "virulent" herbivore genotype, which may have a more effective detoxification system (Haukioja and Hakala, 1975; Georghiou and Saito, 1983; Rhoades, 1985). Still another possibility is the plant's effect on the herbivore's detoxification system. Is it possible, for example, that some changes inherent in the plant's stress response may enhance, in some manner, the functioning of the herbivore's detoxification systems? We hypothesize that it is. Stress brings about changes that enhance the herbivore's detoxification processes more than they enhance the plants defensive properties per se, especially levels of noxious allelochemicals. Is there evidence to support this hypothesis?

There are abundant examples of the fact that the efficacy of pesticides on phytophagous insects varies with the plant cultivar or species consumed by the insect before exposure to pesticides (Potter and Gillham, 1957; Bass and Rawson, 1960; Wood *et al.*, 1981; Yu, 1982; Heinrichs, *et al.*, 1984). Decreased sensitivity to pesticides as a result of dietary experience has been explained in two basic ways. The first is that the diet contains some ingredient(s), an inducer, that increases the amounts and activities of those enzymes in the insects that normally metabolize the pesticides (Terriere, 1984). For example, Brattsten (1979a) reported that armyworms that ingested a diet containing α-pinene (an inducer) were twice as tolerant to nicotine as armyworms on a normal diet. Plant allelochemicals are believed to be a general class of such inducers for many organisms (Brattsten, 1979b). The second explanation is that plant nutrients are more nearly optimal in those plants where insects show reduced pesticide sensitivity. Nutrient levels affect the production and sensitivity of the detoxification enzymes and the conjugate molecular species that are important in the detoxification process (Nutritional Reviews, 1985). In particular, nutritional imbalances have been demonstrated to suppress the actions of the mixed-function oxidases (MFO). For example, Campbell and Hayes (1974) reviewed numerous cases of the depressing effects of deficiencies or imbalances of macronutrients (sugars, lipids, proteins) and micronutrients (vitamins, minerals) on the operation of the MFO systems in vertebrates. Wahl and Ulm (1983) similarly reviewed evidence of inadequate pollen intake (protein and vitamin deficiency) on the increased sensitivity of adult bees to various pesticides. Only sensitivity to pesticides containing manganese was not affected by inadequate pollen intake. In addition, a high level of sucrose intake decreased the sensitivity of bees to $MnSO_4$ and $ZnSO_4$. In general a nutritionally poor or imbalanced diet reduces the actions of detoxifying enzymes. Organisms thus become more sensitive to those pesticides that are toxic in their initial states and less sensitive to pesticides that are rendered more toxic after enzymatic action by the detoxification system. The effects of nutrients on pesticide penetration and target site sensitivity are poorly understood. However, there is reason to suspect that nutrition may also affect these components of pesticide efficacy (Campbell and Hayes, 1974).

We hypothesize that moisture stress may enhance the nutritional properties of plants and in so doing also enhance the effectiveness of the herbivore's detoxification system, particularly the detoxification enzymes. This is in fact very similar to the mechanisms that permit the fungus *Armillaria mellea* to colonize the roots of stressed, deciduous trees. According to Wargo (1984), *A. mellea* succeeds in the roots of such trees because the increased levels of both glucose and amino acids enhance its ability to oxidize the normally inhibitory phenols and even to utilize the products as a source of carbon. Therefore, the concomitant rise in the concentration of many plant allelochemicals (which are potential toxins) following moisture stress is not debilitating because they are more readily detoxified than they were before moisture stress. Moreover, if the insects need to eat less and yet grow larger after water stress because of enhanced plant nutritional value, perhaps the total dosages of allelochemicals they receive is actually less per unit weight of insect tissue than before.

3. *Immunocompetence*

Just as we argued in the preceding section that diet influences the detoxification systems of herbivores, we argue here that it influences the efficacy of pathogenic organisms through direct effects on the pathogen (Merdan *et al.,* 1975) and on the herbivore's immune system. The immune system is critically important in the herbivore's response to infections by microorganisms (protozoans, fungi, bacteria, viruses, etc.) as well as to the eggs and larvae of such macroorganisms as insect and nematode parasitoids. Moreover, there may be critical interactions between the immune and detoxification systems of herbivores. For example, Wahl and Ulm (1983) reported that *Nosema* infections in the honeybee increase its sensitivity to pesticide.

The immune system of insects (or invertebrates in general) is believed to be less complex than that of vertebrates (Maramorosch and Shope, 1975), but because it is not nearly as well studied as the latter, we will use the vertebrate system as the general model for our discussion. The vertebrate immune system is believed to consist of two major parts: nonspecific and specific immunity (Pestka and Witt, 1985). The first consists of general responses: (1) physical barriers, (2) secretions, (3) phagocytosis, and (4) special blood proteins such as interferon. The second consists of highly specific host responses broadly covered as humoral-mediated immunity and cellular-mediated immunity. Humoral immunity is in large part the production of specific antibodies by the host in response to antigens. Cellular immunity largely involves special cells (macrophages, killer cells) that defend against living antigens such as intracellular pathogens.

The link between immunocompetence and nutrition is certain but not yet very well understood. Murray and Murray (1981) argued that undernourishment may actually enhance specific parts of the human immune system and thereby lead to an arrestment of intracellular disease, both malignant and infective. The mechanisms of this arrestment have been speculated to be the promotion of certain growth-inhibiting factors such as nonspecific immune proteins like chalones and interferons. However, nutritional deficiencies and nutritional imbalances are also well known to reduce general immunocompetence. For example, Porter *et al.* (1984) demonstrated that undernourished mice were more sensitive than normal mice to virus, as is generally true with insects. In addition, an interaction between dietary and environmental chemicals reduced immunity. Chandra (1985) reviewed all components of the immune system known to be affected by diet. Protein-energy malnutrition commonly impairs cellular-mediated immunity, humoral-mediated immunity, and nonspecific components. However, according to Chandra (1985) the main suppressing effects of malnutrition are on cellular immunity, phagocyte activity, secretory immunoglobin class IgA antibody responses, and the complementary system of nonspecific blood proteins. In other words, the effects are mainly on the nonspecific immune system, which is believed to predominate in invertebrates (Maramorosch and Shope, 1975). Furthermore, the competence of the immune system also depends on its anatomical and biochemical links to the nervous system (Marx, 1985; Siegel, 1985). Perhaps there are some plant traits that may enhance or inhibit the immunocompetence of insects via the CNS connections.

Our point here is that, early in the development of plant water stress, an insect's immunocompetence against microorganisms and other parasites may be subtly if not markedly enhanced, thereby allowing more individuals to survive and/or produce more gametes. We admit that this is mere speculation, but nevertheless it could be a higly significant factor in outbreak development. We do know that once phytophage outbreaks have occurred there is usually a rapid explosion of disease organisms. Whether this is simply a result of more rapid transmission of disease between healthy and infected animals or also to a concomitant decline in the general immunocompetence of the population due to host plant quality changes or other factors remains to be established.

C. Does Water Stress Affect Regulation Capacity?

To this point we have itemized many of the significant changes that occur in plants as a result of water stress, as well as some of the probable responses of phytophagous insects to many of these changes. We have implied that many of the changes favor consumers. However, we have not addressed whether the plant's ability to regulate consumers is substantially changed owing to water stress.

Probably two basic strategies have evolved in the plant kingdom for the regulation of plant consumers. At one extreme is the near-immune strategy, and at the other is the highly tolerant strategy (Mattson *et al.*, 1987). The first is employed against those insects whose normal damage to the plant is so debilitating that recovery is almost impossible and death usually results quickly. Examples might be the "primary" bark beetles or sucking or other insects that transmit deadly plant pathogens. We hypothesize that, after water stress, plants may become highly susceptible to this class of insects and have no capacity for regulating them because they essentially had none in the first place—they had immunity. Someone might argue that this is actually a perfect form of regulation, to a pinpoint, zero density. Whatever the point of view, we still argue that these plants become highly susceptible after water stress and are left with little or no regulation capacity because the original "defenses" were compromised by the stress. Hence, they are totally defenseless and usually die unless the stressing conditions are reversed. Even then, insects may be able to overwhelm some plants because their density became so high during the stress period that they were able to overcome the plants' restored and normally effective defenses (see Raffa and Berryman, 1983).

At the other extreme in the plant regulation continuum are the defensive strategies of those species of consumers whose damage can be tolerated—that is, the highly tolerant strategy. Because plants can recover to some extent from the attacks of these insects, they can employ a less precise system for regulating them. As a result there are no immune plants; all are susceptible, but not necessarily to the same degree. The mechanisms that evolved for the regulation of this class of consumers include rapidly changing and suboptimal levels of nutritional, physical, and allelochemical properties and facilitation of natural enemies. Inducible defenses such as the *de novo* synthesis of allelochemicals are probably an important subset of such properties. However, other plant properties that change significantly in response to herbivory may be equally or even more important in the regulation process. We hypothesize that water stress changes many important plant properties in a direction that favors insects. Hence, it increases their mean density. At the same time it reduces the plant's regulation capacity because the improved nutritional conditions for the insect favor its faster growth, more effective detoxification capacity, and higher levels of immunocompetence. The higher temperatures of water-stressed plants may also promote the same capacities of phytophagous insects. All of these changes favor "escape" from the additional regulatory power of parasites and/or predators. In the case of plants that normally regulate insects to very low levels, there will also be only a very small number of natural enemies. However, once the host allows insect population expansion, the normal guild of natural enemies may be ill equipped to respond to it functionally and numerically and thereby will have no ability to help regulate the incipient outbreak.

We are proposing that, in the case of plants using the highly tolerant strategy against insects, stress-precipitated outbreaks occasionally occur not so much because actual defenses decline as because plant nutritional quality increases so substantially that it allows insects to overcome some defenses like toxicants and parasites. We are also proposing that elevated microenvironmental temperatures, coupled with stress, may facilitate the insect's abilities. We also recognize that the changed plants or the elevated temperatures may actually contribute to the development of a more "virulent" insect phenotype either through classical natural selection or through shock-induced expression of alternative isozymes that may be more concentrated and have higher catalytic efficiency than standard isozyme (enzyme) variants. Another possibility is the abrupt emergence of a previously uncommon allozyme variant that is better adapted to the higher-temperature regimes and more capable of detoxifying the higher levels of secondary compounds in stressed plants than the previously common allozyme variants.

After a period of sustained population expansion and mounting damage to plants, a rapid decline in insect number eventually sets in owing to several deteriorating conditions. Foremost of these are long-term, high levels of herbivory, which cause severe reductions in plant tissue quantity and quality (see Baltensweiler, 1984; Tuomi *et al.*, 1984; Rhoades, 1985). This reduced quality, growing populations of pathogens and natural enemies, restored moisture contents, and reduced temperatures all interact to bring levels of phytophagous insects back to a more innocuous density where they can be held in "check" by the plant and natural enemies until the next episode of sustained moisture and/or temperature stress.

REFERENCES

Abdel Rahman, A. A., Shalably, A. F., and El Monayeri, M. O. (1971). Effect of moisture stress on metabolic products and ion accumulation. *Plant Soil* **34,** 65–90.

Acevedo, E., Fereres, E., Hsiao, T. C., and Henderson, D. W. (1979). Diurnal growth trends, water potential, and osmotic adjustment of maize and sorghum leaves in the field. *Plant Physiol.* **64,** 476–480.

Ahmad, I., Larher, F., and Stewart, G. R. (1979). Sorbitol, a compatible osmotic solute in *Plantago maritima*. *New Phytol.* **82,** 671–678.

Ahmad, S., ed. (1983). "Herbivorous Insects: Host-Seeking Behavior and Mechanisms." Academic Press, New York.

Ahmad, Z. F., and Fahmy, I. R. (1949). The effect of environment on the growth and alkaloidal content of *Hyoscyamus muticus* L. *J. Am. Pharm. Assoc., Sci. Ed.* **38,** 484–487.

Ahrns, W. (1924). Weitere Untersuchungen über die Abhangigkeit des gegen seitigen Mengenverhaltnisses der Kohlenhydrate in Laubblatt vom Wassergehalt. *Bot. Arch.* **5,** 234–259.

Åkeson, W. R., Gorz, H. J., and Haskins, F. A. (1970). Sweetclover weevil feeding stimulants: Variations in levels of glucose, fructose, and sucrose in *Melilotus* leaves. *Crop Sci.* **10,** 477–479.

Albert, P. J., and Jerrett, P. A. (1981). Feeding preferences of spruce budworm (*Choristoneura fumiferana* Clem.) larvae to some host-plant chemicals. *J. Chem. Ecol.* **7,** 391–401.

Albert, P. J., Cearley, C., Hanson, F., and Parisella, S. (1982). Feeding responses of eastern spruce budworm larvae to sucrose and other carbohydrates. *J. Chem. Ecol.* **8,** 233–239.

Alfaro, R. I., Pierce, H. D., Jr., Borden, J. H., and Oehlschlager, A. E. (1980). Role of volatile and nonvolatile components of Sitka spruce bark as feeding stimulants for *Pissodes strobi* Peck (Coleoptera: Curculionidae). *Can. J. Zool.* **58,** 626–632.

Aloni, B., and Rosenshtein, G. (1984). Proline accumulation: A parameter for evolution of sensitivity of tomato variation to drought stress? *Physiol. Plant.* **61,** 231–235.

Altner, H., and Loftus, R. (1985). Ultrastructure and function of insect thermo- and hygroreceptors. *Annu. Rev. Entomol.* **30,** 273–296.

Amer, F. A., and Williams, W. T. (1958). Drought-resistance in *Pelargonium zonale*. *Ann. Bot.* (*London*) [N.S.] **22,** 369–379.

Baker, E. A. (1974). The influence of environment on leaf wax development in *Brassica oleracea* var. *gemmifera*. *New Phytol.* **73,** 955–966.

Ball, D. M., and Hoveland, C. S. (1978). Alkaloid levels in *Phalaris aquatica* L. as affected by environment. *Agron. J.* **70,** 977–981.

Baltensweiler, W. (1984). The role of environment and reproduction in the population dynamics of the larch bud moth, *Zeiraphera diniana* Gn. (Lepidoptera: Tortricidae). *In* "Advances in Invertebrate Reproduction" (W. Engels *et al.*, eds.), pp. 291–301. Elsevier, Amsterdam.

Barlow, E. W. R., Boersma, L., and Young, J. L. (1976). Root temperature and soil water potential effects on growth and soluble carbohydrate concentration of corn seedlings. *Crop Sci.* **16,** 59–62.

Barlow, E. W. R., Munns, R. E., and Brady, C. J. (1980). Drought responses of apical meristems. *In* "Adaptation of Plants to Water and High Temperature Stress" (N. C. Tuner and P. J. Kramer, eds.), pp. 191–205. Wiley, New York.

Barnett, N. M., and Naylor, A. W. (1966). Amino acid and protein metabolism in Bermuda grass during water stress. *Plant Physiol.* **41,** 1222–1230.

Bartholic, J. F., Namker, L. N., and Wiegand, C. L. (1972). Aerial thermal scanner to determine temperature of soils and of crop canopies differing in water stress. *Agron. J.* **64,** 603–608.

Bass, M. H., and Rawson, T. W. (1960). Some effects of age, premarginal habitat and adult food on the susceptibility of boll weevil to certain insecticides. *J. Econ. Entomol.* **53,** 334–336.

Bates, T. E. (1971). Factors affecting critical nutrient concentrations in plants and their evaluation: A review. *Soil Sci.* **112,** 116–130.

Begg, J. E. (1980). Morphological adaptations of leaves to water stress. *In* "Adaptation of Plants to Water and High Temperature Stress" (N. C. Turner and P. J. Kramer, eds.), pp. 33–42. Wiley, New York.

Begg, J. E., and Torssell, W. R. (1974). Diaphotonastic and parahelionastic leaf movements in *Stylosanthes humilis* H. B. K. (Townsville stylo). *Bull. R. Soc. N. Z.* **12,** 277–283.

Begg, J. E., and Turner, N. C. (1976). Crop water deficits. *Adv. Agron.* **28,** 161–217.

Begon, M. (1983). Grasshopper populations and weather: The effects of insolation on *Chorthippus bruneus*. *Ecol. Entomol.* **8,** 361–370.

Bejer-Peterson, B. (1972). The nun moth, *Lymantria monacha* L., in Denmark (Lep., Lymantriidae). *Entomol. Medd.* **40,** 129–139.

Belesky, D. P., Wilkinson, S. R., and Pallas, J. E., Jr. (1982). Responses of four tall fescue cultivars grown at two nitrogen levels to low soil water availability. *Crop Sci.* **22,** 93–97.

Bell, W. J., and Carde, R. T. (1984). "Chemical Ecology of Insects." Sinauer Assoc., Sunderland, Massachusetts.

Bentley, M. D., Leonard, D. E., Leach, S., Reynolds, E., Stoddard, W., Tomkinson, B., Tomkinson, D., Strunz, G., and Yatagai, M. (1982). Effect of some naturally occurring chemicals and extracts of non-host plants on feeding by spruce budworm larvae, *Chroristoneura fumiferana. Maine Agric. Exp. Stn., Tech. Bull.* **107.**

Berryman, A. A. (1972). Resistance of conifers to invasion by bark beetle–fungal associations. *BioScience* **22,** 598–602.

Berryman, A. A. (1973). Population dynamics of the fir engraver, *Scolytus ventralis* (Coleoptera: Scolytidae). I. Analysis of population behavior and survival from 1964 to 1971. *Can. Entomol.* **105,** 1465–1488.

Bewley, J. D., and Larsen, K. M. (1982). Differences in the responses to water stress of growing and non-growing regions of maize mesocotyls: Protein synthesis on total, free, and membrane-bound polyribosome fractions. *J. Exp. Bot.* **33,** 406–415.

Bible, B. B., Hu, J., and Chong, C. (1980). Influence of cultivar, season, irrigation, and date of planting on thiocyanate ion content in cabbage. *J. Am. Soc. Hortic. Sci.* **105,** 88–91.

Blackman, M. W. (1924). The effect of deficiency and excess in rainfall upon the hickory bark beetle. *J. Econ. Entomol.* **17,** 460–470.

Blanche, C. A., Nebecker, T. E., Hodges, J. D., Karr, B. L., and Schmitt, J. J. (1985). Effect of thinning damage on bark beetle susceptibility indicators in loblolly pine. *U.S. Dep. Agric., For. Serv., Gen. Tech. Rep.* **50–54,** 471–479.

Blohm, H. (1962). "Poisonous Plants of Venezuela." Harvard Univ. Press, Cambridge, Massachusetts.

Blum, A. (1975). Effect of the BM gene on epicuticular wax and the relations of *Sorghum bicolor* L. (Moench). *Isr. J. Bot.* **24,** 50–51.

Borger, G. A., and Kozlowski, T. T. (1972). Effects of water deficits on first periderm and xylem development in *Fraxinus pennsylvanica. Can. J. For. Res.* **2,** 144–151.

Boyd, F. T., Aamodt, O. S., Bohstedt, G., and Truog, E. (1938). Sudan grass management for control of cyanide poisoning. *J. Soc. Agron.* **30,** 569–582.

Bradford, K. J., and Hsiao, T. C. (1982). Physiological responses to moderate water stress. *Encycl. Plant Physiol., New Ser.* **12B,** 263–324.

Brattsten, L. B. (1979a). Ecological significance of mixed-function oxidations. *Drug. Metab. Rev.* **10,** 35–58.

Brattsten, L. B. (1979b). Biochemical defense mechanisms in herbivores against plant allelochemics. *In* "Herbivores: Their Interaction with Secondary Plant Metabolites" (G. A. Rosenthal and D. H. Janzen, eds.), pp. 199–270. Academic Press, New York.

Brattsten, L. B. (1983). Cytochrome *P*-450 involvement in the interaction between plant terpenes and insect herbivores. *ACS Symp. Ser.* **208,** 173–195.

Brattsten, L. B., Samuelian, J. H., Long, K. Y., Kincaid, S. A., and Evans, C. K. (1983). Cyanide as a feeding stimulant for the southern armyworm, *Spodoptera eridania. Ecol. Entomol.* **8,** 125–132.

Braun, S., and Fluckiger, W. (1984). Increased population of the aphid *Aphis pomi* at a motorway. Part 2. The effect of drought and deicing salt. *Environ. Pollut., Ser. A* **36,** 261–270.

Briske, D. D., and Camp, B. J. (1982). Water stress increases alkaloid concentrations in threadleaf groundsel (*Senecio longilobus*). *Weed Sci.* **30,** 106–108.

Bruijn, G. H. (1973). The cyanogenic character of cassava (*Manihot esculenta*). *In* "Chronic Cassava Toxicity" (B. Nestel and R. MacIntyre, eds.), pp. 43–48. Int. Dev. Res. Cent., Ottawa.

Bucks, D. A., Nakayama, F. S., and French, O. F. (1984). Water management for guayule rubber production. *Trans. ASAE* **27,** 1763–1770.

Bunting, E. S. (1963). Changes in the capsule of *Papaver somniferum* between flowering and maturity. *Ann. Appl. Biol.* **51,** 459–471.

Buttery, B. R., and Boatman, S. G. (1976). Water deficits and flow of latex. *In* "Water Deficits and Plant Growth" (T. T. Kozlowski, ed.), Vol. 4, pp. 233–289. Academic Press, New York.

Callahan, P. S. (1975). "Tuning in to Nature." Devin-Adair Co., Greenwich, Connecticut.

Campbell, T. C., and Hayes, J. R. (1974). Role of nutrition in the drug metabolizing enzyme system. *Pharmocol. Rev.* **26,** 171–197.

Carroll, W. J. (1956). History of the hemlock looper, *Lambdina fiscellaria* (Guen.) (Lepidoptera: Geometridae), in Newfoundland, and notes on its biology. *Can. Entomol.* **88,** 587–599.

Cates, R. G., and Alexander, H. (1982). Host resistance and suceptibility. *In* "Bark Beetles of North American Conifers: A System for Study of Evolutionary Biology" (J. B. Mitton and K. B. Sturgeon, eds.), pp. 212–263. Univ. of Texas Press, Austin.

Cates, R. G., Redak, R. A., and Henderson, C. B. (1983). Patterns in defensive natural product chemistry: Douglas fir and western spruce budworm interactions. *ACS Symp. Ser.* **208,** 3–19.

Chambliss, O. L., and Jones, C. M. (1966). Cucurbitacins: Specific insect attractants in Cucurbitaceae. *Science* **153,** 1392–1393.

Chandra, R. K. (1985). Effect of macro- and micronutrient deficiencies and excesses on immune response. *Food Technol.* **39,** 91–93.

Chararas, C. (1979). "Ecophysiologie des Insectes Parasites des Forêts." Printed by the Author, Paris.

Chen, D., Kessler, B., and Monselise, S. P. (1964). Studies on water regime and nitrogen metabolism of citrus seedlings grown under water stress. *Plant Physiol.* **39,** 379–386.

Clark, J. A., and Levitt, J. (1956). The basis of drought resistance in the soybean plant. *Physiol. Plant.* **9,** 598–606.

Clum, H. H. (1926). The effect of transpiration and environmental factors on leaf temperatures. I. Transpiration. *Am. J. Bot.* **13,** 194–216.

Cobbinah, J. R., Morgan, F. D., and Douglas, T. J. (1982). Feeding responses of the gum leaf skeletonizer *Uraba lugens* Walter to sugars, amino acids, lipids, sterols, salts, vitamins and certain extracts of eucalypt leaves. *J. Aust. Entomol. Soc.* **21,** 225–236.

Cook, A. G. (1977). Nutrient chemicals as phagostimulants for *Locusta migratoria* (L.). *Ecol. Entomol.* **2,** 113–121.

Coyne, J. F., and Lott, L. H. (1976). Toxicity of substances in pine oleoresin to southern pine beetle. *J. Ga. Entomol. Soc.* **11,** 297–301.

Craighead, F. C. (1925). Bark-beetle epidemics and rainfall deficiency. *J. Econ. Entomol.* **18,** 577–586.

Cram, W. J. (1976). Negative feedback regulation of transport in cells. The maintenance of turgor, volume, and nutrient supply. *In* "Transport in Plants" (U. Lüttge and M. G. Pitman, eds.), pp. 284–316. Springer-Verlag, Berlin and New York.

Cutler, J. M., and Rains, D. W. (1978). Effects of water stress and hardening on the internal water relations and osmotic constituents of cotton leaves. *Physiol. Plant.* **42,** 262–268.

Cutler, J. M., Rains, D. W., and Loomis, R. S. (1977). Role of changes in solute concentration in maintaining favorable water balance in field-grown cotton. *Agron. J.* **69,** 773–779.

Dadd, R. H. (1960). Observations on the palatability and utilization of food by locusts, with particular reference to the interpretation of performances in growth trials using synthetic diets. *Entomol. Exp. Appl.* **3,** 283–304.
Dethier, V. G. (1941). Chemical factors determining the choice of food plants by *Papilio* larvae. *Am. Nat.* **75,** 61–73.
Dethier, V. G. (1977). The taste of salt. *Am. Sci.* **65,** 744–751.
Diamantoglou, S., and Kull, U. (1982). Die Jahresperiodik der Fettspeicherung und ihre Beziehungen zum Kohlenhydrathaushalt bei immergrunen mediterranen Holzpflanzen. *Acta Oecol.* [*Ser.*]: *Oecol. Plant.* **3,** 231–248.
Diamantoglou, S., and Kull, U. (1984). Kohlenhydratgehalte und osmotische Verhaltnisse bei Blattern und Rinden von *Arbutus unedo* L. und *Arbutus Andrachne* L. im Jahregang. *Ber. Dtsch. Bot. Ges.* **97,** 433–441.
Diamantoglou, S., and Meletiou-Christou, M. S. (1980). Kohlenhydratgehalte und osmotische Verhaltnisse bei Blattern und Rinden von *Pistacia lentiscus, Pistacia terebinthus* und *Pistacia vera* in Jahresgang. *Biochem. Physiol. Pflanz.* **169,** 168–176.
DiCosmo, F., and Towers, G. H. N. (1984). Stress and secondary metabolism in cultured plant cells. *Recent Adv. Phytochem.* **18,** 97–175.
Dina, S. J., and Klikoff, L. G. (1973). Effect of plant moisture stress on carbohydrate and nitrogen content of big sagebrush. *J. Range Manag.* **26,** 207–209.
Drake, B. G. (1976). Estimating water status and biomass of plant communities by remote sensing. *In* "Water and Plant Life" (O. L. Lange, L. Kappen, and E.-D. Shultze, eds.), pp. 432–438. Springer-Verlag, Berlin and New York.
Drake, B. G., and Salisbury, F. B. (1972). After-effects of low and high temperature pretreatment on leaf resistance, transpiration, and leaf temperature in *Xanthium. Plant Physiol.* **50,** 572–575.
Dubetz, S. (1969). An unusual photonastism by drought in *Phaseolus vulgaris. Can. J. Bot.* **47,** 1640–1641.
Eaton, F. M., and Ergle, D. R. (1948). Carbohydrate accumulation in the cotton plant at low moisture levels. *Plant Physiol.* **23,** 169–187.
Ehleringer, J. (1980). Leaf morphology and reflectance in relation to water and temperature stress. *In* "Adaptation of Plants to Water and High Temperature Stress" (N. C. Turner and P. J. Kramer, eds.), pp. 295–308. Wiley, New York.
Ehrler, W. L., and van Bavel, C. H. M. (1967). Sorghum foliar responses to changes in soil water content. *Agron. J.* **59,** 243–246.
Ehrler, W. L., Idso, S. B., Jackson, R. D., and Reginato, R. J. (1978). Wheat canopy temperature: Relation to plant water potential. *Agron. J.* **70,** 251–256.
Evans, W. G., and Kuster, J. E. (1980). The infrared receptive fields of *Melanophila acuminata* (Coleoptera: Buprestidae). *Can. Entomol.* **112,** 211–216.
Feeny, P., Rosenberry, L., and Carter, M. (1983). Chemical aspects of oviposition behavior in butterflies. *In* "Herbivorous Insects: Host-Seeking Behavior and Mechanisms" (S. Ahmad, ed.), pp. 27–76. Academic Press, New York.
Ferrell, G. T., and Hall, R. C. (1975). Weather and tree growth associated with white fir mortality caused by fir engraver and roundheaded fir borer. *U.S., For. Serv., Res. Pap. PSW* **PSW-109.**
Fluck, H. (1963). Intrinsic and extrinsic factors affecting the production of secondary plant products. *In* "Chemical Plant Taxonomy" (T. Swain, ed.), pp. 167–186. Academic Press, New York.
Ford, C. W. (1984). Accumulation of low molecular weight solutes in water-stressed tropical legumes. *Phytochemistry* **23,** 1007–1015.
Ford, C. W., and Wilson, J. R. (1981). Changes in levels of solutes during osmotic adjust-

ment of water stress in leaves of four tropical pasture species. *Aust. J. Plant Physiol.* **8,** 77–91.

Freeman, G. G., and Mossadeghi, N. (1971). Water regime as a factor in determining flavor strength in vegetables. *Biochem. J.* **124,** 61F–62F.

Fukutoku, Y., and Yamada, Y. (1984). Sources of proline-nitrogen in water-stressed soybean (*Glycine max*). II. Fate of ^{15}N-labelled protein. *Physiol. Plant.* **61,** 622–628.

Gates, D. M. (1980). "Biophysical Ecology." Springer-Verlag, Berlin and New York.

Georghiou, G. P., and Saito, T. (1983). "Pest Resistance to Pesticides." Plenum, New York.

Gershenzon, J. (1984). Changes in the levels of plant secondary metabolites under water and nutrient stress. *In* "Phytochemical Adaptations to Stress" (B. N. Timmermann, C. Steelink, and F. A. Loewus, eds.), pp. 273–320. Plenum, New York.

Gershenzon, J., Lincoln D. E., and Langenheim, J. H. (1978). The effect of moisture stress on monoterpenoid yield and composition in *Satureja douglasii. Biochem. Syst. Ecol.* **6,** 33–43.

Gilmore, A. R. (1977). Effects of soil moisture stress on monoterpenes in loblolly pine. *J. Chem. Ecol.* **3,** 667–676.

Gollob, L. (1980). Monoterpene composition in bark beetle-resistant loblolly pine. *Naturwissenchaflen* **67,** 409–410.

Grimalsky, V. I. (1961). The causes of the resistance of pine stands to needle-eating pests. *Zool. Zh.* **40,** 1656–1664.

Grimalsky, V. I. (1966). The role of oleoresin exudation in the pine resistance against pests. *Zool. Zh.* **45,** 551–557.

Haack, R. A., and Benjamin, D. M. (1982). The biology and ecology of the two-lined chestnut borer, *Agrilus biliniatus* (Coleoptera: Buprestidae), on oaks, *Quercus* spp., in Wisconsin. *Can. Entomol.* **114,** 385–396.

Haack, R. A., and Slansky, F., Jr. (1987). Nutritional ecology of wood-feeding Coleoptera, Lepidoptera, and Hymenoptera. *In* "Nutritional Ecology of Insects, Mites, and Spiders" (F. Slansky, Jr., and J. G. Rodriguez, eds.), pp. 449–486. Wiley, New York.

Haack, R. A., Wilkinson, R. C., Foltz, J. L., and Cornell, J. A. (1984). Gallery construction and oviposition by *Ips calligraphus* (Coleoptera: Scolytidae) in relation to slash pine phloem thickness and temperature. *Can. Entomol.* **116,** 625–632.

Haack, R. A., Foltz, J. L., and Wilkinson, R. C. (1985). Effects of temperature and slash pine phloem thickness on *Ips colligraphus* life processes. *U.S., Dep. Agric., Tech. Bull.* **50–56,** 102–113.

Hanson, A. D., and Hitz, W. D. (1982). Metabolic responses of mesophytes to plant water deficits. *Annu. Rev. Plant Physiol.* **33,** 163–203.

Hanson, A. D., and Nelson, C. E. (1978). Betaine accumulation and [^{14}C]formate metabolism in water-stressed barley leaves. *Plant Physiol.* **62,** 305–312.

Hanson, A. D., Nelson, C. E., and Everson, E. H. (1977). Evaluation of free proline accumulation as an index of drought resistance using two contrasting barley cultivars. *Crop Sci.* **17,** 720–726.

Hartt, C. E. (1967). Effect of moisture supply upon translocation and storage of ^{14}C in sugarcane. *Plant Physiol.* **42,** 338–346.

Harvey, G. T. (1974). Nutritional studies of eastern spruce budworm (Lepidoptera: Tortricidae). I. Soluble sugars. *Can. Entomol.* **106,** 353–365.

Haukioja, E., and Hakala, T. (1975). Herbivore cycles and periodic outbreaks. Formulation of a general hypothesis. *Rep. Kevo Subarct. Res. Stn.* **12,** 1–9.

Heinrichs, E. A., Fabellar, L. T., Basilio, R. P., T.-Ch. Wen, and Medrano, F. (1984). Susceptibility of rice plant hoppers, *Nilaparvata ingens* and *Sogatella furcifera* (Homoptera: Delphacidae), to insecticides as influenced by level of resistance in the host plant. *Environ. Entomol.* **13,** 455–458.

Heron, R. J. (1965). The role of chemotactic stimuli in the feeding behavior of spruce budworm larva of white spruce. *Can. J. Zool.* **43,** 247–269.

Hicks, K. L. (1974). Mustard oil glucosides: Feeding stimulants for adult cabbage flea beetles, *Phyllotreta cruciferae* (Coleopt., Chrysomelidae). *Ann. Entomol. Soc. Am.* **67,** 261–264.

Hodges, J. D., and Lorio, P. L., Jr. (1969). Carbohydrates and nitrogen fractions of the inner bark of loblolly pines under moisture stress. *Can. J. Bot.* **47,** 1651–1657.

Hodges, J. D., and Lorio, P. L., Jr. (1971). Comparison of field techniques for measuring moisture stress in large loblolly pines. *For. Sci.* **17,** 220–223.

Hodges, J. D., and Lorio, P. L., Jr. (1975). Moisture stress and composition of xylem oleoresin in loblolly pine. *For. Sci.* **21,** 283–290.

Hoffman, J. J., Kingsolver, B. E., McLaughlin, S. P., and Timmermann, B. N. (1984a). Production of resins by acid-adapted Astereae. *In* "Phytochemical Adaptations to Stress" (B. N. Timmermann, C. Sterlink, and F. A. Loewus, eds.), pp. 251–271. Plenum, New York.

Hoffman, J. J., Kingsolver, B. E., McLaughlin, S. P., and Timmermann, B. N. (1984b). Production of resins by acid-adapted Astereae. *Recent Adv. Phytochem.* **18,** 251–271.

House, H. L. (1966). Effect of temperature on the nutritional requirements of an insect, *Pseudosarcophaga affinis* Auct. nec Fallen (Diptera: Sarcophagidae), and its probable ecological significance. *Ann. Entomol. Soc. Am.* **59,** 1263–1266.

Hsiao, T. C. (1973). Plant responses to water stress. *Annu. Rev. Plant Physiol.* **24,** 519–570.

Hsiao, T. C., Acevedo, E., Fereres, E., and Henderson, D. W. (1976). Stress metabolism. Water stress, growth, and osmotic adjustment. *Philos. Trans. R. Soc. London, Ser. B* **273,** 479–500.

Hsiao, T. H. (1969). Chemical basis of host selection and plant resistance in oligophagous insects. *Entomol. Exp. Appl.* **12,** 777–788.

Hsiao, T. H., and Fraenkel, G. (1968). The influence of nutrient chemicals on the feeding behavior of the Colorado potato beetle, *Leptinotarsa decemlineata* (Coleoptera: Chrysomelidae). *Ann. Entomol. Soc. Am.* **61,** 44–54.

Ives, W. G. H. (1974). Weather and outbreaks of the spruce budworm, *Choristoneura fumiferana*. *Inf. Rep. NOR-X—North. For. Res. Cent.* **NOR-X-118.**

Jager. H. J., and Meyer, H. R. (1977). Effect of water stress on growth and proline metabolism of *Phaseolus vulgaris* L. *Oecologia* **30,** 83–96.

Jones, M. M., Osmond, C. B., and Turner, N. C. (1980). Accumulation of solutes in leaves of sorghum and sunflower in response to water deficits. *Aust. J. Plant Physiol.* **7,** 193–205.

Julander, O. (1945). Drought resistance in range and pasture grasses. *Plant Physiol.* **20,** 573–599.

Kemble, A. R., and MacPherson, H. T. (1954). Liberation of amino acids in perennial rye grass during wilting. *Biochem. J.* **58,** 46–49.

Kemp, W. P., and Moody, V. L. (1984). Relationships between regional soils and foliage characteristics and western spruce budworm (Lepidoptera: Tortricidae) outbreak frequency. *Environ. Entomol.* **13,** 1291–1297.

Kennedy, C. W., and Bush, L. P. (1983). Effect of environmental and management factors on the accumulation of *N*-acetyl and *N*-formyl loline alkaloids in tall fiscue. *Crop Sci.* **23,** 547–552.

Kilmer, V. J., Bennett, O. L., Stahly, V. F., and Timmons, D. R. (1960). Yield and mineral composition of eight forage species grown at four levels of soil moisture. *Agron. J.* **52,** 282–285.

Kolomiets, N. G., Stadnitskii, G. V., and Vorontzov, A. I. (1979). "The European Pine Sawfly." Amerind Publ. Co., Pvt. Ltd., New Delhi (U.S. Dep. Comm. Note, Tech. Inf. Serv., Springfield, Virginia).

Kozlowski, T. T. (1976). Water supply and leaf shedding. *In* "Water Deficits and Plant Growth" (T. T. Kozlowski, ed.), Vol. 4, pp. 191–231. Academic Press, New York.

Kozlowski, T. T. (1982). Water supply and tree growth. Part I. Water deficits. *For. Abstr.* **43,** 57–95.

Kramer, P. J. (1983). "Water Relations of Plants." Academic Press, New York.

Kramer, P. J., and Kozlowski, T. T. (1979). "Physiology of Woody Plants." Academic Press, New York.

Kriedemann, P. E., and Downton, W. J. S. (1981). Photosynthesis. *In* "Physiology and Biochemistry of Drought Resistance in Plants" (L. G. Paleg and D. Aspinall, eds.), pp. 283–314. Academic Press, New York.

Larsson, S., and Tenow, O. (1984). Areal distribution of a *Neodiprion sertifer* (Hym., Diprionidae) outbreak on Scots pine as related to stand condition. *Holarctic Ecol.* **7,** 81–90.

Lechowicz, M. J. (1983). Leaf quality and the host preferences of gypsy moth in the northern deciduous forest. *U.S., Dep. Agric., For. Serv., Gen. Tech. Rep. NC* **NC-85,** 67–82.

Lorio, P. L., Jr., and Hodges, J. D. (1968). Oleoresin exudation pressure and relative water content of inner bark as indicators of moisture stress in loblolly pines. *For. Sci.* **14,** 392–398.

Lorio, P. L., Jr., and Hodges, J. D. (1974). Host and site factors in southern pine beetle infestations. *In* "Proceedings of the Southern Pine Beetle Symposium" (T. L. Payne, R. N. Coulson, and R. C. Thatcher, eds.), pp. 32–34. Tex. Agric. Exp. Stn. and U.S. For. Serv., Washington, D. C.

Ludlow, M. M. (1980). Adaptive significance of stomatal responses to water stress. *In* "Adaptation of Plants to Water and High Temperature Stress" (N. C. Turner and P. J. Kramer, eds.), pp. 123–138. Wiley, New York.

Ma, W. C. (1972). Dynamics of feeding responses in *Pieris brassicae* L. as a function of chemosensory input: A behavioral, ultrastructural and electrophysiological study. *Meded. Landbouwhogesch. Wageningen* **72,** 1–62.

Ma, W. C. (1977). Electrophysiological evidence for chemosensitivity to adenosine, adenine and sugars in *Spodoptera exempta* and related species. *Experientia* **33,** 356–358.

McManus, M. L., and Giese, R. L. (1968). The Columbia timber beetle, *Corthylus columbianus*. VII. the effect of climatic integrants on historic density fluctuations. *For. Sci.* **14,** 242–253.

McMichael, B. L., and Elmore, C. D. (1977). Proline accumulation in water stressed cotton leaves. *Crop Sci.* **17,** 905–908.

Magness, J. R., Regeimbal, L. O., and Degman, E. S. (1932). Accumulation of carbohydrates in apple foliage, bark and wood as influenced by moisture supply. *Proc. Am. Soc. Hortic. Sci.* **29,** 246–252.

Majak, W., McDiarmid, R. E., Van Ryswyz, A. L., Broersha, K., and Bonin, S. G. (1979). Alkaloid levels in reed canarygrass grown on wet meadows in British Columbia. *J. Range Manage.* **32,** 322–326.

Majak, W., McDiarmid, R. E., Hall, J. W., and van Ryswyk, A. L. (1980a). Seasonal variation in the cyanide potential of arrowgrass (*Triglochin maritima*). *Can. J. Plant Sci.* **60,** 1235–1241.

Majak, W., Quinton, D. A., and Broersma, K. (1980b). Cyanogenic glycoside levels in Saskatoon serviceberry. *J. Range Manage.* **33,** 197–199.

Mansfield, T. A., and Davies, W. J. (1981). Stomata and stomatal mechanisms. *In* "Physiology and Biochemistry of Drought Resistance in Plants" (L. G. Paleg and D. Aspinall, eds.), pp. 315–346. Academic Press, New York.

Mansfield, T. A., and Davies, W. J. (1985). Mechanisms for leaf control of gas exchange. *BioScience* **35,** 158–164.

Maramorosch, K., and Shope, R. E. (1975). "Invertebrate Immunity: Mechanisms of Invertebrate Vector–Parasite Relations." Academic Press, New York.

Marx, J. (1985). The immune system "belongs in the body." *Science* **227,** 1190–1192.

Mason, R. R. (1971). Soil moisture and stand density affect oleoresin exudation flow in a loblolly pine plantation. *For. Sci.* **17,** 170–177.

Mason, R. R., and Tigner, T. C. (1972). Forest-site relationships within an outbreak of lodgepole needle miner in central Oregon. *U.S., For. Serv., Res. Pap.* **PNW-146.**

Mattson, W. J. (1980). Herbivory in relation to plant nitrogen content. *Annu. Rev. Ecol. Syst.* **11,** 119–161.

Mattson, W. J., and Addy, N. D. (1975). Phytophagous insects as regulators of forest primary production. *Science* **190,** 515–522.

Mattson, W. J., and Haack, R. A. (1987). The role of drought in outbreaks of plant-eating insects. *BioScience* **37,** 110–118.

Mattson, W. J., and Scriber, J. M. (1987). Feeding ecology of insect folivores of woody plants: Water, nitrogen, fiber, and mineral considerations. *In* "Nutritional Ecology of Insects, Mites and Spiders" (F. Slansky, Jr., and J. Rodriquez, eds.), pp. 105–146. Wiley, New York.

Mattson, W. J., Lorimer, N., and Leary, R. A. (1982). Role of plant variability (trait vector dynamics and diversity) in plant/herbivore interactions. *In* "Resistance to Diseases and Pests in Forest Trees" (H. M. Heybroek, R. B. Stephen, and K. von Weissenberg, eds.), pp. 295–303. Pudoc, Wageningen.

Mattson, W. J., Slocum, S. S., and Koller, C. N. (1983). Spruce budworm performance in relation to foliar chemistry of its host plants. *USDA For. Serv. Gen. Tech. Rep.* **NE-85,** 55–56.

Mattson, W. J., Lawrence, R. K., Haack, R. A., Herms, D. A., and Charles, P. J. (1987). Plant defensive strategies for different insect feeding guilds in relation to plant ecological strategies and intimacy of host association. *In* "Mechanisms of Plant Resistance to Insects" (W. J. Mattson, J. Levieux, and C. Bernard-Dagan, eds.). Springer-Verlag, New York (in press).

Merdan, A., Abdel-Rahman, H., and Soliman, A. (1975). On the influence of host plants on insect resistance to bacterial diseases. *Z. Angew. Entomol.* **78,** 280–285.

Michaels, P. J., Sappington, D. E., and Stenger, P. J. 1985. An automated objective prediction package for the spread of southern pine beetle. *Proc. Conf. Agric. For. Meteorol. 17th, and Conf. Biometeorol. Aerobiol. 7th,* pp. 70–73.

Milborrow, B. V. (1981). Abscisic acid and other hormones. *In* "The Physiology and Biochemistry of Drought Resistance in Plants" (L. G. Paleg and D. Aspinall, eds.), pp. 347–388. Academic Press, New York.

Miller, J. R., and Strickler, K. L. (1984). Finding and accepting host plants. *In* "Chemical Ecology of Insects" (W. J. Bell and R. T. Carde, eds.), pp. 127–157. Sinauer Assoc., Sunderland, Massachusetts.

Mittler, T. E., and Dadd, R. H. (1964). Gustatory discrimination between liquids by the aphids *Myzus persicae* (Sulzer). *Entomol. Exp. Appl.* **7,** 315–328.

Morgan, J. M. (1984). Osmoregulation and water stress in higher plants. *Annu. Rev. Plant Physiol.* **35,** 299–319.

Morris, R. F., and Fulton, W. C. (1970). Models for the development and survival of *Hyphantria cunea* in relation to temperature and humidity. *Mem. Entomol. Soc. Can.* **70.**

Munns, R., and Weir, R. (1981). Contribution of sugars to osmotic adjustments in elongating and expanded zones of wheat leaves during moderate water deficits at two light levels. *Aust. J. Plant Physiol.* **8,** 93–105.

Munns, R., Brady, C. J., and Barlow, E. W. R. (1979). Solute accumulation in the apex and leaves of wheat during water stress. *Aust. J. Plant Physiol.* **6,** 379–389.

Murray, J., and Murray, M. (1981). Toward a nutritional concept of host resistance to malignancy and intracellular infection. *Perspect. Biol. Med.* **24,** 290–301.

Mustaparta, H. (1984). Olfaction. *In* "Chemical Ecology of Insects" (W. J. Bell and R. T. Carde, eds.), pop. 37–70. Sinauer Assoc., Sunderland, Massachusetts.

Nayar, J. K., and Thorsteinson, A. J. (1963). Further investigation into the chemical basis of insect–host plant relationships in an oligophagous insect, *Plutella maculipennis* (Curtis) (Lepidoptera: Plutellidae). *Can. J. Zool.* **41,** 923–929.

Naylor, A. W. (1972). Water deficits and nitrogen metabolism. *In* "Water Deficits and Plant Growth" (T. T. Kozlowski, ed.), Vol. 3, pp. 241–254. Academic Press, New York.

Nelson, C. E. (1953). Hydrocyanic acid content of certain sorghums under irrigation as affected by nitrogen fertilization and soil moisture stress. *Agron. J.* **45,** 615–617.

Nielsen, J. K. (1978). Host plant discrimination within crucifers: Feeding responses of four leaf beetles (Coleoptera: Chrysomelidae) to glucosinolates, cucurbitacins, and cardenolides. *Entomol. Exp. Appl.* **24,** 41–54.

Nutritional Reviews (1985). Drug metabolism and disposition at varying levels of proteins and calories in human subjects. *Nutr. Rev.* **43,** 71–72.

Nuttall, W. F. (1976). Effect of soil moisture tension and amendments on yields and on herbage N, P, and S concentrations of alfalfa. *Agron. J.* **68,** 741–744.

O'Toole, J. C., Turner, N. C., Namuco, O. P., Dingkuhn, M., and Gomez, K. A. (1984). Comparison of some crop water stress measurement methods. *Crop Sci.* **24,** 1121–1128.

Otto, D. (1970). Zur Bedeutung des Zuckergehaltes der Nahrung fur die Entwicklung nadelfressender Kieferninsekten. *Arch. Forstwes.* **19,** 135–150.

Otto, D., and Geyer, W. (1970). Zur Bedeutung des Kiefernnadel harzes und des Kiefernnadeloles fur die Entwicklung nadelfressender Isektem. *Arch. Forstwes.* **19,** 151–167.

Ouellet, M. J., Lailamme, M., and Perron, J.-M. (1983). Effects des sucres sur le comportement alimentaire de *Lambdina fiscellaria*. *Entomol. Exp. Appl.* **34,** 139–142.

Pahlich, E., and Grieb, B. (1983). Turgor pressure and proline accumulation in water stressed plants. *Angew. Bot.* **57,** 295–299.

Pallardy, S. G., and Kozlowski, T. T. (1980). Cuticle development in the stomatal region of *Populus* clones. *New Phytol.* **85,** 363–368.

Parker, J. (1979). Effects of defoliation and root height above a water table on some red oak root metabolites. *J. Am. Soc. Hortic. Sci.* **104,** 417–421.

Parker, J., and Patton, R. L. (1975). Effects of drought and defoliation on some metabolites in roots of black oak seedlings. *Can. J. For. Res.* **5,** 457–463.

Parsons, L. R. (1982). Plant responses to water stress. *In* "Breeding Plants for Less Favorable Environments" (M. N. Christiansen and C. F. Lewis, eds.), pp. 175–192. Wiley, New York.

Perttunen, V. (1957). Reactions of two bark beetle species, *Hylurgops palliatus* Gyll. and *Hylastes ater* Payk. (Col., Scolytidae), to the terpene α-pinene. *Ann. Entomol. Fenn.* **23,** 101–110.

Pestka, J. J., and Witt, M. F. (1985). An overview of the immune function. *Food Technol.* **39,** 83–90.

Pitman, M. G. (1981). Ion uptake. *In* "Physiology and Biochemistry of Drought Resistance in Plants" (L. G. Paleg and D. Aspinall, eds.), pp. 71–96. Academic Press, New York.

Popp, M. (1984). Chemical composition of Australian mangroves. II. Low molecular weight carbohydrates. *Z. Pflanzenphysiol.* **113,** 395–409.

Porter, W. P., Hindsdill, R., Fiarbrother, A., Olson, L. J., Jalger, J., Yuill, T., Bisgaard, S., and Nolan, Y. (1984). Toxicant–disease–environment interactions associated with suppression of immune system, growth, and reproduction. *Science* **224,** 1014–1017.

Potter, C., and Gillham, E. M. (1957). Effects of host plants on the resistance of *Acyrthosiphon pisum* (Harris) to insecticides. *Bull. Entomol. Res.* **48,** 317–322.

Prokopy, J. R., and Owens, E. D. (1983). Visual detection of plants by herbivorous insects. *Annu. Rev. Entomol.* **28,** 337–364.

Puritch, G. S., and Mullick, D. G. (1975). Effect of water stress on the rate of non-suberized impervious tissue formation following wounding in *Abies grandis*. *J. Exp. Bot.* **26,** 903–910.

Quarrie, S. A., and Jones, H. G. (1977). Effects of abscisic acid and water stress on development and morphology of wheat. *J. Exp. Bot.* **28,** 192–203.

Raffa, K. F., and Berryman, A. A. (1983). The role of host plant resistance in the colonization behavior and ecology of bark beetles (Coleoptera: Scolytidae). *Ecol. Monogr.* **53,** 27–49.

Reichenbach, N. G., and Stairs, G. R. (1984). Response of the western spruce budworm (Lepidoptera: Tortricidae) to temperature and humidity: Development rates and survivorship. *Environ. Entomol.* **13,** 611–618.

Renwick, J. A. A. (1983). Nonpreference mechanisms: Plant characteristics influencing insect behavior. *ACS Symp. Ser.* **208,** 199–214.

Rhoades, D. F. (1983). Herbivore population dynamics and plant chemistry. *In* "Variable Plants and Herbivores in Natural and Managed Systems" (R. F. Denno and M. S. McClure, eds.), pp. 155–220. Academic Press, New York.

Rhoades, D. F. (1985). Offensive–defensive interactions between herbivores and plants: Their relevance in herbivore population dynamics and ecological theory. *Am. Nat.* **125,** 205–238.

Roberts, B. R. (1963). Effects of water stress on the translocation of photosynthetically assimilated carbon-14 in yellow poplar. *Diss. Abstr.* **24,** 918 (Order No. 63-66166).

Rodriguez, E. (1983). Cytotoxic and insecticidal chemicals of desert plants. *ACS Symp. Ser.* **208,** 291–302.

Rogers, C. F., and Frykolm, O. C. (1937). Observations on the variation in cyanogenetic power of white clover plants. *J. Agric. Res.* (*Washington, D.C.*) **55,** 535–537.

Routley, D. G. (1966). Proline accumulation in wilted ladino clover leaves. *Crop Sci.* **6,** 358–361.

Rudinsky, J. A. (1966). Scolytid beetles associated with Douglas fir response to terpenes. *Science* **152,** 218–219.

Runckel, W. J., and Knapp, I. E. (1946). Viscosity of pine gum. *Ind. Eng. Chem., Ind. Ed.* **38,** 555–556.

St. George, R. A. (1929). Weather, a factor in outbreaks of the hickory bark beetle. *J. Econ. Entomol.* **22,** 573–580.

St. George, R. A. (1930). Drought affected and injured trees attractive to bark beetles. *J. Econ. Entomol.* **23,** 825–828.

Saxena, K. N. (1965). Control of the orientation and feeding behavior of red cotton bug, *Dysdercus koenigii* (F.), by chemical constituents of plants. *Proc. Int. Congr. Entomol., 12th, 1964,* p. 294.

Schwenke, W. (1968). Neve Hinweise Auf einer Abhaengigkeit der Vermehrung blattund nadelfressender Forstinsekten vom Zuckergehalt ihrer Nahrung. *Z. Angew. Entomol.* **61,** 365–369.

Scriber, J. M. (1984). Host plant suitability. *In* "Chemical Ecology of Insects" (W. J. Bell and R. T. Carde, eds.), pp. 159–202. Sinauer Assoc., Sunderland, Massachusetts.

Scriber, J. M., and Lederhouse, R. C. (1983). Temperature as a factor in the development and feeding of tiger swallowtail caterpillars, *Papilio glaucus* (Lepidoptera). *Oikos* **40,** 95–102.

Siegel, H. S. (1985). Immunological responses as indicators of stress. *World's Poult. Sci. J.* **41,** 36–44.

Singh, T. N., Paleg, L. G., and Aspinall, D. (1973a). Stress metabolism. III. Variation in response to water deficit in the barley plant. *Aust. J. Biol. Sci.* **26,** 65–76.

Singh, T. N., Aspinall, D., and Paleg, L. G. (1973b). Stress metabolism. IV. The influence of (2-chloroethyl)trimethylammonium chloride and gibberellic acid on the growth and proline accumulation of wheat plants during water stress. *Aust. J. Biol. Sci.* **26,** 77–86.

Skoss, J. D. (1955). Structure and composition of plant cuticle in relation to environmental factors and permeability. *Bot. Gaz. (Chicago)* **117,** 155–172.

Smith, J. A. C., and Milburn, J. A. (1980). Osmoregulation and the content of phloem-sap composition in *Ricinus communis* L. *Planta* **148,** 28–34.

Smith, R. H. (1963). Toxicity of pine resin vapors to three species of *Dendroctonus* bark beetles. *J. Econ. Entomol.* **56,** 827–831.

Smith, R. H. (1975). Formula for describing effect of insect and host tree factors on resistance to western pine beetle attack. *J. Econ. Entomol.* **68,** 841–844.

Sokolov, V. S. (1959). The influence of certain environmental factors on the formation and accumulation of alkaloids in plants. *Symp. Soc. Exp. Biol.* **13,** 230–257.

Stadler, E. (1984). Contact chemoreception. *In* "Chemical Ecology of Insects" (W. J. Bell and R. T. Carde, eds.), pp. 3–35. Sinauer Assoc., Sunderland, Massachusetts.

Stark, J. C., and Wright, J. L. (1985). Relationship between foliage temperature and water stress in potatoes. *Am. Potato J.* **62,** 57–68.

Stewart, C. R. (1971). Effect of wilting on carbohydrates during incubation of excised bean leaves in the dark. *Plant Physiol.* **48,** 792–794.

Stewart, C. R., Morris, C. J., and Thompson, J. F. (1966). Changes in amino acid content of excised leaves during incubation. II. Role of sugar in the accumulation of proline in wilted leaves. *Plant Physiol.* **41,** 1585–1590.

Stoszek, K. J., Mika, P. G., Moore, J. A., and Osborne, H. L. (1981). Relationships of Douglas-fir tussock moth defoliation to site and stand characteristics in northern Idaho. *For. Sci.* **27,** 431–442.

Tanner, C. B. (1963). Plant temperatures. *Agron. J.* **55,** 210–211.

Terriere, L. C. (1984). Induction of detoxification enzymes. *Annu. Rev. Entomol.* **29,** 71–88.

Thomson, A. J., and Shrimpton, D. M. (1984). Weather associated with the start of mountain pine beetle outbreaks. *Can. J. For. Res.* **14,** 255–258.

Todd, G. W. (1972). Water deficits and enzymatic activity. *In* "Water Deficits and Plant Growth" (T. T. Kozlowski, ed.), Vol. 3, pp. 177–216. Academic Press, New York.

Treichel, S., Brinckmann, E., Scheitler, B., and von Willert, D. J. (1984). Occurrence and changes of proline content in plants in the southern Namib Desert in relation to increasing and decreasing drought. *Planta* **162,** 236–242.

Tsiropoulos, G. J. (1981). Effects of varying the nitrogen to carbohydrate ratio upon the biological performance of adult *Daucus olea*. *Arch. Int. Physiol. Biochim.* **89,** 101–105.

Tuomi, J., Niemela, P., Haukioja, E., Siven, S., and Neuvonen, S. (1984). Nutrient stress:

An explanation for plant anti-herbivore responses to defoliation. *Oecologia* **61,** 208–210.

Turner, N. C. (1979). Drought resistance and adaptation to water deficits in crop plants. *In* "Stress Physiology in Crop Plants" (H. Mussell and R. C. Staples, eds.), pp. 343–372. Wiley, New York.

Turner, N. C., and Jones, M. M. (1980). Turgor maintenance by osmotic adjustment: A review and evaluation. *In* "Adaptations of Plants to Water and High Temperature Stress" (N. C. Turner and P. J. Kramer, eds.), pp. 87–103. Wiley, New York.

Turrell, F. M., and Turrell, M. E. (1943). The effect of the great drought of 1934 on the leaf structure of certain Iowa plants. *Proc. Iowa Acad. Sci.* **50,** 185–194.

Tyree, M. T., and Jarvis, P. G. (1982). Water in tissues and cells. *Encycl. Plant Physiol., New Ser.* **12B,** 35–77.

van Bavel, C. H. M. (1953). Chemical composition of tobacco leaves as affected by soil mosture condition. *Agron. J.* **45,** 611–614.

Vassiliev, I. M., and Vassiliev, M. G. (1936). Changes in carbohydrate content of wheat plants during the process of hardening for drought resistance. *Plant Physiol.* **11,** 115–125.

Viets, F. G., Jr. (1972). Water deficits and nutrient availability. *In* "Water Deficits and Plant Growth" (T. T. Kozlowski, ed.), Vol. 3, pp. 217–239. Academic Press, New York.

Visser, J. H. (1983). Differential sensory perception of plant compounds by insects. *ACS Symp. Ser.* **208,** 215–230.

Vite, J. P. (1961). The influence of water supply on oleoresin exudation pressure and resistance to bark beetle attack in *Pinus ponderosa. Contrib. Boyce Thompson Inst.* **21,** 37–66.

Vite, J. P., and Wood, D. L. (1961). A study on the applicability of the measurement of oleoresin exudation pressure in determining susceptibility of second growth ponderosa pine to bark beetle infestation. *Contrib. Boyce Thompson Inst.* **21,** 67–78.

Wahl, O., and Ulm, K. (1983). Influence of pollen feeding and physiological condition on pesticide sensitivity of the honey bee, *Apis mellifera carnica. Oecologia* **59,** 106–128.

Wainwright, C. M. (1977). Sun-tracking and related leaf movements in a desert lupine (*Lupinus arizonicus*). *Am. J. Bot.* **64,** 1032–1041.

Waldbauer, G. P., Cohen, R. W., and Friedman, S. (1984). Self selection of an optimal nutrient mix from defined diets by larvae of the corn earworm, *Heliothis zea* (Boddie). *Physiol. Zool.* **57,** 590–597.

Walgenbach, R. P., Martin, G. C., and Blake, G. R. (1981). Release of soluble protein and nitrogen in alfalfa. I. Influence of growth temperature and soil moisture. *Crop Sci.* **21,** 843–849.

Waller, G. R., and Nowacki, E. K. (1978). "Alkaloid Biology and Metabolism in Plants." Plenum, New York.

Wample, R. L., and Bewley, J. D. (1975). Proline accumulation in flooded and wilted sunflower and the effects of benzyladenine and abscisic acid. *Can. J. Bot.* **53,** 2893–2896.

Wardlaw, I. F. (1968). The control and pattern of movement of carbohydrates in plants. *Bot. Rev.* **34,** 79–105.

Wargo, P. M. (1984). How stress predisposes trees to attack by *Armillaria mellea*—a hypothesis. *In* "Root and Butt Rots of Forest Trees" (G. A. Kile, ed.), pp. 115–121. CSIRO, Melbourne.

Weete, J. D., Leek, G. L., Peterson, C. M., Currie, H. E., and Branch, W. D. (1978). Lipid and surface wax synthesis in water-stressed cotton. *Plant Physiol.* **62,** 675–677.

Wellington, W. G. (1950). Climate and spruce budworm outbreaks. *Can. J. Res.* **28,** 308–331.
Wensler, R. J. D. (1962). Mode of host selection by an aphid. *Nature (London)* **195,** 830–831.
Wensler, R. J. D., and Dudzinksi, A. E. (1972). Gustation of sugars, amino acids and lipids by larvae of the scarabaeid, *Sericesthis geminata* (Coleoptera). *Entomol. Exp. Appl.* **15,** 155–165.
White, T. C. R. (1969). An index to measure weather-induced stress of trees associated with outbreaks of psyllids in Australia. *Ecology* **50,** 905–909.
White, T. C. R. (1974). A hypothesis to explain outbreaks of looper caterpillars, with special reference to populations of *Selidosema suavis,* in a plantation of *Pinus radiata* in New Zealand. *Oecologia* **16,** 279–301.
White, T. C. R. (1976). Weather, food and plagues of locusts. *Oecologia* **22,** 119–134.
White, T. C. R. (1984). The abundance of invertebrate herbivore in relation to the availability of nitrogen in stressed food plants. *Oecologia* **63,** 90–105.
Wiegand, C. L., and Namken, L. N. (1966). Influences of plant moisture stress, solar radiation, and air temperature on cotton leaf temperature. *Agron. J.* **58,** 582–586.
Williams, J. D. (1972). Effects of time of day, moisture stress, and frosting on the alkaloid content of *Phalaris tuberosa. Aust. J. Agric. Res.* **23,** 611–621.
Witanachchi, J. P., and Morgan, F. D. (1981). Behavior of the bark beetle, *Ips grandicollis,* during host selection. *Physiol. Entomol.* **6,** 219–223.
Wood, D. L. (1962). Experiments on the interrelationship between oleoresin exudation pressure in *Pinus ponderosa* and attack by *Ips confusus* (Lec.) (Coleoptera: Scolytidae). *Can. Entomol.* **94,** 473–477.
Wood, K. A., Wilson, B. H., and Graves, J. B. (1981). Influence of host plants on the susceptibility of the fall armyworm to insecticides. *J. Econ. Entomol.* **74,** 96–98.
Woodhams, D. H., and Kozlowski, T. T. (1954). Effects of soil moisture stress on carbohydrate development and growth in plants. *Am. J. Bot.* **41,** 316–320.
Wyn Jones, R. G. (1984). Phytochemical aspects of osmotic adaptation. *In* "Phytochemicals Adaptation to Stress" (B. N. Timmermann, C. Steelink, and F. A. Loewus, eds.), pp. 55–78. Plenum, New York.
Yamamoto, R. T., and Fraenkel, G. (1960). Assay of the principal gustatory stimulant for the tobacco hornworm, *Protoparce sexta,* from solanaceous plants. *Ann. Entomol. Soc. Am.* **53,** 499–503.
Yancey, P. H., Clark, M. E., Hand, S. C., Bowlus, R. D., and Somero, G. N. (1982). Living with water stress: Evolution of osmolyte systems. *Science* **217,** 1214–1222.
Yu, S. J. (1982). Induction of microsomal oxidases by host plants in the fall armyworm, *Spodoptera frugiperda* (J. C. Smith). *Pestic. Biochem. Physiol.* **17,** 59–67.
Zahner, R. (1968). Water deficits and growth of trees. *In* "Water Deficits and Plant Growth" (T. T. Kozlowski, ed.), Vol. 2, pp. 191–254. Academic Press, New York.
Zeiger, E. (1983). The biology of stomatal guard cells. *Annu. Rev. Plant Physiol.* **34,** 441–475.

Part IV

EVOLUTIONARY CONSEQUENCES

Chapter 16

Insect Population Dynamics and Induction of Plant Resistance: The Testing of Hypotheses

ERKKI HAUKIOJA and SEPPO NEUVONEN

Laboratory of Ecological Zoology
Department of Biology
University of Turku
SF-20500 Turku 50, Finland

I. INTRODUCTION

It is a relatively recent idea that plants may defend themselves actively and that phenotypic plasticity in plant quality may vastly modify insect performance (Rudnew, 1963). Even the most complete population studies of defoliating insects (Morris, 1963; Klomp, 1966; Varley *et al.*, 1973; Wellington, 1977; Schwenke, 1978) have usually neglected variable food quality as a factor in population fluctuation (but see Benz, 1974; Baltensweiler *et al.,* 1977). Methods such as life table analysis have been the prevailing paradigm in the study of insect population dynamics. Such approaches can show only the stage of insect life cycle when mortality best corresponds to changes in population size. A life table analysis does not reveal, for example, whether food quality can affect both mortality and fecundity.

There is an increasing number of studies demonstrating either chemical changes in plant composition that may affect herbivores or changes in some measure of herbivore performance as a bioassay of plant quality. Still, such studies seldom elucidate the effects on herbivore populations. Consequently, existing data cannot be easily used to test hypotheses arguing that variable food quality plays a role in the population dynamics of outbreak or cyclic insects (Rudnew, 1963; Benz, 1974; White, 1974; Haukioja and Hakala, 1975; Rhoades, 1979, 1985; Haukioja, 1980; Haukioja *et al.,* 1983). Our main aim in this chapter is to investigate the demands the data must satisfy and the kind of data that could actually be applied to such questions.

Food may play both a density-independent and a density-dependent role in the dynamics of insect cycles. A density-independent increase in food availability can cause an increase in population density and create the possibility of an outbreak. Good synchrony of leaf flush with a sensitive stage of the insect life cycle (Feeny, 1970; Varley *et al.,* 1973) or stress (Lewis, 1979) are examples. The latter has been claimed to involve an increase in available nitrogen (White, 1974, 1984) and/or carbohydrates (Schwenke, 1963, 1968) or a decrease in defenses (Rudnew, 1963; Grimalsky, 1966, 1974; Rudnew and Smeljanez, 1978) or a general change in allocation of carbon to nondefensive use (Pitman *et al.,* 1982). The release may also be density dependent with time lags, so that a new peak can be initiated only after relaxing of induced defenses. If herbivores can manipulate host plants to make them more suitable (Norris, 1979), major increases in pest density occur after they reach a threshold density.

Food may contribute to the cessation of an outbreak or cycle. This may come about because of an absolute deprivation of food (a density-dependent phenomenon; for a review, see Dempster, 1983) or for density-independent reasons such as relaxation of stress with an accompanied decrease in food quality or a poor synchrony between budbreak and hatching of eggs. What is most interesting is that plant quality may deteriorate in a density-dependent manner as a consequence of an insect attack and thus may be important in the population dynamics of herbivores. In this chapter we concentrate on the effects of inducible changes in plant quality because these may be promising and neglected causal factors in the fluctuation of herbivore populations (Haukioja *et al.*, 1983). For this discussion the crucial question is: Does foliage quality affect insect performance in a way that is relevant to fluctuation in the density of herbivore populations? We do not discuss the chemical and/or physical basis of these reactions (for the chemical basis of resistance, see Rosenthal and Janzen, 1979).

II. INDUCIBLE RESISTANCE

Insect-inducible changes in food quality may be physical, as in the timing of budbreak (Heichel and Turner, 1976) and early leaf abscission (Faeth *et al.*, 1981; but see Kahn and Cornell, 1983), or physicochemical, as in the occurrence of resinosis (Karban, 1983; McClure, 1983), an increase in fiber (Benz, 1974) or phenolics (Niemelä *et al.*, 1979; Schultz and Baldwin, 1982), and/or a decrease in nitrogen (Benz, 1974; McClure, 1980; Tuomi *et al.*, 1984).

Inducible responses in host plants depend on pest density; they can create negative feedbacks and, consequently, have the potential to regulate herbivore density. Their effects on herbivore population dynamics depend on how rapidly they are triggered, how strongly they affect the herbivore, and how long it takes for them to subside. These characteristics together determine whether induced responses of plants tend to stabilize or destabilize herbivore densities.

The impact of inducible defenses falls between two extremes. If the triggering takes place rapidly and the decay time of the response is shorter than the generation time of the herbivore, the effects are experienced by the herbivore generation doing the damage. We classify such a response as a rapid (short-term) inducible resistance (RI). The RI may be triggered in minutes (Carroll and Hoffmann, 1980) or hours (Green and Ryan, 1972; Wratten *et al.*, 1984). When the time before the decay of inducible resistance is longer than the generation time of the herbivore, the negative effects on herbivore performance last well after the pest generation that triggered the response. This type of response is the long-term inducible resistance (LI), and it may take several years to relax (Benz, 1974; Haukioja, 1982). Because our classification of inducible resistance is based on a time scale of herbivore generations, the same response of a plant may be experienced as either a RI or an LI by different insects.

The effects of the RI on an insect population depend, in part, on the ability of the insects to search for uninduced tissues. As population density increases, this becomes more difficult and herbivore performance (survival, fecundity) may be impaired. The RI may cause negative feedback on insect population growth and hence tends to stabilize herbivore populations.

The LI may enhance the decline of an insect population in a delayed density-dependent manner. Because its relaxation time may be years, it may affect insects even after an outbreak subsides, that is, at low densities. Thus, the LI produces a negative feedback and a time-lag component into the population dynamics of the pest. It tends to destabilize the density of herbivore populations. The LI offers a simple mechanism both for the decline in density (even before apparent food shortage) and for the long intervals between successive peaks in density.

III. STUDYING THE INFLUENCE OF PLANT QUALITY ON INSECT POPULATIONS: MEASURING THE EFFECTS OF FOOD ON HERBIVORES

Unambiguous experimental results from several plant–insect systems should precede the formulation of generalizations on the importance of plants in herbivore population dynamics. Unfortunately, comparable data are rare owing to the many differences in the methods of studying the effect of food quality on field populations.

A. Making Measurements

The effects of food quality on growth rate, consumption rate, energetic parameters, length of the larval period, survival, final larval (or pupal or adult) weight, and fecundity can be demonstrated by simple bioassays.

The easiest indices of insect performance to obtain—consumption and growth rates achieved by insects on different diets—are the least useful. They may or may not be correlated with each other, for example, final weight and fecundity (Haukioja and Hanhimäki, 1985). Another index that is easy to measure, duration of the feeding stage, may also be of little relevance. In insects and mites with overlapping generations, delayed maturation considerably decreases the growth rate of populations (Dixon and Barlow, 1979; see also Karban and Carey, 1984). In univoltine species a prolonged larval period as such may not necessarily be detrimental. However, if a longer feeding period renders larvae susceptible to parasites and predators or a slow growth delays development of the animal in relation to maturation of the foliage, slow growth may increase mortality.

Survival and fecundity are directly related to the growth rate of a population. Female reproductive potential in nonfeeding adults is often linearly related to size (e.g., Haukioja and Neuvonen, 1985b). Therefore, pupal and adult weights can be converted to fecundities in many cases. The dependence of male reproductive potential on size in folivorous insects is not well understood. In the geometrid *Epirrita autumnata* (Borkhausen) the lowest weight classes of males (in a range characteristic of insects on poor natural diets) have a reduced fertilization ability (Haukioja and Neuvonen, 1985b).

Detrimental effects on insect performance (e.g., slow growth, long larval periods, low pupal weights, lower survival) often are correlated. An increased mortality on one diet biases estimates of the effects on other parameters, like the pupal weight. These can be measured only for surviving individuals. Weight gain and other indices of performance in dead individuals presumably would have been worse.

For comparison involving different diets and different studies we recommend computing responses on a per generation basis by multiplying the effects on survival and fecundity (Young and Wrensch, 1981; Haukioja *et al.,* 1985; see also Tables 2–4). Fecundity is best measured directly but can be estimated from the weights of animals. Extrapolating even these indices to field populations may be risky because, in a natural situation, other density-dependent mortality factors may compensate for those caused by variable food quality.

B. Quality of Experimental Animals

The outcome of a feeding experiment depends not only on the quality of the food but also on the nature of the organism (e.g., its detoxification capacity) (Batzli and Cole, 1979). Different species of pests or even individuals of the same species may respond differently to certain diets. This places extra demands on the planning of experiments and interpretation of results. Consequently, in experiments on insect outbreaks, the population from which the insect originates (outbreak or decline and endemic populations) may be profoundly important. This possibility has received practically no attention, although some studies (Moran, 1981; Service and Lenski, 1982; Blau and Feeny, 1983; Service, 1984; Haukioja *et al.,* 1985) indicate interactions between food quality and herbivore quality. At least the following possibilities should be considered:

1. Poor food quality may reduce the quality of offspring (i.e., it may have intergenerational cumulative effects): Poor nutritional status in one generation may deter possibilities for normal development in the next. Studies by Kovasevic (1956), Wellington (1965), and Morris (1967) indicate such presumably nongenetic effects in the gypsy moth (*Lymantria dispar*), the western caterpillar (*Malacosoma pluviale*), and the fall webworm (*Hyphantria cunea*).

2. Poor food quality may select for insect genotypes that are able to exploit poor-quality diets. It is difficult to find relevant and significant results that unambiguously show this alternative to be true. However, in the larch budmoth, *Zeiraphera diniana,* Day and Baltensweiler (1972) found a change in the proportion of the two color morphs, which were dominant at different phases of a gradation when reared on different diets. Jeker (1981) studied the fecundity of *Agelastica alni* on alders in relation to defoliation in the previous year. When he used animals from an area with no or low levels of damage, their fecundity (obtained by multiplying the mean number of clutches per female by the clutch size; Tables 19 and 20 in Jeker, 1981) was about 35% lower on defoliated than on control trees. In insects from an outbreak area there was no difference between control and moderately or heavily defoliated trees. The statistical significance of the interaction between insect origin and food quality was not reported.

We reared the autumnal moth, *Epirrita autumnata,* for two generations in netting enclosures on branches of control birches and of birches defoliated in the previous year. Larvae from parents reared on foliage in trees defoliated in the previous year, with accompanying higher mortality, were less sensitive to inducible responses than were larvae coming from parents reared on control trees (Table 1).

For practical reasons we had to use on each tree an identical mixture of broods (one mixture from parents on control trees and another mixture from defoliated trees). Therefore, it is impossible to strictly test the effect of parental diet. However, the result is consistent with a hypothesis that poor food may select for individuals with better ability to process poor-quality diets.

If the above trends are typical, effects of food quality on field populations are underestimated by bioassays using animals from outbreak popu-

TABLE 1

Influence of Rearing Experience on Susceptibility to Induced Defenses as Reflected in the Proportion of Larvae Able to Pupate

	Parents reared on	
Offspring reared on	Control birches, % (*n*)	Defoliated birches, % (*n*)
Control birches	31 (180)	35 (180)
Defoliated birches	6 (100)	26 (100)

lations. They may or may not be overestimated when insects from endemic populations are used.

The density of the insects in an experiment may also modify the apparent suitability of different diets. Solitary larvae of *Epirrita autumnata* on good diets grew better than their crowded siblings, but when the larvae were on poor diets the difference was negligible or reversed (E. Haukioja, unpublished, 1980). Consequently, growth retardation on poor-quality foliage was smaller with crowded than with solitary larvae.

C. Demonstrating Inducible Defenses in the Field

The results of field experiments and observations of various kinds have been used as evidence for or against the proposition that foliage damage induces higher resistance (whether short term, long term, or both) in host plants, which in turn plays a role in the population dynamics of pests. We distinguish three kinds of approaches:

1. Chemical properties of plants or the success of herbivores are compared on naturally attacked and unattacked hosts (e.g., Thielges, 1968; Myers and Williams, 1984; Williams and Myers, 1984). Such studies may be impossible to interpret because plants are not randomly assigned to treatments, but instead herbivores are allowed to select the plants on the basis of their feeding preferences.

2. Plant quality is monitored for several years, with different phases of fluctuation in a local pest population (Benz, 1974), or different out-of-phase host stands, that is, attacked and unattacked ones, are compared (e.g., Jeker, 1981; Schultz and Baldwin, 1982). Also in this case a local outbreak may be caused by a difference in food quality, and so measured differences in quality may cause rather than result from the different levels of herbivory.

When hypotheses about the existence of inducible resistance are considered, studies of types 1 and 2 are especially prone to type II errors (acceptance of a false null hypothesis), since the attacked plants or stands may be originally of higher quality (by an amount X). Consequently, instead of the null hypothesis H_O: $-I = 0$ (in which $-I$ is the reduction in plant quality caused by inducible resistance), the H_O: $X - I = 0$, is tested.

3. Actual experiments in which plants are randomly assigned to control and experimental groups (e.g., Haukioja and Niemelä, 1977; Wallner and Walton, 1979; Werner, 1979; Karban and Carey, 1984; Haukioja *et al.*, 1985; Raupp and Denno, 1984) are easiest to interpret. These experiments are controlled against biases potentially confusing the previous types of studies.

Other potential sources of error are worth controlling:

1. Plants must not be affected by pretreatment inducible effects. In practice this is difficult to control except by choosing a study area without heavy previous or current-year herbivory.
2. The relative merits of using either insect-made foliage damage or artificial damage must be carefully considered. Insect-made damage may trigger stronger inducible responses than artificial damage (Jeker, 1981; Haukioja and Neuvonen, 1985a), but it is difficult to exclude insect-spread diseases as causal factors for poor performance of animals on induced diets (Fowler and Lawton, 1985).
3. The insect instar(s) used, the duration of the feeding period, and the duration of the experiment depend on the goal of the experiment. For practical reasons it is often preferable to use large (e.g., last-instar) larvae. If the effect of plant quality on insect population dynamics is the main focus, the outcome of the experiment may be too mild. Youngest larvae are usually regarded as most sensitive (White, 1974). Consequently, effects on insect performance are probably monitored most relevantly over the whole feeding stage of the insect.

Many studies on inducible resistance suffer from pseudoreplication (Fowler and Lawton, 1985). This means that, when there are multiple observations from each experimental unit (e.g., several insects from a tree or a site), the treatment effects are tested against an inappropriate error term (Hurlbert, 1984). Field tests of "communication" in red alder offer an example. Rhoades (1983) claimed to have shown the existence of "communication" among trees, partially because larvae grew better on a far-control site (far from damaged trees) than on a near-control site. In contrast, Myers and Williams (1984) and Williams and Myers (1984) did not find differences between the performance of larvae on far- versus near-control sites and thus claimed to have falsified the existence of "communication." In all these studies there was only one true replicate (site) for each treatment. Accordingly, statistical inferences about treatment effects (near versus far) were unwarranted.

The possible existence of "communication" creates serious difficulties in tests of induced responses. If a treatment (damage) to one plant also affects adjacent plants, and treated and control plants are close to one another, the treatment effect is systematically underestimated. If an induced response is found when damaged and undamaged plants (or parts of the same plant) are compared by bioassays, "communication" cannot be extremely effective. More important, its effects must be studied if something other than demonstrating the existence of an induced response is the objective of the experiment.

The easiest and, at the same time, a statistically sound way to test the occurrence of "communication" in the field is to damage the foliage of several widely separated plants in an area without high herbivory and to measure the performance of insects on untouched plants in relation to the distance to the closest damaged one. Plant-specific mean values should be used as independent observations. Haukioja *et al.* (1985) found by using this approach that the performance of *E. autumnata* larvae reared on control trees correlated positively with the distance to the closest trees defoliated during the previous year.

In summary, testing in a valid way the hypothesis that previous foliage damage induces higher resistance (RI or LI or both) requires (1) choosing an area without high densities of folivores before and during the experimental period, (2) choosing plants beyond a minimum distance from one another (the supposed "communication" distance), (3) assigning plants randomly to control and experimental treatments, (4) damaging the foliage of experimental plants either mechanically or by using larvae, and (5) comparing the success of larvae when reared through the whole feeding stage on both types of plants.

Demonstrating the existence and, especially, quantifying the effects of RI require some additional qualifications. A good test should minimize the probability of induction in control foliage, effectively simulate the efficacy (e.g., duration) of insect damage, and allow statistically sound analysis of the data. To our knowledge no one has executed a field experiment on RI satisfying all these criteria. The following are some ways in which RI has been or should be studied (Neuvonen and Haukioja, 1985):

1. Every insect gets all its food from one plant, either from a previously damaged one or from an undamaged control. In this case individual plants are true replicates. The data can be analyzed either by a nested ANOVA or by plant-specific means. The experiment should last only a short period to avoid induction of the control plants. This is the most serious drawback of the method. It presumably does not yield realistic estimates of the effects on parameters relevant to population dynamics.
2. Larvae can be reared through the whole larval period by feeding control and experimental larvae from the same individual plants but rotating leaf picking so that control animals always get the leaves before the damage and experimental animals afterward (Haukioja and Niemelä, 1977). Unfortunately, this design does not simulate the effects of repeated insect feeding.
3. It is possible to choose a reasonable number of plants (e.g., 20–30) for each treatment and to feed each insect with randomized leaf samples from experimental and control plants (Haukioja, 1982; see also Williams

and Myers, 1984). This method allows rearing of insects through their whole feeding period while minimizing the damage (and the possibility of inducing resistance) on control trees. Although the method may control for most of the intertree variation, the true degrees of freedom probably cannot be estimated and thus statistical tests on the basis of a single experiment may be unwarranted. It is possible to test the significance of the results by repeating the experiment with different sets of plants.

4. The density of insects in individual trees can be artificially manipulated (cf. Eisenberg, 1966). For example, insects should be reared through their whole larval period on living trees at two densities. In the low-density treatment they should cause "insignificant" damage. Although rapidly induced defenses may be triggered by feeding, larvae can avoid such effects by moving to intact leaves. In the high-density treatment the damage by larvae should be adjusted to a level not causing quantitative food shortage but forcing the insects to feed on damaged and/or adjacent leaves. Difficulties in this design arise from the need to control for other density-dependent factors (e.g., "group effects"; see Iwao, 1968; parasites, predators, "communication") and to have reasonable sample sizes. Individual plants are true replicates.

The existence and efficacy of long-term inducible responses are easier to demonstrate. In the year(s) following damage to the foliage of experimental plants, larvae are reared both on them and control plants (e.g., Haukioja *et al.*, 1985). In this case RI, if expressed, is experienced by animals both on control and on previously defoliated plants. A difference in the performance between experimental and control insects gives an estimate of the effect of LI. Means of individual plants are true replicates. It is not wise to use the same plants as controls in successive years since even limited insect damage may induce defensive responses in the undamaged parts of the same plant (Haukioja and Neuvonen, 1985a; unpublished).

Another research strategy is to manipulate confined subpopulations of insects by causing them to be out of phase with the normal (surrounding) density fluctuation. Myers (1981) kept an isolated population of tent caterpillars (on Imrie Island) at a low level during an ongoing outbreak. She found that this population did not build up when other populations crashed. This suggested that a lack of recent defoliation was not sufficient to cause a population increase in the tent caterpillar. Auer *et al.* (1981) distributed a large number of larch budmoth pupae into a postpeak population. They found that a new increase did not follow. These two experiments exemplify potentially very useful approaches that are directly relevant to the population dynamics of the insect. Such designs should be used more frequently, although they necessarily bear more uncontrolled factors than the simple experiments ordinarily used to demonstrate the existence or absence of inducible responses in plants.

IV. EXPERIMENTAL EVIDENCE OF RAPID INDUCIBLE RESISTANCE

From the viewpoint of insect population dynamics, the following questions are relevant to RI: (1) Is there any RI capable of affecting insect performance in a specific plant–herbivore system? (2) What are the actual and potential quantitative effects of RI on insect performance? (3) How much temporal variation is there in the expression of RI?

Rapid chemical changes following an actual or simulated insect attack are widespread among plants (Walker-Simmons and Ryan, 1977; Niemelä *et al.*, 1979; Carroll and Hoffmann, 1980; Baldwin and Schultz, 1983; Wink, 1983). Their relevance to herbivores can most easily be studied by bioassays in the laboratory. A typical qualitative proof in a bioassay is a low growth rate, which may produce a longer larval period and/or lower pupal weight (e.g., Haukioja and Niemelä, 1977; Haukioja, 1982). Changes in the palatability of foliage are also easily monitored (Edwards and Wratten, 1982; Wratten *et al.*, 1984). Slower growth or reduced palatability as such does not indicate effects on the population dynamics of the herbivore, but may be useful indices of, for example, interspecific (Edwards and Wratten, 1982) and seasonal (Haukioja and Niemelä, 1979; Edwards and Wratten, 1982; Wratten *et al.*, 1984) differences in the expression of RI.

The RI also occurs at low insect densities (Edwards and Wratten, 1983, 1985; Haukioja and Hanhimäki, 1985). Its effects may therefore be widespread and, paradoxically, difficult to show. In addition to factors treated in Section III, methodological problems may arise from differences in the responses of detached and growing leaves (e.g., Haukioja, 1980, Fig. 4; Haukioja *et al.*, 1983, Fig. 1). Consequently, the relevance of laboratory studies to a natural situation is difficult to evaluate. In addition, the effects of RI on mortality are easily underestimated in laboratory studies, since the environment is benign for larvae. However, because of practical difficulties we must generally be satisfied with laboratory estimates of RI.

Table 2 reviews quantitative estimates of the efficacy of RI on selected parameters of insect population dynamics. We include only cases in which fecundity was reported or it was possible to estimate fecundity from females' weights. The tabulation indicates that RI may in some cases have considerable effects on the potential rate of increase of insect popu-

TABLE 2

Effects of Rapid Inducible Resistance on the Performance[a] of Some Herbivorous Insects[b]

Plant	Herbivore	A	B	C	D	Ref.[c]
Betula tortuosa	*Epirrita autumnata*	b, s	d	0.91	**/ns	1
	E. autumnata	b, s	d	0.78	?	2
	E. autumnata	b, s	d	1.02	ns	3
	E. autumnata	b	d	0.83–0.89	*	4
B. populifolia	*Lymantria dispar*	b, s	g	0.59–0.83	?	5
Quercus velutina	*L. dispar*	b, s	g	0.68–0.75	?	5
Salix alba	*Plagiodera versicolora*	a	d	0.82	**	6
S. babylonica	*P. versicolora*	a, s	d	0.67	**	6
Rumex obtusifolius	*Gastrophysa viridula*	a, s	g	0.16	**	7

[a] Performance relative to insects on control plants (column C).

[b] Column A: Effects on fecundity (a, actually measured; b, estimated from female pupal weights) and larval survival (s). Column B: Reared on (d) detached leaves (g, reared on living plants). Column C: Performance (see column A) on induced foliage as a proportion of that on control foliage. Column D: Statistical inferences (?, tests suffer from sacrificial pseudoreplication; *, **, significant differences with $p < .05$ or $p < .01$, respectively, in fecundity; ns, differences not significant; **/ns, variable results).

[c] References: 1. Haukioja (1977); Haukioja and Niemelä (1977) (experiment was begun with half-grown larvae). 2. Haukioja and Niemelä (1979). 3. Niemelä *et al.* (1979). 4. Haukioja and Hanhimäki (1985). 5. Wallner and Walton (1979); Valentine *et al.* (1983). 6. Raupp and Denno (1984). 7. Jeker (1981).

lations. The experimental design may not even fully reveal the efficacy of RI, for example, because "communication" was not controlled in any of the studies. The results with the gypsy moth (Wallner and Walton, 1979; Valentine *et al.*, 1983) may underestimate the efficacy of RI because the experiment was begun with the third-instar larvae. Among cases excluded from the table because of insufficient information for estimating per generation effects, however, there are cases in which no effects were obvious (Haukioja and Niemelä, 1979; Myers and Williams, 1984; Niemelä *et al.*, 1984). Furthermore, these and most other studies designed to ascertain whether RI exists suffer from pseudoreplication (Fowler and Lawton, 1985).

Temporal variation in RI has received little attention. However, it is crucial because a temporal weakening of RI may be a prerequisite for an increase in population density. Studies performed with mature foliage (Haukioja and Niemelä, 1979; Edwards and Wratten, 1982) have not demonstrated the existence of RI. Annual variation in the efficacy of RI has been demonstrated in the birch–autumnal moth system (Haukioja and Hanhimaki, 1985).

V. EXPERIMENTAL EVIDENCE OF THE EFFECTS OF LONG-TERM INDUCIBLE RESISTANCE

A. Efficacy of Long-Term Inducible Resistance

Long-term inducible resistance is especially important for our topic because it offers a potential mechanism for creating herbivore cycles. Its effects have been demonstrated in several defoliating insects. Table 3 demonstrates examples selected by the same criteria as in Table 2.

In the Alps, LI in the larch (*Larix decidua*) caused a strong (62%) reduction in the performance of the larch budmoth. Benz (1974) estimated that the survival of larvae decreases linearly with an increase in the fiber

TABLE 3

Effects of Long-Term Inducible Resistance on the Performance[a] of Some Herbivorous Insects[b]

Plant	Herbivore	A	B	C	D	E	Ref.[c]
Larix decidua	*Zeiraphera diniana*	b, s	d	0.38	*	N (100)	1
Alnus incana	*Melasoma aenea*	a, s	d	0.80	?	N (40–100)	2
	Agelastica alni	a, s	d, g	0.41	?	N (40–79)	2
	A. alni	a, s	d, g	0.53	?	N (80–100)	2
Betula pubescens ssp. *tortuosa*	*Epirrita autumnata*	b, s	g	0.27	**	A (845), F	3
	E. autumnata	a, s	g	0.34	*	A (45 d to 100)	4
B. populifolia	*Lymantria dispar*	b, s	g	0.75	?	A, r (100+)	5
Quercus velutina	*L. dispar*	b, s	g	0.63	?	A, d (100+)	5
Betula resinifera	*Rheumaptera hastata*	a, s	d	0.63	?	A, r (100)	6
Populus tremuloides	*Choristoneura conflictana*	a, s	d	0.71	?	A, r (100)	6

[a] Performance relative to insects on control plants (column C).

[b] Columns A–D as in Table 2. Column E: Type of foliage damage (N, natural defoliation; A, artificial defoliation; d, leaves damaged; r, leaves removed; F, *Epirrita* frass to soil) and intensity (%; + indicates ongoing additional defoliation). In all cases the defoliation took place during the previous year.

[c] References: 1, Benz (1974). 2, Jeker (1981); see also Jeker (1983). 3, Haukioja *et al.* (1985). 4, Haukioja (1980); Neuvonen and Haukioja (1984). 5, Valentine *et al.* (1983). 6, Werner (1979, 1981).

and a decrease in the protein content of larch foliage. Both of these changes in foliage quality resulted from defoliation and consequently could be regarded as delayed density-dependent alterations. Whether just these characteristics were causally connected with the poor success of larvae is not known.

Jeker (1981, 1983) compared the performance of two chrysomelid beetles on defoliated versus nondefoliated alders. The performance of *Melasoma aenea* was reduced by 20% in damaged trees. In *Agelastica alni* (animals from a low-density population; see Section III,B) the reduction in performance was 47–59%.

Wallner and Walton (1979) tested the efficacy of RI (current-year defoliation) + LI (previous-year defoliation) on gypsy moth feeding on black oak and gray birch. The performance was 32–36% (oak) to 18–41% (birch) poorer on defoliated trees than on controls. Note that they had reared gypsy moth larvae on the control trees also in the previous year, which may have reduced their quality relative to unattacked trees. Actually, in the third study year the larvae managed better on defoliated gray birches than on their control trees, on which insects had been reared in the two previous years (Valentine *et al.*, 1983).

Simulated natural defoliation (damage + frass or fertilizer subsidy to soil beneath the trees) reduced the performance of *Epirrita autumnata* (from egg to egg) by 70% or more, when compared with larvae on control birches (Haukioja *et al.*, 1985). The reduction in *Epirrita* performance may be even greater because the control trees were close enough to treated trees to be affected by "communication." Furthermore, neither male-biased sex ratios in the strongest treatments nor possible reduced fecundity in males (Haukioja and Neuvonen, 1985b) were corrected for. Control trees also exhibit RI (Haukioja and Niemelä, 1977; Haukioja and Hanhimäki, 1985), which leads to a greater underestimation of the combined effect of induced defenses (LI and RI).

To demonstrate the total effect of inducible resistance (LI + RI) on *Epirrita* performance requires unbiased estimates of the performance of animals on control trees, that is, on trees in which only constitutive resistance functions. Obtaining such a figure in a strict sense may be impossible at present. However, we can estimate the performance of larvae reared, by a method similar to that in the above mentioned experiments, on birches that are most suitable for the defoliator. We can compare this figure with the performance of *Epirrita* in defoliation experiments by Haukioja *et al.* (1985). This yields an estimate of the magnitude of the effect of phenotypic plasticity in birch foliage on *Epirrita* performance. Figure 2 in Haukioja *et al.* (1983) demonstrated that survival and pupal weights of *Epirrita* depended on the distance to the native site of the birch provenance. Reproductive performance calculated per first-instar female larva (same method as in Haukioja *et al.*, 1985) was 82 eggs for larvae reared on mountain birches growing 1000 km from their native site. Larvae reared on trees expressing LI (Haukioja *et al.*, 1985) had 93% lower performance. This figure does not necessarily show the effect of induced resistance alone because trees growing far from their native sites may have been stressed (although they had been grown from seedlings at the study sites) and, consequently, may have had low constitutive resistance. However, it demonstrates how great a reduction in *Epirrita* performance can be caused by phenotypic differences in the foliage quality within the same birch provenance.

Table 4 and Fig. 1 include results from our work with *Epirrita* and the mountain birch. They reveal a great deal of variation even in one system. There may be yearly variation in the efficiency of LI, although the major differences are caused by different experimental techniques (use of enclosures in trees versus laboratory rearing, degree of defoliation and the manner of defoliation—tearing versus removing leaves—and fertilization treatment).

B. Threshold Values for Damage and Relaxation Times of Long-Term Inducible Resistance

The effects of LI vary among different host plants as a consequence of amount and duration of foliage damage (Fig. 1). In the mountain birch one

TABLE 4

Fecundity of *Epirrita autumnata* Moths Reared on Foliage from Mountain Birches Defoliated 1–4 Year(s) Earlier as a Proportion of Fecundity in Insects on Control Foliage[a]

Type of damage	Time since defoliation				Ref.
	1 year	2 years	3 years	4 years	
A, pr (100)	0.86	0.78	—	—	Haukioja (1977), Haukioja and Niemelä (1977)
A, d (50)	—	0.62	0.84	0.79	Haukioja (1982)

[a] All experiments with detached leaves. Type of damage as in column E in Table 3, but pr denotes removal of leaves from single twigs; i.e., the trees were only partially defoliated. Tests suffer from sacrificial pseudoreplication.

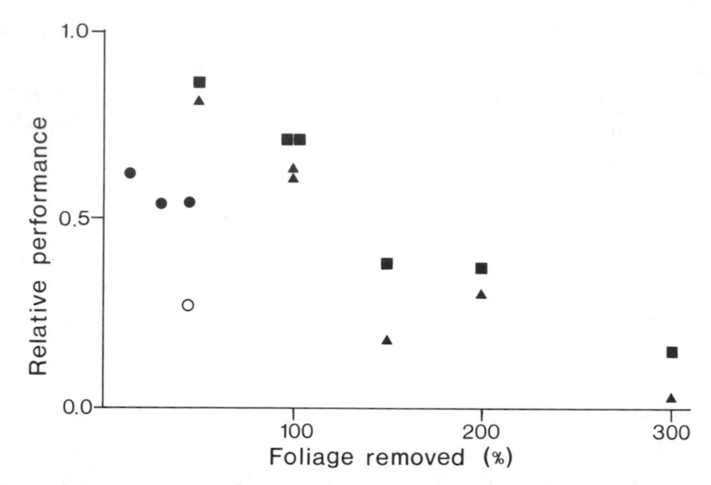

Fig. 1. Performance of some moths (success on defoliated/success on control trees) in relation to cumulative artificial foliage losses in the previous year(s). ■, *Choristoneura conflictana* on *Populus tremuloides* (Werner, 1981); ▲, *Rheumaptera hastata* on *Betula resinifera* (Werner, 1981); ●, *Epirrita autumnata* on *Betula pubescens* spp. *tortuosa* (E. Haukioja, J. Suomela, and S. Neuvonen, unpublished); ○, defoliation + *Epirrita* frass in the soil (Haukioja *et al.*, 1985).

partial defoliation increased resistance (Table 4). Around 15–45% leaf removal caused indistinguishable responses (Fig. 1). In this species the insect outbreaks generally do not last more than a couple of years (Tenow, 1972).

In another species with short outbreaks, the gypsy moth, resistance of two hosts (black oak and gray birch) was also induced rapidly (Wallner and Walton, 1979). A third defoliation increased resistance in black oak only slightly, whereas gray birches died after two successive defoliations (Valentine *et al.*, 1983).

In some defoliators (e.g., the spruce budworm) the outbreaks may last several (>5) years (Morris, 1963). In such cases it is obvious that there is no LI, the insects adapt to it, or it reaches its full potential very slowly. The same is obvious in sawflies defoliating the Scots pine (Niemelä *et al.*, 1984).

The relaxation time and rate of LI may contribute to minimum intervals between successive peaks in insect density. Environmental factors like nutrient availability may modify the relaxation rate of LI (Tuomi *et al.*, 1984). Unfortunately, there are very few estimates on the relaxation of LI.

In the larch–larch budmoth system (Benz, 1974), the quality of larch foliage may have been poor 3–4 years after defoliation. Studies with the mountain birch (Table 4) performed with excised leaves showed decreased performance of *Epirrita* larvae on trees defoliated 1–4 years earlier (Haukioja, 1982). However, more estimates of the relaxation time in different host species, and at different sites, are needed before any generalizations about its effects on the cyclic behavior of defoliator populations can be made.

VI. CONCLUSIONS

The first verbal models in the middle and late 1970s hypothesized that induced plant defenses play a role in insect cycles. Those models, based on inference and vague information, suggested that the mechanism sought after for so long might exist. Their merit lay in their explanations of the most problematic phenomena in cycles: the decrease in density and the long latent phases after a decline of a peak. Furthermore, depending on the triggering and decay times of induced defenses, these models indicated that induced resistance may either stabilize or destabilize herbivore populations. At present we are still far from understanding the role of inducible plant resistance in insect cycles. However, we now know that there are both plant–herbivore systems in which induced defenses play a role and systems in which they have not been shown to occur.

Insect cycles do not occur in a plant–herbivore vacuum. Consequently, all factors affecting density, by definition, must be taken into account in explaining density changes (Haukioja *et al.*, 1983). Indirect effects of food quality are difficult to assess but may still be important. Poor nutrition has been claimed to render larvae more vulnerable to diseases or parasitism (Rudnew, 1963; Price *et al.*, 1980; Wallner, 1983; Schultz, 1983). Increased plant resistance, however, may also have negative effects on the natural enemies of herbivores (Price *et al.*, 1980; Barbosa *et al.*, 1982). There is an urgent need for experimental study of the effects of foliage quality on the incidence of diseases and the success of parasitoids and predators of herbivores.

Simulation models have not yet been used extensively to study the effects of plant quality on herbivore cycles. Fischlin and Baltensweiler (1979) demonstrated that models taking into account only changes in foliage quality caused by defoliation can realistically simulate the cyclic behavior of a moth population. Such simulations based on correlations and approaches with other correlations in the same systems have yielded different but also positive results (van den Bos and Rabbinge, 1976; Anderson and May, 1980).

Analytical models of population dynamics are traditionally based on numbers of individuals in the interacting populations. This is not a fruitful approach to describing insect herbivores exploiting plants that are both much larger and much more long lived than they and may vary greatly in quality (Caughley and Lawton, 1981).

Most experimental confirmation so far concerns phenomena that are relevant at the end of an outbreak. In some systems the role of induced plant resistance is strong enough to contribute to density fluctuations in herbivores. With regard to the initiation of outbreaks, there have been very few experimental tests demonstrating that loss of plant vigor or RI or the relaxation of previous inducible resistance is crucial. There is a great deal of circumstantial evidence that this is a fruitful direction for future research. The knowledge of interactions between variable insect and plant quality at different phases of the cycles is very scanty. So far all demonstrations of inducible defenses with long relaxation times have concerned deciduous plants.

ACKNOWLEDGMENTS

T. B. Jeker, H. T. Valentine, W. E. Wallner, and R. A. Werner kindly supplied detailed information about their research. P. Barbosa, J. Schultz, and S. F. MacLean helped with the manuscript, and Terttu Laurikainen drew the figure. We thank all of them.

REFERENCES

Anderson, R. M., and May, R. M. (1980). Infectious diseases and population cycles of forest insects. *Science* **210,** 658–661.

Auer, C., Roques, A., Goussard, F., and Charles, P. J. (1981). Effects de l'accroissement provoque du niveau de population de la tordeuse du mélèze Zeiraphera diniana Guenee (Lep., Tortricidae) au cours de la phase de regression dans un massif forestier du Brianconnais. *Z. Angew. Entomol.* **92,** 286–302.

Baldwin, I. T., and Schultz, J. C. (1983). Rapid changes in tree leaf chemistry induced by damage: Evidence for communication between plants. *Science* **221,** 277–279.

Baltensweiler, W., Benz, G., Bovey, P., and Delucchi, V. (1977). Dynamics of larch bud moth populations. *Annu. Rev. Entomol.* **22,** 79–100.

Barbosa, P., Saunders, J. A., and Waldvogel, M. (1982). Plant-mediated variation in herbivore suitability and parasitoid fitness. *Proc. Int. Symp. Insect-Plant Relationships, 5th,* pp. 63–71.

Batzli, G. O., and Cole, F. C. (1979). Nutritional ecology of microtine rodents: Disgestibility of forage. *J. Mammal.* **60,** 740–750.

Benz, G. (1974). Negative Rückkoppelung durch Raum- und Nahrungskonkurrenz sowie zyklische Veränderungen der Nahrungsgrandlage als Regelprinzip in der Populationsdynamik des Grauen Lärchenwicklers, *Zeiraphera diniana* (Guenee) (Lep., Tortricidae). *Z. Angew. Entomol.* **76,** 196–228.

Blau, W. S., and Feeny, P. (1983). Divergence in larval responses to food plants between temperate and tropical populations of the black swallowtail butterfly. *Ecol. Entomol.* **8,** 249–257.

Carroll, C. R., and Hoffmann, C. A. (1980). Chemical feeding deterrent mobilized in response to insect herbivory and counteradaptation by *Epilachna tredecimnotata. Science* **209,** 414–416.

Caughley, G., and Lawton, J. H. (1981). Plant–herbivore systems. *In* "Theoretical Ecology" (R. M. May, ed.), pp. 132–166. Blackwell, Oxford.

Day, K. R., and Baltensweiler, W. (1972). Change in proportion of larval colour types of the larchform *Zeiraphera diniana* when reared on two media. *Entomol. Exp. Appl.* **15,** 287–298.

Dempster, J. P. (1983). The natural control of populations of butterflies and moths. *Biol. Rev. Cambridge Philos. Soc.* **58,** 461–481.

Dixon, A. F. G., and Barlow, N. D. (1979). Population regulation in the lime aphid. *Zool. J. Linn. Soc.* **67,** 225–237.

Edwards, P. J., and Wratten, S. D. (1982). Wound-induced changes in palatability in birch (*Betula pubescens* Ehr. ssp. *pubescens*). *Am. Nat.* **120,** 816–818.

Edwards, P. J., and Wratten, S. D. (1983). Wound-induced defenses in plants and their consequences for patterns of insect grazing. *Oecologia* **59,** 88–93.

Edwards, P. J., and Wratten, S. D. (1985). Induced plant defenses against grazing: Fact or artefact? *Oikos* **44,** 70–74.

Eisenberg, R. M. (1966). The regulation of density in a natural population of the pond snail, *Lymnaea elodes. Ecology* **47,** 889–906.

Faeth, S. H., Conner, E. F., and Simberloff, D. (1981). Early leaf abscission: A neglected source of mortality for folivores. *Am. Nat.* **117,** 409–415.

Feeny, P. P. (1970). Seasonal changes in oak leaf tannins and nutrients as a cause of spring feeding by winter moth caterpillars. *Ecology* **51,** 565–581.

Fischlin, A., and Baltensweiler, W. (1979). System analysis of the larch budmoth system. Part I. The larch–larch budmoth relationship. *Mitt. Schweiz. Entomol. Ges.* **52,** 273–289.

Fowler, S. V., and Lawton, J. H. (1985). Rapidly induced defenses and talking trees: The devil's advocate position. *Am. Nat.* **126,** 181–195.

Green, T. R., and Ryan, C. A. (1972). Wound-induced proteinase inhibitor in plant leaves: A possible defense mechanism against insects. *Science* **175,** 776–777.

Grimalsky, V. I. (1966). The role of oleoresin exudation in the pine resistance against pests. *Zool. Zh.* **45,** 1656–1664 (in Russian, with English summary).

Grimalsky, V. I. (1974). Resistance of tree stands against needle- and leaf-eating pests with respect to the trophic theory of dynamics of insect numbers. *Zool. Zh.* **53,** 189–198 (in Russian, with English summary).

Haukioja, E. (1977). The mechanism of *Oporinia autumnata* cycles. *Proc. Circumpolar Conf. North. Ecol.* Vol. 1, pp. 235–242.

Haukioja, E. (1980). On the role of plant defenses in the fluctuation of herbivore populations. *Oikos* **35,** 202–213.

Haukioja, E. (1982). Inducible defenses of white birch to a geometrid defoliator, *Epirrita autumnata. Proc. Int. Symp. Insect-Plant Relationships, 5th,* pp. 199–203.

Haukioja, E., and Hakala, T. (1975). Herbivore cycles and periodic outbreaks. Formulation of a general hypothesis. *Rep. Kevo Subarct. Res. Stn.* **12,** 1–9.

Haukioja, E., and Hanhimäki, S. (1985). Rapid wound-induced resistance in white birch (*Betula pubescens*) foliage to the geometrid *Epirrita autumnata:* A comparison of trees and moths within and outside the outbreak range of the moth. *Oecologia* **65,** 223–228.

Haukioja, E., and Neuvonen, S. (1985a). Induced long-term resistance of birch foliage against defoliators: Defensive or incidental? *Ecology* **66,** 1303–1308.

Haukioja, E., and Neuvonen, S. (1985b). The relationship between size and reproductive potential in male and female *Epirrita autumnata* (Lep., Geometridae). *Ecol. Entomol.* **10,** 267–270.

Haukioja, E., and Niemela, P. (1977). Retarded growth of a geometrid larva after mechanical damage to leaves of its host tree. *Ann. Zool. Fenn.* **14,** 48–52.

Haukioja, E., and Niemela, P. (1979). Birch leaves as a resource for herbivores: Seasonal occurrence of increased resistance in foliage after mechanical damage of adjacent leaves. *Oecologia* **39,** 151–159.

Haukioja, E., Kapiainen, K., Niemelä, P., and Tuomi, J. (1983). Plant availability hypothesis and other explanations of herbivore cycles: Complementary or exclusive alternatives? *Oikos* **40,** 419–432.

Haukioja, E., Suomela, J., and Neuvonen, S. (1985). Long-term inducible resistance in birch foliage: Triggering cues and efficacy on a defoliator. *Oecologia* **65,** 363–369.

Heichel, G. H., and Turner, N. C. (1976). Phenology and leaf growth of defoliated hardwood trees. *In* "Perspectives in Forest Entomology" (J. F. Anderson and H. K. Kaya, eds.), pp. 31–40. Academic Press, New York.

Hurlbert, S. H. (1984). Pseudoreplication and the design of ecological field experiments. *Ecol. Monogr.* **54,** 187–211.

Iwao, S. (1968). Some effects of grouping in lepidopterous insects. *Colloq. Int.* C. N. R. S. **173,** 185–211.

Jeker, T. B. (1981). Durch Insektenfrass induzierte, resistenzähnliche Phänomene bei Pflanzen. Wechselwirkungen zwischen Grauerle, *Alnus incana* (L.) und den Erlenblattkäfern *Agelastica alni* L. und *Melasoma aenea* L. sowie zwischen stumpflattrigem Ampfer, *Rumex obtusifolius* L. und Ampferblattkafer, *Gastrophysa viridula* Deg. Dissertation No. 6895. Eidgem. Technische Hochschule, Zürich.

Jeker, T. B. (1983). Einfluss einer Defoliation im Vorjahr und des Blattalterns auf die Larvenentwicklung von *Melasoma aenea* L. (Coleoptera: Chrysomelidae). *Mitt. Schweiz. Entomol. Ges.* **56,** 237–244.

Kahn, D. M., and Cornell, H. V. (1983). Early leaf abscission and folivores: Comments and consideration. *Am. Nat.* **122,** 428–432.

Karban, R. (1983). Induced responses of cherry trees to periodical cicada oviposition. *Oecologia* **59,** 226–231.

Karban, R., and Carey, J. R. (1984). Induced resistance of cotton seedlings to mites. *Science* **225,** 53–54.

Klomp, H. (1966). The dynamics of a field population of the pine looper (*Bupalus piniarius* L.) (Lepidoptera: Geometridae). *Adv. Ecol. Res.* **3,** 207–305.

Kovasevic, Z. (1956). Die nahrungswahl und das auftreten der pflanzenschadlinge. *Anz. Schaedlingskd.* **24,** 97–101.

Lewis, A. C. (1979). Feeding preference for diseased and wilted sunflower in the grasshopper *Melanopus differentialis*. *Entomol. Exp. Appl.* **26,** 202–207.

McClure, M. S. (1980). Foliar nitrogen: A basis for host suitability for elongate hemlock scale, *Fiorinia externa* (Homoptera: Diaspididae). *Ecology* **61,** 72–79.

McClure, M. S. (1983). Population dynamics of a pernicious parasite: Density-dependent vitality of red pine scale. *Ecology* **64,** 710–718.

Moran, N. (1981). Intraspecific variability in herbivore performance and host quality: A field study of *Uroleucon caliqatum* (Homoptera: Aphididae) and its *Solidago* hosts (Asteraceae). *Ecol. Entomol.* **6,** 301–306.

Morris, R. F., (1963). The dynamics of epidemic spruce budworm populations. *Mem. Entomol. Soc. Can.* **31,** 1–332.

Morris, R. F. (1967). Influence of parental food quality on the survival of *Hyphantria cunea*. *Can. Entomol.* **99,** 24–33.

Myers, J. H. (1981). Interactions between western tent caterpillars and wild rose: A test of some general plant herbivore hypotheses. *J. Anim. Ecol.* **50,** 11–25.

Myers, J. H., and Williams, K. S. (1984). Does tent caterpillar attack reduce the food quality of red alder foliage? *Oecologia* **62,** 74–79.

Neuvonen, S., and Haukioja, E. (1984). Low nutritive quality as defense against herbivores: Induced responses in birch. *Oecologia* **63,** 71–74.

Neuvonen, S., and Haukioja, E. (1985). How to study induced plant resistance? *Oecologia* **66,** 456–457.

Niemela, P., Aro, A. M., and Haukioja, E. (1979). Birch leaves as a resource for herbivores. Damaged-induced increase in leaf phenols with trypsin-inhibiting effects. *Rep. Kevo Subarctic Res. Stn.* **15,** 37–40.

Niemela, P., Tuomi, J., Mannila, R., and Ojala, P. (1984). The effect of previous damage on the quality of Scots pine foliage as food for diprionid sawflies. *Z. Angew. Entomol.* **98,** 33–43.

Norris, D. M. (1979). How insects induce disease. *In* "Plant Disease" (J. G. Horsfall and E. B. Cowling, eds.), Vol. 4, pp. 239–255. Academic Press, New York.

Pitman, G. B., Larsson, S., and Tenow, O. (1982). Stem growth efficiency: An index of susceptibility to bark beetle and sawfly attack. *In* "Carbon Uptake and Allocation in Subalpine Ecosystems as a Key to Management" (R. H. Waring, ed.), pp. 52–56. IUFRO Proc., Corvallis, Oregon.

Price, P. W., Bouton, C. E., Grass, P., McPheron, B. A., Thompson, J. N., and Weis, A. E. (1980). Interactions among three trophic levels: Influence of plants on interactions between insect herbivores and natural enemies. *Annu. Rev. Ecol. Syst.* **11,** 41–65.

Raupp, M. J., and Denno, R. F. (1984). The suitability of damaged willow leaves as food for the leaf beetle, *Plagiodera versicolora*. *Ecol. Entomol.* **9,** 443–448.

Rhoades, D. F. (1979). Evolution of plant chemical defenses against herbivores. *In* "Herbivores: Their Interaction with Secondary Plant Metabolites" (G. A. Rosenthal and D. H. Janzen, eds.), pp. 3–54. Academic Press, New York.

Rhoades, D. F. (1983). Responses of alder and willow to attack by tent caterpillars and webworms: Evidence for pheromonal sensitivity of willows. *ACS Symp. Ser.* **208,** 55–68.

Rhoades, D. F. (1985). Offensive–defensive interactions between herbivores and plants: Their relevance in herbivore population dynamics and ecological theory. *Am. Nat.* **125,** 205–238.

Rosenthal, G. A., and Janzen, D. H., eds. (1979). "Herbivores: Their Interaction with Secondary Plant Metabolites." Academic Press, New York.

Rudnew, D. F. (1963). Physiologischer Zustand der Wirtspflanze und Massenvermehrung von Forstschadlingen. *Z. Angew. Entomol.* **53,** 48–68.

Rudnew, D. F., and Smeljanez, W. P. (1978). Pest resistance of plants. *Zh. Obshch. Biolo.* **39,** 414–421 (in Russian, with English summary).

Schultz, J. C. (1983). Impact of variable plant defensive chemistry on susceptibility of insects to natural enemies. *ACS Symp. Ser.* **208,** 37–54.

Schultz, J. C., and Baldwin, I. T. (1982). Oak leaf quality declines in response to defoliation by gypsy moth larvae. *Science* **217,** 149–151.

Schwenke, W. (1963). Über die Beziehungen zwischen dem Wasserhaushalt von Bäumen und der Vermehrung blattfressender Insekten. *Z. Angew. Entomol.* **51,** 371–376.

Schwenke, W. (1968). Neue Hinweise auf eine Abhengigkeit der Vermehrung batt-und nadelfressender Forstinsekten vom Zuckerhalt ihrer Nahrung. *Z. Angew. Entomol.* **61,** 365–369.

Schwenke, W., ed. (1978). "Die Forstschädlinge Europas." Schmetterlinge. Parey, Berlin.

Service, P. (1984). Genotypic interactions in an aphid–host plant relationship: *Uroleucon rudbeckiae* and *Rudbeckia laciniata*. *Oecologia* **61,** 271–276.

Service, P., and Lenski, R. E. (1982). Aphid genotypes, plant phenotypes, and genetic diversity: A demographic analysis of experimental data. *Evolution* (*Lawrence, Kans.*) **36,** 1276–1282.

Tenow, O. (1972). The outbreaks of *Oporinia autumnata* Bkh. and *Operophtera* spp. (Lep., Geometridae) in the Scandinavian mountain chain and northern Finland 1962–1968. *Zool. Bidr. Uppsala, Suppl.* **2,** 1–107.

Thielges, B. A. (1986). Altered polyphenol metabolism in the foliage of *Pinus sylvestris* associated with European pine sawfly attack. *Can. J. Bot.* **46,** 724–725.

Tuomi, J., Niemela, P., Haukioja, E., Siren, S., and Neuvonen, S. (1984). Nutrient stress: An explanation for plant anti-herbivore responses to defoliation. *Oecologia* **61,** 208–210.

Valentine, H. T., Wallner, W. E., and Wargo, P. M. (1983). Nutritional changes in host foliage during and after defoliation, and their relation to the weight of gypsy moth larvae. *Oecologia* **57,** 298–302.

van den Bos, J., and Rabbinge, R. (1976). "Simulation of the Fluctuations of the Grey Larch Budmoth." Cent. Agric. Publ. Doc., Wageningen.

Varley, G. C., Gradwell, G. R., and Hassell, M. P. (1973). "Insect Population Ecology: An Analytical Approach." Blackwell, Oxford.

Walker-Simmons, M., and Ryan, C. A. (1977). Wound-induced accumulation of trypsin inhibitor activities in plant leaves. Survey of several plant genera. *Plant Physiol.* **59,** 437–439.

Wallner, W. E. (1983). Gypsy moth host interactions: A concept of room and board. *USDA For. Serv. Gen. Tech. Rep. NE* **NE-85,** 5–8.

Wallner, W. E., and Walton, G. S. (1979). Host defoliation: A possible determinant of gypsy moth population quality. *Ann. Entomol. Soc. Am.* **72,** 62–67.

Wellington, W. G. (1965). Some maternal influences on progeny quality in the western tent caterpillar, *Malacosoma pluviale* (Dyar). *Can. Entomol.* **97,** 1–14.

Wellington, W. G. (1977). Returning the insect to insect ecology: Some consequences for pest management. *Environ. Entomol.* **6,** 1–8.

Werner, R. A. (1979). Influence on host foliage on development, survival, fecundity and oviposition of the spear-marked black moth, *Rheumaptera hastata* (Lepidoptera, Geometridae). *Can. Entomol.* **111,** 317–322.

Werner, R. A. (1981). Advantages and disadvantages of insect defoliation in the taiga ecosystem. *Proc. Alaska Sci. Conf., 32nd.*

White, T. C. R. (1974). A hypothesis to explain outbreaks of looper caterpillars with special reference to populations of *Seliosema suavis* in a plantation of *Pinus radiata* in New Zealand. *Oecologia* **16,** 279–301.

White, T. C. R. (1984). The abundance of invertebrate herbivory in relation to the availability of nitrogen in stressed food plants. *Oecologia* **63,** 90–105.

Williams, K. S., and Myers, J. H. (1984). Previous herbivore attack of red alder may improve food quality for fall webworm larvae. *Oecologia* **63,** 166–170.

Wink, M. (1983). Wounding-induced increase of quinolizidine alkaloid accumulation in lupin leaves. *Z. Naturforsch.* **38,** 905–909.

Wratten, S. D., Edwards, P. J., and Dunn, I. (1984). Wound-induced changes in the palatability of *Betula pubescens* and *B. pendula*. *Oecologia* **61,** 372–375.

Young, S. S. Y., and Wrensch, D. L. (1981). Relative influence of fitness components on total fitness of the two-spotted spider mite in different environments. *Environ. Entomol.* **10,** 1–5.

Chapter 17

Geographic Invasion and Abundance as Facilitated by Differential Host Plant Utilization Abilities

J. MARK SCRIBER[1] and JOHN H. HAINZE[2]

Department of Entomology
University of Wisconsin
Madison, Wisconsin 53706

I. INTRODUCTION

We suggest that the term "outbreak" has no technical or scientific definition but that in common entomological parlance it refers to a perceived increase in numbers of insects that are a nuisance or have an economic impact. Since we are concerned here with insect outbreaks, we must then examine those factors that may impinge on insect population dynamics and result in measurable population increases in pest species. It

[1] Present address: Department of Entomology, Michigan State University, East Lansing, Michigan 48824.
[2] Present address: S. C. Johnson and Son, Inc., Racine, Wisconsin 53402.

INSECT OUTBREAKS

may also be instructive, however, to examine similar factors in noneconomic species. To this end we describe two cases in which changes in host utilization preference and abilities in Wisconsin have resulted in significant enough changes in population dynamics to be classified as genuine outbreaks and contrast them with the population dynamics of two noneconomic species of phytophagous Lepidoptera (the tiger swallowtail butterfly, *Papilio glaucus* L., and the promethea silk moth, *Callosamia promethea* Drury) in relation to differential host plant utilization patterns (particularly across Wisconsin). These comparisons are made in order to illustrate the effect of food plants and thermal ecology on the behavioral, physiological, and ecological adaptations moderating population dynamics.

Both the red pine shoot moth, *Dioryctria resinosella* Mutuura, which feeds on red pine shoot and cones, and the hopvine borer, *Hydraecia immanis* Guenee, which feeds on the underground stem and root portions of corn plants, have undergone a shift in host plant preference and utilization, which has directly or indirectly resulted in significant economic damage in Wisconsin and adjacent states. The implications of a host plant shift for insect population dynamics include changes in plant spatial distribution (particularly density) and variability in host plant quality. These differences in distribution and quality were encountered by the red pine shoot moth as a result of human activities and by the hopvine borer as a result of a host shift that was probably enhanced by human activities.

Subtle differences in host preference and/or the ability to survive on various hosts may have very significant implications in the population dynamics and evolutionary ecology of insects (see reviews by Scriber, 1983; Diehl and Bush, 1984; Futuyma and Peterson, 1985). Differential adaptations of entomophagous and phytophagous insect species to their various hosts are generally accompanied by a number of other behavioral, physiological, morphological, and ecological adaptations, including temporal variation in activity patterns; differential voltinism patterns; color, shape, and size differences; migration, dispersal, and mating strategies; differences in susceptibility to pathogens or synthetic insecticides; predator and parasite defense behaviors (Denno and Dingle, 1981; Dingle and Hegmann, 1982).

Several of the above-mentioned adaptations have had the term "biotype" applied to them. Diehl and Bush (1984) proposed the following classifications to help standardize terminology in the future: (1) *nongenetic polyphenisms* (also called ecomorphs or phenocopies), in which the same genotype produces various phenotypes in different environments; (2) *polymorphic* and/or *polygenic* variation within populations (i.e., discontinuous or continuous variation within a freely breeding population with a genetic basis); (3) *geographic races,* which are geographically isolated biotypes (e.g., semispecies or subspecies); (4) *host races* as "a population of a species that is partially reproductively isolated from other conspecific populations as a direct consequence of adaptation to a specific host" (whether due to isolation by host preference, host-associated allochronic isolation, or some other form of assortative mating arising as a direct result of different host use); (5) *species* as "natural populations that are reproductively isolated from one another and that follow distinct and independent evolutionary paths" (with sibling species morphologically so similar that recognition requires careful studies of biochemical, cytological, or behavioral traits). Since these five categories of organisms are not necessarily mutually exclusive, we do find biotypes at several stages of evolutionary divergence amid various processes of speciation. The four insect taxa we will describe have different (and unknown) degrees of genetic differentiation in their abilities to survive and grow on their food plants. The objective is to elucidate the food plant use patterns and environmental effects in order to relate them to observed geographic limits of "outbreaks" and thereby enhance our understanding of the fluctuating relationships between distribution and abundance.

II. RED PINE SHOOT MOTH

Insect outbreaks are not always accompanied by such spectacular occurrences as frass dropping like rainfall from the forest canopy or highways slickened by columns of voracious larvae searching for food. Often increases in insect numbers occur in relative obscurity, where there are no entomologists to observe them or where their activities are of little human (economic) interest. In the case of the red pine shoot moth, *Dioryctria resinosella* Mutuura (Lepidoptera: Pyralidae), the increase in number in central Wisconsin was, in fact, not observed by entomologists, but its effects were first noticed by foresters.

Foresters observed a gradual decline in tree form in the mid-1970s, in red pine plantations in Wisconsin's central sand plains. Tree crowns were changing from a straight-stemmed and conical aspect to a bushy and flat-topped appearance. The involvement of the shoot moth in this decline was subtle enough that the foresters termed the change "stagnation" and attributed it, variously, to soil nutrient deficiency, moisture stress, overstocking, or premature senescence. The difficulty of determining the cause of the decline led to the involvement of entomologists, who found an insect feeding in the new growing shoots of the trees. This insect was first identified as the Zimmerman pine moth, *Dioryctria zimmermani* (Grote).

The Zimmerman pine moth feeds on a variety of *Pinus* species, attacking trees at branch whorls or at injuries along the main stem or occasionally feeding on shoots (Carlson and Wilson, 1967). Presently, its preferred hosts appear to be imported species, Scots pine, *Pinus sylvestris* L., Austrian pine, *Pinus nigra* L., and Japanese red pine, *Pinus densiflora* Sieb. and Zucc. However, it has been speculated that its distribution follows that of white pine, *Pinus strobus* L. (Mutuura, 1982), which suggests that white pine was probably the most important host for the Zimmerman pine moth before the introduction of the imported species. Early observers of the shoot feeding by this insect in the outbreak area in Wisconsin found that, in mixed plantations, neither the main stem, the shoots, nor the cones of jack, *Pinus banksiana* Lamb., or white pine were attacked; the shoots and cones, but not the main stem, of adjacent red pine were heavily damaged (Hainze and Benjamin, 1983). Subsequently, taxonomic investigations determined that the insect of concern was actually a sibling species, the specialized red pine shoot moth, *Dioryctria resinosella* (Mutuura, 1982). The red pine shoot moth appears to be a native species, distributed in generally low numbers throughout the range of red pine (Fig. 1). Preliminary host range studies and field observations indicate that, in contrast to Zimmerman pine moth, the red pine shoot moth does not feed on white, jack, or Scots pine and remains limited to red pine (Hainze and Benjamin, 1983). In Wisconsin, populations reached very high levels in the middle to late 1970s and then rapidly declined. The shoot moth outbreak in Wisconsin seems to have resulted from a complex interaction of human activities, the abiotic environment, and biotic factors, which combined to produce conditions analogous to those experienced by an insect after a host shift.

The genus *Dioryctria* is of holarctic distribution. It has been divided on the basis of adult morphology into seven species groups, six of which are present in North America (Mutuura and Munroe, 1972; Fig. 2). Members of the genus feed on a variety of conifers, including pines, spruce, fir, larch, and bald cypress. Larvae feed on shoots, cones, the cambium layer beneath the bark of the trunk, or fungus-caused galls on host trees (Table 1). As a group, then, they have evolved to feed in all above-ground areas of concentrated meristematic tissue in conifers. Some of the species are very host specific, while others may feed on as many as five coniferous genera. There has been a great divergence within the *Dioryctria* involving specializations to fill available niches on most species of North American pines. The result is a number of host species-specific *Dioryctria* that are limited in distribution to their host plant distribution and exist in allopatric

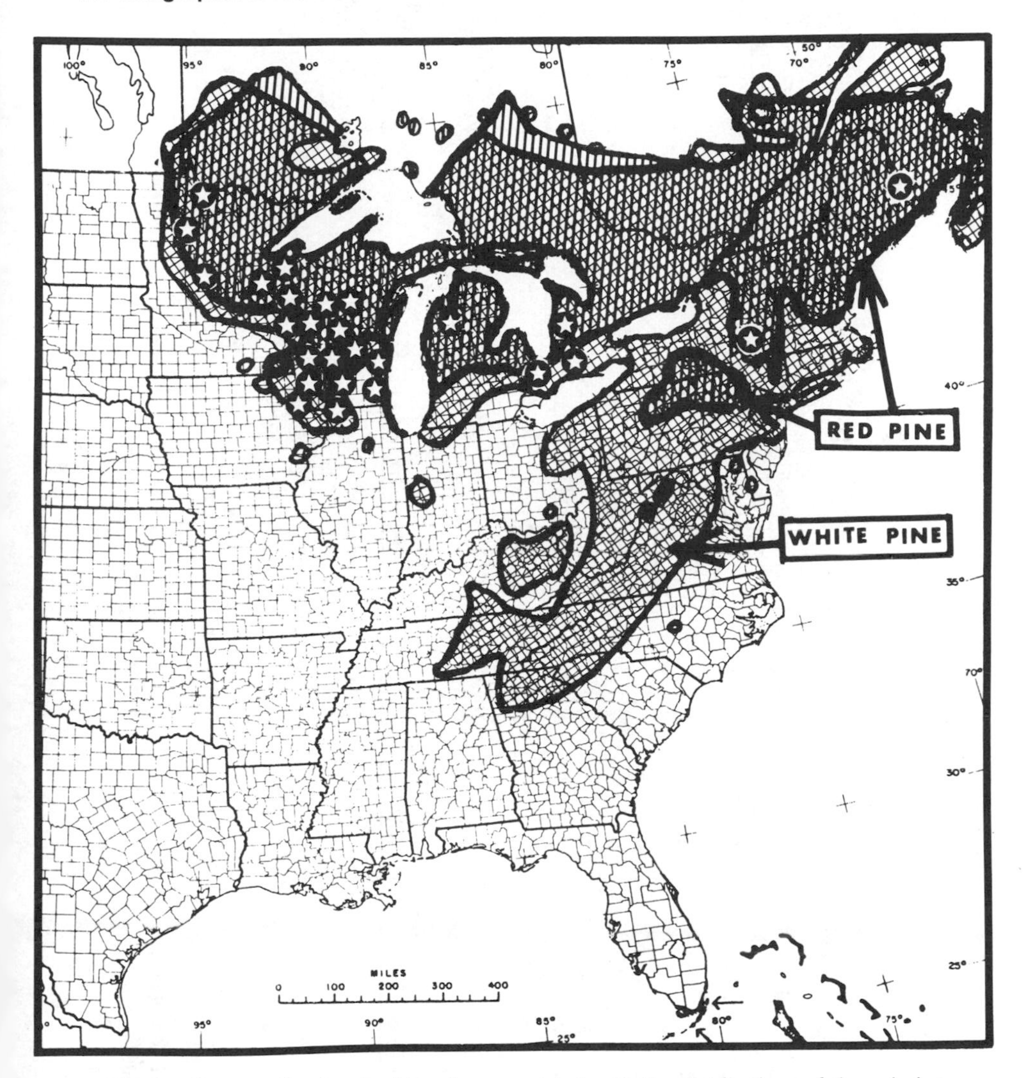

Fig. 1. Ranges of red and white pine associated with the distributions of the red pine shoot moth and Zimmerman pine moth, respectively. Stars indicate reports of the red pine shoot moth.

distributions (Fig. 3) or, like *D. resinosella* and *D. zimmermani*, in sympatry. The host shifts and resulting specific divergences have enabled the *Dioryctria* to invade all areas of North America where conifers are present. Much remains to be learned about this group, and new North American species remain to be described. The difficulties of distinguishing between *D. zimmermani* and *D. resinosella* are typical of the taxo-

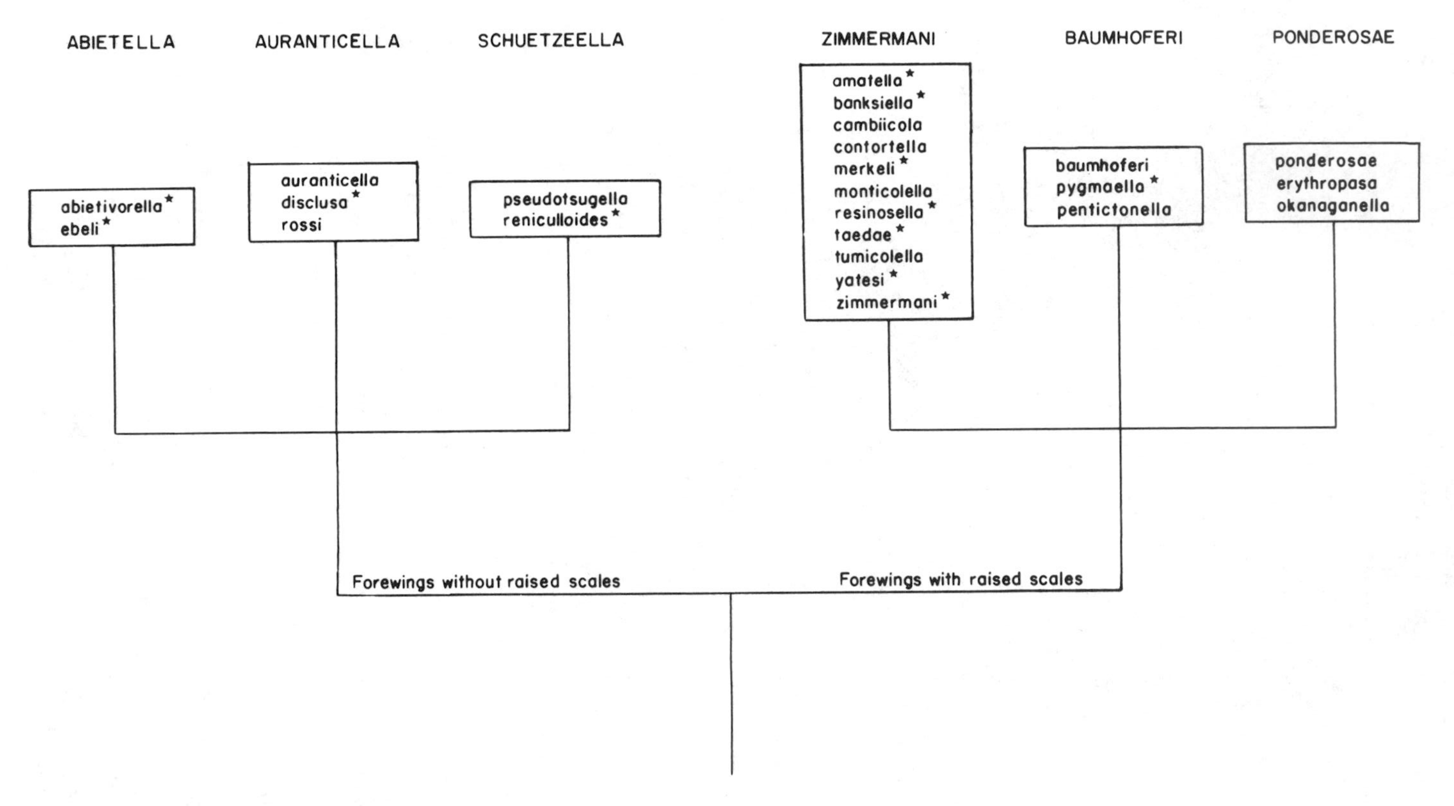

Fig. 2. Phylogeny of *Dioryctria* species groups. Note that *D. albovitella* and *D. clarioralis* have not been assigned to species groups and that current taxonomic work by A. Mutuura (personal communication) will likely result in additional new species descriptions. Asterisks indicate eastern species (see Table 1). Adapted from Mutuura and Munroe (1972).

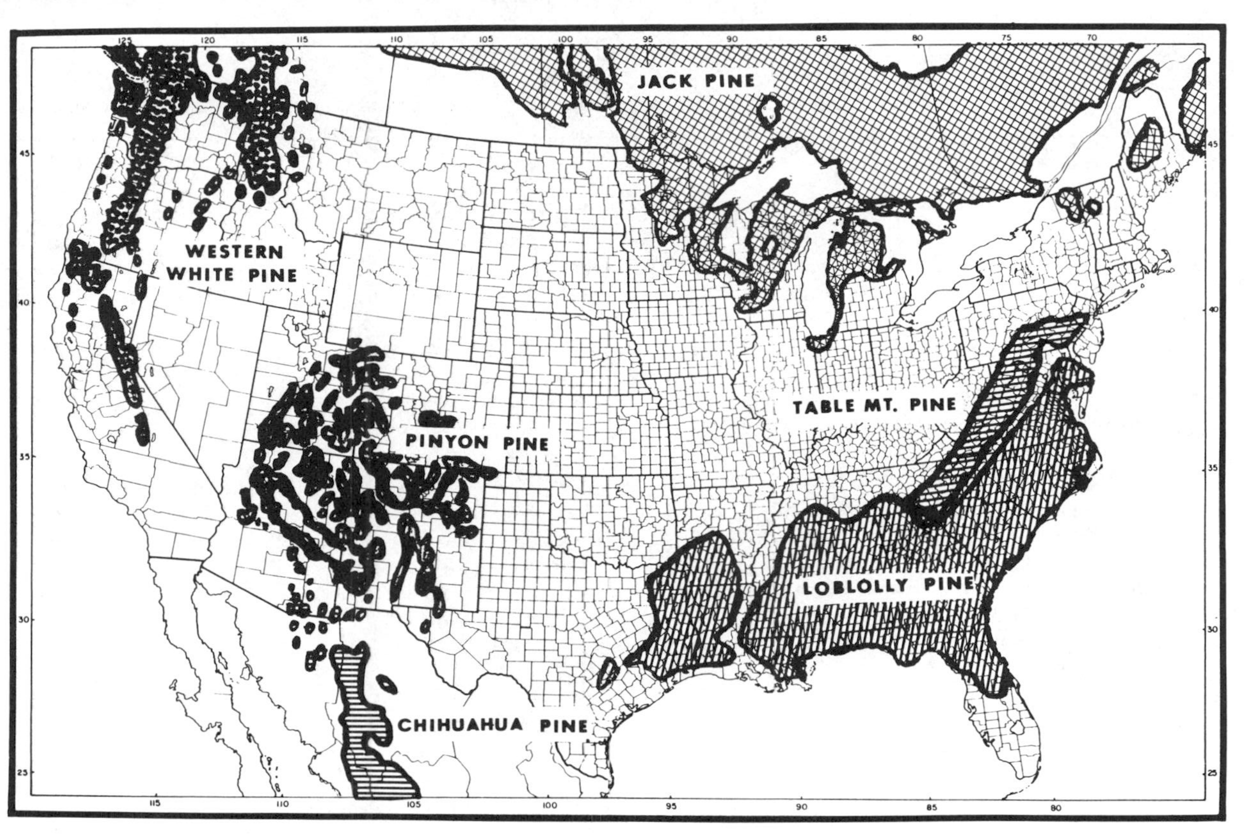

Fig. 3. Distributions of six host tree species for six *Dioryctria* specialists: western white pine for *D. monticolella,* pinyon pine for *D. albovitella,* chihuahua pine for *D. erythropasa,* jack pine for *D. banksiella,* table mountain pine for *D. yatesi,* and loblolly pine for *D. taedae*. *Dioryctria pygmaella* is a specialist on bald cypress that has a similar distribution to loblolly pine except that it extends up the Mississippi River to Illinois.

TABLE 1

***Dioryctria* Species in Eastern North America North of Mexico**

Species	Primary native hosts	Feeding sites	References
D. resinosella Mutuura	Old red pine[a]	Shoots, cones	Mutuura (1982); Hainze and Benjamin (1983)
D. zimmermani (Grote)	Young red pine, white pine, ornamental imports	Trunk, shoots	Heinrich (1956); Carlson and Butcher (1967)
D. amatella (Hulst)	Loblolly, longleaf, slash pines	Flowers, shoots, cones	Heinrich (1956); Coulson and Franklin (1970)
D. banksiella Mutuura and Munroe	Jack pine[a]	Blister rust galls	Mutuura *et al*. (1969)
D. merkeli Mutuura and Munroe	Loblolly pine and others	Flowers, shoots, cones	Mutuura and Munroe (1979)
D. taedae Schaber and Wood	Loblolly pine[a]	Shoots, cones	Schaber and Wood (1971); Schaber (1981)
D. yatesi Mutuura and Munroe	Table mountain pine[a]	Cones	Mutuura and Munroe (1979)
D. abietivorella (Grote)	Douglas fir, firs, spruces, pines	Cones, trunk, shoots	Munroe (1959); Lyons (1957)
D. ebeli Mutuura and Munroe	Loblolly, longleaf, slash pines, pond cypress	Cones	Mutuura and Munroe (1979); Ebel (1965)
D. clarioralis (Walker)	Loblolly, longleaf pines	Flowers, shoots, cones	Heinrich (1956); Ebel (1965)
D. reniculelloides Mutuura and Munroe	Douglas fir, firs, hemlock, pine, spruces	Shoots, cones	Mutuura and Munroe (1973); MacLeod and Daviault (1963)
D. disclusa Heinrich	Jack, loblolly, longleaf, pitch, red, shortleaf, Virginia	Pines, spruce, cones	Farrier and Tauber (1953)
D. pygmaella Ragonot	Bald cypress,[a] pond cypress	Cones	Heinrich (1956); Merkel (1982)

[a]Specialized (probably true monophagous species).

nomic problems in the species complexes of this genus. The previously defined distribution and host range of several other species have also been reduced as our understanding of the systematics has increased.

Species of *Dioryctria* are regarded as pests of forest trees throughout the Northern Hemisphere. However, some *Dioryctria* species by virtue of their specialized feeding habits are not considered pests. *Dioryctria banksiella* Mutuura and Munroe is such an insect, feeding only in rust galls on jack pine. Perhaps the best-known pests are the cone-feeding *Dioryctria* species, which are particularly troublesome in the southeastern United States, where they damage high-value pine seed crops. Coniferous cone crops are quite variable from year to year, and so the number of cone-feeding *Dioryctria* may often be limited in years of low cone production. However, many cone-feeding *Dioryctria* feed on the cones of several different host species, and still others may feed on shoots as well as cones, so they are less dependent on the vagaries of cone production. The development of seed orchards to produce a large quantity of high-quality seed has also provided these moths with a much more predictable resource and thus a greater potential for population increases. As a result *Dioryctria* species have become a vexing problem for seed orchard managers. Outbreaks of *Dioryctria* involve an increase in population levels, although a relatively small number of larvae. Feeding sites (e.g., cones, shoots, branch whorls, or galls) are limited for these insects, and a single larva generally feeds on more than one shoot or cone.

The life history of the red pine shoot moth in Wisconsin is somewhat typical of the genus (Hainze and Benjamin, 1983). It is univoltine, and oviposition occurs in late July and August. Eggs are laid singly beneath bark scales on branches and along the main stem. Neonate larvae eclose from middle to late August. There are five larval instars. The larvae overwinter as first instars in silken hibernacula beneath bark scales on branches and along the main stem, and molt to second instar in the hibernacula. Third-instar larvae generally initiate shoot feeding in late May, and larvae may mine two or three shoots or cones. A single larva feeds on a shoot, and the point of attack, identified by a pitch mass composed of frass and resin, usually occurs near the tip of the shoot. Pupation most often takes place within a shoot or cone, but pupae also are found under bark scales and in the litter. Pupae were first seen in early July in Wisconsin, and the pupal stage lasts about 12 days. Adults emerge in mid-July and August and oviposition begins about a week after emergence.

Although the buildup of shoot moth populations in central Wisconsin was not documented directly, it was reflected indirectly in the dramatic decline in annual height growth in affected forest stands (Fig. 4). Height growth is reduced when shoot moth larvae feed on and destroy the tree's

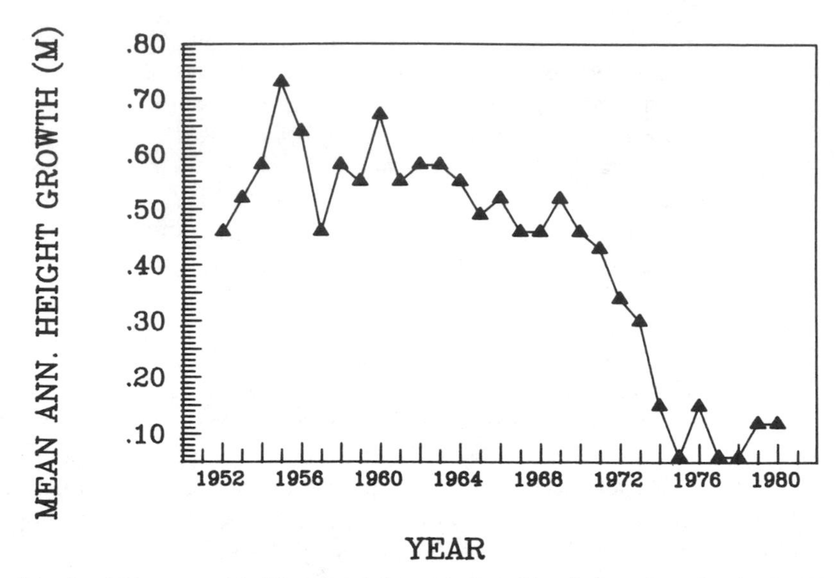

Fig. 4. Mean annual height growth in a 36-year-old red pine plantation in the Black River Falls State Forest, Wisconsin.

terminal shoot. Height growth, in fact, was reduced by 38 to 65% over the outbreak period in infested red pine plantations (Hainze and Benjamin, 1984). The outbreak in central Wisconsin appears to have been the most severe of any that can be attributed to the red pine shoot moth. There were other reports of similar damage attributed to other *Dioryctria* species before the description of the red pine shoot moth in Wexford County, Michigan (Anonymous, 1957), and the Lake Simcoe region of Ontario (Rose and Lindquist, 1973), for example, but none to the same degree. If this species is distributed throughout the natural range of red pine and is a native species, what factor or factors led to the significant outbreak in Wisconsin in the 1970s?

One of the most readily apparent aspects of the shoot moth outbreak in Wisconsin was its localized nature. The outbreak was generally limited to the central part of the state, an area known as the central sand plains. Most of the central plains of Wisconsin are occupied by sandy, excessively drained soils. These soils originated from deposition into Glacial Lake Wisconsin and glacial outwash. Glacial Lake Wisconsin was formed by the blockage of glacial drainage during the Cary substage of the Wisconsin glaciation, about 14,000 years ago. The former lake bottom covers a large portion of the central plain in Wisconsin. Before settlement this area was covered by prairie and oak barrens in its southern portion and pine barrens in the northern portion. The droughty soils caused severe crop losses for early farmers in the area during dry years, and much of the land was abandoned for agriculture by the 1930s. Subsequently, these areas were included in Wisconsin's reforestation effort and large acreages were planted with red pine. It is these plantings, in part, that were subject to the red pine shoot moth outbreak in the 1970s. Although survey results indicate that the shoot moth is present wherever red pine occurs in the state, outbreak populations occurred only on the sandy soils of the central plains and along the floodplain of the Wisconsin River. The specific relationship between the sandy soils and shoot moth abundance is as yet unclear. Forest stands suffering from high shoot moth populations had exhibited moderate to excellent growth before the outbreak, so the effect of the soil on the trees, if any, is not obvious. A nutrient cycling study in a red pine plantation in the central sand plain, however, indicated that over a period of years, the available potassium in the sandy soil is bound up in red pine biomass and plantations are then subject to a potassium deficit (Bockheim *et al.*, 1983). A marked deficiency of potassium reduces photosynthesis and increases respiration, resulting in lowered production of carbohydrates necessary for growth or production of some defensive compounds (Kramer and Kozlowski, 1979). In fact, in some forest trees there may be a trade-off in the utilization of carbohydrates for growth or defensive compounds, so that limitations on carbohydrate production may result in limited production of defensive compounds (Wright *et al.*, 1979). The excessively drained soils may also have caused water stress in dry years and thus predisposed the stands to the shoot moth outbreak. It has been suggested that shoot moth populations were lower in some areas with a high water table.

In studies of shoot moth biology, we observed that younger trees were rarely attacked whereas heavy populations occurred on older trees. Significant differences were found in levels of attack on adjacent stands less than 20 years of age and greater than 20 years of age (Hainze and Benjamin, 1983). In 1970, at the beginning of the shoot moth outbreak, there were approximately 84,000 acres of red pine plantations greater than 5 acres in size in Wisconsin. In 1950, there were only about 1500 acres of 5-acre or greater red pine plantations in the state (Thorne, 1980). A reforestation boom occurred in Wisconsin and the other Lake States beginning in the 1930s with the efforts of the CCC and peaking in the late 1950s and early 1960s. The result is that by the late 1980s there will be some 290,000 acres of red pine plantations larger than 5 acres in size and greater than 20 years of age. This represents an enormous expansion of the food base for the red pine shoot moth. Furthermore, red pine represents around 70% of the coniferous plantations on a statewise basis. Within the central sand plain counties in particular, red pine plantations account for 88% of all coniferous plantations. Therefore, red pine shoot moth resources of the critical age are rapidly increasing and are particularly concentrated in the very geographic locale in which the moth seems to be most successful.

It is apparent that conditions were conducive to a shoot moth population increase in the late 1960s and early 1970s, but it is difficult to surmise what triggered the outbreak. Climate is a possibility; however, the years preceding the outbreak seem to have been normal; they were not exceedingly dry or wet, nor were there unusually mild winters for the survival of overwintering larvae. Another possibility involves the bimodal feeding pattern of the shoot moth. Shoot moth larvae usually leave their initial shoot and attack an additional shoot or a second-year cone. Cone feeding generally is undertaken by the later instars. Cones vary in their availability, however. Cone crops are very irregular, and a large cone crop may occur only once every 10 or even 20 years. Perhaps this is why some *Dioryctria* cone feeders have evolved the ability to feed on cones of several different species or why some are able to feed on both cones and shoots, which are much more predictable resources. Another red pine cone-feeding species, *Conophthorous resinosae* Hopkins, generally attacks only cones but will switch to shoots when cone crops are low (Mattson, 1971). Cones may be nutritionally more advantageous to shoot moth larvae and/or may afford greater protection from natural enemies. In either case, with greater fecundity or survival, a large cone crop might lead to outbreak populations if the other conditions are right. Another possibility that might increase the availability of cones to the shoot moth is a reduction in competition from other species feeding on second-year cones. In the central sands area of Wisconsin during our studies of shoot moth biology and population dynamics, two other species utilized second-year red pine cones: *Dioryctria disclusa* Heinrich and *Eucosma monitorana* Heinrich. These two species fed on a large number of cones and provided significant competition for the shoot moth. *Dioryctria disclusa* larvae enter cones in late spring or early summer long before the red pine shoot moth. *Eucosma monitorana* larvae feed almost concurrently with shoot moth larvae or attack cones just a little later. A reduction in number of either or both of these species would result in a greater availability of cones for the red pine shoot moth.

Parasitoids may have played an important role in the Wisconsin shoot moth outbreak as well. Several of the most important red pine shoot moth parasitoids also parsitize the European pine shoot moth, *Rhyacionia bouliana* (Schiff). The European pine shoot moth primarily attacks young red pine, and it was in outbreak in the 1950s and 1960s in Wisconsin. Populations of its parasitoids would also have built up during that period and then declined as host populations declined in the mid-1960s. The higher parasitoid populations may also have suppressed red pine shoot moth populations during that period. The decline of parasitoid numbers may have released red pine shoot moth populations, resulting in the outbreak population levels in the 1970s.

Since about 1980 red pine shoot moth populations have been in decline; they reached endemic levels in much of the central sand plains by 1983 (Fig. 5). Partial life tables were constructed for red pine shoot moth larvae during 1982 and 1983 in order to determine some of the factors involved in the decline. Very high mortality of the fifth instar occurred in the stands where populations were declining (Table 2); one of the major mortality factors was the parasitoid *Bracon rhyacioniae* (Muesebeck), which was also a parasitoid of the European pine shoot moth. Also during this period of decline, populations of *Dioryctria disclusa* and *Eucosma monitorana* were very high, reducing the availability of the cone resource for the red pine shoot moth. The possibility of induced chemical resistance in the plants (Rhoades, 1983) has not been investigated.

Finally, it appears likely that the specialized shoot and cone feeder *Dioryctria resinosella* arose in a host shift from the more polyphagous *Dioryctria zimmermani*. The latter feeds on the cambium layer below the bark of several *Pinus* species and may also occasionally feed on shoots. It does not, however, feed on cones. Its primary native host apparently is white pine, although it is also able to feed on young red pine. Recorded

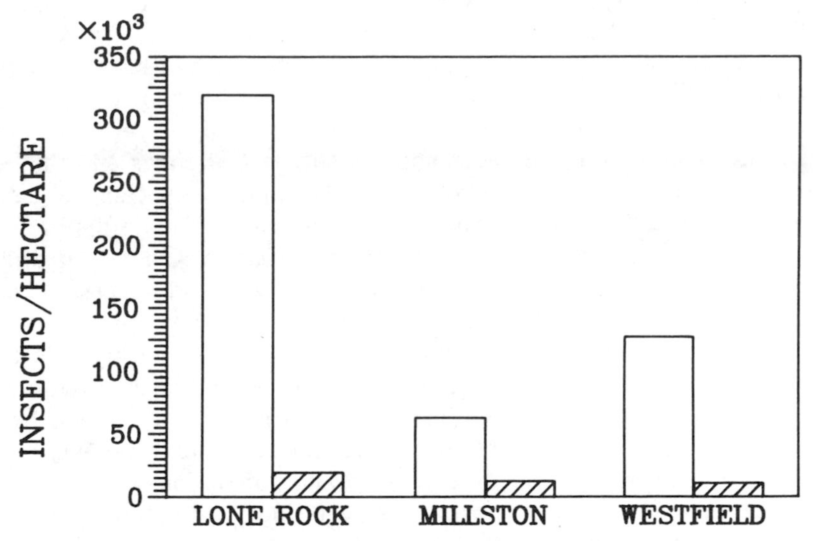

Fig. 5. Red pine shoot moth population levels (number of insects per hectare) at three Wisconsin study sites (open columns: 1982; striped columns: 1983).

TABLE 2

Condensed Partial Life Table for *Dioryctria resinosella* at Lone Rock, Wisconsin, 1982[a]

x	lx	dx	d x F	100qx	Sx
Early instars	319,448	58,202	*Bracon rhyacioniae*	18.2	
		30,595	Other factors	9.6	
		88,797		27.8	.722
Fifth instar	230,651	119,046	*Bracon rhyacioniae*	51.6	
		78,176	Unidentified parasitoids	33.9	
		13,544	*Hyssopus rhyacioniae*	5.9	
		17,624	Other factors	7.6	
		228,390		99.0	.010
Pupae	2,261	0			1,000

[a] The lx and dx columns represent number of insects per hectare. Adapted from Hainze and Benjamin (1985).

shoot feeding by *D. zimmermani,* in association with its primary mode of attack along the trunk, suggests how such a host shift might have occurred. The adaptation to utilize cones would seem an easy step from a primary dependence on shoots because of their similar location. Numbers of shoots and cones are greater, they may be nutritionally more advantageous, and they may have a different chemical profile in older as opposed to younger trees. This shift to shoots and cones might later involve a preference for older trees. *Dioryctria zimmermani* larvae have not been recorded feeding on the trunks of older red pine trees, the result perhaps of chemical or physical differences between these and the younger trees, which are readily attacked. So a shift to shoots and cones and thus to older trees might result in a loss of the ability to utilize the cambial tissue below the bark on the trunk. Such an evolutionary scenario fits the observed present-day feeding biologies of *D. resinosella* and *D. zimmermani.* Their very close morphological similarity would suggest recent speciation, and their sympatric distribution indicates that a host shift was involved. This host shift in past evolutionary time was probably followed by a gradual expansion in the range of the insect within the distribution of red pine. However, no population increases (or outbreaks) have been recorded until now. Human activities in recent ecological time may have resulted in a general increase in *D. resinosella* populations across northern states as its food base has expanded, with localized outbreak populations occurring where the further interaction of biotic and abiotic factors were favorable.

III. HOPVINE BORER

The hopvine borer (HVB), *Hydraecia immanis* Guenee, is a stem-feeding caterpillar (Lepidoptera: Noctuidae) that has been causing severe localized damage to corn in large portions of Wisconsin, Minnesota, Iowa, and Illinois only since 1985 (Giebink *et al.,* 1984). Since its description by Guenee in 1852, economic problems caused by the HVB have exclusively involved hop (*Humulus lupulus* L.) (Bethune, 1873; Comstock, 1883; Howard, 1897; Brittain, 1915; Hawley, 1918; Godfrey, 1981, for a review). The HVB's common name is derived from the larval stage, which feeds in and damages the head, vine, and root of the hop plant (Dodge, 1882). The larvae has also been referred to as the hop grub (Comstock, 1883; Smith 1884) and the collar worm of hop (Fletcher, 1883).

This insect caused severe economic problems in the once extensive hops industry in Wisconsin and was commonly responsible for 25–50% losses of hops crops (Dodge, 1882). Thus, the HVB was in large part to blame, along with downy mildew of hop (*Pseudoperonospora humuli* Miyb. and Tak.) and the damson hop aphid (*Phorodon humuli* Schr.), for the shift of hop cultivation to the west coast of the United States by the early twentieth century (Schwartz, 1973; see also Campbell, 1983). Wild hop currently occurs in at least 28 states, from California to Maine (Gross, 1900; Davis, 1957; Small, 1981).

From about 1920 to 1975, populations of this noctuid moth in New York and the Midwest apparently remained very localized and attracted essentially no attention. This situation has changed significantly since the mid-1970s. Long thought to be restricted to grasses and hops, the insect has successfully switched to corn, and local populations are currently causing significant economic damage (Scriber, 1980; Giebink, 1983; Giebink *et al.,* 1984). Since the first heavy localized damage was first detected in a few southwestern Wisconsin cornfields, serious damage to young corn has been reported in more than 50 counties in four major corn-growing states in the Midwest: Wisconsin, Illinois, Iowa, and Minnesota (Fig. 6). The distribution, phenology, and developmental biology in relation to the major change in feeding habits are discussed in detail by Giebink *et al.* (1984). A brief review is in order here, however.

As indicated by its common name, the HVB had been found in close association with hop plants. Wild hop, found from the east coast of the United States to the Rocky Mountains, presumably served as the primary host for the insect in North America (Forbes, 1954). Hop is also a favored host for the congeneric, but more polyphagous potato stem borer (PSB), *H. micacea* Esper, in Europe, which has also damaged corn and potatoes

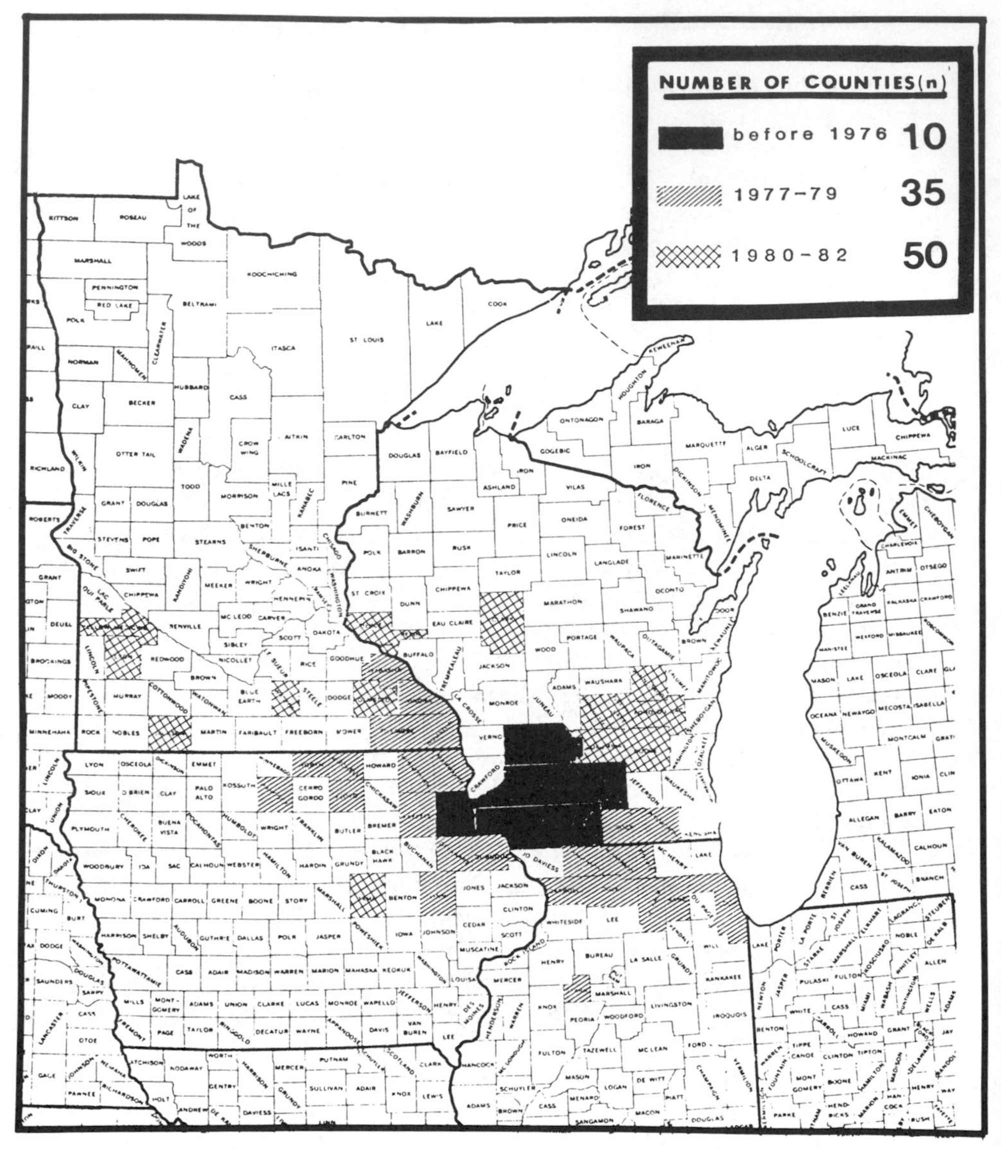

Fig. 6. Geographic invasion of Wisconsin and adjacent states by the hopvine borer, *Hydraecia immanis*, from its discovery in 1975 to the present. After Giebink *et al.*, (1984).

as well as many other crops such as sugar beet, rhubarb, onions, tomatoes, strawberries, and raspberries in Scandanavia, Europe, the United Kingdom, Russia, and Canada (Nordstrom *et al.*, 1941; Jobin, 1963; Kondakov and Nogina, 1968; Seppanen, 1970; French *et al.*, 1973; Chawla *et al.*, 1975; Deedat, 1981; Deedat *et al.*, 1983). Whether HVB has the same capacity for polyphagy as the PSB is unknown, although it is of obvious significance (e.g., to the seed potato industry in northern Wisconsin).

Since hops are a key flavoring ingredient in beer, and because beer has been such a critical socioeconomic factor since the time of the early European-American settlers, the commercial production of hops in the United States has always been intense. New York State and Wisconsin became the leading hop producers throughout the latter half of the nineteenth century, and at one point in the 1860s the work force in a single Wisconsin county (Sauk) was greater than 30,000 persons, and the hop crop was 4 million pounds (Hibbard, 1904). Climate, plant disease, and insect problems contributed to a decline in production toward the end of the nineteenth century. The west coast states (Washington, Oregon, Idaho, and northern California) had surpassed New York and Wisconsin in production by 1900 and today represent the only commercial production sites in the United States (approximately 60% of current world production) (Burgess, 1964; Schwartz, 1973; A. Haunold, personal communication). To date, however, neither the HVB nor PSB has been detected as a pest. One final, unsuccessful attempt to revive commercial hop production in Wisconsin after the prohibition era was attempted in Sauk County in the 1930s, perhaps providing a crucial resource (food) for the HVB during their decline years (Fig. 7).

In Wisconsin, hop production was concentrated in several southern counties, including Sauk, Richland, Grant, Iowa, Lafayette, and Dane (Schwartz, 1973), where, in the spring of 1975, the first reports of serious HVB larval damage to corn originated. Although there has been essentially no hop production east of the Rocky Mountains since the 1930s, small patches of wild or escaped hop, along roadsides and drainage ditches near heavily damaged cornfields, have been located and HVB larvae found feeding on below-ground portions of these plants (Giebink *et al.*, 1984). Thus, it appears that isolated endemic populations have continued to exist even in the absence of the hop industry and that the HVB has been able to make the transition from its grass and hops feeding habit to a quackgrass and corn feeding pattern in these areas (Fig. 7). Since hop is a perennial and corn an annual, this transition in larval feeding has certainly been favored by continuous corn production, which has made corn a dependable resource locally since the 1940s. It is tempting to speculate that the response of the insect populations is related directly to the in-

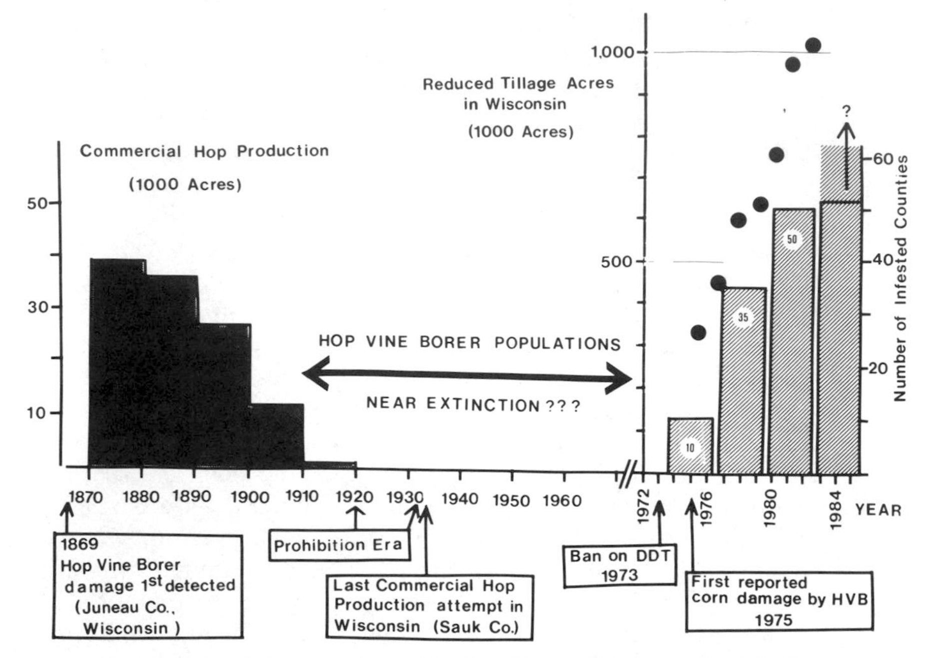

Fig. 7. Relationship between hop production, tillage practices, and hopvine borer populations over the past century. As Wisconsin and New York hop production was declining after the "Hop Crash of 1867," California, Oregon, and Washington took over. Tillage data from the Wisconsin Soil Conservation Service; Hop production data from Schwartz (1973).

creased carrying capacity caused by this change in agronomic practices in areas adjacent to pockets of hop-feeding HVBs.

The reasons for the lack of any outbreaks before 1975 are unknown, but it is possible that the general use of chlorinated hydrocarbons (DDT, aldrin, dieldrin, etc.) in the two decades before 1970 had a general and complete suppressive effect on HVB. It is also possible that small population pockets have only recently been forced onto corn with the removal of perennial hop plants or other unknown potential broadleaf hosts along field edges as a result of changing agronomic practices (e.g., enlarging fields by removing fence rows between fields and/or increased herbicide use). In any case, poor weed control, continuous corn, no till, conservation tillage, and reduced tillage favor increasing insect population densities and increase the potential for further spread in the corn belt (Phillips and Phillips, 1984; Scriber, 1984c).

Except for its much more stenophagous feeding behavior, the native HVB is very similar to the introduced PSB in many respects. The biology and damage to corn are very similar for both species (see Deedat, 1981).

Both species are relatively difficult to control by current conventional chemical methods (Deedat *et al.*, 1982; Giebink, 1983). As yet, effective and legally acceptable chemical controls are essentially lacking. Furthermore, the HVB apparently has few natural enemies, skunks being the most significant (Dodge, 1882; Howard, 1897).

In view of the fact that the HVB has few natural enemies and effective chemical means of control are lacking, this insect could become an even greater economic concern to corn producers in the Midwest. Then there is the strong possibility that the HVB in the Midwest may follow a trend similar to that of the PSB on corn in Canada. The PSB has been well-established in Canada since the turn of the century (Muka, 1976) and has since appeared as far west as Manitoba (C. R. Ellis, personal communication) and more recently (1982) in Manitowoc and Kewaunee Counties of Wisconsin. In 1984, economically damaging levels of PSB on corn were detected in several locations of Calumet County, Wisconsin (Fig. 1). On the basis of its similar life cycle, habits, and damage to corn (Deedat and Ellis, 1983), we could anticipate that the HVB might also increase its densities and/or range throughout the Midwest (see Muka, 1976; Rings and Metzler, 1982).

The total range of acceptable plants for oviposition is unknown for both the HVB and the PSB; however, grasses are generally preferred (Deedat *et al.*, 1983; Giebink *et al.*, 1984). The range of subsequent food plants that are suitable for larval feeding and growth of the HVB is also unknown and untested. In addition to the current economic concerns that these two insects create among corn growers across Canada and the northern United States, the basic ecological and evolutionary interactions between these herbivores and their potential host plants are intriguing. The recent range expansion and overlap of these two species provide an excellent natural experimental system in which to monitor population biology during the outbreak stage and the subsequent colonization and spread of a native insect and its congeneric European introduction.

The implications of hybridization of the HVB with its more polyphagous relative (the PSB) may be significant. For example, some Lepidoptera hybrids exhibit increased survival, growth, and reproduction on atypical plants (Scriber, 1982, 1983; Collins, 1984). Furthermore, hybridization can significantly alter larval developmental threshold temperatures (Ritland and Scriber, 1985). At current rates of spread, we expect that the opportunity for natural hybridization between HVB and PSB may have already occurred this year in Wisconsin, since the two species are currently separated (as indicated by their economic damage to corn) by only a single county, Calumet, in central Wisconsin and are likely to be sympatric in this area. The degree of reproductive isolation between HVB and the PSB is currently unknown, but the existence of hybrids has been suspected (e.g., Forbes, 1954; Smith, 1899), and males in the presence of conspecifics will mate with females of the other species (B. L. Giebink, unpublished; but see also Teal *et al.*, 1983). The southern limits on the geographic spread of either of these *Hydraecia* species through the Midwest Cornbelt are also unknown. A change in the voltinism patterns, as has occurred with the European corn borer (Beck and Apple, 1961; Showers, 1979), might also be mediated by hybridization and could accelerate the geographic spread throughout other corn-growing regions of the Midwest and southward (Fig. 8).

In summary, undetermined factors have led to increased local densities of hopvine borers, *Hydraecia immanis,* in the Midwest. Economic damage of corn is intense where populations occur, yet these insects have never been listed as even a minor pest of corn in any economic entomology textbooks of the twentieth century. This change in feeding behavior from grass and hops to grass and corn has led to increased HVB abundance and a geographic range expansion of noticeable economic damage to corn from 5 to 6 counties in Wisconsin to more than 50 counties in Wisconsin and three adjacent states. A simultaneous but even more rapid range expansion of the introduced potato stem borer, *Hydracea micacea,* has occurred westward across Canada and into Wisconsin with economically significant damage to corn.

IV. TIGER SWALLOWTAIL BUTTERFLY

The larval stages of leaf-feeding tiger swallowtail butterflies are virtually unknown except for sporadic host plant observations over the past century. Survival of eggs and pupae is favored by solitary breeding behavior on a varied and widely dispersed array of non-economically important deciduous trees (Scriber, 1975; West and Hazel, 1982). Lepidopterists are sometimes aware of focal food plant relationships; however, generally only the large, showy adult butterflies attract any interest. Despite very noticeable adult populations across North America (Fig. 9), "outbreak" levels have never (to our knowledge) been reported. Densities of male adults at puddling sites may reach $100/m^2$; however, these "puddle clubs" are extremely sporadic, short-lived, and very likely to remain largely unnoticed in the forest clearings across most of the continent (Norris, 1936; Arms *et al.*, 1974; Scriber and Lintereur, 1982).

Whereas millions of swallowtail butterflies blanket the southern half of North America during late spring through most of the summer with multiple generations [e.g., *Papilio glaucus glaucus* L. and *P. g. australis* (May-

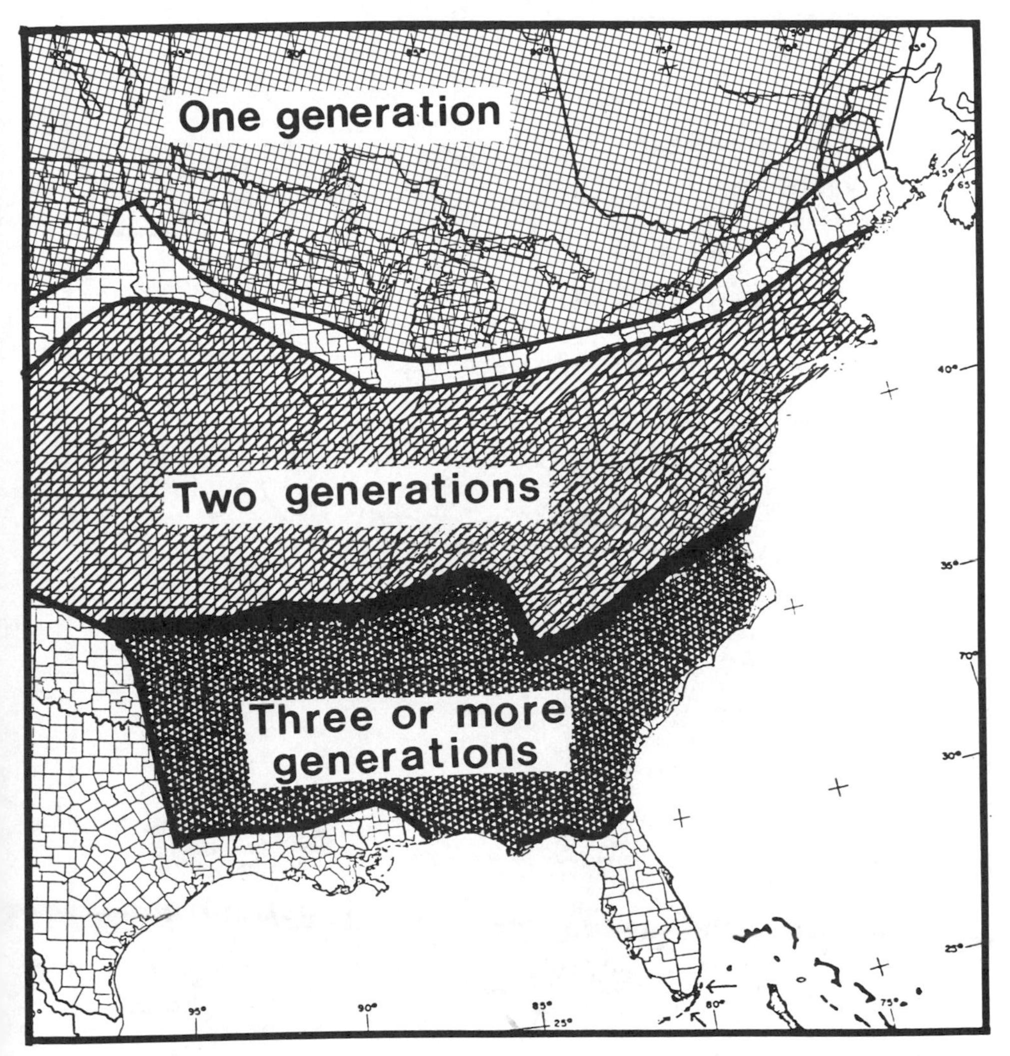

Fig. 8. Voltinism patterns observed for the European corn borer, *Ostrinia nubilalis*, across the eastern United States. After Showers (1979, 1981).

nard)], the northern half of the continent will experience only a single June flight from the putative subspecies *Papilio glaucus canadensis* R & J (Scriber, 1982). Consequently, there are no adults "on the wing" after mid-July across all of Canada and the northern United States.

We now know that significant differences in food plant utilization abilities exist not only between the western and eastern species of the *P.*

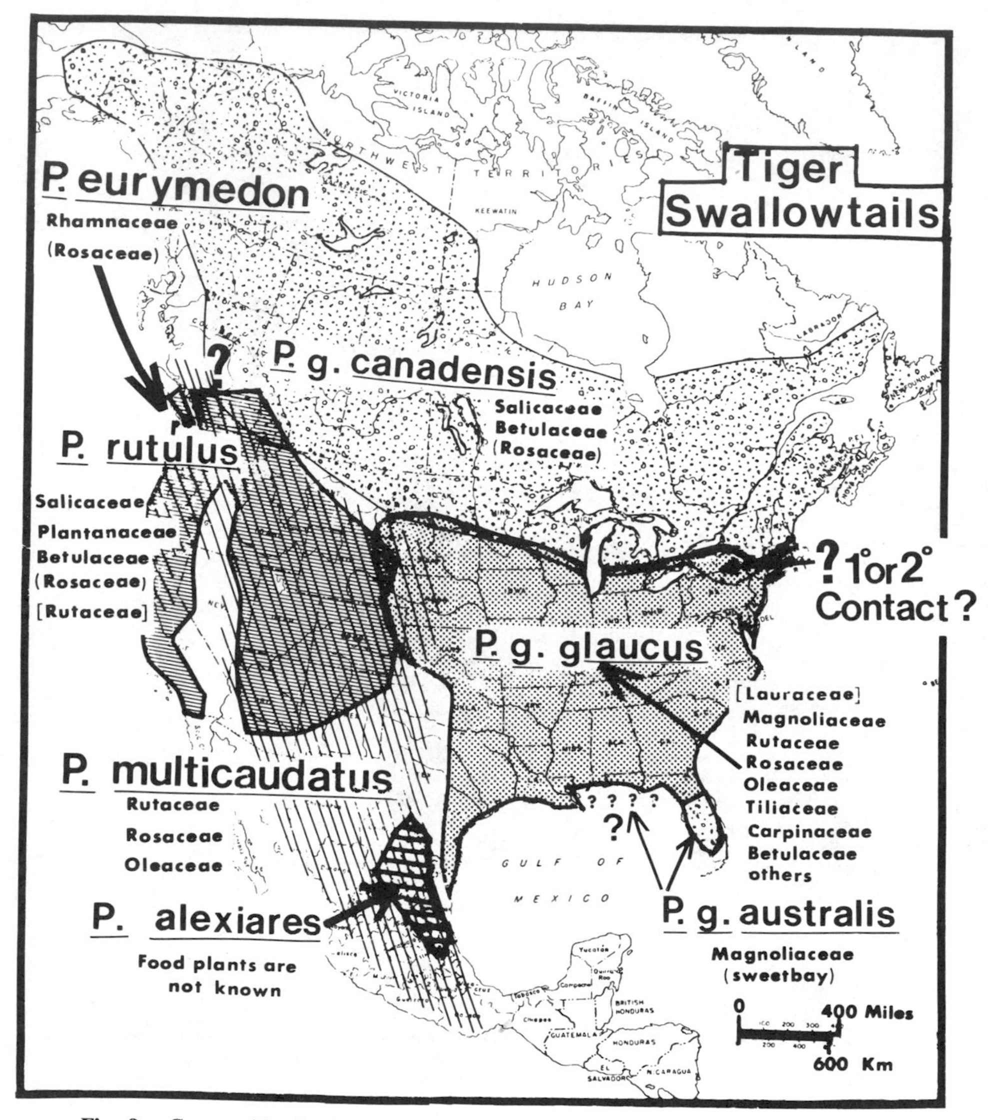

Fig. 9. Geographic distribution and favorite food plant families of the three putative subspecies of the eastern tiger swallowtail butterfly, *Papilio glaucus*. *Papilio rutulus* is sympatric with *P. eurymedon* throughout most of its range, *P. multicaudatus* exists throughout most of Mexico, and the Mexican swallowtail (*P. alexiares*) is relatively localized and poorly studied.

glaucus complex, but also among the subspecies *P. g. canadensis, P. g. glaucus,* and *P. g. australis* (e.g., see Scriber *et al.,* 1982; Scriber, 1986). In fact, the primary factor that has enabled *P. g. canadensis* to inhabit Canada may be the ability to utilize food plants of the Salicaceae (e.g., *Populus tremuloides* Michx., *P. balsamifera* L., *P. grandidentata* Michx., and various *Salix* spp.) and of the Betulaceae (e.g., *Betula papyrifera* Marsh and *Alnus* spp.). These are the only suitable food plants for *P. glaucus* available at latitudes north of 50° (Scriber, 1984a). The southern subspecies, *P. g. glaucus,* does not survive well on plants of these two families (Scriber, 1983), which may be part of the explanation of its northern distribution limits.

More significant perhaps is the observation that *P. g. glaucus* does not appear to exist farther north than where two generations can be completed (Scriber, 1982). Furthermore, it is noteworthy that the latitude at which two generations can be completed does not depend on the heat units (degree-days) accumulated, but is significantly altered by the food plant that the larva is feeding on. For example, we know that *P. g. glaucus* larvae require approximately 340 degree-days (above a base 10°C threshold) to complete their growth on black cherry leaves, *Prunus serotina* L. Combined with the egg (90 degree-days) and pupal stages (270 degree-days), this results in an absolute minimum of approximately 700 degree-days for one generation. With the egg and larval stages of a second generation conservatively requiring an additional (90 + 340) 430 degree-days (not including adult mating and oviposition delays), we observe that successful completion of the second generation (i.e., reaching the over-wintering pupal stage) realistically requires a total minimal degree-day accumulation of 1130 units above a base temperature of 10°C (Ritland and Scriber, 1985). This, of course, assumes that the black cherry leaves are at their best nutritional quality throughout this period of time, which we know is not the case (see Scriber, 1984b). In fact, it would not be unreasonable to add 150–200 degree-day units to the second larval generation because of the reduced-quality late-season leaves and 50–100 degree-days for eclosion, mating, and oviposition delays in the adults.

The northernmost observations of bivoltine populations for *P. glaucus* in Wisconsin occur at just about the same geographic location as the plant transition (tension) zone in the center of the state (Fig. 10). The northernmost limit to the bivoltine potential is defined by degree-days required on black cherry or tulip tree (the two "best" foods for most rapid growth). On many other food plants, however, a second generation would not be possible except in areas farther to the south, which have a greater total seasonal degree-day accumulation. In fact, the presumed second brood in central New York is really a "false second flight," which is not derived

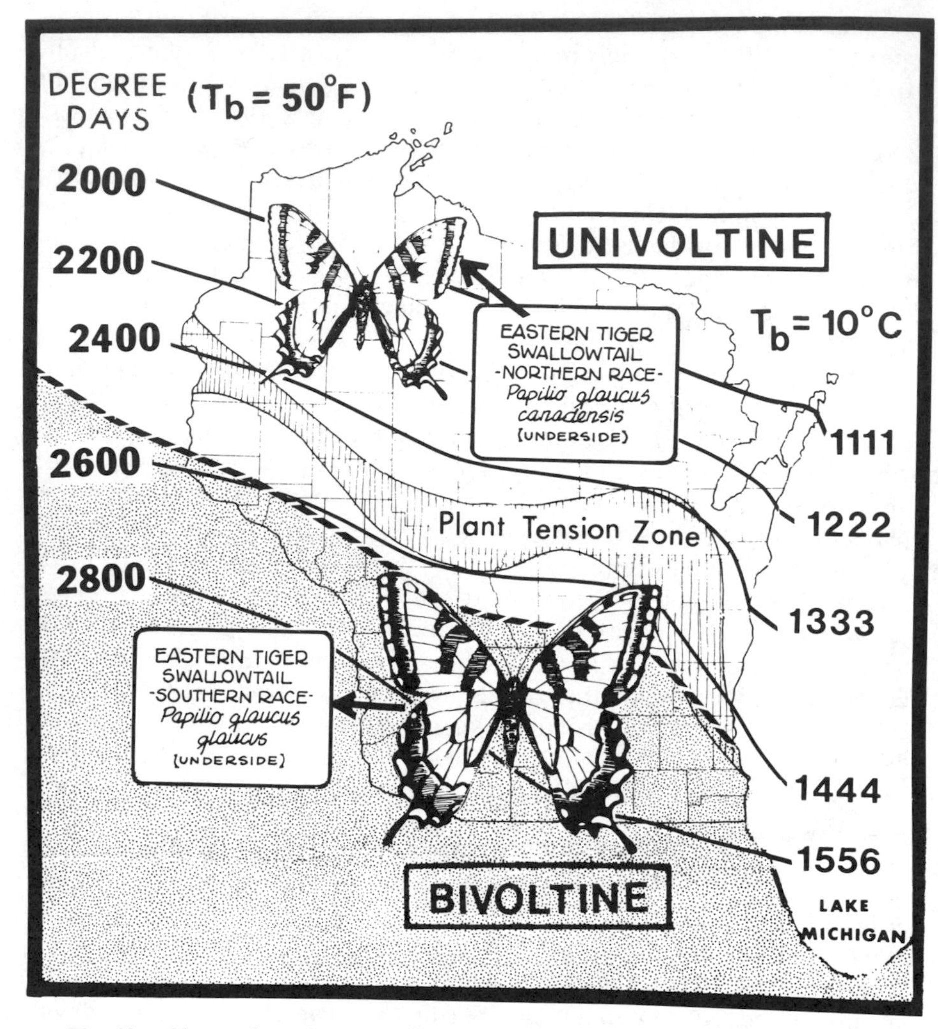

Fig. 10. Observed northernmost limits of bivoltine populations of *Papilio glaucus* in Wisconsin (indicated by shading) in relation to the total seasonal degree-day accumulation above a base temperature of 50°F (and 10°C). After Ritland and Scriber (1985).

from the first generation (Scriber, 1975; Scriber and Lederhouse, 1983; Hagen and Lederhouse, 1985). Whether a second flight occurs at all in Wisconsin depends on the food plant choice. Some food plant choices such as white ash, *Fraxinus americana* L., may require 200–240 more degree-days than black cherry to complete larval development. This is equivalent to a total required accumulation of 1530 to 1670 degree-days for completion of a second generation. Therefore, a second generation is feasible on white ash only in the warmer Mississippi Valley region of southwestern Wisconsin 200 miles farther south than for black cherry (Fig. 11).

Across this entire 40–45° north latitudinal zone, the potential for a second brood of *Papilio glaucus* depends both on the weather (variable seasonal heat unit accumulation) and on the choice of food plant. Similar developmental thresholds (9–10°C) and a similar facultative diapause in the European corn borer, *Ostrinia nubilalis* Hubner (Beck and Apple, 1961), result in an intriguingly similar voltinism pattern at the same latitudes (Showers, 1981). As with *Papilio glaucus,* the zone of thermal unit accumulation of 1389 to 1500 degree-day units above a base 10°C (2500–2700 degree-day units above a base 50°F) represents the zone in which

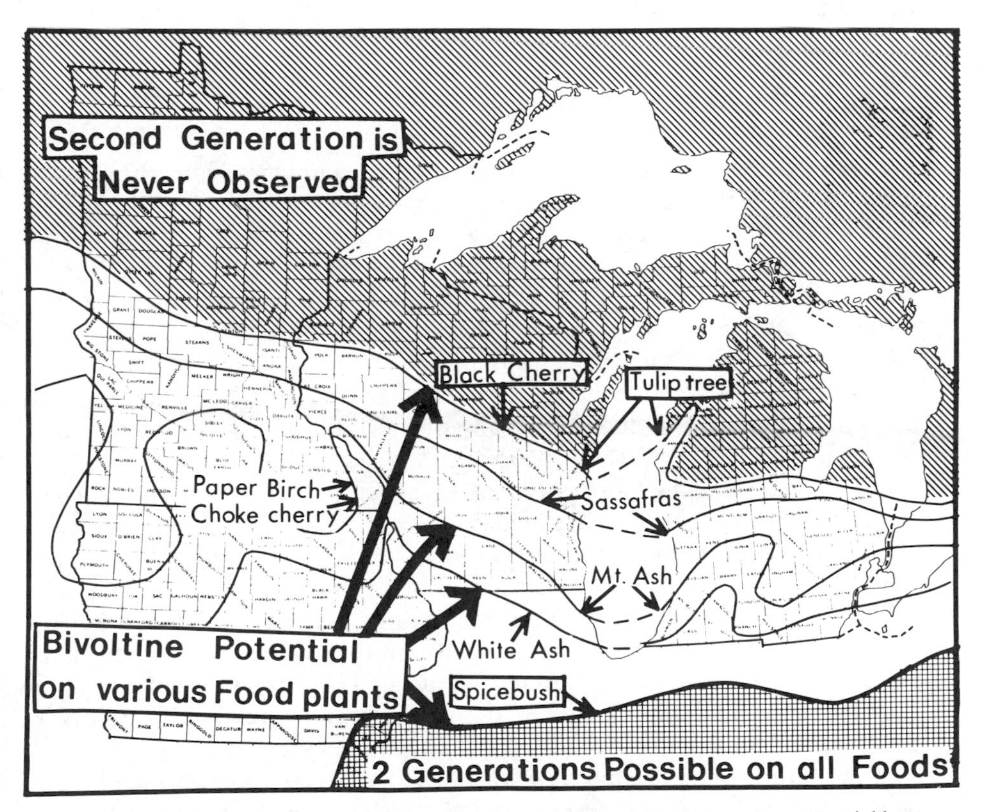

Fig. 11. Northermost limits of bivoltine potential of *P. glaucus* on various potential host plants. Calculated from various rearing studies (see text for details).

European corn borer populations vacillate between one and two generations per year (see Showers, 1981; also compare Fig. 8).

Corresponding with the HVB, the northernmost observations of the European corn borer are not expected to transcend significantly the geographic range of maize production, even with diapause capabilities that would permit a much greater northward distribution in the northern United States and Canada. In contrast, the distribution of *Papilio glaucus* includes most of Canada because the northern subspecies not only has 10 necessary facultative and/or obligate diapause capabilities, but also has the ability to locate, accept, and grow on Canadian food plants (primarily species in the Salicaceae and Betulaceae families). This ability is critical to the successful habitation of these latitudes (50°N to 65°N) in North America since no other suitable food plant species are available (Fig. 12; Scriber, 1984b). Thus, it appears that a shift in host plant utilization ability has enabled *P. glaucus canadensis* to increase its geographic range and overall abundance greatly.

V. PROMETHEA SILK MOTH

A phytogeographic relationship is observed with the promethea silk moth group in the genus *Callosamia* (Lepidoptera: Saturniidae) that is similar to that of the *Papilio glaucus* group. However, unlike *P. glaucus,* the promethea silk moth remains only an oviposition behavior away from successful colonization of Canada. In the southeastern United States, the sweetbay silk moth, *Callosamia securifera* (Maassen), feeds only on sweetbay, *Magnolia virginiana* L. (much as do the populations of *Papilio glaucus australis* Maynard; Fig. 13). Another congeneric silk moth, *C. angulifera* (Walker), is found virtually only within the range of its food plant, tulip tree, *Liriodendron tulipifera* L. (see Scriber, 1983, for additional discussion). The more polyphagous species *Callosamia promethea* (Drury) exhibits a geographic range (Fig. 13) that extends beyond that of sweetbay and tulip tree to include white ash, *Fraxinus americana* L., and black cherry, *Prunus serotina* Ehrhart. As with *Papilio glaucus glaucus* butterflies, *Callosamia promethea* moths locate and accept these food plants as favorites and grow successfully on their leaves in Wisconsin. Like the *P. g. glaucus* butterflies, larvae of *C. promethea* all die on quaking aspen (and other members of the Salicaceae family) in laboratory no-choice situations (Scriber, 1983; Manuwoto, 1984).

Paper birch, *Betula papyrifera* Marshall, has never been verified as a food plant for *Callosamia promethea* (Wagner *et al.,* 1981; Scriber, 1983). However, in laboratory no-choice bioassays, certain populations of *C.*

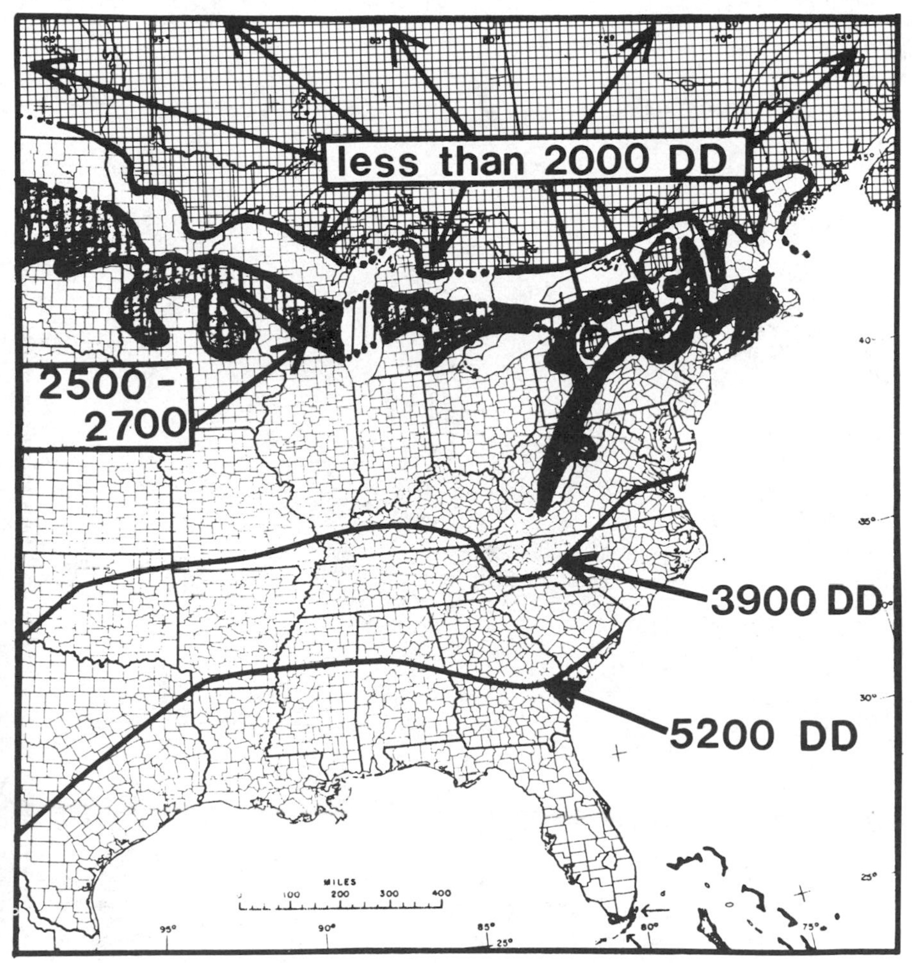

Fig. 12. The 20-year average of seasonal growing degree-day (thermal) accumulations above a base temperature (T_b) of 50°F as reported by the National Meteorological Reporting Service (NOAA), summarized from individual states. The 2500–2700 degree-day (DD) zone (1389–1500°C) corresponds to the northernmost limits of bivoltine populations of *P. glaucus* and also the area of facultative diapause in *O. nubilalis* (the European corn borer). The 3900°F DD (2167°C) zone corresponds to the northernmost limits of trivoltine populations and 5200 DD to the northernmost limits of four generations per year (compare Fig. 8).

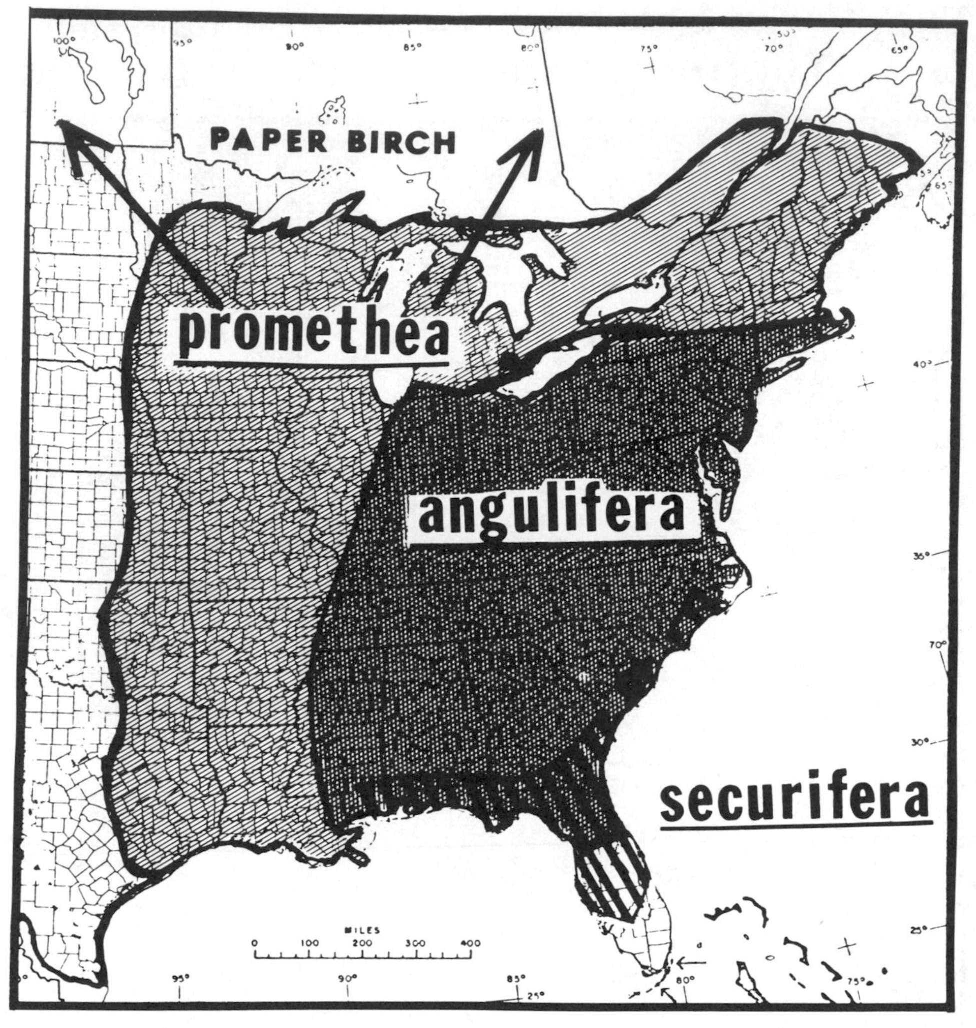

Fig. 13. Geographic range of *Callosamia securifera* (the sweetbay silkmoth), *C. angulifera* (the tulip tree silk moth), and *C. promethea* (a generalist on several additional food plants, including black cherry and white ash in the northernmost western parts of its range).

promethea not only survived the critical neonate stage in birch but grew rapidly and produced very large healthy cocoons and adults. The lowest first-instar mortality was observed for populations from areas that are sympatric with (or very near) paper birch (e.g., central Wisconsin survival was 94%, southern Wisconsin survival 85%, Maine survival 86%, and North Carolina mountain population survival 95% compared with values of only 0 to 70% for populations allopatric to paper birch in Indiana, Illinois, Ohio, Kentucky, Alabama, Tennessee, and New Jersey; see Fig. 14).

Given these facultative and obligate diapause capabilities and the ability of larvae of northern populations to detoxify and convert paper birch

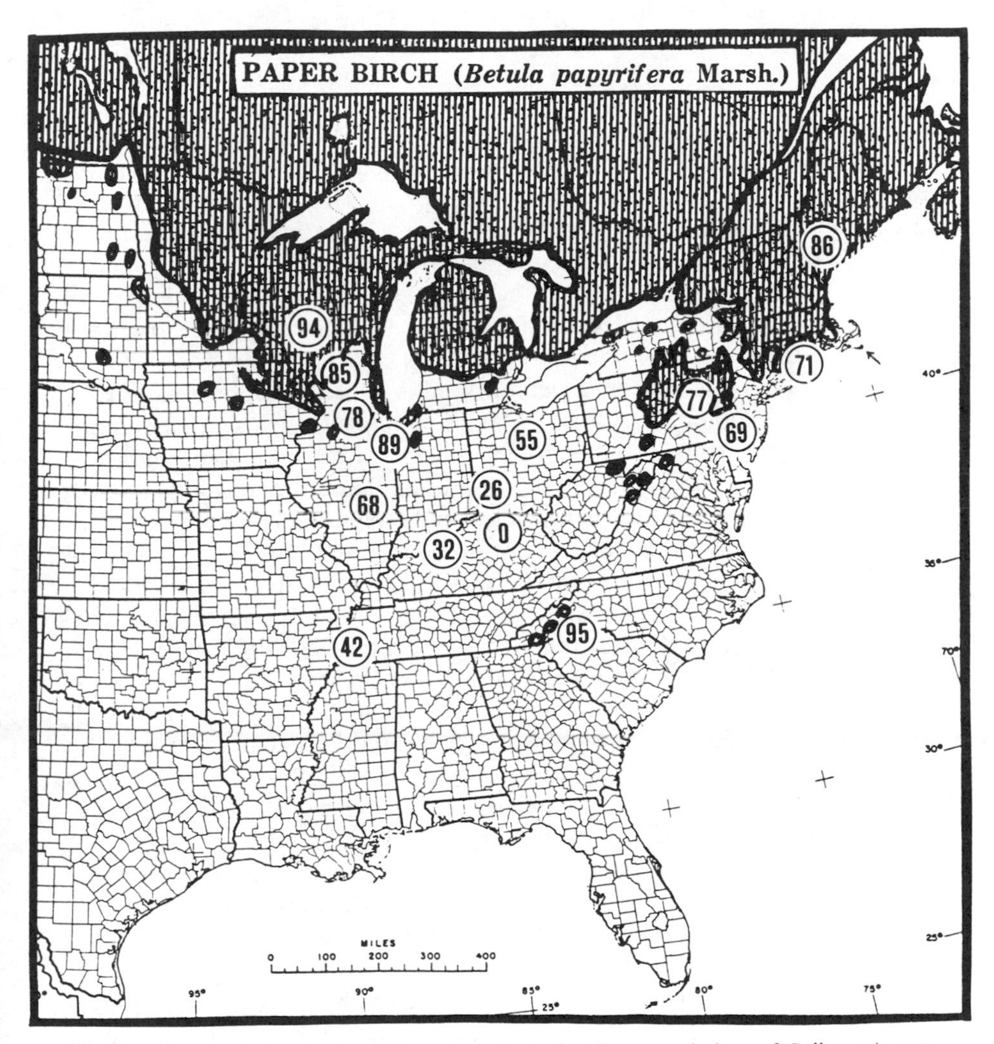

Fig. 14. First-instar (neonate) larval survival of various populations of *Callosamia promethea* on leaves of paper birch, *Betula papyrifera*, in laboratory no-choice feeding bioassays. Survival is presented as a percentage of initial larvae from each geographic source in relation to the natural range of the food plant. A total of 2007 larvae from 107 different females are represented here (Madison, Wisconsin, 1980–1983).

leaves into healthy adults, we might expect occupation of the entire range of paper birch by *C. promethea* moths. Apparently the reason this has not yet occurred is that adult females do not (or at best rarely) oviposit on paper birch leaves. Oviposition avoidance of paper birch may thus be the only barrier to the geographic invasion of most of Canada by *C. promethea.*

VI. CONCLUSIONS

Our understanding of the evolutionary significance of food plants in the speciation of phytophagous insects is surprisingly poor. The behavioral, biochemical, genetic, and environmental processes involved are of concern for agronomists and silviculturalists as well as evolutionary ecologists (see also Gould, 1983). We have described one agronomic pest of corn and one silvicultural pest of red pine that have reached outbreak levels in Wisconsin. In both cases a change of host preference underlies these two economically significant outbreaks. Differential host plant utilization abilities, then, may have a marked impact on insect population dynamics. By virture of a relatively recent (in evolutionary time) host shift, the red pine shoot moth was able to exploit an increasing resource base and develop to outbreak-level populations. Thus abundance may have been favored by conspecific differences in red pines grown on poor soils, representing an even greater effect on insect abundance of differential utilization abilities. Furthermore, human activities in this instance produced conditions that stimulated a recent host shift (in ecological time). That is, the red pine shoot moth was presented with a new and expanding resource; red pines reaching a susceptible age on poor soils (qualitatively different) were present at a greater density than in the natural state. Similarly, the HVB is presently expanding its range, perhaps beyond its original limits as dictated by the presence of hops, as a result of an apparently recent host shift to corn. Along with this range expansion, the utilization of a new and more abundant host has allowed localized population increases or "outbreaks." Incidentally, both the HVB and the red pine shoot moth illustrate how changes in agricultural or silvicultural practices may influence insect population dynamics.

In addition, we have described two species of noneconomic foliage-feeding Lepidoptera that reach an ecologically significant distributional limit in Wisconsin. In one case, successful "invasion" of Canada was accomplished by a subspecies that has both the appropriate ovipositional behavior and the appropriate larval detoxification system for paper birch; in the other case, the larval detoxification system is operational (along with the requisite diapause capabilities), but the adults prefer not to oviposit on suitable food plants and thus are unable to take advantage of this widespread food plant resource across Canada.

The interaction of the nutritional suitability of food plants and temperature-related voltinism patterns has not been well studied. We have presented evidence that both the geographic limits on the insect distribution and the number of generations feasible during a given growing season depend on the plant a population is feeding on. In *Papilio glaucus*, the northernmost limits on a successful second generation for any given seasonal total thermal unit accumulation varies for up to 600 miles (in latitude), depending on the plant species its larvae feed on. Thus, we suggest that food plant nutritional quality is at least as important a consideration in developing models of insect population dynamics as is temperature. More significant is the obvious fact that when appropriate (acceptable and suitable) food plants are absent for a phytophagous insect, the population distribution will end (i.e., zero abundance).

The examples presented herein illustrate the significant role that food plant utilization abilities may play in the population dynamics of phytophagous insects. Changes in host distribution and quality, whether the result of a major ecological host shift, as in the HVB, or of human activities, as in the red pine shoot moth, evidently provide sufficient impetus for major population increases. These conditions may also contribute to the success of insects introduced into new geographic locales. Furthermore, short-term changes in food plant quality, often the result of climatic changes, and subsequently the insect's utilization abilities have often been implicated in insect outbreaks. Examples of noneconomic species further demonstrate the generality of this theme. Although population levels of such insects were not measured, again we are impressed by the way that differential food plant utilization abilities may limit or facilitate their distribution and thus directly affect their population dynamics. It follows that models of insect population dynamics must account for variations in plant quality and in host plant utilization abilities as well as their interaction with temperature effects if we are to have the ability to predict insect outbreaks. This area of study, which has such broad implications for both evolutionary entomology and insect ecology, merits continued and increasing efforts to develop a continuity in our understanding of the processes involved.

ACKNOWLEDGMENTS

This research was supported in part by the National Science Foundation (DEB 7921749 and BSR 8306060), the Graduate School and the College of Agriculture and Life Sciences (Hatch 5134) of the University of Wisconsin, and the Wisconsin Department of Natural

Resources and in part by NC-105 and NC-180 Regional Research. We thank Dan Benjamin, Mark Evans, Bruce Giebink, and John Wedberg for their assistance with various aspects of this work. Discussions with Michael Collins and Robert Lederhouse have been helpful throughout the development of our ideas over the past couple of years.

REFERENCES

Anonymous (1957). "Michigan Forest Pest Detection Program Report for 1957." Mich. Dep. Conserv. For. Div., East Lansing, MI.

Arms, K., Feeny, P., and Lederhouse, R. C. (1974). Sodium stimulus for puddling behavior of tiger swallowtail butterflies, *Papilio glaucus*. *Science* **185,** 372–374.

Beck, S. D., and Apple, J. W. (1961). Effects of temperature and photoperiod on voltinism of geographic populations of the European corn borer, *Pyrausta nubilalis*. *J. Econ. Entomol.* **54,** 550–558.

Bethune, C. J. S. (1873). Insects affecting the hop. *In* "Report of the Entomological Society of Ontario, 1882, Toronto," pp. 27–34. Entomol. Soc. Ont.

Bockheim, J. G., Lee, S. W., and Leide, J. E. (1983). Distribution and cycling of elements in a *Pinus resinosa* plantation ecosystem, Wisconsin. *Can. J. For. Res.* **13,** 609–619.

Brittain, W. H. (1915). *Hydroecia micacea* as a garden pest. *N. S. Entomol. Soc. Proc.* **2,** 96–97.

Burgess, A. H. (1964). "Hops." Wiley (Interscience), New York.

Campbell, C. A. M. (1983). Antibiosis in hop (*Humulus lupulus*) to the damson-hop aphid, *Phorodon humuli*. *Entomol. Exp. Appl.* **33,** 57–62.

Carlson, R. B., and Butcher, J. W. (1967). Biology and behavior of Zimmerman pine moth, *Dioryctria zimmermani,* in Michigan. *Can. Entomol.* **99,** 529–536.

Carlson, R. B., and Wilson, L. F. (1967). Zimmerman pine moth. *USDA For. Serv. Pest Leafl.* **106,** 1–6.

Chawla, S. S., Perron, J. M., and Cloutier, M. (1975). Preliminary observations on an artificial diet for *Hydroecia micacea* (Esp.) (Lepidoptera: Noctuidae). *Ann. Soc. Entomol. Que.* **20,** 8–10.

Collins, M. M. (1984). Genetics and ecology of a hybrid zone in *Hyalophora* (Lepidoptera: Saturniidae). *Univ. Calif. Publ. Entomol.* **104,** 1–93.

Comstock, J. H. (1883). The hop-vine borer, or hop grub. *Am. Agric.* **42,** 275.

Coulson, R. N., and Franklin, R. T. (1970). The biology of *Dioryctria amatella* (Lepidoptera: Phycitidae). *Can Entomol.* **102,** 697–684.

Davis, E. L. (1957). Morphological complexes in hops (*Humulus lupulus* L.) with special reference to the American race. *Ann. Mo. Bot. Gard.* **44,** 271–294.

Deedat, Y. D. (1981). The life history and control of the potato stem borer (Esper) (Lepidoptera: Noctuidae) *Hydraecia immanis* on corn in Ontario. M.S. Thesis, University of Guelph, Ontario.

Deedat, Y. D., and Ellis, C. R. (1983). Damage caused by potato stem borer (Lepidoptera: Noctuidae) to field corn. *J. Econ. Entomol.* **76,** 1055–1060.

Deedat, Y. D., Ellis, C. R., and Elmhist, J. (1982). Studies on the control of potato stem borer, *Hydraecia micacea* (Lepidoptera: Noctuidae), in field corn. *Proc. Entomol. Soc. Ont.* **113,** 43–51.

Deedat, Y. D., Ellis, C. R., and West, R. J. (1983). Life history of the potato stem borer (Lepidoptera: Noctuidae) in Ontario. *J. Econ. Entomol.* **76,** 1033–1037.

Denno, R. F., and Dingle, H. (1981). "Insect Life History Patterns: Habitat and Geographic Variation." Springer-Verlag, Berlin and New York.

Diehl, S. R., and Bush, G. L. (1984). An evolutionary and applied perspective of insect biotypes. *Annu. Rev. Entomol.* **29,** 471–504.

Dingle, H., and Hegmann, J. P. (1982). "Evolution and Genetics of Life Histories." Springer-Verlag, Berlin and New York.

Dodge, C. R. (1882). The hop-vine borer. *Can. Entomol.* **14,** 93–96.

Ebel, G. H. (1965). The *Dioryctria* coneworms of North Florida pines (Lepidoptera: Phycitidae). *Ann. Entomol. Soc. Am.* **58,** 623–630.

Farrier, M. H., and Tauber, O. C. (1953). *Dioryctria disclusa* Heinrich, n. sp. (Phycitidae) and its parasites in Iowa. *Iowa State Coll. J. Sci.* **27,** 495–507.

Fletcher, J. (1883). The hop-vine borer, "the collar-worm of the hop." *In* "Report of the Entomologist and Botanist," Can. Exp. Farms. Rep. 1882, pp. 149–151.

Forbes, W. T. M. (1954). Lepidoptera of New York and neighboring states. Noctuidae. Part 3. *Mem—N.Y., Agric. Exp. Stn.* (*Ithaca*) **329,** 1–443.

French, N., Ludlam, F. A. B., and Wardlow, L. R. (1973). Biology, damage and control of the rosy rustic moth, *Hydraecia micacea* (Esp.) on hops. *Plant Pathol.* **22,** 58–64.

Futuyma, D. J., and Peterson, S. C. (1985). Genetic variation in the use of resources by insects. *Annu. Rev. Entomol.* **30,** 217–238.

Giebink, B. L. (1983). Biology, phenology, and control of the hop-vine borer (*Hydraecia immanis* Guenee) in Wisconsin corn. M.S. Thesis, University of Wisconsin, Madison.

Giebink, B. L., Scriber, J. M., and Wedberg, J. L. (1984). Biology and phenology of the hop vine borer, *Hydraecia immanis* Guenee, and detection of the potato stem borer, *H. micacea* (Esper) (Lepidoptera: Noctuidae), in Wisconsin. *Environ. Entomol.* **13,** 1216–1224.

Godfrey, G. L. (1981). Identification and descriptions of the ultimate instars of *Hydraecia immanis* (hop-vine borer) and *H. micacea* (potato stem borer) (Lepidoptera: Noctuidae). *Biol. Notes* (*Ill. Nat. Hist. Surv.*) **114,** 1–8.

Gould, F. (1983). Genetics of plant–herbivore systems: Interactions between applied and basic study. *In* "Variable Plants and Herbivores in Natural and Managed Systems" (R. F. Denno and M. S. McClure, eds.), pp. 599–653. Academic Press, New York.

Gross, E. (1900). "Hops in Their Botanical, Agricultural and Technical Aspect and as an Article of Commerce." Scott-Greenwood, London.

Guenée, A. (1852). *Hydrecia immanis. In* "Histoire naturelle des insectes," Lepidopteres 5, Noctuelities I, pp. 128–129.

Hagen, R. H., and Lederhouse, R. C. (1985). Polymodal emergence of the tiger swallowtail, *Papilio glaucus* (Lepidoptera: Papilionidae): Source of a false second generation in central New York State. *Ecol. Entomol.* **10,** 19–28.

Hainze, J. H., and Benjamin, D. M. (1983). Bionomics and pattern of attack of the red pine shoot moth, *Dioryctria resinosella* (Lepidoptera: Pyralidae), in Wisconsin. *Can. Entomol.* **115,** 1169–1175.

Hainze, J. H., and Benjamin, D. M. (1984). Impact of the red pine shoot moth, *Dioryctria resinosella* (Lepidoptera: Pyralidae), on height and radial growth in Wisconsin red pine plantations. *J. Econ. Entomo.* **77,** 36–42.

Hainze, J. H., and Benjamin, D. M. (1985). Partial life tables for the red pine shoot moth, *Dioryctria resinosella* (Lepidoptera: Pyralidae), in Wisconsin red pine plantations. *Environ. Entomol.* **14,** 545–551.

Hawley, I. M. (1918). Insects injurious to the hop in New York with special reference to the hop grub and the hop redbug. *Mem.—N.Y., Agric. Exp. Stn.*(*Ithaca*) **15,** 141–224.

Heinrich, C. (1956). American moths of the subfamily Phycitinae. *Bull.—U.S. Natl. Mus.* **207,** 1–581.

Hibbard, B. H. (1904). Hops. *In* "History of Agriculture in Dane County, Wisconsin," Wis. Bull. Econ. & Political Sci., Ser. 1, 67–214, pp. 149–154.

Howard, L. O. (1897). The hop plant borer. In some insects affecting the hop plant. *U.S. Div. Entomol. Bull.* **7,** 40–44.

Jobin, L. J. (1963). Etude preliminaire de la biologie due perce-tige de la pomme de terre *Hydraecia micacea* Esp. (Lepidoptera: Noctuidae). *Ann. Entomol. Soc. Que.* **8,** 111–120.

Kondakov, N. I., and Nogina, L. A. (1968). The potato moth on hops. *Zashch. Rast. (Leningrad)* **13,** 47.

Kramer, P. J., and Kozlowski, T. T. (1979). "Physiology of Woody Plants." Academic Press, New York.

Lyons, L. A. (1957). Insects affecting seed production in red pine. II. *Dioryctria disclusa* Heinrich, *D. abietella* (D & S), and *D. cambiicola* (Dyar) (Lepidoptera: Phycitidae). *Can. Entomol.* **89,** 70–79.

MacLeod, J. M., and Daviault, L. (1963). Notes on the life history and habits of the spruce coneworm, *Dioryctria reniculella* (Grt.) (Lepidoptera: Pyralidae). *Can. Entomol.* **95,** 309–316.

Manuwoto, S. (1984). Feeding and growth of three Lepidoptera species as influenced by natural and altered nutrient and allelochemical concentration in their diet. Ph.D. Thesis, University of Wisconsin, Madison.

Mattson, W. J., Jr. (1971). Relationship between cone crop size and cone damage by insects in red pine seed-production areas. *Can. Entomol.* **103,** 617–621.

Merkel, E. P. (1982). Biology of the baldcypress coneworm, *Dioryctria pygmaella* Ragonot (Lepidoptera: Pyralidae), in North Florida. *J. Ga. Entomol. Soc.* **17,** 13–19.

Muka, A. (1976). A new corn pest is south of border. *Hoard's Dairyman* **121,** 688.

Munroe, E. (1959). Canadian species of *Dioryctria zeller* (Lepidoptera: Pyralidae). *Can. Entomol.* **91,** 65–72.

Mutuura, A. (1982). American species of *Dioryctria* (Lepidoptera: Pyralidae). VI. A new species of *Dioryctria* from eastern Canada and northeastern United States. *Can. Entomol.* **114,** 1069–1076.

Mutuura, A., and Munroe, E. (1972). American species of *Dioryctria* (Lepidoptera: Pyralidae). III. Grouping of species: Species of the *Auranticella* group, including the Asian species, with the description of a new species. *Can. Entomol.* **104,** 609–625.

Mutuura, A., and Munroe, E. (1973). American species of *Dioryctria* (Lepidoptera: Pyralidae). IV. The *schuetzeela* group and the taxonomic status of the spruce cone moth. *Can. Entomol.* **105,** 653–668.

Mutuura, A., and Munroe, E. (1979). American species of *Dioryctria* (Lepidoptera: Pyralidae). V. Three new cone-feeding species from the southeastern United States. *J. Ga. Entomol. Soc.* **14,** 290–304.

Mutuura, A., Munroe, E., and Ross, D. A. (1969). American species of *Dioryctria* (Lepidoptera: Pyralidae). I. Western Canadian species of the *zimmermani* group. *Can. Entomol.* **101,** 1009–1023.

Nordstrom, F., Wahgren, E., and Tullgren, A. (1941). *Hydroecia*. *In* "Storfjarilar Macrolepidoptera," pp. 187–188. Nordisk-Fumiljeboks Forlags Aktiebolag, Stockholm, Sweden.

Norris, M. J. (1936). The feeding habits of the adult Lepidoptera, Heteroneura. *Trans. R. Entomol. Soc.* **85,** 61–90.

Phillips, R. E., and Phillips, S. H. (1984). "No-tillage Agriculture: Principals and Practices." Van Nostrand-Reinhold, Princeton, New Jersey.

Rhoades, D. F. (1983). Herbivore population dynamics and plant chemistry. *In* "Variable Plants and Herbivores in Natural and Managed Systems" (R. F. Denno and M. S. McClure, eds.), pp. 155–220. Academic Press, New York.

Rings, R. W., and Metzler, E. W. (1982). Two newly detected noctuids (*Hydraecia immanis* and *Hydraecia micacea*) of potential economic importance in Ohio. *Ohio J. Sci.* **82,** 299–302.

Ritland, D. B., and Scriber, J. M. (1985). Larval developmental rates of three putative subspecies of tiger swallowtail butterflies, *Papilio glaucus,* and their hybrids in relation to temperature. *Oecologia* **65,** 185–193.

Rose, A. H., and Lindquist, O. H. (1973). Insects of eastern pines. *Can. For. Serv., Publ.* **1313,** 1–127.

Schaber, B. D. (1981). Description of the immature stages of *Dioryctria taedae* Schaber and Wood, with notes on its biology and that of *D. disclusa* Heinrich (Lepidoptera: Pyralidae). *Proc. Entomol. Soc. Wash.* **83,** 680–689.

Schaber, B. D., and Wood, F. E. (1971). A new species of *Dioryctria* infesting loblolly pine (Lepidoptera: Pyralidae). *Proc. Entomol. Soc. Wash.* **73,** 215–223.

Schwartz, B. W. (1973). History of hops in America. *In* "Steiner's Guide to American Hops" (Louis Steiner Gumbel), 3rd ed., pp. 37–71. Hopsteiner, S. S. Steiner, Inc., New York.

Scriber, J. M. (1975). Comparative nutritional ecology of herbivorous insects: Generalized and specialized feeding strategies in the Papilionidae and Saturniidae (Lepidoptera). Ph.D. Thesis, Cornell University, Ithaca, New York.

Scriber, J. M. (1980). The potato stem borer and hop-vine borer: New problems in corn? *Proc. Wis. Aglime, Fert., Pest. Manage. Conf.* **19,** 23–31.

Scriber, J. M. (1982). Food plants and speciation in the *Papilio glaucus* group. *Proc. Int. Symp. Insect-Plant Relationships, 5th,* pp. 307–314.

Scriber, J. M. (1983). The evolution of feeding specialization, physiological efficiency, and host races. *In* "Variable Plants and Herbivores in Natural and Managed Systems" (R. F. Denno and M. S. McClure, eds.), pp. 373–412. Academic Press.

Scriber, J. M. (1984a). Larval foodplant utilization by the World Papilionidae (Lepidoptera): Latitudinal gradients reappraised. *Tokurana* (*Acta Rhaloceralogica*) **2,** 1–50.

Scriber, J. M. (1984b). Host-plant suitability. *In* "Chemical Ecology of Insects" (W. J. Bell and R. J. Carde, eds.), pp. 159–202. Sinauer Assoc., Sunderland, Massachusetts.

Scriber, J. M. (1984c). Nitrogen nutrition of plants and insect invasion. *In* "Nitrogen in Crop Production" (R. D. Hauck, ed.), pp. 441–460. ASA-CSSA-SSSA, Madison, Wisconsin.

Scriber, J. M. (1986). Origins of regional feeding abilities in the tiger swallowtail butterfly: Ecological monophagy and the *Papilio glaucus australis* subspecies in Florida. *Oecologia.* **71,** 94–103.

Scriber, J. M., and Lederhouse, R. C. (1983). Temperature as a factor in the development and feeding ecology of tiger swallowtail caterpillars, *Papilio glaucus* (Lepidoptera). *Oikos* **40,** 95–102.

Scriber, J. M., and Lintereur, G. L. (1982). A melanic aberration of *Papilio glaucus canadensis* from northern Wisconsin. *J. Res. Lepid.* **21,** 199–201.

Scriber, J. M., Lintereur, G. L., and Evans, M. H. (1982). Foodplant suitabilities and a new oviposition record for *Papilio glaucus canadensis* (Lepidoptera: Papilionidae) in northern Wisconsin and Michigan. *Great Lakes Entomol.* **15,** 39–46.

Seppanen, E. J. (1970). *Hydroecia.* In "Suuperhostoukkien Ravintokasuit," pp. 81–82. Werner Soderstrom Osakeyhtio, Helsinki, Finland.

Showers, W. B. (1979). Effect of diapause on the migration of the European corn borer into the southwestern United States. *In* "Movement of Highly Mobile Insects: Concepts and Methodology in Research'' (R. L. Rabb and G. G. Kennedy, eds.), pp. 420–430. North Carolina State Univ. Press, Raleigh.

Showers, W. B. (1981). Geographic variation of the diapause response in the European corn borer. *In* ''Insect Life History Patterns'' (R. F. Denno and H. Dingle, eds.), pp. 97–111. Springer-Verlag, Berlin and New York.

Small, E. (1981). A numerical analysis of morpho-geographic groups of cultivars of *Humulus lupulus* based on samples of cones. *Can. J. Bot.* **59,** 311–324.

Smith, J. B. (1884). The hop grub. *U.S. Div. Entomol. Bull.* **4,** 34–39.

Smith, J. B. (1899). Contributions toward a monograph of the Noctuidae of boreal North America. Revision of the *Genus Hydroecia. Trans. Am. Entomol. Soc.* **26,** 1–48.

Teal, P., West, R. J., and Laing, J. E. (1983). Identification of a blend of sex pheromone components of the potato stem borer (Lepidoptera: Noctuidae) for monitoring adults. *Proc. Entomol. Soc. Ont.* **114,** 15–19.

Thorne, H. W. (1980). ''Wisconsin Coniferous Plantation Survey—1979.'' Wis. Dep. Nat. Res., Div. For. Rec., Bur. For.

Wagner, W. H., Hansen, M. K., and Mayfield, M. R. (1981). True and false foodplants of *Callosamia promethea* (Lepidoptera: Saturniidae) in Southern Michigan. *Great Lakes Entomol.* **14,** 159–165.

West, D. A., and Hazel, W. N. (1982). An experimental test of natural selection for pupation site in swallowtail butterflies. *Evolution* (*Lawrence, Kans.*) **36,** 152–159.

Wright, L. C., Berryman, A. A., and Gurusiddiak, S. (1979). Host resistance to the fir engraver beetle, *Scolytus ventralis* (Coleoptera: Scolytidae). 4. Effect of defoliation on wound monoterpene and innerbark carbohydrate concentrations. *Can. Entomol.* **111,** 1255–1262.

Chapter 18

Phenotypic Plasticity and Herbivore Outbreaks

PEDRO BARBOSA

Department of Entomology
University of Maryland
College Park, Maryland 20742

WERNER BALTENSWEILER

Institut für Phytomedizin
ETH-Zentrum/CLS
CH-8092 Zurich, Switzerland

The more diversified the descendents from any one species become in structure, constitution and habits by so much will they be better able to seize on many and widely diversified places in the policy of nature and so be enabled to increase in numbers.

Charles Darwin, 1859

I. INTRODUCTION

In 1961 B. P. Uvarov noted that "it is the physiological variability and the adaptability of a temporary, reversible kind closely linked with equally temporary environmental changes, which are important in population dynamics." In this chapter we reemphasize and support Uvarov's assertion. That is, variability among individuals of a population, particularly among insect herbivores, is of importance for population dynamics.

In this chapter we propose that insect species existing in heterogeneous environments exhibit phenotypic variability that is manifested at the population level as phenotypic plasticity. Phenotypic plasticity allows existence and persistence in environments that are highly variable in time, within a season or from year to year, and in location. It facilitates adaptation to environmental heterogeneity and uncertainty and ensures survival and reproduction of at least some proportion of the population under any given set of environmental conditions. Although all variability is ultimately genetically based, what we are suggesting is that in heterogeneous habitats there are adaptive advantages to genomes that allow for environmentally induced expression of many phenotypes (i.e., for great phenotypic variation) within a single population. This phenotypic plasticity may range from continuous variation in one or more traits among individuals of a population to more predictable fixed sets of traits or polymorphisms. These environmentally induced changes may be transferred from generation to generation by various mechansisms like maternal influence and short-term differential survival.

In species that exhibit phenotypic plasticity, populations consist of one or more phenotypes, each providing near maximal fitness for a given set of mean environmental conditions. No one phenotype persist due to the alteration of sets of environmental conditions. The multiple-phenotype condition persists because of the continuous alteration of environmental conditions or biologically relevant factors (Fig. 1). Indeed, implicit in the concept of phenotypic plasticity is that polymorphism represents an adaptation to unpredictable but recurrent instabilities (in time and space) in the onset, duration, and intensity of environmental factors. Phase polymorphism is an example of the most advanced form of phenotypic plasticity. In such a polymorphism, there is a set number of morphs but environmental triggers still determine the proportion of each morph and whether females of a given morph produce progeny of their own morph. Phase polymorphism reflects the basic features of phenotypic plasticity and makes a species (or the progeny of certain females) able to survive and reproduce, in spite of the unstable conditions that may exist in its usual environment.

We further propose that the capacity of at least some proportion of a population to survive and increase in abudance because of its ability to adjust to changing environmental conditions makes species exhibiting phenotypic plasticity more likely to become outbreak species. The concept of phenotypic plasticity can provide testable hypotheses of some underlying causal relationships affecting an insect's ability to reach out-

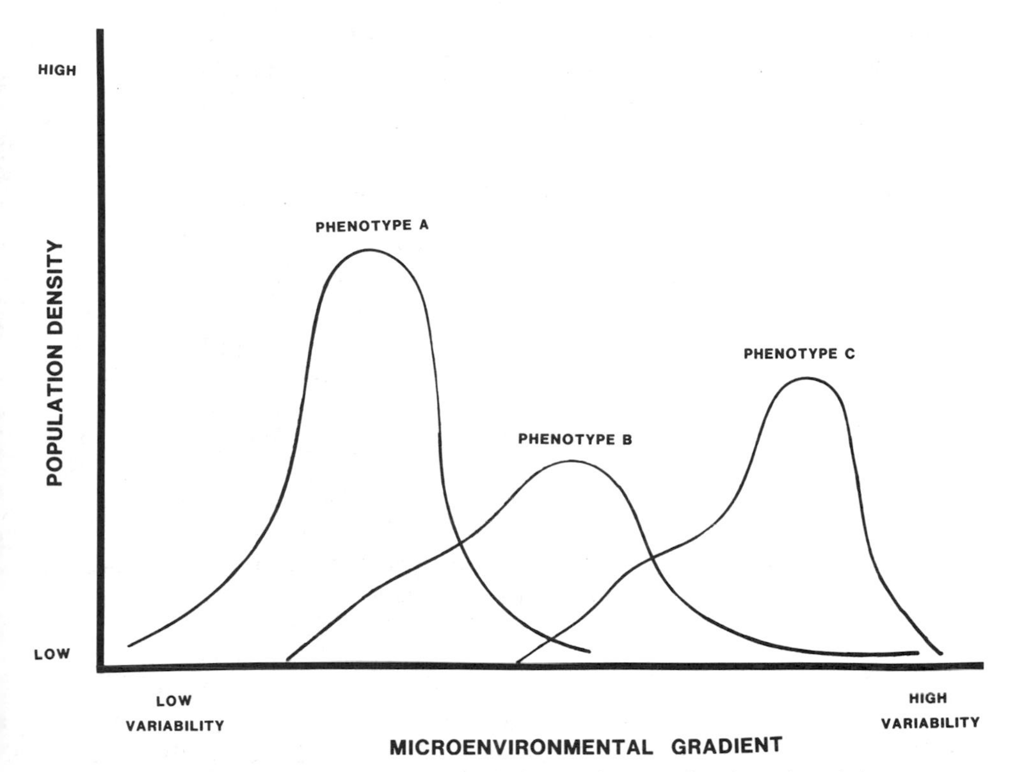

Fig. 1. Hypothetical abundance and relative persistence of variants (morphs) among individuals of an insect population.

break levels. There are also numerous consequences implicit in the concept of phenotypic plasticity that are important for understanding the management of insect pests.

In the following sections we define phenotypic plasticity and relate it to other similar concepts. This is followed by a consideration of selected theoretical issues aimed at answering the following questions. Are all life history traits subject to environmentally induced variation? How is phenotypic plasticity maintained? What is the relationship between phenotypic plasticity and insect outbreaks? Can species be categorized as phenotypically plastic based on the capacity of the environment to support population growth and the ability of insect species to respond accordingly? Following these theoretical considerations we provide illustrations of the mechanisms that can trigger (i.e., induce) the expression of phenotypic plasticity and brief but specific examples of phenotypic plasticity. The penultimate section provides a discussion of the relationship between phenotypic plasticity and numerical increase. The final section provides the reader with some insight into the nature of species that tend to exhibit phenotypic plasticity. The chapter concludes with suggested testable hypotheses that might guide further research.

A. Phenotypic Plasticity and Related Concepts

We propose to use the term "phenotypic plasticity" to describe environmentally induced adaptive behavioral, physiological, and/or ecological variation among individuals of a population. The term "population quality" has been used in a multitude of ways, ranging from any change in growth and development to the occurrence of complex polymorphism (e.g., Hirata, 1957, 1963; Wellington, 1957, 1960, 1962, 1964, 1965; Nakamura, 1966; Iwao and Wellington, 1970; Danthanarayana, 1975; Myers, 1978; Capinera, 1979; Sahota and Thomson, 1979; Barbosa *et al.,* 1981; Moran *et al.,* 1981). In fact, it has been used in so many ways that it perhaps has lost its usefulness. To avoid confusion it is essential to relate the concept of phenotypic plasticity to similar or related concepts such as population quality, population vigor, and geographic variation.

The term "population quality" is rarely specifically defined and often for any given paper one must analyze the examples given in order to decipher the author's meaning of the term (see references just cited). One of the few specific definitions is that of Sahota and Thomson (1979). They define the term as a "measure of reproductive efficiency and survival capability of the organisms that comprise the population." This definition is somewhat related to the concept of phenotypic plasticity, but it is restrictive in that it excludes critical behavioral and physiological factors that also vary in response to environmental heterogeneity and alter the course of population growth and development. A second related phenomenon is population vigor. The latter can be described as the state or condition of individuals of a population that influence their response to and/or their vulnerability to inimical environmental factors. It represents primarily physiological or physiologically based behavioral functions that vary in response to a wide array of factors including but not limited to nutritional quality, nonnutritional chemicals, physical factors, parasitism, artificial selection, and so on. This condition has sometimes been referred to as population quality (Talerico, 1981), although it differs from the concept defined by Sahota and Thomson (1979). A final related concept, geographic variation, is distinct from phenotypic plasticity because the former involves interpopulation variation, whereas the latter is an intrapopulational phenomenon.

B. Phenotypic Plasticity: Elements of a Concept

How is population plasticity maintained? Why is there not selection for a fixed set of morphs? Charnov (1979) states that an evolutionary stable strategy may consist of a set of actions with a particular one appropriate for a particular patch. This concept is somewhat analogous to population plasticity, in that one has a set of morphs, with particular ones appropriate for particular patches or habitat types. No one morph persists or dominates because no one patch or habitat type persists or dominates in time and space.

What traits exhibit phenotypic plasticity? The traits that characterize population plasticity may vary with the nature of environmental heterogeneity. For example, the cotton stainer bug, *Dysdercus bimaculatus,* is a colonizing insect living on seed crops that are short-lived and that mature at various times throughout a season. Although characteristics like the timing of the first clutch and the interclutch interval are under direct genetic control, other traits like the size of the first clutch or the number of clutches and the total number of eggs produced are influenced by directional changes in the environment and are of great importance in numerical changes in the population. As seen in this example phenotypic plasticity may be expressed only in some traits. Variation in these traits should reflect the patterns of certainties and uncertainties in the insect's environment (Derr, 1980).

Phenotypic variation that allows for the survival of individuals in populations in highly heterogeneous environments can be viewed as simply the realized manifestation of multiple phenotypes within a population. The degree of phenotype plasticity required to provide a range of variation sufficient to track a variable environment can occur only if there exists sufficient genetic variation. Most characteristics of natural populations are probably under stabilizing selection or some combination of stabilizing and fluctuating directional selection that reduces genetic variance. However, one counterdirectional force may be mutations that maintain genetic variability in polygenic characters. This genetic variability facilitates phenotypic plasticity. For example, Lande (1977) cites studies on *Drosophila* demonstrating that observed amounts of mutation in polygenic characters are sufficient to maintain heritable variation even under strong stabilizing selection. Indeed, failure to maintain heritable variation may be counteradaptive for many species. Plasticity of a given character can itself be considered a trait and may be under genetic control (Caswell, 1983; Bradshaw, 1965). Thus, genetic heterogeneity and its concomitant phenotypic plasticity reduce the likelihood of local population extinctions. An alternative view is that of Lees (1961, 1966), who concluded that, although nuclear genes participate in the realization of polymorphic characters, genetic mechanisms are not known to influence the choice of alternative developmental pathways involved in the differentiation of a polymorphic organism.

Finally, not even the "most adaptive" trait is expressed consistently and unerringly. We suggest that variation from mean behavior, physiology, and so on is not an evolutionary error but an adaptive paradigm. That is, one may often find variance in a behavior or physiological process that appears fully adaptive. For example, in any given cohort of larvae the vast majority may reject an inappropriate host, whereas a few accept the inappropriate host. Pronounced individual variation does occur in insect feeding behavior and host selection (Schoonhoven, 1977; Prokopy *et al.*, 1982). Why should such variance be maintained? Ostensibly adaptive, fixed behavior may not be adaptive at all times. When given a choice, female butterflies do not always prefer to oviposit on the species or variety of food plant on which their larvae will grow and survive the best (Chew, 1980; Jones and Ives, 1979; Wiklund, 1975). In such circumstances the above-noted variation (i.e., feeding on a novel host) allows survival of some proportion of a female's progeny.

Is there a relationship between the concept of phenotypic plasticity and insect outbreak theory? An affirmative answer is provided by models that consider environmental heterogeneity. Such a model is that of Whittaker and Goodman (1979), which proposes three strategies for adaptations to environmental adversity and resource fluctuations as paradigms emerging from modeling the effects of microhabitat heterogeneity on population fluctuations and growth. These proposed strategies result from *adversity selection* (which involves survival in a predominantly unfavorable environment), *exploitation selection* (which relates to the utilization of intermittently favorable and unpredictable environments), and *saturation selection* (which involves interactions and survival in a predominantly favorably and fully occupied environment). Although this model deals with the evolution of population-level strategems, it does provide an example of the relevance of phenotypic plasticity to population phenomena.

Individuals in populations under certain selective pressures are probably more likely than other individuals to exhibit phenotypic plasticity. Individuals of exploitation-selected species are probably the most likely to exhibit phenotypic plasticity. The adversity-selected species is one for which the environment's carrying capacity is usually at a minimum with occasional occurrences of significantly increased carrying capacity. For exploitation-selected species, whose environment is unpredictable and intermittently favorable, carrying capacity fluctuates broadly around an intermediate range. Finally, for saturation-selected species, carrying capacity is usually near a maximum with occasional episodes of significantly reduced carrying capacity. The opportunities for rapid increase are infrequent in adversity-selected species since the model would predict that they are under the most pressure to evolve the greatest efficiency of survival during unfavorable periods. Whittaker and Goodman further suggest that the capacity for population outbreaks may persist but only coincidentally or as a result of some minimal cost-compromise tactic.

Because exploitation-selected species exist in variable and unpredictable environments, one would expect them to be opportunists or generalists and thus be able to adapt to existence at various densities and to various degrees of environmental stress. There should be frequent periods of strong selection for rapid increase during favorable times and frequent periods of strong selection for the capacity to dampen fast decrease in unfavorable times (Whittaker and Goodman, 1979). Thus, although species exhibiting the exploitation or saturation strategies may both be outbreak species, exploitation-selected species are perhaps most likely to fall in that category. In addition, species in this group would be ones in which various levels of phenotypic plasticity would most likely be expressed, since it is these species whose environment is unpredictable and intermittent.

To the extent that adaptations for fast growth, for example, may be disadvantageous during an episode of population decline (but advantageous at other times), this and other characters should be reduced or made facultative. Otherwise an adaptation that increases the sustainable population density at one end of the gradient may occur at the expense of survival and development at another portion of the environmental gradient (Fig. 1). Thus, the exploitation-selected species can be distinguished by (1) its adoption of the simplest form of population plasticity, that is, continuous variation in life history and demographic traits, (2) some intermediate form of phenotypic plasticity, or (3) its adoption of a fixed polymorphism (discontinuous variation). Saturation-selected species are likely to be specialized with respect to density and therefore reflect adaptations to the pressures of interactions of individuals with one another. Susceptibility to other environmental factors is thus essentially irrelevant. This is merely one demonstration of the way a model can describe and predict a set of life history strategies that include the concept of phenotypic plasticity and can potentially increase our understanding of insect outbreaks.

C. Mechanisms Inducing Phenotypic Plasticity

Changes in the age, stability, resource variability, floral and faunal diversity, and other characteristics of habitats can result in changes in the types of individuals (morphs) and/or their proportion within a population

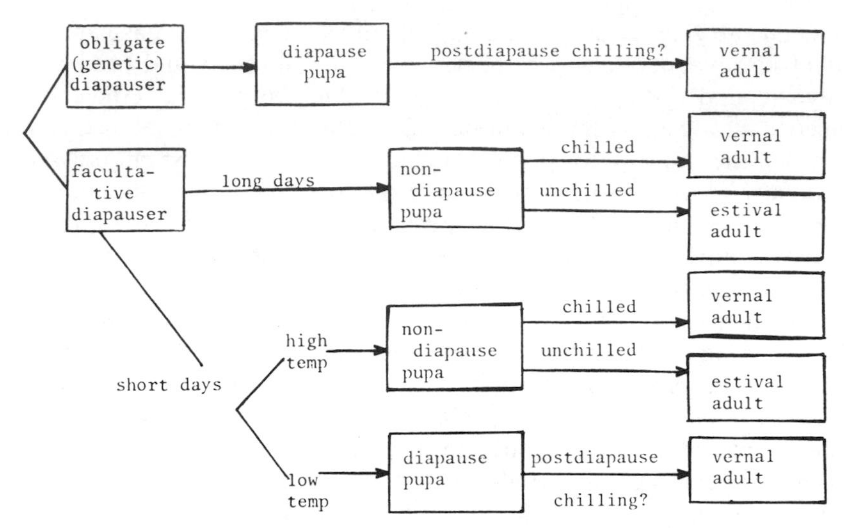

Fig. 2. Schematic representation of the developmental and phenotypic options available to *Pieris napi* under different environmental conditions. From Shapiro (1978).

(Fig. 2). Such changes in what might be termed habitat quality can be an important trigger of phenotypic plasticity. The quality of a habitat or environment may be reflected in its capacity to support survival and reproduction. Although in some circumstances the capacity of an environment to support survival and reproduction may be an all-or-none proposition, in many cases the habitat provides a mosaic of resource units that influence survival and reproduction to different degrees. Phenotypic plasticity can provide an adaptive solution to habitat quality heterogeneity by maximizing resource utilization.

The following are examples of this advanced form of phenotypic plasticity. In these populations there exists a specific number of morphs (green versus brown pupae, short versus long-winged individuals, etc.), but the frequency of their occurrence or the dominance of a particular morph are environmentally mediated. The "success" of each morph is not fixed but reflects the direction of environmental variation from season to season, habitat to habitat, or within a habitat. The environmental cues that facilitate the expression of specific phenotypes can include the quantity and quality of food, the quantity or quality of other resources, variation in physical factors, and so on. This may depend on which of the stages is subject to the most significant mortality from, for example, predation. Whether pupal color is polymorphic or monomorphic in certain Lepidoptera species can be linked to the nature and, more important, to the variability of pupation substrates. In some *Papilio* species, the success of a morph depends on the relative abundance of brown versus green pupation substrates. Pupae of *Papilio glaucus,* which occur very close to the ground in the litter, are brown (West and Hazel, 1979). In contrast, *Papilio polyxenes,* which pupates on grasses, weed stalks, posts, and so on, has both green and brown pupae. The abundance and frequency of morphs depend in large part on the type, frequency, and abundance of substrate types.

Critical environmental factors may not be variable within one location but may change over time. Thus, adaptive features in insects living in such a habitat may also vary over time. An illustration is provided by four curculionid species, *Apion virens* Herbst, *Sitona hispidulus* F., *S. sulcifrons* Thumb., and *Phytonomus nigrirostris* F., in which adult morphology changes with habitat age. The percentage of brachypterous individuals within a population has been shown to increase over 6 years, after the invasion of a newly seeded meadow. Thus, increasing habitat age (and presumably stability) requires lower proportions of dispersing individuals (Stein, 1977).

Climate, weather, or so-called microclimate may also impose itself on organisms in a population, in different ways, throughout the habitat or from season to season. The length of the elytra of adults of the grasshopper, *Zonocerus variegatus* (L.), varies among the individuals of a population depending on the severity of the season. Although individuals with differing elytra lengths occur in dry-season populations, there is nevertheless a clear size dimorphism such that two morphs dominate: a long-winded and a short-winged form. In contrast, no such dimorphism exists in wet-season populations, and individuals are comparable to the dry-season short-winged type (Chapman *et al.,* 1978). More significantly, long-winged individuals have well-developed flight muscles and short-winged types do not. The flight capacity of long-winged individuals would enhance survival under the limitations of the dry season by enabling dispersal from a degenerating environment. Dimorphism in flight capability in response to a deteriorating environment has been well documented in various species (Denno, 1983; Steffan, 1973) and found to be mediated by proximate environment cues rather than existing as inherited traits.

The quality of an insect's food rarely remains constant. Even when the quality is optimal, the amount of food available is often unpredictable and thus important. Variations in parameters affected by food availability, like body size, may have immediate demographic consequences because of their correlation with patterns of reproduction and survival (Palmer, 1985). In the absence of compensatory polymorphism (e.g., size morphs) species may evolve toward maximal utilization of their food resources. However, the need for efficient utilization of resources is often achieved by the development of polymorphism.

Polymorphism in the parasitic wasp *Melittobia chalybii* illustrates this relationship. Like other species of the genus, *M. chalybii* is a parasitoid of both larval bees and wasps as well as their hymenopterous and dipterous parasitoids. When attacking bees or wasps, females gnaw their way into the nest or cell of the host, over its adult life of about 60 to 75 days. Larvae from the first parasitoid eggs to hatch (i.e., the first 12–20 larvae) feed and develop at a rate that is very different from that of the normal ("original") morph of the species (Table 1). When the former individuals become adults, they do not leave the cocoon or cell of their host but mate and deposit their own eggs. The remaining larvae, emerging from eggs laid by the original female, and the larvae from eggs laid by early-morph females exhibit the more typical extended development of the original morph. Morphological differences between the types are also quite dramatic. The wings of early-morph females are short and crumpled and their abdomens are enlarged and distended relative to the original morph. In addition, segments of the early-morph antennae are fused. These and other structural differences are associated with equally dramatic physiological and behavioral differences (Schmieder, 1933).

The polymorphism exhibited by *M. chalybii* may constitute an adaptation providing maximal utilization of each available host and as a consequence the maximization of progeny production (Schmieder, 1933). The original female is minute (<2 mm) relative to its host, and the four to five eggs she lays per day are too few to produce sufficient offspring to consume the entire host. The early morph is adapted only to life in the cocoon or cell, has a female-biased sex ratio, and oviposits from 40 to 60 eggs per female. These eggs plus the original ones laid may produce a total of about 500 to 800 larvae. Original-morph individuals and the progeny of early-

TABLE 1

Comparison of Development in Two Morphs of *Mellitobia chalybii*[a]

	Mean duration of developmental stage (days)	
Developmental stage	Early morph	Original morph
Embryo	4	4
Larva	7	71
Pupa	3	15
Total	14	90
Longevity	2–30	60–75

[a] Based on Schmieder (1933).

morph females are adapted to escaping the cocoon or cell, dispersing, and seeking new hosts.

Transfer experiments also demonstrate that all eggs of both morphs are potentially alike and capable of developing into adults of either morph. In all cases the first few larvae to feed on the host (mainly on hemolymph) give rise to early-morph adults, while all remaining larvae (feeding mainly on remaining tissues) become original-morph adults.

The food consumed by individuals in a population may be a critical triggering mechanism. For a generalist herbivore species polymorphism may be dependent on which of its many potential plant hosts is available at a given time or place or the variability in quality of hosts in one locale. For other generalists and for specialists the absolute amount of food available, the concentration of an essential nutrient, or the changes in essential nutrients (or allelochemicals) over time may be important. For still others, whether food has been utilized and thus presumably altered by other herbivores may influence the degree of variation in their responses to the resource. Polymorphism in *Acheta domesticus* (L.) is associated with food quality, although in several other cricket species it is associated with photoperiod. The incidence of the macropterous form of this species decreases as the protein content of its diet decreases, from 78% macroptery on 30% protein diets to 39% on 15% diets (Patton, 1975). Similarly, nutrients in host plants of aphids like *Myzus persicae* can regulate morph determination (Harrewijn, 1978).

Phenotypic plasticity may also be expressed as continuous variation among the individuals of a population. Thus, clear distinctions in behavior, physiology, or ecology are not easily perceived unless individuals at opposite extremes of the continuum are observed or evaluated. Two examples are noted below where a continuous range of sizes of (gypsy moth or tent caterpillar) eggs as well as associated behaviors among individuals of a given population are highlighted by measurements and observations of the largest and the smallest individuals. As in previous examples the triggers for observed changes can be habitat quality, variation in physical factors, food, and so on.

Studies of the light brown apple moth, *Epiphyas postvittana* (Walk), provide an illustration of the variation in fecundity and size (as well as size-related flight capacity) that occurs as a result of changes in the quality of available food and temperature. Continuous size change cued by environmental effects, primarily temperature and nutrition (larval food quality), ensures maximal fitness during highly variable and suboptimal periods. During the latter, smaller individuals are produced that possess significantly lower wing loadings than large individuals and thus are more adapted for dispersal. Large individuals occur during the cool wet months of the year (Danthanarayana, 1975, 1976).

Variation in a particular physical factor rather than in overall climate may be all that is required to provide the conditions under which phenotypic plasticity is adaptive. *Colias eurytheme* (alfalfa caterpillar) adults can vary in wing color from primarily yellow to predominantly orange at the same site through the year. These changes appear to be under environmental control. Exposure of larvae to higher temperatures (26–32°C) results in the darker forms of the adults, whereas lower temperatures (18°C) result in adults of the lighter coloration (Tuskes and Atkins, 1973). Similarly, the amount of dark pigment in the color pattern of the ambush bug, *Phymata americana,* is strongly negatively correlated with the temperature during the egg stage (Mason, 1976). In both of these examples the implications of the adaptations were not elaborated, but there is nevertheless considerable evidence from many other studies of the importance of pigmentation in mating, thermoregulation, crypsis, and so on. Temperature-induced changes in color patterns are not uncommon and have been noted in many other insects (Sabath *et al.,* 1973; Watabe, 1977; Mariath, 1982).

Adaptation to changing temperatures experienced in a natural environment may be enhanced by phenotypic plasticity. In the lampyrid *Luciola cruciata,* size-related differences in the viability of eggs, associated with temperature differences, maximize the survival value of all the eggs of an individual female. Mean egg weight decreases with female age and seasonally through the oviposition period. Heavy, middle, and light eggs show high hatchability at low, middle, and high temperatures, respectively. When eggs develop at a favorable temperature, a maximum larval size is achieved, and these larvae exhibit enhanced tolerance of starvation when newly hatched (Yuma, 1984).

Phenotypic plasticity in response to heterogeneity and unpredictability of food quantity and quality is expressed in various Lepidoptera species. Like most environmental factors, the quantity and quality of food rarely act independently of other environmental factors. Thus, the amount and/or quality of food may exert an influence while, concurrently, other equally unpredictable and varied environmental requisites exert an influence on the individuals of a population. In species like *Malacosoma californicum pluviale* (western tent caterpillar), *M. neustria* (L.) (lackey moth), *Lymantria dispar* (gypsy moth), and others, phenotypic plasticity is reflected in egg size and the behavior and physiology of larvae from each egg type. In these species behavioral and developmental characteristics are correlated with differences in nutrition and associated unequal partitioning of ovarian substrate during egg production (Wellington and Maelzer, 1967). In the gypsy moth, large eggs have significantly greater yolk reserves (per unit weight) available than small eggs during embryological development and for postdiapause consumption (Capinera *et al.*, 1977). Of various biotic and abiotic factors including virus infection, crowding, and insecticidal treatment, only the host plant species consumed altered the mean egg size of egg masses (Capinera and Barbosa, 1977). Variation in egg size reflects a morphological and behavioral polymorphism within the population. Larvae from small and large eggs vary accordingly (Table 2).

Dispersal behavior also differs among individuals of a population. Although all first instars disperse, the tendency toward repeated dispersal is greatest among larvae emerging from larger eggs (Fig. 3). The rate of movement of larvae and their orientation to light also differ significantly, as is best illustrated by the behavior of small and large larvae (Barbosa *et al.*, 1981; Tables 3 and 4). In the gypsy moth, females are flightless, and phenotypic plasticity is expressed in a critical life history parameter: first-instar dispersion. This adaptive strategy may enable a female's progeny to colonize new sites within a habitat well before overexploitation of currently utilized sites within a habitat (Barbosa, 1978b; Lance and Barbosa, 1981; Lance, 1983). Thus, these differences in dispersal capacity may affect the numerical increase of both resident (established) and peripheral (expanding) populations and as such have important applied consequences.

The impact of phenotypic plasticity on the biology and population dynamics of the western tent caterpillar have been well documented (Wellington, 1957, 1960, 1964, 1965; but see Myers, 1978). The polymorphism observed in this species is also reflected in egg size and associated behav-

TABLE 2

Emergence and Morphometrics of *Lymantria dispar* (Gypsy Moth) First Instars Originating from Small and Large Eggs

Parameter evaluated[a]	Egg size of origin	
	Large	Small
Mean emergence from egg (days)	2.9 (±1.20)	3.3 (±0.73)
Head capsule width (mm)	0.63 (±0.02)	0.58 (±0.01)
Seta (mm)		
A	1.16 (±0.06)	1.10 (±0.10)
B	2.06 (±0.10)	1.88 (±0.14)
C	1.48 (±0.06)	1.39 (±0.11)
Body length (mm)	3.12 (±0.11)	2.84 (±0.11)
Body weight (mg)	0.725 (±0.075)	0.503 (±0.117)

[a] All parameters evaluated are significantly different. Based on Barbosa and Capinera (1978).

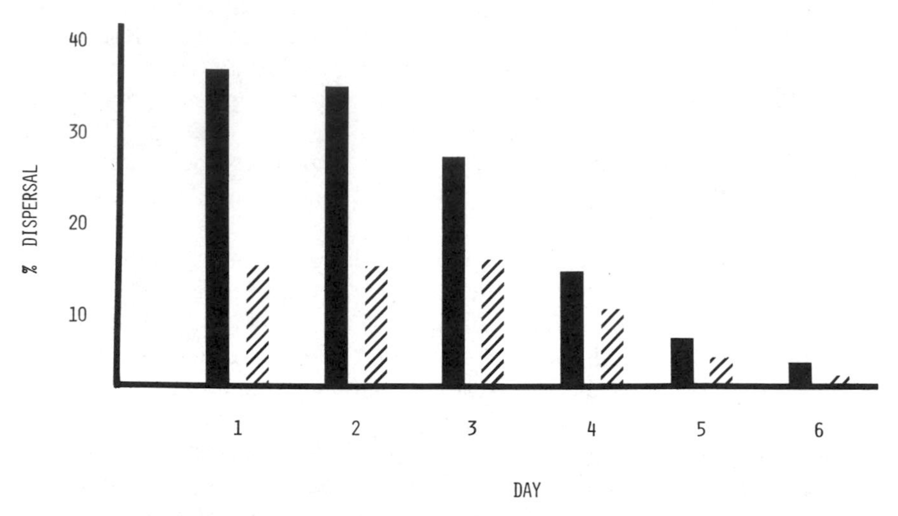

Fig. 3. Dispersal of first-instar gypsy moths from large and small eggs in the presence of acceptable food. Solid columns and striped columns denote larvae that hatched from large (>1.20 mm) and small (<1.17 mm) eggs, respectively. Based on data from Barbosa and Capinera (1978).

ioral and ecological characteristics of larvae. Indeed, egg size is an important property that produces qualitative differences in several characters among individuals of species in a variety of taxa (Capinera, 1979). In the western tent caterpillar, colonies are composed of morphs differing in relative levels of activity. Colonies vary in the proportional composition of each morph. Individuals from the largest eggs exhibit higher levels of mobility, the capacity for directed independent orientation in the production of silk trails, and thus in the capacity to find new food sources

TABLE 3

Percentage of First-Instar Gypsy Moths (*Lymantria dispar*) Not Responding to a Light Stimulus within 300 sec[a]

Egg size	Day tested			No. of larvae tested
	1	2	3	
Small	45.5	18.2	30.0	33
Medium	25.7	25.7	29.4	35
Large	26.7	6.7	10.3	30

[a] From Barbosa *et al.* (1981).

TABLE 4

Velocity of Larvae from Small, Medium, and Large Eggs of the Gypsy Moth, *Lymantria dispar*[a]

	Day tested			
Egg size	1	2	3	No. of larvae tested
Small	0.27 (±0.14)	0.28 (±0.12)	0.26 (±0.23)	29
Medium	0.28 (±0.11)	0.29 (±0.11)	0.26 (±0.12)	18
Large	0.33 (±0.13)	0.39 (±0.27)	0.30 (±0.17)	46

[a] Velocity expressed as centimeters per second (± standard deviation). Values for days 1 and 2 are significantly different at $p < .07$ and $p < .04$, respectively. Values for day 3 are not significant. From Barbosa *et al.* (1981).

readily. These individuals suffer less early-instar mortality, have fewer stadia, and develop faster than the more sluggish individuals. As one would expect (from the definition of phenotypic plasticity) the sluggish individuals also exhibit characteristics of importance for survival under certain environmental conditions. Indeed, sluggish individuals produce better silken tents than larger active individuals, and larger individuals cannot survive without some sluggish individuals in the colony.

The occurrence of active and sluggish individuals in response to heterogeneity of environmental stresses and its relationship to egg size is relatively common. In the migratory locust, large egg size is associated with the gregarious form of the species (Uvarov, 1961). Similarly, active and sluggish individuals have been reported in *Malacosoma neustria, M. disstria* (forest tent caterpillar), *Adelges piceae* (balsam wooly aphid), *Spodoptera exempta* Wlk. (African armyworm), *Callosobruchus chinensis* L. (azuki bean weevil), and several other species (Laux, 1962; Greenblatt and Witter, 1976; Edwards, 1966; Nakamura, 1966; Rose, 1975). There is often a close association between density and polymorphism, and it is expressed in a variety of traits (Tables 5 and 6). Flightlessness and flight-competent morphs have been reported in many species including planthoppers, aphids, and milkweed bugs (Denno, 1983). Even among those species that are flight competent like *Dendroctonus pseudotsugae* and *Ips paraconfusus,* there may be considerable behavioral variability, for example, in their flight response to host tree odors and/or pheromones (Atkins, 1966, 1975; Hagen and Atkins, 1975), or the dispersive behavior of larvae of flight-competent species (Mariath, 1984). Not enough is known about the ecology and population dynamics of some of these species to understand the adaptive advantages of their phenotypic plasticity. For other species, observed differences in fecundity among individuals and associ-

TABLE 5

Density-Associated Changes Manifested in Morphs of Selected Species[a]

State affected	Change	Species	Reference
Adult	Initiation of polymorphism	*Dysaphis devecta*	Forrest (1974)
		Leucania unipuncta	Iwao (1959a)
		Chaetopsiphon fragaefolii	Judge and Schaefers (1971)
		Brevicoryne brassicae	Kawada (1964)
		Megoura viciae	Lees (1967)
		Myzus persicae	Mittler and Kunkel (1971)
		Aphis fabae	Shaw (1970a,b,c)
		Acyrthosiphon pisum	Sutherland (1969)
		Therioaphis maculata	Toba *et al.* (1967)
		Callosobruchus maculata	Utida (1972, 1976)
		Various species in the genera *Aedes, Culex,* and *Anopheles*	Nayar and Sauerman (1970b)
	Lower wing loading	Various species in the genera *Aedes, Culex,* and *Anopheles*	Nayar and Sauerman (1970a,b)
	Delayed emergence	*Drosophila* sp.	Mori (1954)
	Decreased life span of female	*Melanoplus sanquinipes*	Smith (1972)
	Fecundity	*Callosobruchus chinensis*	Nakamura (1969)
	Coloration, body and wing size, reproductive maturity, fat content, response to biotic and abiotic factors	*Callosobruchus maculatus*	Utida (1972)
	Decreasing proportion of females with increasing density	*Laspeyresia pomonella*	MacLellan (1972)
		Various species of Lepdioptera	Anderson (1961)
Immature	Darker coloration	*Mamestra brassicae*	Hirata (1957)
		Dysdercus fasciatus	Hodjat (1969)
		Leucania unipuncta	Iwao (1959a,b)

TABLE 5 (*Continued*)

State affected	Change	Species	Reference
Immature	Hormonal control of coloration	*Leucania separata*	Ogura (1975)
	Higher proportion of extra-instar individuals	*Leucania unipuncta*	Iwao (1959b)
		Lymantria dispar	Leonard (1968)
	Fewer instars	Several species of Lepidoptera	Long (1953)
	Reduced respiratory levels	*Aedes aegypti*	Barbosa and Peters (1972, 1973)
	Wider food tolerance	*Leucania unipuncta*	Iwao (1959c)
	Differential food ingestion by instars	*Mamestra brassica*	Hirata (1963)
	Greater tolerance to starvation	*Leucania separata*	Iwao (1967)
	Increased incidence of disease	*Trichnoplusia ni*	Jacques (1962)
		Various species in the genera *Culex, Anopheles,* and *Psorophora*	Nayar and Sauerman (1973)
		Various insect species	Steinhaus (1958)
	Higher fat content and lower water content	Various species of Lepidoptera	Long (1953)
	Increased egg mortality	*Callosobruchus chinensis*	Nakamura (1969)

[a] Modified from Peters and Barbosa (1977).

ated traits can be expected to have a significant impact on numerical change (Nakamura, 1966; Zera, 1984).

The importance of food quality in the induction of phenotypic plasticity is evident not only within a habitat but over a season. Seasonal changes in host plants can also be a mechanism of variation among individuals of an herbivore population. Morris (1967) studied the survival of the fall webworm, *Hyphantria cunea,* reared on early (E), midseason (M), and late (L) foliage collected from the same apple trees. Although no significant differences in larval and pupal survival were exhibited by E- and M-type individuals, in the L type both larval and pupal mortality were higher. Egg to adult survival in L-type individuals was about one-half that of the other two types. Significant differences in fecundity were exhibited. Fecundity in L-type females was dramatically lower since they retained

TABLE 6

Examples of Behavioral Changes Associated with Shifts in Population Density[a]

Type of change	Species	Reference
Activity	*Lecuania separata*	Atsuhiro (1969)
	Dysdercus fasciatus	Hodjat (1969)
	Callosobruchus chinensis	Nakamura (1966)
	Blattella germanica	Ebeling and Reierson (1970)
	Leucania unipuncta	Iwao (1959a)
	Several species of Lepidoptera	Long (1953)
Orientation to light	*Hodotermes mossambicus*	Hewitt and Nel (1969)
	Leucania separata	Iwao (1967)
Flight behavior	*Mamestra brassica*	Hirata (1956)
	Aedes taeniorhynchus	Nayar and Sauerman (1969)
	Various mosquitoes in the genera *Culex, Anopheles,* and *Psorophora*	
Ability to fly	*Callosobruchus maculatus*	Utida (1965)
Migratory activity or enhanced flight tendency	*Schistocerca gregaria*	Michel (1971)
	Aedes taeniorhynchus	Nayar and Sauerman (1975)
	Aphis fabae	Shaw (1970c)
	Various mosquitoes in the genera *Aedes, Culex,* and *Anopheles*	
Flight–reproduction relationship	*Aphis fabae*	Shaw (1970c)
Adult biting rate	*Anopheles quadrimaculatus*	Terzian and Stahler (1955)
Larval feeding behavior	*Trichoplusia ni*	Jacques (1962)
Song length	*Meimuna opalifera*	Nakao (1958)
Emission of sex pheromone	*Tenebrio molitor*	Happ and Wheeler (1969)
Learning ability	Various insect species	Allee (1934)
Mating behavior	*Callosobruchus maculatus*	Utida (1972)

[a] Modified from Peters and Barbosa (1977).

about 70% of their oocytes. The effect of food-related stress (poor late-season foliage) was manifested in the progeny of females of each type. When F_1 individuals were themselves stressed (by poor food) the L-type individuals exhibited the greatest negative effects. L-type individuals were disoriented, wandered off their food, and died as first instars. Survival of M-type individuals, from egg to adult, was less than one-fourth that of E-type individuals even though no such difference was exhibited in the parental generation. All M-type adults were too deformed to mate. Thus, for species whose critical resource (e.g., food) deteriorates over the season, phenotypic plasticity may be exhibited over time, particularly if the resource deterioration is discontinuous and asynchronous and may have considerable impact on numerical increase.

It is clear from such results that selection per se may be needed to ensure the presence of various morphs in some populations, but maternal influences can be more important in others. For *M. pluviale* the nutritional experience of one generation has demonstrable effects on individual viability and activity in the next generation. Thus, although the effects are not heritable in the usual sense, they are transmissible. Brief starvation during the earliest part of the maternal larval stage has no effects on the behavior of parents but does effect subsequent progeny. Reduced progeny vigor and ability to take in food leads to subsequent decreased egg production. The effects can be cumulative over several generations (Wellington, 1965). Phenotypic plasticity can be maintained and transmitted over time when a population is composed of individuals whose parents have experienced different developmental conditions as might exist in variable and unpredictable habitats.

Thus, a variety of environmental constraints such as habitat quality, physical factors, and food, and others, like crowding and predation, may all create sufficient heterogeneity in survival requisites to induce and maintain phenotypic plasticity.

II. PHENOTYPIC PLASTICITY AND NUMERICAL INCREASE

Any analysis of the role of environmentally induced variation in the survival, persistence, and increase of a population must take into account the specific environmental adversities confronting individuals of a population and the mechanism they have to cope with these problems. Phenotypic plasticity may enable a population to survive the initial impact of adversity, to endure the ensuing period, and to recuperate (Thomson *et al.*, 1976). We would argue that survival and population growth in highly variable and/or transitory habitats can be enhanced by the existence of individuals varying in morphology, behavior, and physiology. Such plasticity would manifest itself in various critical traits like adaptive reproductive, or dispersive capabilities, which in turn could maximize numerical increase or spread the population into new habitats. We have already provided some illustrations in the previous section, but others might be useful.

Biotypes of *Acyrthosiphon pisum* (Harris) are designated by the difference in survival and rates of population increase among sympatric individuals feeding on broom, bean, and clover. Indeed, the sympatric biotypes were believed to represent aphids that had adapted to particular host plants the previous spring (Frazer, 1972). Laboratory transfer experiments showed that adaptation was facultative; thus, adaptation to one host did not limit the acceptability of other hosts. One could argue that host-adapted biotypes, developed each spring, would result in more efficient use of plant resources than a more homogeneous population of aphids (Frazer, 1972).

Divergence from the previously held presumption of traditional population dynamics models that practically all individuals within a population are identical has led to the development of models that consider one or more manifestations of population plasticity to be a key in the regulation of populations. Lomnicki (1978), for example, suggests that animal density can be regulated by emigration differences among individuals, particularly nongenetic differences. The specific cause-and-effect mechanisms that drive changes in numerical change may be quite varied. Numerous factors in any given situation can be the critical reflection of the phenotypic plasticity that regulates numerical change. The relationship between phenotypic plasticity and the influence of natural enemies is one illustration.

Phenotypic plasticity may provide situations in which individuals of a population are differentially affected by natural enemies. Significant but different interactions occur among the various larval morphs of the western tent caterpillar, *Malacosoma pluviale,* and its natural enemies. Death caused by nuclear polyhedrosis virus and spore-forming *Bacillus* sp. occurred significantly more frequently among individuals from compact colonies than among those from elongate colonies (Wellington, 1962). Sluggish (type IIb and IIc) larvae predominate in compact colonies, whereas active type I and IIa larvae are heavily represented in elongate colonies. Similarly, more than a twofold difference can be demonstrated in the mortality of active and sluggish larvae of *M. neustria,* sluggish larvae being more susceptible to death due to *Bacillus* infection (Laux, 1962). A braconid parasitoid, *Rogas* sp., attacks both kinds of colonies with the same intensity but, within a colony, parasitizes more of the clustering type IIb larvae than the type I larvae. In contrast, parasitism by five tachinid parasitoids tends to be higher among individuals from compact colonies (Iwao and Wellington, 1970). All of these interactions can significantly influence outbreak development.

What is perhaps most important in terms of the plasticity of the population and the maintenance of morphs is that even in relation to parasitism no one morph has a total selective advantage. Instead, maximal fitness varies with different environmental subsets. The effectiveness of tachinid parasitoids is a function of the differential phenology of parasitoid and morph development. *Tachinomyia similis* preferentially parasitizes colonies inhabited by third- and fourth-instars. At any given point, weather affects host (colony) and parasitoid development differentially. In general, elongate colonies develop faster than compact colonies. Under unfavorable conditions when host (colony) development is delayed in relation to the fly's emergence period, the first flies available will attack elongate instead of compact colonies because of the former's faster development. During especially favorable spring periods, however, host development is accelerated and larvae may be in their (prepupa) wandering phase when adult flies are active. Under these conditions slower-developing compact colonies bear the brunt of parasitoid activity (Iwao and Wellington, 1970).

Examples of other natural enemies similarly illustrate the role of phenotypic plasticity in studies showing the relationship between fat content of *Ips paraconfusus* individuals and their dispersive tendencies (responsiveness). Hagen and Atkins (1975) found that there was considerable variation in the level of infection of each morph by nematodes. Nonresponders (i.e., less dispersive individuals) were generally less heavily parasitized than responders.

Finally, the effectiveness of the generalist predator *Podisus maculiventris* is a function of the movement of its prey. Thus, differential activity of active and sluggish *M. pluviale* larvae results in different levels of predation on each morph. Preference for one morph or another varied with prey age and the changes in their behavior (Iwao and Wellington, 1970). Thus, predator–prey interactions and the regulation of populations by natural enemies can be subject to the influence of phenotypic plasticity.

In summary, phenotypic plasticity can provide sufficient variation to allow close tracking of changing environmental resources and, in turn, enhance a population's potential for numerical increase. At the same time under certain circumstances, differential mortality of specific morphs may enhance the decline of population numbers. One would predict that many of the species exhibiting phenotypic plasticity should also exhibit dramatic shifts in density.

TABLE 7

Characteristics of Selected Lepidoptera

	Feeding type (species/family)	Variability in density[b]	Voltinism[c]	Pupal weight[d] (mg)	Fecundity[e]	Eggs/(mg) weight[f]	Polymorphism[a]
Conifer feeders							
1	*Hyloicus pinastri*/Sphing.	66	1+	2,427	334	0.14	
2	*Dendrolimus pini*/Lasio.	1,195	1+	2,325	201	0.09	Hibernation
3	*Lymantria monacha*/Lym.		1	527	158	0.30	Adult color?
4	*Panolis flammea*/Noct	114	1	265	141	0.53	Larval color?
5	*Bupalis piniarius*/Geom.	2,240	1	152	160	1.05	
6	*Choristoneura fumiferana*	7,405	1	106	156	1.47	Egg weight, diapause
6a	*C. fumiferana*/Tortr.			173	322	1.86	
7	*C. occidentalis*/Tortr.		1	80	111	1.39	
8	*C. murinana*		1	60	105	1.75	
9	*Zeiraphera diniana*/Tortr.	35,270	1	29	134	4.62	Ecotypes
9a	*Z. diniana*			33	261	7.91	
10	*Acleris variana*/Tortr.	156	1	20	53	2.65	
11	*Argresthia fundella*/Ypon.		1	0.02	20	1000	
Broad leaf tree feeders							
12	*Lymantria dispar*/Lym.		1	1,359	499	0.37	Dispersal/development/flight
13	*Malacosoma neustria*/Lasio		1	502	223	0.44	
14	*M. pluviale*/Lasio.		1	422	184	0.44	Maternal effects
15	*Ennomos subsignarius*/Geom.		1	280	248	0.89	Larval color
16	*Aporia crataegi*/Pier.		1	258	172	0.67	
17	*Hyphantria cunea*/Arct.		2–3	149	470	3.15	Ecotypes
18	*Epirrita autumnata*/Geom.		1	70	105	1.50	

19	*Tortrix viridana*/Tortr.	1	58	55	0.95	
20	*Alsophila pometaria*/Geom.	1	49	117	2.39	♀ Aptery/parthenogenesis
21	*Adoxophyes orana*/Tortr.	2	35	216	6.17	
21a	*A. orana*		23	126	5.48	
22	*Hyponomeuta padellus*/Ypono.	1	33	150	4.55	
23	*Operophthera brumata*/Geom.	1	31	173	5.58	♀ Aptery
24	*Phyllonorycter harrisella*/Grac.	2	1.1	26	23.6	
Herb feeders						
25	*Mamestra brassicae*/Noct.	2–3	457	1,575	3.45	Phase polymorphism
26	*Leucania separata*/Noct.	3–4	411	1,232	3.00	Phase polymorphism
27	*L. loreyi*/Noct.	3–4	384	1,309	3.41	Phase polymorphism
28	*Heliothis virescens*/Noct.	2	288	864	3.00	
29	*Spodoptera littoralis*/Noct.	1–7	278	1,191	4.28	Phase polymorphism
30	*Trichoplusia ni*/Noct.	5	222	611	2.75	
31	*S. frugiperda*/Noct.	2–3	143	714	4.99	
32	*S. exigua*/Noct.		114	541	4.75	
33	*Naranga aenescens*/Noct.	4–5	29	241	8.31	
33a	*N. aenescens*		24	188	7.83	Diapause/polymorphism

[a] Polymorphic or species-specific traits; ? indicates that the ecological significance was not evaluated. All data based on data from Ali (1934), Altwegg (1971), Badr *et al.* (1983), Barbosa (1978a), Barbosa and Greenblatt (1979), Blunck and Wilbert (1962), Brandt (1936), Brewer *et al.* (1985), Campbell (1962), Cuming (1961), Drooz (1970, 1971), Feeny (1968), Franz (1940), Gerrits-Heybroek *et al.* (1978), Graf (1974), Greenbank (1956), Gruys (1970), Harvey (1983), Haukioja and Neuvonen (1985), Henneberry and Kishaba (1966), Hirata (1956, 1962), Hoy (1978), Injac *et al.* (1973), Iwao (1962), Janssen (1958), Klomp (1966), Kopec (1924), Leonard and Doane (1966), Lynch (1981), Magnoler (1970), Maksimovic (1958), Mehmet (1935), Miller (1957, 1966), Mitter *et al.* (1979), Morris (1967), Morris and Fulton (1970), Mors (1942), Oldiges (1959), Opalicki (unpublished, 1985), Rudelt (1935), Sattler (1939), Schmid (1973), Schneider (1980), Schneider-Orelli (1917), Schutte (1957), Schwenke (1953, 1978), Schwerdtfeger (1953), Shorey and Hale (1965), Sivcec (1983), Statelow (1935), Tamaki (1966), Thalenhorst (1938), Vaclena (1977), Varley *et al.* (1973), van Salis (1974), Vasiljevic and Injac (1971), Wellington and Maelzer (1967), West (1985), and Zwolfer (1934).

[b] Variability in density: antilog 3 SD of population densities (see text).

[c] Voltinism equals the number of generations per year; + indicates prolonged diapause.

[d] Pupal fresh weight (milligrams).

[e] Fecundity: generally eggs laid.

[f] Conversion index: eggs per milligram pupal weight.

III. Characteristics of Species Exhibiting Phenotypic Plasticity

It is often suggested that organisms that produce spectacular mass outbreaks, such as lepidopteran forest pests or some nonforest pests like armyworms, are inherently more plastic than species with less extreme fluctuations in abundance. Although this hypothesis would seem easy to verify, such an analysis is not straightforward. Accurate information on the variability of populations is available for only a few insect species. Moreover, plasticity expresses itself in many ways but has not been observed and recorded in a sufficiently consistent fashion to be subject to analysis.

We have argued that species that populate intermittently favorable and unpredictable environments exhibit phenotypic plasticity, are able to adapt to existence at various densities, and thus are potential outbreak species. A general comparison among Lepidoptera species feeding on conifers, deciduous trees, and herbs can provide insight into the nature of species exhibiting phenotypic plasticity and the relationship between variation in population density and rate of increase. We have characterized these species by comparing the following specific parameters: density variability, fecundity (eggs laid) or absolute fecundity (eggs laid and eggs developed in the ovaries), pupal wet weight, number of generations, and any known manifestation of phenotype variation (Table 7). The comparison of relative population variability is based on the range of population densities as reflected in the antilogarithm of three times the standard deviation (Williamson, 1984). Dividing pupal weight of a female by her fecundity provides a "reproductive index" (Campbell, 1962; Capinera and Barbosa, 1977), which is a measure of food conversion to reproduction and can be used to compare individuals within a species that vary in size. This conversion index is used here for interspecies comparisons (Fig. 4). The species-specific parameters are means calculated from various sources, and data on pupal weight and fecundity are from rearings on natural food and/or artificial media. A large index value means an efficient food conversion, which leads to a large number of offspring. Conversely, a relatively low index means a low reproductive effort (i.e., a small number of progeny). The selection of species for this analysis is somewhat arbitrary because it is dependent on the availability of relevant data.

It is obvious that the variability of population density results from a great number of influences; nevertheless, the reproductive potential is an important condition for fast increase in population density. Therefore, the conversion index (i.e., the number of eggs per milligram of pupal weight) is thought to provide a means for comparing and classifying various spe-

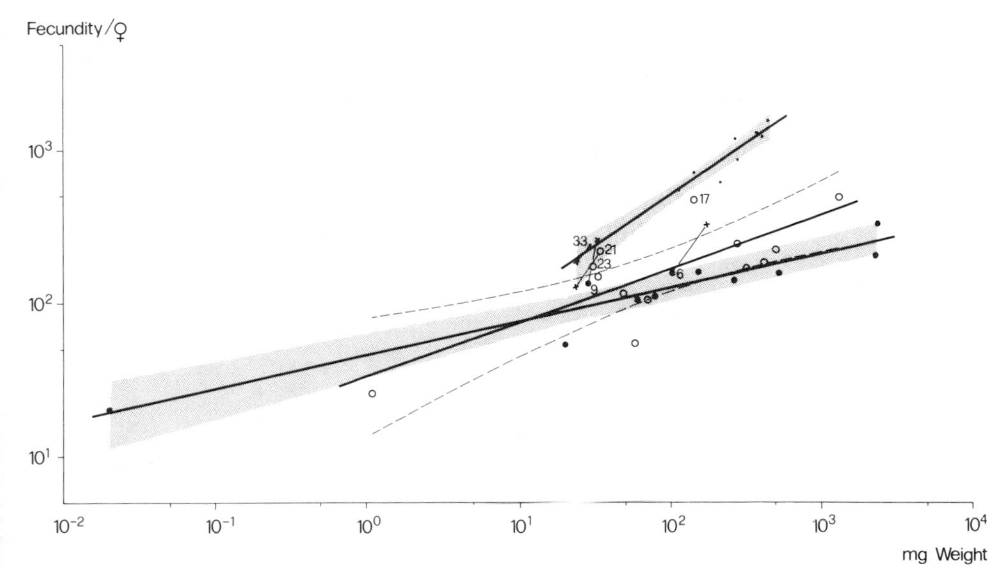

Fig. 4. Relationship between fecundity and weight for Lepidoptera feeding on host plants of different growth form: conifers, broadleaf trees, and herbs; shaded area: 95% significance for the conifer and herb groups; dashed lines, 95% significance for the group on broadleaf trees; for significance of "17" and +, see Table 7.

cies and allow us to ask whether population variability and a high conversion index are correlated. This analysis must be restricted to the conifer-feeding species since it is only for these species that we have data on both parameters. A regression analysis comparing density variability and conversion indices for all conifer-feeding species results in an r^2 of .71. If these species are separated into two groups, one consisting of species exhibiting a high degree of population variability and one with low variability, the r^2 is increased. A very significant result was that the species with the higher conversion indices also manifested the greatest variability in population density (r^2 = .97) (Table 8). Unfortunately, the lack of appropriate data does not allow a similar comparison of the species feeding on hardwoods and herbs. Nevertheless, an analysis of the relationship between fucundity and pupal weight indicates that these two parameters are most closely correlated among herb-feeding species. The r^2 is highest for the herb-feeding species, although that for conifer-feeding species is similarly high (Table 8). The rapid increases in fecundity per unit pupal weight among herb-feeding species is also reflected in their high conversion index. It is about 2.3 times greater than that of the conifer-feeding species. As noted previously, there are many examples in the literature of

TABLE 8

Comparison of Lepidoptera Living on Plants of Different Growth Forms

Relationship analyzed	Species or growth form evaluated	Statistic	r^2
		Regression	
Between population variability y and the conversion index x of species on conifers	All species	$y = 3398 + 6657x$	.71
	Low variability in density species	$y = 83 + 28x$	.77
	High variability in density species	$y = 3021 + 8049x$	.97
Between fecundity y and pupal weight x of species on conifers, broadleaf trees, or herbs	Conifer feeders	$\log y = 1.65 + 0.22x$	.89
	Broadleaf feeders	$\log y = 1.52 + 0.35x$	.62
	Herb feeders	$\log y = 1.38 + 0.66x$	.95
		Mean (±SD)	Coefficient of variation (%)
Conversion indices	Conifer group (without species 11)	1.98 ± 2.28	114.14
	Broadleaf group (without species 24)	2.51 ± 2.21	88.05
	Herb group	4.58 ± 1.99	43.45

species in which there is an association between the occurrence of high densities and polymorphism (Tables 5 and 6).

Finally, variability in population density clearly depends not only on the reproductive potential of a species but also on the length of time over which that potential is expressed. The coupling of a large conversion index with multiple generations within a growing season greatly enhances variability in numbers. The coupling of traits is exhibited by herb-feeding species. One would expect the existence of a mechanism allowing populations to cope with large and fast changes in density. It is perhaps not surprising that phase polymorphism, the most advanced form of phenotypic plasticity, has evolved in this group of species. For example, phase polymorphism in *Leucania separata* allow this species to become well adapted to environments that induce severe and rapid changes in density (Iwao, 1968). The morphofunctional changes manifested in solitary and ''crowded'' morphs facilitate rapid responses to changing environments: changes that may occur in the subsequent instar or that may even be reversible if the larvae are subjected to an additional density changes.

IV. CONCLUSIONS

The impact of phenotypic plasticity on outbreak development may be important, although short term (see Chapter 19). We have provided arguments and illustrations suggesting that plasticity among individuals of a population is a trait of evolutionary and ecological importance and specifically has consequences for the dynamics of outbreaks. A great deal of the difficulty in establishing the role of phenotypic plasticity has been due to the lack of formulation and testing of concepts against which specific data sets can be evaluated. Indeed, one of the greatest challenges of future research is to distinguish adaptive variation from the unavoidable affects of the environment. A second important challenge is to determine the relative importance of phenotypic changes as a means of adaptation and indeed the degree to which they are interdependent (Caswell, 1983).

Some of the research and concepts that have been discussed here await experimental evaluation. For example, in habitats that do not remain favorable for a much longer period than an insect's generation time, one would expect a higher likelihood of evidence of population plasticity. A corollary of this proposition would be that species living under temporary resource availability should exhibit greater phenotypic plasticity than those in long-lived habitats. Another idea that merits consideration is the proposition that phenotypic plasticity should be a particularly important adaptation in colonizing insect species, as it appears to be in colonizing species of plants (Bradshaw, 1965; Moran *et al.*, 1981). These and other predictions can provide the basis for further research that hopefully will elaborate the many ways in which phenotypic plasticity may influence outbreak development.

REFERENCES

Ali, M. (1934). Experimentelle untersuchungen uber den einfluss von temperatur und luftfeuchtigkeit auf die entwicklung des schwammspinners, *Porthetria dispar* L. *Z. Angew. Entomol.* **20,** 354–381.

Allee, W. C. (1934). Recent studies in mass physiology. *Biol. Rev. Cambridge Philos. Soc.* **9,** 1–40.

Altwegg, P. (1971). Ein semisynthetisches nahrmedium und ersatzsubstrate fur die oviposition zur von der jahreszeit unabhangigen zucht des grauen larchenwicklers, *Zeiraphera diniana* (Gn.) (Lepidoptera, Tortricidae). *Z. Angew. Entomol.* **69,** 135–170.

Anderson, F. S. (1961). Effect of density on animal sex ratio. *Oikos* **12,** 1–16.

Atkins, M. D. (1966). Laboratory studies of the Douglas-fir beetle *Dendroctonus pseudotsugae* Hopkins. *Can. Entomol.* **98,** 953–991.

Atkins, M. D. (1975). On factors affecting the size, fat content and behavior of a scolytid. *Z. Angew. Entomol.* **78,** 209–218.

Atsuhiro, S. (1969). The locomotive activity of *Leucania separata* larvae in relation to rearing densities. *Jpn. J. Ecol.* **19,** 73–75.

Badr, N. A., Moawad, G. M., and Salem, I. E. M. (1983). Host plant shifting affects the biology of *Spodoptera littoralis*(Boisd.) (Lepidoptera: Noctuidae). *Meded. Fac. Landbouwwet. Rijksuniv. Gent* **48,** 369–374.

Barbosa, P. (1978a). Population quality, dispersal and numerical change in the gypsy moth, *Lymantria dispar* (L.). *Oecologia* **36,** 203–209.

Barbosa, P. (1978b). Host plant exploitation by the gypsy moth, *Lymantria dispar* L. *Entomol. Exp. Appl.* **24,** 228–237.

Barbosa, P., and Capinera, J. L. (1978). Population quality, dispersal and numerical change in the gypsy moth, *Lymantria dispar*. *Oecologia* **36,** 203–209.

Barbosa, P., and Greenblatt, J. (1979). Suitability, disgestibility and assimilation of various host plants of the gypsy moth, *Lymantria dispar* L. *Oecologia* **43,** 111–119.

Barbosa, P., and Peters, T. M. (1972). The effect of larval overcrowding on pupal respiration in *Aedes aegypti*. *Can. J. Zool.* **50,** 1179–1181.

Barbosa, P., and Peters, T. M. (1973). Some effects of overcrowding on the respiration of larval *Aedes aegypti*. *Entomol. Exp. Appl.* **16,** 146–156.

Barbosa, P., Cranshaw, W., and Greenblatt, J. A. (1981). Influence of food quantity and quality on polymorphic dispersal behaviors in the gypsy moth, *Lymantria dispar*. *Can. J. Zool.* **59,** 293–297.

Blunck, H., and Wilbert, H. (1962). Der baumweissling, *Aporia crataegi* (L.) (Lep., Pieridae) und sein massenwechsel. *Z. Angew. Entomol.* **50,** 166–221.

Bradshaw, A. D. (1965). Evolutionary significance of phenotypic plasticity in plants. *Adv. Genet.* **13,** 115–155.

Brandt, H. (1936). Puppengewicht, puppengroesse und eizahl beim kiefernspanner, *Bupalus piniarius*. *Mitt. Forstwirtsch. Forstwiss.* **4,** 413–427.

Brewer, J. W., Capinera, J. L., Deshon, R. E., Jr., and Walmsley, M. L. (1985). Influence of foliar nitrogen levels on survival, development, and reproduction of western spruce budworm, *Choristoneura occidentalis* (Lepidoptera: Tortricidae). *Can. Entomol.* **117,** 23–32.

Campbell, I. M. (1962). Reproductive capacity in the genus *Choristoneura* Led. (Lepidoptera: Tortricidae). I. Quantitative inheritance and genes as controllers of rates. *Can. J. Genet. Cytol.* **4,** 272–288.

Capinera, J. (1979). Qualitative variation in plants and insects: Effects of propagule size on ecological plasticity. *Am. Nat.* **114,** 350–361.

Capinera, J. L., and Barbosa, P. (1976). Dispersal of first-instar gypsy moth larvae in relation to population quality. *Oecologia* **26,** 53–60.

Capinera, J. L., and Barbosa, P. (1977). Influence of natural diets and larval density on gypsy moth, *Lymantria dispar* (Lepidoptera: Orgyiidae), egg mass characteristics. *Can. Entomol.* **109,** 1313–1318.

Capinera, J. L., Barbosa, P., and Hagedorn, H. H. (1977). Yolk and yolk depletion of gypsy moth eggs: Implications for population quality. *Ann. Entomol. Soc. Am.* **70,** 40–42.

Caswell, H. (1983). Phenotypic plasticity in life history traits: Demographic effects and evolutionary consequences. *Am. Zool.* **23,** 35–46.

Chapman, R. F., Cook, A. G., Mitchell, G. A., and Page, W. W. (1978). Wing dimorphism and flight in *Zonocerus variegatus* (L.) (Orthoptera: Pyrgomorphidae). *Bull. Entomol. Res.* **68,** 229–242.

Charnov, E. L. (1979). The genetical evolution of patterns of sexuality: Darwinian fitness. *Am. Nat.* **113,** 465–480.

Chew, F. A. (1980). Food plant preferences of *Pieris* caterpillars (Lepidoptera). *Oecologia* **46,** 347–353.

Cuming, F. G. (1961). The distribution, life history and economic importance of the winter moth, *Operophtera brumata* (L.) (Lepidoptera: Geometridae) in Nova Scotia. *Can. Entomol.* **18,** 135–142.

Danthanarayana, W. (1975). Factors determining variation in fecundity of the light brown apple moth, *Epiphyas postvittana* (Walker) (Tortricidae). *Aust. J. Zool.* **23,** 439–451.

Danthanarayana, W. (1976). Environmentally cued size variation in the light-brown apple moth *Epiphyas postvittana* (Walk.) (Tortricidae), and its adaptive value in dispersal. *Oecologia* **26,** 121–132.

Darwin, C. (1859). "The Origin of Species." Murray, London.

Denno, R. F. (1983). Tracking variable host plants in space and time. *In* "Variable Plants and Herbivores in Natural and Managed Systems" (R. F. Denno and M. McClure, eds.), pp. 291–342. Academic Press, New York.

Derr, J. A. (1980). The nature of variation in life history characters of *Dysdercus bimaculatus* (Heteroptera: Pyrrhocoridae), a colonizing species. *Evolution (Lawrence, Kans.)* **34,** 548–557.

Drooz, A. T. (1970). The elm spanworm (Lepidoptera: Geometridae): How several natural diets affect its biology. *Ann. Entomol. Soc. Am.* **63,** 391–397.

Drooz, A. T. (1971). The elm spanworm (Lepidoptera: Geometridae): Natural diets and their effect on the F2 generation. *Ann. Entomol. Soc. Am.* **64,** 331–333.

Ebeling, W., and Reierson, D. A. (1970). Effect of population density on exploratory activity and mortality rate of German cockroaches in choice boxes. *J. Econ. Entomol.* **63,** 350–355.

Edwards, D. K. (1966). Observations on the crawlers of the balsam woolly aphid, *Adelges piceae* (Ratz.). *Can. For. Serv. Bi-Mon. Res. Notes* **22,** 4–5.

Feeny, P. P. (1968). Effect of oak leaf tannins on larval growth of the winter moth, *Operophtera brumata*. *J. Insect Physiol.* **14,** 805–817.

Forrest, J. M. S. (1974). The effects of crowding on morph determination of the aphid *Dyaphis devecta*. *J. Entomol.* **48,** 171–175.

Franz, J. (1940). Der tannentriebwickler, *Cacoecia murinana* (Hb.). Beitrage zur bionomie ubd oekologie. *Z. Angew. Entomol.* **27,** 585–620.

Frazer, B. D. (1972). Population dynamics and recognition of biotypes in the pea aphid (Homoptera: Aphididae). *Can. Entomol.* **104,** 1729–1733.

Gerrits-Heybroek, E. M., Herrebout, W. M., Ulenberg, S. A., and Weibes, J. T. (1978). Host plant preference of five species of small ermine moths (Lepidoptera: Yponomeutidae). *Entomol. Exp. Appl.* **24,** 16–168.

Graf, E. (1974). Zur biologie und gradologie des brauen larchenwickers, *Zeiraphera diniana,* Gn. (Lep., Tortricidae), im schweizerischen mittelland. *Z. Angew. Entomol.* **76,** 347–379.

Greenbank, D. O. (1956). The role of climate and dispersal in the initiation of outbreaks of the spruce budworm in New Brunswick. *Can. J. Zool.* **34,** 453–476.

Greenblatt, J. A., and Witter, J. A. (1976). Behavioral studies on *Malacosoma disstria* (Lepidoptera: Lasiocampidae). *Can. Entomol.* **108,** 1225–1228.

Gruys, P. (1970). Growth of *Bupalus piniarius* (Lepidoptera: Geometridae) in relation to larval population density. *Res. Inst. Nat. Manage., Verh.* **1,** 1–127.

Hagen, B. W., and Atkins, M. D. (1975). Between generation variability in the fat content and behavior of *Ips paraconfusus* Lanier. *Z. Angew. Entomol.* **79,** 169–172.

Happ, G. M., and Wheeler, J. (1969). Bioassay, preliminary purification and effect of age, crowding and mating on the release of sex pheromone by female *Tenebrio molitor*. *Ann. Entomol. Soc. Am.* **62,** 846–851.

Harrewijn, P. (1978). The role of plant substances in polymorphism of the aphid *Myzus persicae*. *Entomol. Exp. Appl.* **24,** 198–214.

Harvey, G. T. (1983). Environmental and genetic effects of mean egg weight in spruce budworm (Lepidoptera: Tortricidae). *Can. Entomol.* **115,** 1109–1117.

Haukioja, E., and Neuvonen, S. (1985). The relationship between size and reproductive potential in male and female *Epirrita autumnata* (Lep., Geometridae). *Ecol. Entomol.* **10,** 267–270.

Henneberry, T. J., and Kishaba, A. N. (1966). Pupal size and mortality, longevity and reproduction of cabbage loopers reared at several densities. *J. Econ. Entomol.* **59,** 1490–1493.

Hewitt, P. H., and Nel, J. J. C. (1969). The influence of group size on the sarcosomal activity and the behavior of *Hodotermes mossambicus* alate termites. *J. Insect Physiol.* **15,** 2169–2177.

Hirata, S. (1956). Influence of larval density upon the variations observed in the adult stage on the phase variation of cabbage armyworm (*Mamestra brassicae* L. IV. Some regulating mechanisms of development in the crowded population. *Jpn. J. Appl. Entomol. Zool.* **1,** 204–208.

Hirata, S. (1957). On the phase variation of the cabbage armyworm *Barathra brassicae* L. IV. Some regulating mechanisms of development in the crowded population. *Jpn. J. Appl. Entomol. Zool.* **1,** 204–208.

Hirata, S. (1962). Local variations in certain characters of the cabbage moth, *Mamestra brassicae* L. *Jpn. J. Ecol.* **12,** 133–140.

Hirata, S. (1963). On the phase variation of the cabbage armyworm *Mamestra* (*Barathra*) *brassicae* (L.). VII. Effect of larval crowding on food consumption and body weight. *Jpn. J. Ecol.* **13,** 125–127.

Hodjat, S. H. (1969). The effects of crowding on the survival, rate of development size, color, and fecundity of *Dysdercus fasciatus* Sign. in the laboratory. *Bull. Entomol. Res.* **58,** 487–504.

Hoy, M. (1978). Selection for a non-diapausing gypsy moth: Some biological attributes of a new laboratory strain. *Ann. Entomol. Soc. Am.* **71,** 75–80.

Injac, M., Veyrunes, J.-C., and Kuhl, G. (1973). L'élévage de *Hypantria cunea* sur milieu artificiel en conditions aseptiques. *Zast. Bilja* **24,** 111–117.

Iwao, S. (1959a). Comparisons of the manifestation of density-dependent variabilities in *Leucania unipuncta* Haworth, *L. loreyi* Dopunchel and *L. placida* Butler. *Jpn. J. Ecol.* **9,** 32–38.

Iwao, S. (1959b). Some analyses on the effect of population density on larval coloration and growth in armyworm *Leucania unipuncta* Haworth. *Physiol. Ecol.* **8,** 107–116.

Iwao, S. (1959c). Phase variation in the armyworm, *Leucania unipuncta* Haworth. IV. Phase difference in the range of food tolerance of the final instar larva. *Jpn. J. Appl. Entomol. Zool.* **3,** 164–171.

Iwao, S. (1962). Studies on the phase variation and related phenomena in some lepidopterous insects. *Mem. Coll. Agric., Kyoto Univ.* **84,** 1–80.

Iwao, S. (1967). Resistance to starvation of pale and black larva of the armyworm *Leucania separata* Walker. *Botyu-Kagaku* **32,** 44–46.

Iwao, S. (1968). Some effects of grouping in lepidopterous insects. *Colloq. Int. C. N. R. S.* **173,** 185–212.

Iwao, S., and Wellington, W. G. (1970). The western tent caterpillar: Qualitative differences and the action of natural enemies. *Res. Popul. Ecol.* **12,** 81–99.

Jacques, R. P. (1962). Stress and nuclear polyhedrosis in crowded populations of *Trichoplusia ni*. *J. Insect Pathol.* **4,** 1–22.

Janssen, M. (1958). Uber biologie, massenwechsel und bekampfung von *Adoxophyes orana* Fischer von Roeslerstamm (Lepidoptera: Tortricidae). *Beitr. Entomol.* **8,** 291–324.

Jones, R. E., and Ives, P. M. (1979). The adaptiveness of searching and host selection behavior in *Pieris rapae* (L.). *Aust. J. Ecol.* **4,** 75–86.

Judge, F. D., and Schaefers, G. A. (1971). Effects of crowding on alary polymorphisms in the aphid *Chaetosiphon fragaefolii*. *J. Insect Physiol.* **17,** 143–148.

Kawada, K. (1964). The development of winged forms in the cabbage aphid *Brevicoryne brassicae* L. *Ber. Ohara Inst. Landwirtsch. Biol., Okayama Univ.* **12,** 189–195.

Klomp, H. (1966). The dynamics of a field population of the pine looper, *Bupalus piniarius* L. (Lep., Goem.). *Adv. Ecol. Res.* **3,** 207–305.

Kopec, S. (1924). Studies on the influence of inanition on the development and the duration of life in insects. *Biol. Bull.* (*Woods Hole, Mass.*) **66,** 1–34.

Lance, D. R. (1983). Host-seeking behavior of the gypsy moth: The influence of polyphagy and highly apparent host plants. *In* "Herbivorous Insects: Host-Seeking Behavior and Mechanisms" (S. Ahmad, ed.), pp. 201–224. Academic Press, New York.

Lance, D., and Barbosa, P. (1981). Host tree influences on the dispersal of first instar gypsy moths, *Lymantria dispar*. *Ecol. Entomol.* **6,** 411–416.

Lande, R. (1977). The influence of the mating system on the maintenance of genetic variability in polygenic characters. *Genetics* **86,** 485–498.

Laux, W. (1962). Individual differences in the behavior and capabilities of the tent caterpillar, *Malacosoma neustria* L. *Z. Angew. Zool.* **49,** 465–524.

Lees, A. D. (1961). Clonal polymorphism in aphids. *Symp. R. Entomol. Soc. London* **1,** 67–79.

Lees, A. D. (1966). The control of polymorphism in aphids. *Adv. Insect Physiol.* **3,** 207–277.

Lees, A. D. (1967). The production of the apterous and alate forms in the aphid *Megoura viciae* Buckton, with special reference to the role of crowding. *J. Insect Physiol.* **13,** 289–318.

Leonard, D. E. (1968). Effects of density of larvae on the biology of the gypsy moth (*Porthetria dispar*). *Entomol. Exp. Appl.* **11,** 291–304.

Leonard, D. E., and Doane, C. C. (1966). An artificial diet for the gypsy moth, *Porthetria dispar* (Lepidoptera: Lymantridae). *Ann. Entomol. Soc. Am.* **59,** 462–464.

Lomnicki, A. (1978). Individual differences between animals and the natural regulation of their numbers. *J. Anim. Ecol.* **47,** 461–475.

Long, D. B. (1953). Effects of population density on larvae of Lepidoptera. *Trans. R. Entomol. Soc. London* **104,** 543–585.

Lynch, R. E. (1981). Effects of "coastal" bermudagrass fertilization levels and age of regrowth on fall armyworm (Lepidoptera: Noctuidae): Larval biology and adult fecundity. *J. Econ. Entomol.* **74,** 122–123.

MacLellan, C. R. (1972). Sex ratio in three stages of field collected codling moth. *Can. Entomol.* **104,** 1661–1664.

Magnoler, A. (1970). A wheat germ medium for rearing of the gypsy moth, *Lymantria dispar* L. *Entomophaga* **15,** 401–406.

Maksimovic, M. (1958). Experimental researches on the influences of temperature upon the development and the dynamics of population of the gypsy moth. *Biol. Inst. Srbije* **3,** 1–115.

Mariath, H. A. (1982). Experiments on the selection against different color morphs of a twig caterpillar by insectivorous birds. *Z. Tierpsychol.* **60,** 135–145.

Mariath, H. A. (1984). Factors affecting the dispersive behavior of larvae of an Australian geometrid moth. *Entomol. Exp. Appl.* **35,** 159–167.

Mason, L. G. (1976). Habitat and phenetic variation in *Phymata americana* Melin (Heteroptera: Phytmatidae). II. Climate and temporal variation in color pattern. *Syst. Zool.* **25,** 123–128.

Mehmet, B. (1935). Experimentelle undetsuchunen uber den einfluss von temperatur und luftfeuchtigkeit auf die sterblichkeit und entwicklung des ringelspinners, *Malacosoma neustria* L. *Z. Angew. Entomol.* **21,** 534–522.

Michel, R. (1971). Influence du groupement en essaim artificiel sur la tendence au vol du criquet Pelerin (*Schistocerca gregaria* Forsk). *Behaviour* **39,** 58–72.

Miller, C. A. (1957). A technique for estimating the fecundity of natural populations of the spruce budworm. *Can. J. Zool.* **35,** 1–13.

Miller, C. A. (1966). The black-headed budworm in eastern Canada. *Can. Entomol.* **98,** 592–613.

Mitter, C., Futuyma, D. J., Schneider, J. C., and Hare, D. J. (1979). Genetic variation and host plant relations in a parthenogenetic moth. *Evolution* (*Lawrence, Kans.*) **33,** 777–790.

Mittler, T. E., and Kunkel, H. (1971). Wing production by grouped and isolated apterae of the aphid *Myzus persicae* on artificial diet. *Entomol. Exp. Appl.* **14,** 83–92.

Moran, G. F., Marshall, D. R., and Muller, W. J. (1981). Phenotypic variation and plasticity in the colonizing species *Xanthium strumarium* L. (Noogoora Burr). *Aust. J. Biol. Sci.* **34,** 639–648.

Mori, S. (1954). Population effect on the daily periodic emergence of *Drosophila*. *Mem. Coll. Sci., Univ. Kyoto., Ser. B* **21,** 49–54.

Morris, R. F. (1967). Influence of parental food quality on the survival of *Hyphantria cunea*. *Can. Entomol.* **99,** 24–33.

Morris, R. F., and Fulton, W. C. (1970). Models for the development and survival of *Hyphanthria cunea* in relation to temperature and humidity. *Mem. Entomol. Soc. Can.* **70,** 1–60.

Mors, H. (1942). Untersuchungen zur nonenprognose wellensteins und die bedeutung gradologischer merkmale. *Monogr. Z. Angew. Entomol.* **15,** 535–553.

Myers, J. H. (1978). A search for behavioral variation in first and last laid eggs of western tent caterpillar and an attempt to prevent a population decline. *Can. J. Zool.* **56,** 2359–2363.

Nakamura, H. (1966). The activity types observed in the adult of *Callosobruchus chinensis* L. *Jpn. J. Ecol.* **16,** 236–241.

Nakamura, H. (1969). The effect of density on progeny population in *Callosobruchus chinensis* L. from different localities. *Jpn. J. Ecol.* **19,** 92–96.

Nakao, S. (1958). On the diurnal variation of a song length of *Meimuna opalifera* Walker and the effects of population density upon it. *Kontyu* **26,** 201–209.

Nayar, J. K., and Sauerman, D. M., Jr. (1969). Flight behavior and phase polymorphism in the mosquito *Aedes taeniorhynchus*. *Entomol. Exp. Appl.* **12,** 365–375.

Nayar, J. K., and Sauerman, D. M., Jr. (1970a). A comparative study of growth and development in Florida mosquitoes. Part 1. Effects of environmental factors on ontogenetic timings, endogenous diurnal rhythm and synchrony of pupation and emergence. *J. Med. Entomol.* **7,** 163–174.

Nayar, J. K., and Sauerman, D. M., Jr. (1970b). A comparative study of growth and development in Florida mosquitoes. Part 3. Effects of temporary crowding on larval aggregation formation, pupal ecdysis and adult characteristics at emergence. *J. Med. Entomol.* **7,** 521–528.

Nayar, J. K., and Sauerman, D. M., Jr. (1973). A comparative study of growth and development of Florida mosquitoes. Part 4. Effects of temporary crowding during larval stages on female flight activity patterns. *J. Med. Entomol.* **10,** 37–42.

Nayar, J. K., and Sauerman, D. M., Jr. (1975). Flight and feeding behavioral of autogenous and anautogenous strains of the mosquito *Aedes taeniorhynchus*. *Ann. Entomol. Soc. Am.* **68,** 791–796.

Ogura, N. (1975). Hormonal control of larval coloration in the armyworm, *Leucania separata*. *J. Insect Physiol.* **21,** 559–576.

Oldiges, H. (1959). Der einfluss der temperatur auf stoffwechsel und eiproduktion von Lepidopteren. *Z. Angew. Entomol.* **44,** 115–166.

Palmer, J. O. (1985). Life-history consequences of body size variation in the milkweed leaf beetle, *Labidomera clivicolis* (Coleoptera: Chrysomelidae). *Ann. Entomol. Soc. Am.* **78,** 603–608.

Patton, R. L. (1975). Wing polymorphism in *Acheta domesticus* (Orthoptera: Gryllidae). *Ann. Entomol. Soc. Am.* **68,** 852–854.

Peters, T. M., and Barbosa, P. (1977). Influence of population density on size, fecundity, and developmental rate of insects in culture. *Annu. Rev. Entomol.* **22,** 431–450.

Prokopy, R. J., Averill, A. L., Cooley, S., Roitberg, C. A., and Kallet, C. (1982). Variation in host acceptance pattern in apple maggot flies. *Proc. Intern. Symp. Insect Plant Relationships, 5th* pp. 123–129.

Rose, D. J. W. (1975). Field development and quality changes in successive generations of *Spodoptera exempta* Wlk., the African armyworm. *J. Appl. Ecol.* **12,** 727–739.

Rudelt, J. (1935). Uber die bezichungen zwischen puppengewicht und eiproduktion beim kiefernspinner (*Dendrolimus pini* L.). *Anz. Schaedlingskd.* **11,** 1–6.

Sabath, M. D., Richmond, R. C., and Torrela, R. M. (1973). Temperature-mediated seasonal color changes in *Drosophila putrida*. *Am. Midl. Nat.* **90,** 509–512.

Sahota, T. S., and Thomson, A. J. (1979). Temperature induced variation in the rates of reproductive processes in *Dendroctonus rufipennis* (Coleoptera: Scolytidae): A new approach to detecting changes in population quality. *Can Entomol.* **111,** 1069–1078.

Sattler, H. (1939). Die entwicklung der nonne, *Lymantria monacha,* ihrer abhangigkeit von der nahrungsqualitat. *Z. Angew. Entomol.* **25,** 543–587.

Schmid, A. (1973). Beitrag zur mikrobiologischen bekampfung des grauen larchenwicklers, *Zeiraphera diniana* (Gn.). Ph.D. Dissertation, No. 5045. ETH-Zurich. 1–73.

Schmieder, R. G. (1933). The polymorphic forms of *Melittobia chalybii* Ashmead and the determining factors involved in their production (Hymenoptera: Chalcidoidea, Eulophidae). *Biol. Bull.* (*Woods Hole, Mass.*) **65,** 338–354.

Schneider, J. C. (1980). The role of parthenogenesis and female aptery in microgeographic, ecological adaptation in the fall cankerworm, *Alsophila pometaria* Harris (Lepidoptera: Geometridae). *Ecology* **61,** 1082–1090.

Schneider-Orelli, O. (1917). Zur biologie und bekampfung des frostpanners *Operophtera brumata* L. *Z. Wiss. Insekt. Biol.* **13,** 192–197.

Schoonhoven, L. M. (1977). On the individuality of insect feeding behavior. *Proc. K. Ned. Akad. Wet. Ser. C* **80,** 341–350.

Schutte, F. (1957). Untersuchungen uber die populatindynamik des eichenwicklers (*Tortrix viridana* L.). *Z. Angew. Entomol.* **40,** (1), 1–136.

Schwenke, W. (1953). Beitrage zur bionomie der kiefernspanner *Bupalus piniarius* L. und *Semiothisa liturata* Cl. auf biozonotischere Grundlage. *Beitr. Entomol.* **3,** 168–206.

Schwenke, W. (1978). Familienreihe Yponomeutidae gespinstmotten-ahnliche. *In* "Die Forstschadlinge Eurpas" (W. Schwenke, ed.), Vol. 3. Parey, Berlin.

Schwerdtfeger, F. (1953). Untersuchungen uber den eisernen bestand von kiefernspanner (*Bupalus piniarius* L.), forleule (*Panolis flammea* Schiff.) und kiefernschwarmer (*Hyloicus pinastri* L.). *Z. Angew. Entomol.* **34,** 216–283.

Shapiro, A. M. (1978). The evolutionary significance of redundancy and variability in phenotypic-induction mechanisms of pierid butterflies (Lepidoptera). *Psyche* **85,** 275–283.

Shaw, M. J. P. (1970a). Effects of population density on alienicolae of *Aphis fabae* Scop. I. The effect of crowding on the production of alatae in the laboratory. *Ann. Appl. Biol.* **65,** 191–196.

Shaw, M. J. P. (1970b). Effects of population density on alienicolae of *Aphis fabae* Scop. II. The effects of crowding on the expression of migratory urge among alatae in the laboratory. *Ann. Appl. Biol.* **65,** 197–203.

Shaw, M. J. P. (1970c). Effects of population density on alienicolae of *Aphis fabae* Scop. III. The effect of isolation on the development of form and behavior of alatae in a laboratory clone. *Ann. Appl. Biol.* **65,** 205–212.

Shorey, H. H., and Hale, R. L. (1965). Mass-rearing of the larvae of nine noctuid species on a simple artificial medium. *J. Econ. Entomol.* **58,** 522–524.

Sivcec, I. (1983). A contribution to the rearing of *Mamestra brassicae* L. (Lep. Noctuidae) with two kinds of semi-synthetic food. *Zast. Bilja* **164,** 275–285.

Smith, D. S. (1972). Crowding of grasshoppers. II. Continuing effects of crowding on subsequent generations of *Malanoplus sanguinipes*. *Environ. Entomol.* **1,** 314–317.

Statelow, N. (1935). Experimentelle untersuchungen zur oekologie des basumweisslings *Aporia crategi* L. *Z. Angew. Entomol.* **21,** 523–546.

Steffan, W. A. (1973). Polymorphism in *Plastosciara perniciosa*. *Science* **182,** 1265–1266.

Stein, W. (1977). Dil beziehung zwischen biotop-alter und auftreten der kurzflugeligkeit bei populationen dimorpher russelkafer-arten (Col., Curculionidae). *Z. Angew. Entomol.* **83,** 37–39.

Steinhaus, E. A. (1958). Crowding as a possible stress factor in insect disease. *Ecology* **39,** 503–514.

Sutherland, O. R. W. (1969). The role of crowding in the production of winged forms by two strains of the pea aphid, *Acyrthosiphon pisum*. *J. Insect Physiol.* **15,** 1385–1410.

Talerico, R. L. (1981). Methods of gypsy moth detection and evaluation. *U.S., Dep. Agric., Tech. Bull.* **1584,** 31–34.

Tamaki, J. (1966). Mass rearing of the smaller tea tortrix, *Adoxophyes orana* Fischer von Roeslerstamm, on a simplified artificial diet for successive generations (Lepidoptera: Tortricidae). *Appl. Entomol. Zool.* **1,** 120–124.

Terzian, L. A., and Stahler, N. (1955). The effects of larval population density on some laboratory characteristics of *Anopheles quadrimaculatus* Say. *J. Parasitol.* **35,** 487–495.

Thalenhorst, W. (1938). Die puppengewicht-eizahlrelation der forleule (*Panolis flammea* Schiff.). *Anz. Schaedlingskd.* **14,** 105–112.

Thomson, W. A., Cameron, P. J., Wellington, W. G., and Vertinsky, I. B. (1976). Degrees of heterogeneity and the survival of an insect population. *Res. Popul. Ecol.* **18,** 1–13.

Toba, H. H., Paschke, J. D., and Friedman, S. (1967). Crowding as a primary factor in the production of the agamic alate form of *Therioaphis maculata*. *J. Insect Physiol.* **13,** 381–396.

Tuskes, P. M., and Atkins, M. D. (1973). Effect of temperature on occurrence of color phases in the alfalfa caterpillar (Lepidoptera: Pieridae). *Environ. Entomol.* **2,** 619–622.

Utida, S. (1965). "Phase" dimorphism in the laboratory population of the cowpea weevil *Callosobruchus maculatus*. IV. The mechanism of induction of the flight form. *Jpn. J. Ecol.* **15,** 193–199.

Utida, S. (1972). Density dependent polymorphism in the adult *Callosobruchus maculatus*. *J. Stored Prod. Res.* **8,** 111–126.

Utida, S. (1976). Polymorphism in the adult *Callosobruchus maculatus*. *Proc. J. U.S.-Jpn. Semin. Stored Prod. Pests,* pp. 174–185.

Uvarov, B. P. (1961). Quantity and quality in insect populations. *Proc. R. Entomol. Soc. London, Ser. C* **25,** 52–59.
Vaclena, C. (1977). Untersuchungen zur dispersionsdynamik des grauen larchenwicklers (Lep., Tortricidae). 1. Morpho- und biometrische untersuchungen des puppen- und falterstadiums. *Mitt. Schweiz. Ges.* **50,** 107–134.
Varley, G. C., Gradwell, G. R. F., and Hassell, M. P. (1973). "Insect Population Ecology." Blackwell, Oxford.
Vasiljevic, L. J., and Injac, M. (1971). Artificial diet for the gypsy moth (*Lymantria dispar*). *Zast. Bilja* **22,** 389–396.
von Salis, G. (1974). Beitrag zur oekologie des puppen- und falterstadiums des grauen larchenwicklers, *Zeiraphera diniana* (Gn.). Ph.D. Dissertation, No. 5265. ETH-Zurich. 1–77.
Watabe, H. (1977). *Drosophila* survey of Kokkaido. XXXIV. Seasonal variations of body color of *Drosophila testacea*. *J. For. Sci. Hokkaido Univ., Ser. 6: Zool.* **21,** 21–30.
Wellington, W. G. (1957). Individual differences as a factor in population dynamics: The development of a problem. *Can. J. Zool.* **35,** 293–323.
Wellington, W. G. (1960). Qualitative changes in natural populations during changes in abundance. *Can. J. Zool.* **38,** 289–314.
Wellington, W. G. (1962). Population quality and the maintenance of nuclear polyhedrosis between outbreaks of *Malacosoma pluviale* (Dyar). *J. Insect Pathol.* **4,** 285–305.
Wellington, W. G. (1964). Qualitative changes in populations in unstable environments. *Can. Entomol.* **96,** 435–451.
Wellington, W. G. (1965). Some maternal influences on progeny quality in the western tent caterpillar, *Malacosoma pluviale* (Dyar). *Can. Entomol.* **17,** 1–14.
Wellington, W. G., and Maelzer, D. A. (1967). Effects of farnesyl methyl ether on the reproduction of the western tent caterpillar, *Malacosoma pluviale:* Some physiological, ecological and practical implications. *Can. Entomol.* **99,** 249–263.
West, C. (1985). Factors underlying the late seasonal appearance of the lepidopterous leaf-mining guild on oak. *Ecol. Entomol.* **10,** 111–120.
West, D. A., and Hazel, W. N. (1979). Natural pupation sites of swallowtail butterflies (Lepidoptera: Papilionidae): *Papilio polyxenes* Fabr., *P. glaucus* L., and *Battus philenor* (L.). *Ecol. Entomol.* **4,** 387–392.
Whittaker, R. H., and Goodman, D. (1979). Classifying species according to their demographic strategy. I. Population fluctuations and environmental heterogeneity. *Am. Nat.* **113,** 185–200.
Wiklund, C. (1975). The evolutionary relationship between adult oviposition preferences and larval host plant range in *Papilio machaon* L. *Oecologia* **18,** 185–197.
Williamson, M. (1984). The measurement of population variability. *Ecol. Entomol.* **9,** 239–241.
Yuma, M. (1984). Egg size and viability of the firefly, *Luciola cruciata* (Coleoptera, Lampyridae). *Kontyu* **52,** 615–629.
Zera, A. J. (1984). Differences in survivorship, development rate and fertility between the longwinged and wingless morphs of the waterstrider, *Limnoporus canaliculatus*. *Evolution (Lawrence, Kans.)* **38,** 1023–1032.
Zwolfer, W. (1934). Studien zur oekologie, insbesondere zur bevolkerungslehre der nonne, *Lymantria monacha*. *Z. Angew. Entomol.* **20,** 1–50.

Chapter 19

Genetic Change and Insect Outbreaks

CHARLES MITTER

Department of Entomology
University of Maryland
College Park, Maryland 20742

JOHN C. SCHNEIDER

Department of Entomology
Mississippi State University
Mississippi State, Mississippi 39762

I. INTRODUCTION

This chapter addresses the problem of outbreaks from the standpoint of evolutionary genetics, asking how genetic change affects the size and dynamics of insect populations. The possibility of such effects in ecological time has been repeatedly raised by population ecologists (reviews in

Berry, 1979; Birch, 1971; Krebs, 1978; Wilson, 1967). Nevertheless, the literature on the subject has been fragmentary and inconclusive, and the question lacks even a firm theoretical foundation. We have therefore taken a broad, exploratory approach, surveying the types of genetic effects on population dynamics that have been or might be sought in order to give the problem a better definition.

For our purposes it is useful to distinguish between short- and long-term genetic effects on population size. At one end of this continuum are reversible changes associated with individual outbreak episodes, as in the "Chitty hypothesis" discussed in Section II,C. We examine the possibility of such short-term changes in detail, asking what characters they might involve, how these are inherited, how their microevolution could influence numerical change, and how such postulates might be tested.

At the other temporal extreme lie long-term shifts in average abundance or outbreak patterns (Chapter 1, this volume), occasioned by directional evolutionary change. There are several ways one might test the importance of such changes in population dynamics. Strong evidence would be provided by consistent association across taxa of outbreak behavior with some syndrome of inherited life history traits. The search for such correlations is a central concern of evolutionary ecology but lies outside the more narrowly genetic outlook adopted here. It has been treated elsewhere in this volume (see Chapter 3).

Evidence on this question may also come, however, from the many outbreaks originating in recent agricultural systems, for which it may sometimes be possible to reconstruct both genetic and ecological antecedents. In the final section we examine the histories of several species that have attained pest status within the past two centuries, asking whether genetic change was a major cause of their increased abundance (see also Chapter 17).

II. THEORY

A. Scarcity of Relevant Models

There has been almost no theoretical work directed specifically at genetic causation of population fluctuation (but see Krebs, 1978; Stenseth, 1977, 1978). Recent efforts to combine genetic and population growth models have focused mainly on the role of population dynamics in natural selection, not the converse. These models are worth mentioning, however, because some of them do predict a dependence of population size on gene frequency.

The best-developed body of population genetic theory incorporating ecological complexity is that ascribing genetic variability to "multiple-niche polymorphism," that is, specialization of different genotypes to different subsets of the population's environment (reviewed in Mitter and Futuyma, 1983). The assumptions of some early models (e.g., Levene, 1953; Dempster, 1955) included constant population size, separate for each "niche" and independent of genetic composition. In other versions of this hypothesis, however, the coexistence of genotypes specialized to different resources may yield a population larger than any monomorphic one (Ludwig, 1950; Matessi and Jayakar, 1976; Roughgarden, 1972, 1979), each variant allowing exploitation of a resource otherwise unavailable to the population.

The theory of multiple-niche polymorphism intergrades with that of adaptive geographic variation. Polymorphism, or geographic differentiation, requires that selection favoring niche- or locality-specific differentiation not be overwhelmed by gene flow. Suppose now that the density of populations is determined partly by specific adaptation to particular environments. Then one contributor to the "favorableness" of a site, and hence to the probability of "gradient" outbreaks (Chapter 1), will be its protection from immigrants carrying locally maladaptive genes. This resembles Mayr's (1963) postulate that a species' range can be limited by the balance at its margin between adaptation to the new environment and gene flow from the center of the range.

Whether gene flow ever reduces abundance by thwarting local adaptation is essentially unknown. Mitter *et al.* (1979) reasoned that dispersal in zones of contact between differently adapted populations should produce locally depressed densities, due to frequent mismatch of genotype to resource, if population size is sufficiently sensitive to genetic composition. They presented a small amount of evidence for reduced numbers of the parthenogenetic geometrid moth *Alsophila pometaria* along an ecotone between stands of two host species occupied by different, specifically adapted herbivore genotypes.

A class of models potentially more relevant to this volume treats fitness as a density-dependent function of r and K in the logistic equation (e.g., Clarke, 1972; Charlesworth and Giesel, 1972a,b; Roughgarden, 1971, 1979). The primary focus of these studies, however, has been the equilibrium frequency of genes determining different demographic "strategies." Few have treated the reciprocal effect of equilibrium population size (e.g., Roughgarden, 1979), and as pointed out by Stenseth (1977), there is little consideration of dynamics.

A final class of genetic models that may eventually be applicable to outbreaks are those invoking simultaneous evolution of interacting populations, such as parasites and hosts. The question of whether some existing models could be extended to account for outbreak cycles is reviewed briefly by Stenseth (1977). Coevolutionary theory is in its infancy, however, and most work so far has focused on whether such systems will persist and whether they will exhibit polymorphism (reviewed in May and Anderson, 1983a; Roughgarden, 1979). The interaction of variable host resistance and population fluctuation is considered briefly by May and Anderson (1983b) and Anderson and May (1981), who note that under some conditions hosts and parasites may evolve at similar rates; changing resistance might then affect the course of outbreaks.

In summary, there is no body of general theory on which the genetic study of insect outbreaks can be grounded. The only conclusion that might be drawn on theoretical grounds is that genetic change in the insect seems, in general, an unlikely primary cause of population fluctuation, given the lack of plausible genetic (as opposed to purely ecological) models and the tremendous selection coefficients required to produce substantial gene frequency change over the rapid course typical of outbreaks (Auer, 1978). Thus, our focus is on the ways in which genetic change might modulate numerical change subject to multiple causes. The subject at present consists of a collection of case-specific hypotheses, most of which have yet to be formally modeled. The value of examining such models, as argued by Stenseth (1977), is to ascertain whether the verbal arguments are logically complete and internally consistent.

B. Gene Frequency–Population Fluctuation Models: Necessary Elements

The influence of genetic change on the course of an outbreak could be contributory, retardant, or nonexistent in either the increase or decline phases, with all combinations of these imaginable. Following the lead of Stenseth (1978), however, we can recognize a set of elements basic to any model of such effects, around which our subsequent discussion of both hypothesis and empirical test can be organized. These are as follows:

1. The nature of the variation. What is the phenotypic variation putatively influencing population growth, and how is it inherited?
2. The causes of gene frequency change. How and why, if at all, does fitness of the variants specified above depend on population density? Are density-independent selective forces important as well?
3. The effect of gene frequency (or mean character value) on population growth. In effect, one has to specify how the terms of a general population growth function vary with population composition.

Some proposals lack one or more of these elements and, therefore, seem logically incomplete (Stenseth, 1977). The hypothesis of Lorimer (1979), for example, attributes population increase to natural selection due to change in "environmental conditions" but specifies neither the nature of the microevolutionary event nor the mechanism by which it affects population size.

These elements should be subject to independent tests, with a difficulty inversely related to the order in which we have listed them. Thus, it is relatively easy to document heritability in a trait of interest; determining the effect of that heritability on population growth, as we shall see, is much harder.

C. Chitty Hypothesis

A hypothesis that, although not stated formally, has generated more discussion of the role of genctic variation in population fluctuation than any other is that of Chitty (1960; reviewed by Krebs, 1978). Designed originally to explain the decline phase in cyclic populations of microtine rodents, Chitty's hypothesis postulates alternating selection for aggressive and nonaggressive behavioral types. The nonaggressive types have higher net reproductive rates under low-density conditions and greater resistance to physical stress, but the aggressive types outcompete them for space under crowded conditions.

Stenseth (1978) showed how the alternative "strategies" postulated by Chitty correspond to some existing life history theory. Whether such polymorphism can by itself lead to cyclic fluctuations is, however, in doubt. Stenseth (1977, 1978) argued that under a logistic growth model there will be no cycle, but that cycles could be expected with the addition of even irregularly occurring environmental stress to which the aggressive types were differentially sensitive. Krebs (1978), in contrast, reports simulation models from a thesis by Anderson in which cyles were observed in a stable external environment.

III. SHORT-TERM CHANGE OF GENE FREQUENCY AND NUMBERS: EVIDENCE

There have been almost no attempts to assess the importance of genetic effects on the short-term dynamics of wild insect populations, and there is almost no evidence that such effects exist. We will nonetheless consider how such an attempt might best proceed, as a way of organizing the relevant if inconclusive observations at hand and of identifying the most critical observations still to be made. Two approaches are suggested by the preceding discussion of theory.

A. First Approach: Beginning with a Heritable Marker

If one had a well-developed theory of genetic effects on population growth, predicting widespread interaction of polymorphism and population size (Clarke, 1972), one might begin with a broad search for correlations between gene frequencies and population fluctuations, using a sample of whatever markers were easiest to assay. If the initial survey yielded a suggestive pattern, one could proceed to measure such demographic properties of the genotypes as were invoked by the theory and determine how well these accounted for the observed dynamics.

Such logic underlies attempts to test the Chitty hypothesis by tracking allozyme frequencies through population cycles in various small mammals (Mihok *et al.,* 1983, and references therein; Krebs *et al.,* 1973). Associations between population phase and allele frequencies were established in some studies. However, attempts to correlate these variants with Chitty's behavioral phenotypes have been unsuccessful, as have attempts to manipulate population dynamics experimentally by controlling the gene frequency. The wane of interest in such surveys parallels the consensus that allozyme variation cannot be readily associated with ecological differences, except in highly specific functional studies.

There do not seem to have been analogous surveys on outbreak-exhibiting insects. However, investigation of one of the best-studied cases of genetic change associated with an outbreak, that of the larch budmoth, began in a similar way, the initial observation of an obvious marker (color) leading to the search for a mechanistic connection with change in number.

Case Study: Larch Budmoth

The larch budmoth (*Zeiraphera diniana*) is a tortricid defoliator of conifers in the Palaearctic. Cyclic population fluctuations of this species with a period of about 8 years have been recorded since 1850 in larch forests of southeastern Switzerland (Auer, 1968; Baltensweiler *et al.,* 1977). Fluctuations of similar magnitude, some aperiodic, occur elsewhere as well.

The causes of these fluctuations are not fully known. Several factors contributing to their decline phase have been identified, including larval parasitism, disease (Anderson and May, 1981), and reduced food quality. Baltensweiler (1971), however, argued that populations continue to decrease even after these forces have receded. He proposed that this effect could be due to observed changes in population frequencies of several characters associated with larval coloration. If the morphs were heritable and showed negatively correlated viabilities under optimal versus stress conditions, selection for the stress-adapted types could impose a delay on recovery from the decline phase. We shall examine the elements of this hypothesis in turn.

Zeiraphera diniana as currently defined includes sympatric populations on larch and *Pinus cembra,* which differ in a number of ways (Bovey and Maksymov, 1959). There is quantitative variation in larval color, lighter forms predominating on cembran pine, dark to intermediate ones on larch. Differences in spring egg hatch date correspond to the difference between hosts in the timing of new foliage growth. The two forms survive equally well on larch, but the larch form does very poorly on *P. cembra.* Finally, the forms appear to produce and respond to different female pheromones (Guerin *et al.,* 1984).

Differences of this kind, even when accompanied by interfertility in the laboratory, have frequently turned out to distinguish host-specific sibling species (reviewed in Mitter and Futuyma, 1983). Because the present evidence does not rule out a similar verdict for the host-associated "ecotypes" of *Z. diniana,* we will restrict our discussion to the variation occurring within populations on larch. However, the status of these forms, which might be resolved by allozyme analysis, may bear on the issues raised below: It is possible that gene flow between the "races" contributes to the variability in the larch form (Baltensweiler *et al.,* 1977; Guerin *et al.,* 1984).

Cyclic populations on larch show continuous variation in the coloration of several body regions of the fifth-instar larva, ranging from "dark" (black) to "intermediate" (orange) by comparison with the "light" (yellow) color frequent in collections from cembran pine. Although there are few published data, the color variation seems to show heritability. Baltensweiler *et al.* (1977) state that selection for dark anal plate, but not for black head capsule or light anal plate, produced true-breeding lines; the variation is also said to be sensitive to temperature, but not to crowding. The work of Day and Baltensweiler (1972), though not presented as a heritability experiment, showed consistent differences between groups of larvae derived from dark versus intermediate parents. The nature of the inheritance is unknown. There is some evidence for correlation of color with other traits. Although published data are few (but see Baltensweiler *et al.,* 1977), eggs of the dark form are said to have a longer development time and suffer higher mortality from elevated temperature than those of the intermediate form, but the two morphs were said not to differ in fecundity (Baltensweiler, 1971).

Day and Baltensweiler (1972) tested the hypothesis of differential tolerance to nutritional stress by rearing progenies of dark versus intermediate parentage on optimal (i.e., larch needle) versus suboptimal (i.e., artificial medium) diets. Average survival within progenies on larch was higher for dark than for intermediate parents, supporting the postulate of low fitness for lighter morphs under optimal conditions. On artificial medium, however, mean survivorship by family was independent of parental color. By comparing color frequencies of surviving larvae pooled across families, these authors concluded that the intermediate form showed greater stress tolerance. This contrast, however, confounded the effect of color with large among-family differences in survival. Thus, it is not certain from this experiment that the intermediate form is more fit under nutritional stress.

Indirect evidence for greater stress tolerance in the intermediate form comes from field observation. An insecticide spray program was said to have increased the frequency of this form (Baltensweiler, 1971), as was an unusual weather period resulting in lowered food quality (Baltensweiler *et al.*, 1977).

In summary, there is suggestive but not definitive evidence for the contentions that budmoth larval color variation is heritable and that the "intermediate" forms show lower fitness under optimal conditions, but higher fitness under stress of various kinds, than the dark morphs.

Data from a single cycle in two Swiss populations suggest that population expansion is accompanied by an increase in the frequency of darker larvae, followed by a shift toward more intermediate coloration, beginning at about the point of maximum density (Baltensweiler, 1971). These shifts were not entirely genetic, since color changes measured in the field were not always tracked by differences between successive samples taken from the same location but reared under standard conditions. A similar phenomenon is said to occur elsewhere (Baltensweiler *et al.*, 1977). Further data on this correlation would be desirable.

The impact of these changes on the trajectory of numbers has been variously interpreted (Auer, 1978; Baltensweiler, 1978; Krebs, 1978). One difficulty is that there has been no clear, comprehensive statement of what effects might be expected. Day and Baltensweiler (1972) argued that selection for the "intermediate" morphs at high density could cause populations to decline even after relief of the defoliation-induced nutritional stress, due to the low survivorship of the intermediate forms under good conditions. Elsewhere, Baltensweiler *et al.* (1977) suggest that selection for intermediate morphs by density-independent stress can initiate population decline, presumably for similar reasons.

Other effects are also imaginable. Selection for the intermediate form might prolong the period of high density as well as that of decline by raising population resistance to density-induced stress. This postulate is in accord with Baltensweiler's characterization of genetic change as a strategy for coping with environmental variability (Baltensweiler, 1971;

Baltensweiler *et al.*, 1977) and his conjecture (Baltensweiler, 1971) that a population composed solely of dark morphs would decline to extinction after reaching peak density. Selection for the dark phase under favorable conditions might be expected to accelerate the increase phase.

How might such hypotheses be tested? Because of the vast extent of populations and migration between them, direct manipulation of genetic composition is difficult in this system. As with other phenomena recalcitrant to controlled experiment, two kinds of indirect tests seem available. The first depends on "natural" manipulation—in this case of gene frequencies by density-independent selection. Thus, Baltensweiler *et al.* (1977) cite unpublished observations by C. Auer on an episode of lowered food quality caused by unusual weather. The budmoth population, a typically cyclic one, "stagnated" at an intermediate density, a fact they attribute to a concomitant increase in frequency of the intermediate types from 3 to 24%. Given that lower food quality results in reduced absolute survival for all color types, however (Day and Baltensweiler, 1972), it is difficult to disprove that the change in trajectory of numbers was independent of that in population composition.

In an alternative approach, exemplified by Day and Baltensweiler (1972), one asks whether a general model describing population growth shows realistic behavior under experimentally derived estimates of some of its components. Thus, life table data of Auer (1968), when analyzed by key-factor analysis (Varley and Gradwell, 1970), indicated a large shift in mortality from late to early instars during the decline phase, which was unexplained by such factors as parasitism and disease. Day and Baltensweiler sought to determine whether concomitant shifts in genetic composition could account for this "residual" mortality. Given the general lack of evidence for genetic effects on any outbreak system, it makes sense to concentrate the search for those effects on just such "unexplained" observations. As more sophisticated models of budmoth cycles are developed (Baltensweiler *et al.*, 1977), incorporating full documentation of the genetic basis and demographic properties of the color morphs (see Auer, 1978), it may be possible to make strong inferences about the dependence of dynamics on population composition. In summary, although we do not regard any conclusion about genetic effects on these cycles to be well founded at present, the budmoth remains a candidate for a genetic influence on outbreak dynamics.

B. Second Approach: Beginning with a Life History Trait

The contrasting approach to the "markers-first" strategy would be to begin with a hypothesis as to the traits in which variation should have the greatest influence on population dynamics. Although such a strategy has nowhere been followed through, we can point to some examples for which preliminary evidence is available.

Raffa and Berryman (1983) found that the reproductive success of female *Dendroctonus ponderosae* bark beetles varies strongly with the resistance characteristics of individual host trees and with the number of beetles attacking a particular tree at the same time. The beetles seem to be more selective in their choice of trees for oviposition at low beetle densities than during outbreaks. Raffa and Berryman hypothesized that selection would favor selectivity for susceptible hosts at low densities, when the number of susceptible trees was not limiting. At epidemic densities, initially all susceptible hosts will have been taken, but there will be enough simultaneous colonists to overcome the defenses of previously resistant trees, which, being more vigorous, will now support higher beetle reproduction than the weaker susceptible trees. Under these conditions, relatively nonselective oviposition behavior should be favored. Raffa and Berryman suggested that positive feedback could thus occur between beetle density and increase in frequency of genotypes determining relatively indiscriminant host selection. Although this hypothesis seems disconcertingly complex, a first test of it would be to screen for heritable variation in beetle selectivity.

In general, assessing which traits are most likely to show genetic–population dynamic effects will require detailed bionomic information as in the example just cited. There are, however, several examples of general syndromes common to a variety of unrelated species that seem reasonable candidates even in the absence of detailed ecologies. For instance, a number of insect species, particularly in the Orthoptera (Uvarov, 1928; Key, 1957) and Lepidoptera (Iwao, 1968), exhibit a syndrome of altered development when the immature stages are crowded. Recurrent elements of this "kentromorphic" response (Key and Day, 1954) include increased melanization, faster development, and lowered adult weight and fecundity.

Neither the adaptive significance nor the population consequences of this syndrome are understood (reviewed in Chapter 18). Speculation about the effects of heritable variation in its threshold is therefore premature, but tempting. Thus, Lea (1968; summarized in Krebs, 1978) proposed that gregarious swarms of migratory locusts should consist of genotypes undergoing phase transformation at relatively low density, resulting in high frequencies of density-insensitive types in endemic source areas.

Although Krebs (1978) regarded Lea's postulate as a variation on the Chitty hypothesis, he did not detail how the putative polymorphism would affect the origination or subsidence of "plagues." Perhaps the genetic shift would affect the temporal spacing of swarms originating from any one source, necessitating a "recovery period" before the frequency of sensitive types was high enough to yield a new swarm. A first test of this idea would be a search for genetic varation in density sensitivity. There is some evidence for heritability (possibly maternal) in the locust phase transformation (reviewed in Krebs, 1977). In the lepidopteran *Alsophila pometaria* there are consistent differences among asexual genotypes in degree of kentromorphism when density is held constant (Futuyma *et al.*, 1981). An adaptive basis in demography has been proposed for the genetically based dispersal polymorphisms known in a number of insects (e.g., Harrison, 1980), but the effect of these polymorphisms on numerical change has been little studied.

Disease resistance is a final example of a character potentially important for outbreak dynamics (e.g., Anderson and May, 1981) whose variation among individuals and over time has been little investigated in insects. There is, however, considerable evidence for heritability in resistance of vector insects to vertebrate diseases such as malaria (e.g., Graves and Curtis, 1982).

Sex Ratio Conditions

The picture of our subject thus far drawn is discouraging. Most of the genetic effects on abundance heretofore postulated are complex and subtle and their implications not fully worked out. The phenotypes they invoke are difficult to measure and likely to have complex inheritance. Rapid progress on the question seems unlikely, if we regard the well-known outbreak systems as the primary experimental material.

Given the shadowy status of genetic hypothesis in population dynamics, an alternative approach makes good sense. Suppose we could find a class of discrete, easily scored heritable variants whose potential influence on population growth was both dramatic and easily modeled. Suppose further than no effects on population dynamics were found for such variants. Then the prospects for detecting more suble influences would look dim indeed and the general hypothesis of genetic effects might be efficiently laid to rest.

One candidate for such a critical test is the class of variants that affect sex ratio. Many such conditions are known in insects, involving both chromosomal and extrachromosomal inheritance and producing discrete ratios strongly biased in both directions (Werren *et al.*, 1981; Stenseth *et al.*, 1985). Some possess a selective advantage that, other things being equal, should result in population extinction as the condition spreads.

Our discussion concentrates on the best-studied set of sex ratio conditions capable of causing local extinction, namely, those suppressing the production of males without either eliminating the requirement for mating or lowering fecundity. At least two types of such strains are known in insects. In the butterflies *Acraea encedon* (Chanter and Owen, 1972) and *Danaus chrysippus* (Smith, 1975), all-female strains apparently result from meiotic drive favoring the Y chromosome, female Lepidoptera being heterogametic. More common causes are forms of parthenogenesis or pseudogamy, in which one or more steps of the insemination process are necessary for the production of fertile eggs but result in no sperm contribution to the zygote genotype. Mixed populations containing strains of this description have been found in, for example, the psychid moth *Luffia lapidella* (Narbel-Hofstetter, 1963), the geometrid moth *Alsophila pometaria* (Mitter and Futuyma, 1977), the delphacid planthopper *Muellerianella fiarmairei* (Drosopoulos, 1976), the ptinid beetle *Ptinus clavipes* (Moore *et al.*, 1956), and several other species of the scolytid genus *Ips* (Lanier and Oliver, 1966).

Many authors have noted that population size in such systems should be unstable (Stenseth *et al.*, 1985, and references therein). In the model of Kiester *et al.* (1981), for example, the assumptions of equal survivorship and fecundity of sexual and gynogenetic females, coupled with males of finite mating capacity who do not discriminate between the strains, guarantees continual increase of the all-female strain until the population goes extinct for want of males.

Two kinds of mechanism can prevent global extinction of populations containing such strains. One is the evolution of traits (e.g., mating discrimination against gynogenetic females, or meiotic drive suppressors) that ameliorate the reproductive disadvantage of normal females. The second is a balance between the establishment of new populations by the normal form and their extinction following invasion by all-female strains. Simulation models of population subdivision with Y-driving strains were investigated by Heuch (1978) and Heuch and Chanter (1982).

If an extinction–immigration balance is mainly what prevents the extinction of population complexes harboring all-female strains, the dynamics of individual demes should be strongly governed by the frequency of those variants. The expectation of extinction preceded by a decrease in the proportion of normal females has already been discussed. The dynamics of such populations soon after invasion by all-female strains has not been modeled, but moderate frequencies of these, before males reach limiting supply, should augment the rate of population increase. Thus, sex ratio changes could contribute markedly to both increase and decrease phases of population history.

Does this occur in nature? The sparse evidence does not suggest a marked pattern of localized "boom and bust." In the best-studied case, *Acraea encedon* (Owen and Chanter, 1969; Chanter and Owen, 1972), 10 local populations were monitored closely over periods of six to nine generations. In most of these, strongly female biased sex ratios fluctuated with no apparent trend. Populations with approximately normal sex ratios were also found, as expected under the extinction–recolonization hypothesis. However, populations of similar sex ratio were clustered geographically. Coupled with the observation that females in biased-ratio populations showed an aggregation behavior not seen in populations of normal ratios, this suggests that among-population differences in sex ratio are stable and do not represent stages of a single type of numerical fluctuation. Decisive rejection of the "boom-and-bust" hypothesis for this system, however, would require long-term monitoring of local populations through episodes of extinction and reestablishment.

Spatial and temporal patterns of sex ratio similar to those just described also characterize the gynogenetic delphacid *Mullerianella fairmairei* in the Netherlands. Local populations sampled over about six generations exhibited mostly stable female-biased sex ratios (Booij and Guldemond, 1984), which showed no average trend. Among-population variation in sex ratio, again seen mainly on a scale much larger than the distance between neighboring populations, is in this case well explained by climatic factors.

Short-term stability of sex ratio seems also to hold in *Alsophila pometaria,* although the data for this species are limited. Samples from six populations on Long Island, New York, contained from 1.1 to 6.7% males in 1981 (Harshman, 1982). A population lying within about the 30-km radius spanned by the 1981 samples contained 1.3% males in 1985 (Mitter and Futuyma, 1977).

In all three of these species there is at least some evidence of a mating advantage for normal over sonless females (Chanter and Owen, 1972; Harshman, 1982), which may help to stabilize mixed populations (Stenseth *et al.,* 1985). In *M. fairmairei* there is direct evidence, in one population, of a higher insemination rate for normal females (Booij and Guldemond, 1984).

Even if the existence of local populations containing all-female strains is stabilized by individual adaptations, the dynamics of such populations could be strongly influenced by sex ratio. For example, if the reproductive types differ in weather tolerance, as suggested for *M. fairmairei,* selection could indirectly alter reproductive potential and subsequent population size by changing the sex ratio. Thus, the sex ratio fluctuations observed in some mixed populations should still provide a strong test of the capacity of genetic change to determine abundance.

The observations made by Owen and Chanter (1969) on the "Lubya" population of *Acraea encedon* provide the most clear-cut suggestion among insects that population fluctuation is influenced by genetic change. The sex ratio in this continuously reproducing deme fluctuated between 0 and 13% males over the six-generation study period, with a mean of 2.5%. Circumstantial evidence suggested that ratios in this range resulted in many females going unmated. It is thus perhaps not surprising that sex ratio (measured at 10-day intervals) was an excellent predictor of population size measured one generation (60 days) later.

Danaus chrysippus populations with sex ratio distorters seem to be stabilized by an autosomal suppressor locus linked to a color pattern polymorphism (Smith, 1975). There is consistent seasonal variation in sex ratio, female bias being strongest at the start of the rainy-season-dependent population flush. The sex ratio changes appear to be driven, at least in part, by fluctuating sexual selection on color pattern, but further work is needed to separate their contribution to numerical change from the direct effects of seasonality.

In summary, there is no evidence so far, at least among insects, for the local outbreak–extinction cycles that all-female strains are potentially able to cause. Further search for this phenomenon is desirable. Less extreme fluctuations of sex ratio within populations harboring sex ratio conditions, however, show promise as a test case for the population dynamic effect of genetic variation.

IV. GENETIC CHANGE AND THE HISTORICAL ORIGIN OF OUTBREAK STATUS

In this section we ask whether genetic change in the pest has been an important cause of insect outbreaks originating in the recent history of agriculture. The alternative hypothesis is that a rise to pest status is due entirely to changes in the insect's environment.

Human activity within the past few millennia has resulted in innumerable novel associations between insects and plants (Hawkes, 1983). Introductions of exotic plant species into North America in the past 500 years, for example, have accounted for 20% of our current flora (Fernald, 1950). Extensive cultivation has stripped the eastern seaboard of the United States of nearly all of its original forests and the central states of nearly all of their original prairies, replacing or modifying these communities over large areas with crop plants of exotic origin, such as soybean (*Glycine max*), originally from China, sorghum (*Sorghum bicolor*), native to Africa, and cotton (*Gossypium hirsutum*), originally from Central America.

It is not surprising that this flux of plant and insect communities should result in new associations of insects with crop species or varieties thereof, leading in some cases to major epidemics. Pest problems accompanied by ecological shift, which may often be associated with genetic change, seem an especially likely place to find evolutionary effects on population size. It is by no means a foregone conclusion, however, that change in the pest has been a prerequisite of such outbreaks; the pest may simply be pre-adapted to a newly available host.

Direct manipulations of pest genetic constitution are the ideal way to demonstrate a genetic effect on population size. However, the prospects of examining such data are especially remote in a historical analysis such as this one. For our purposes, evidence for an important contribution to the origin of outbreaks from genetic as opposed to purely environmental change is the following:

1. The existence of heritable differences between the outbreak population and its presumed progenitor, especially in characters whose functional relevance to a host shift can be strongly inferred;
2. The absence of associated environmental changes sufficient in themselves to explain the outbreak.

We shall examine the histories of several outbreaks originating since the European settlement of North America to see how well these criteria are met. The associations to be scrutinized include *Schizaphis graminum* on sorghum, *Mayetiola destructor* on wheat, *Cydia pomonella* on walnut, and *Heliothis virescens* on cotton. These cases were chosen because the evidence on them is more extensive than for most; a similar analysis could be applied to many other insects (see, e.g., Diehl and Bush, 1984).

A. Greenbug

One of the strongest cases for outbreak being triggered by genetically mediated host shift involves the greenbug (*Schizaphis graminum*) on sorghum. This aphid, which feeds on wheat and a great many other graminaceous species, was known to feed on sorghum as early as 1863 in Italy and 1882 in the United States (Harvey and Hackerott, 1969), but was not a problem on that crop in North America. In 1968, however, it became a severe pest of sorghum across a seven-state area (Harvey and Hackerott, 1969; Daniels, 1977). The greenbug was present in low numbers on sorghum in the Texas high plains for at least 2 years preceding the outbreak (Daniels, 1977). Sorghum production had been increasing gradually since the late 1940s but was slightly lower in 1968 than in 1967 (Moore and Majors, 1970; U.S. Department of Agriculture, 1970).

There is strong circumstantial evidence that the outbreak on sorghum was due to the proliferation of a new, specifically adapted "biotype." As detailed below, this putative new genetic entity differs in several characteristics from its presumed progenitor on wheat.

Sticky-trap data collected over twenty-four years (Daniels, 1977) indicate that the outbreak in 1968 was accompanied by a marked shift in greenbug phenology; this was the primary evidence for the origin of a new biotype (Harvey and Hackerott, 1969). Before 1968 the aphids were caught only in spring and fall, originating presumably from wheat. After 1972 they were caught primarily in spring and summer. Between these years, which correspond to the peak densities recorded on sorghum, they were caught predominantly in August, which is when sorghum, but not wheat, is at the developmental stage susceptible to greenbug attack.

As might be expected from the difference in phenology, the new population on sorghum showed higher temperature tolerance than populations on wheat (Wood *et al.,* 1972). It also exhibited a greater preference for sorghum than did the wheat-adapted strains, though immatures of the two strains did not differ in ability to utilize the two hosts (Harvey and Hackerott, 1969). Finally, the new strain on sorghum was lighter in pigmentation than the wheat-adapted strain (Harvey and Hackerott, 1969).

Since these differences are maintained through several generations of rearing under identical conditions (Wood *et al.,* 1972), they clearly seem to be heritable. Since no crosses have been performed, however, the genetic basis of these "biotypes," as of most others, is unknown (Diehl and Bush, 1984; Mitter and Futuyma, 1983). It would be of great interest to know whether the new biotype is genetically homogeneous, suggesting that the widespread outbreak was due to the dispersal of a single new genotype.

Whatever the precise nature of the new biotype proves to be, however, the absence of any apparent correlated change in cultural practice or other environmental factors on sorghum suggests strongly that this outbreak was facilitated by directional genetic change.

B. Hessian Fly

The Hessian fly (*Mayetiola destructor,* Cecidomyiidae) is a second pest for which historical evidence links outbreaks to genetic change. The Hessian fly was probably introduced into the United States from southern Europe or Turkey, where it had long been a pest of wheat and other small grains (Packard, 1883). Beginning with the first serious outbreak in 1779, the history of the Hessian fly in North America has been one of accelerating alternation between control by the introduction of resistant strains of wheat and outbreaks mediated by the appearance of new, specifically adapted pest strains (see, e.g., Painter, 1930; critical review in Diehl and

Bush, 1984). Although Packard (1883) noted that the Underhill cultivar had shown useful levels of resistance for "nearly a century," resistant varieties have proved less durable since the mid-1950s. The release of new cultivars by Purdue University in 1955 and 1972 has been followed within 20 generations by high frequencies of counteradapted genotypes (Sosa, 1981).

This last observation is striking, given that virulence is recessive, and leads to speculation as to whether the unusual breeding system of *M. destructor* contributes to its capacity for rapid breaking of resistance. Male Hessian flies transmit only the maternally derived chromosomes to their sperm. Except for loci subject to frequent crossing over (limited to the male in this species), this markedly reduces the effective breeding size of Hessian fly populations.

How might this effective inbreeding accelerate the evolution of virulence? In areas where resistant wheat is grown, the frequencies of corresponding virulence alleles seem to average .90 or greater (Hatchett and Gallun, 1968; Sosa, 1981). Survival data in Sosa (1981) suggest that selection coefficients for virulent alleles on resistant wheat range between .14 and .45. Selection for a recessive at even the greatest of these intensities (Crow and Kimura, 1970, p. 173) can produce a gene frequency of .90 after 20 generations only from a starting frequency of .10 or more. If the virulent allele were even slightly deleterious on susceptible wheat and its rate of origin were as low as that typically reported for recessive mutants, large populations would very infrequently harbor such frequencies. To explain the rapid loss of control one might then suppose that virulence spread by dispersal from a few small demes in which sampling error had resulted in high preexisting mutant frequencies. The expected number of such demes is proportional to the among-population variance of gene frequency, which is in turn strongly dependent on effective population size. The degree to which the Hessian fly breeding system might magnify the expected number of populations with high initial virulence frequency over a normal segregation mechanism has yet to be modeled, but may be considerable.

Whether this argument is relevant, of course, depends on the distribution of allele frequencies that precedes the introduction of a new resistant wheat variety. Such evidence as exists suggests that some fly populations on susceptible wheat harbor appreciable frequencies (>.10) of virulence alleles (Hatchett and Gallun, 1968).

Indirect evidence for the importance of the breeding system in the evolution of virulence is provided by the linkage relationships of "virulence" loci. Loci that are far from the centromere will experience high crossover with respect to the centromere, resulting in appreciable rates of incorporation of paternally derived alleles into sperm and hence in less effective inbreeding. If the contribution of effective inbreeding to the evolution of virulence is substantial, "virulence" alleles at loci near the centromere should have a better chance of spreading. One such allele mapped, H_3, is close to the centromere (Gallun, 1978); one might predict a similar finding for the H_5 allele, which has risen to high frequencies in the field (Sosa, 1981).

C. Codling Moth

The examples in this and the following section illustrate the much more ambiguous evidence typically available on the role of microevolution in the origin of outbreaks.

The codling moth (*Cydia pomonella*), long known as a pest of rosaceous orchard crops, has within this century become an important pest of walnut in both California and South Africa. Persian walnuts (*Juglans regia*) had been grown extensively for about 20 years in southern California before the first observations of their infestation by the codling moth were published (Smith, 1912; Foster, 1910). The codling moth had been introduced to California in about 1870 (Essig, 1931, and references therein). Infested apple orchards were scattered through the main walnut-producing regions, but the major areas for apple production were to the north, near San Francisco (Quayle, 1926; Gould and Andrews, 1917).

The initial records of codling moth on walnuts in California came from Contra Costa County near San Francisco, where a close search revealed "general but light infestation" (Foster, 1910). Infestations in the region of first discovery remained light, but sporadic outbreaks occurred from 1918 through 1925 in walnut groves of southern California, which were treated with lead arsenate (Quayle, 1926).

Because of the walnut blight and increasing demand for land for citrus production in southern California, walnut production began to shift northward and into the central valley in the early 1900s (Smith, 1912). Important infestations of codling moth on walnuts in the interior regions occurred in 1931 (Boyce, 1935).

In South Africa, the codling moth was first noted on walnuts in 1908, one year earlier than in California (Mally, 1916, cited in Quayle, 1926). It occurred in outbreak numbers in 1914–1915, several years before it was economically important in California walnuts.

One reason to doubt that evolutionary change was a prerequisite of outbreaks of codling moth on walnut in California or South Africa is that the moth had apparently been known from this host within its native Europe as early as 1859 (reported by Quayle, 1926; no reference given).

Howard (1887) disputed rearing records of *C. pomonella* on walnuts in France in the 1870s, arguing that these probably represented a sibling species, *C. putaminana*. The latter, however, has since been treated as merely a variety (Kennel, 1921).

The conflicting literature on inherited differences between walnut and apple populations of codling moth in California is difficult to interpret. Quayle (1926) reported no differences in oviposition preference or host-specific larval fitness between samples collected from these hosts, but gave no numbers. In contrast, he also cited anecdotal observations of walnuts unattacked despite their growing in close proximity to infested apple trees.

Phillips and Barnes (1975) reported quantitative differences in oviposition preference and phenology among apple, walnut, and plum "races." Their conclusions on preferences of "wild" populations are difficult to evaluate because no sample sizes or statistical analyses were presented and were undermined by their concomitant report of "conditioning" of preference by continuous rearing on apple in the laboratory. Their comparison of "wild" parental "race" preferences with those of lab-reared F_1 hybrids seem also to be confounded with these differences in rearing conditions. The clear differences in phenology they describe may be genetic, but the experiments are not described in detail. In summary, the question of heritable life history differences of walnut populations and of codling moth from their presumed progenitors on apple merits further investigation.

Although the evidence just cited does not rule out a contribution from genetic change in *C. pomonella,* the history of walnut production itself seems capable of explaining the development of outbreaks on this crop. The parallel outbreaks in California and South Africa could easily have resulted from similar expansions of production and shared cultural practices communicated through the literature. The use of lead arsenate, for example, could have resulted in secondary outbreaks by the disruption of natural-enemy complexes, as documented for a number of other pests (see Section IV,D and Chapter 12).

D. Tobacco Budworm

The native North American noctuid moth *Heliothis virescens,* the tobacco budworm, has become a pest of cotton in the United States in just the past several decades (Pfrimmer *et al.,* 1981). This is despite the fact that cotton had been planted extensively since about 1810 in the southeastern United States (Handy, 1896). The earliest unambiguous published record of *H. virescens* feeding on cotton in this region dates from 1934

(Folsom, 1936). Consequently, it has generally been assumed that outbreaks of *H. virescens* on cotton result from a recent host shift. Although there are some data to support this view, we shall argue that present evidence is equally compatible with an explanation based on changing cultural practice acting on long-standing low-level infestations rather than genetic change in the host.

A suggestion that the association of *H. virescens* with cotton considerably predates 1934 comes from records of the braconid wasp *Cardiochiles nigriceps*. Since *H. virescens* is the host of this parasite on cotton (Krombein *et al.*, 1979; Kimball, 1965; Lewis and Brazzell, 1966), records of *C. nigriceps* from cotton, even with the host unidentified, suggest attack on that crop by *H. virescens*. A survey by one of us (J. S.) of 14 museum collections (see Acknowledgments) yielded 11 pre-1934 records (totaling 13 specimens) of *C. nigriceps*. Three of these (27%) are associated with cotton. Of these, two are field-collected adults, but one specimen, in the U.S. National Museum, is labeled "25 July 1912, Opelousas, Louisiana, bred ex cotton square." There is also an unpublished rearing record of *H. virescens* itself from cotton bolls in Texas in the fall of 1930 (H. Burke, personal communication; determination by R. A. Vickery). Finally, Riley (1885) refers to the presence of *Aspila* (= *Heliothis*) on cotton in Georgia in 1878. Thus, it is possible that low, economically insignificant levels of *H. virescens* infestation had occurred on cotton for much of that plant's history in the United States.

Has there been genetic change associated with the recent abundance of *H. virescens* on cotton? Indirect evidence is provided by a comparison of populations from Mississippi to ones from the Virgin Islands (Schneider and Roush, 1987). Cotton was produced commercially in the Virgin Islands only from 1908 until 1927, when falling prices and infestation by the pink bollworm (*Pectinophora gossypiella*) made it unprofitable (Ricks, 1932; Briggs, 1933). Cotton has been feral in the Virgin Islands for about as long as it has been cropped in the southeastern United States, but neither tobacco budworms native to the Virgin Islands nor released Mississippi stocks oviposit on cotton under field conditions (M. L. Laster, personal communication), implying a strong preference for alternative, locally common hosts such as *Cajanus cajan* (pigeon pea) or *Bastardia viscosa* (Snow *et al.*, 1974).

In Mississippi, by contrast, cotton has been grown extensively for about 170 years and is the only abundant host of *H. virescens* during the summer (Snow and Brazzell, 1965). If genetic change in *H. virescens* was an important element in the recent abundance of this pest on cotton, one might expect to see greater adaptation to this host in Mississippi than in Virgin Island populations. Schneider and Roush found a markedly greater oviposition preference for cotton, as against geraniaceous hosts, in Mississippi than in Virgin Island strains. Mississippi populations also showed a higher larval growth rate on cotton, standardized against rates on artificial diet. All strains were reared under identical conditions, and the F_1 hybrids were intermediate between the parental values.

These results can be taken as estimates of the degree of adaptation to cotton that has taken place in Mississippi *H. virescens*. Whether such quantitative differences could account for the budworm's rise from economic insignificance to dominant pest status on cotton is unclear. A reason for doubting that they were solely responsible is the close correlation of these outbreaks with the rise of organic pesticide usage.

Heliothis virescens has been abundant on cotton for more than 35 years (Pfrimmer *et al.*, 1981), starting shortly after the beginning of commercial DDT production in 1946 (Metcalf, 1980). About half the insecticide produced in the United States in that time has been used on cotton. Pest outbreaks as a result of insecticide application are a widely recognized problem and have been well documented with cotton (e.g., Ewing and Ivy, 1943). Outbreaks of *H. virescens* due to the death of natural enemies from DDT application have been demonstrated in Peruvian valleys (Wille, 1951). In the United States, a shift from organochlorine to organophosphate pesticides, the latter being more toxic to natural enemies, led to especially large outbreaks of the *Heliothis* complex (Bottrell and Adkisson, 1977).

In summary, although there is evidence that *H. virescens* has undergone some genetic adaptation to the use of cotton, its recent outbreaks on that crop are also plausibly interpreted as resulting simply from the destruction of its natural-enemy complex by pesticides (but see Chapter 5).

V. SUMMARY AND CONCLUSIONS

At present the subject of genetic effects on numerical change consists largely of ill-defined hypotheses and scraps of inconclusive evidence. One aim of this chapter has been to provide a broad sample of relevant conjectures and observations from the literature, placed in a framework that might guide further development of the subject.

Population fluctuation per se has received scant attention in recent genetic–ecological modeling, and the evolutionary literature provides only tangentially relevant theorizing. We have attempted to outline the components that would exist in a complete model of gene frequency–population growth interaction. Measured against this standard, the existing, mainly case-specific genetic hypotheses about outbreaks remain preliminary and logically incomplete. In contrast to the search for environmental determinants of gene frequency, the study of short-term genetic effects on population dynamics is guided little by theory.

There is very little evidence on genetic change associated with outbreak episodes and certainly no conclusive demonstration that such a change influences population growth or decline. Although evidence of genetic effects could be sought in the well-known outbreak systems, the urgency of such an effort is debatable; this question is probably in the background for good reasons, both theoretical and practical. It is not generally clear what characters should be examined, nor how they might be expected to interact with population growth, though features such as disease resistance, density-induced "kentromorphism," and wing polymorphism seem like reasonable places to look. The plausibility of genetic causation of rapid fluctuation is undermined by the drastic selective differences required to produce major gene frequency change in just one or a few generations, though microevolution could still have important modulating effects on outbreak dynamics. The inheritance of most characters of ecological interest is likely to be complex, making controlled rearing necessary for the quantification of heritable change. Finally, the quantities directly affecting population growth, such as fecundity and developmental rate, often show environmentally (including density-) induced variation great enough to overshadow most genetic influences (Chapter 18). In sum, the search for genetic effects on outbreak dynamics, in the absence of suggestive evidence like that available for the larch budmoth, is likely to be disappointing.

Given these difficulties, it may be more profitable to attempt to discredit the general hypothesis of short-term genetic influence on population fluctuations by identifying apt model systems. In the ideal test case, the character of interest will show discrete, strongly heritable variation. The potential effect of this variation on population growth will be strong, obvious, and easily modeled. If variation even of this kind cannot be shown to affect population dynamics, we may assign low priority to the search for more subtle effects in species without these unusual traits.

One class of variants that meet the "test case" criteria above are those affecting sex ratio. Other things being equal, for example, the condition known as pseudogamy (Stenseth *et al.*, 1985) can drive local populations to extinction. Although much more evidence is needed, this does not seem to happen very often in insect populations harboring pseudogamy, which generally seem to be stable mixtures of reproductive types. The moderate temporal fluctuations of sex ratio sometimes seen in such populations, nevertheless, deserve scrutiny as a plausible and easily studied genetic influence on subsequent population growth. If the potential twofold reproductive advantage of pseudogamy can be balanced by other factors, we may expect fewer dramatic effects on abundance from variants of more subtle effect. In sum, sex ratio conditions of various kinds should be ideal material for studying the interaction of population dynamics with gene frequencies.

The evidence for long-term or historical contribution of genetic change to outbreak density appears to be both easier to gather and more compelling than that for short-term effects. An important role for microevolution is apparent in some cases of recently originating outbreaks; it may be significant that these species have atypical breeding systems. Further genetic study on historical cases like those discussed seems justified from both academic and practical points of view. The effect of microevolution on population size is of intrinsic interest to anyone concerned with distribution and abundance. Beyond this, the knowledge that the emergence of a new pest reflects a change in the pest itself might help direct the search for the right management strategy and help predict where new outbreaks will occur (J. Werren, personal communication).

ACKNOWLEDGMENTS

This chapter was greatly improved by the painstaking criticisms of J. Coyne, J. Werren, and the editors; important references were provided by A. Dobson, F. Gould, and J. Werren. We were probably wrong in not taking all of their advice.

J. S. wishes to thank the following people for help in locating records of *Carodochiles nigriceps* reared from hosts on cotton before 1934: P. H. Arnaud, California Academy of Sciences; H. R. Burke, Texas A & M University; C. Carlton, University of Arkansas; J. B. Chapin, Louisiana State University; W. E. Clark, Auburn University; J. P. Donahue, Natural History Museum of Los Angeles County; S. W. Hamilton, Clemson University; C. Parron, North Carolina State University; R. O. Schuster, University of California, Davis; M. Sharkey, Agriculture Canada; S. Shaw, U.S. National Museum; C. L. Smith, University of Georgia; L. A. Stange, Division of Plant Industry, FL; F. W. Werner, University of Arizona.

REFERENCES

Anderson, R. M., and May, R. M. (1981). The population dynamics of microparasites and their invertebrate hosts. *Philos. Trans. R. Soc. London Ser. B* **291,** 451–524.

Auer, C. (1968). Erste Ergebnisse einfacher stochasticher Modelluntersuchungen ueber die Ursachen der Populationsbewegung des grauen Laerchenwicklers *Zeiraphera diniana* Gn. (=*Z. griseana* Hb.) im Oberengadin, 1949/66. *Z. Angew. Entomol.* **62,** 202–235.

Auer, C. (1978). Ursache oder Wirkung? Kritische Betrachtung zum Aufsatz ("Colour-polymorphism and dynamics of larch budmoth populations."). *Mitt. Schweiz. Entomol. Ges.* **51,** 255–260.

Baltensweiler, W. (1971). The relevance of changes in the composition of larch budmoth populations for the dynamics of its numbers. *Proc. Adv. Study Inst. Dyn. Numbers Popul., 1970,* pp. 208–219.

Baltensweiler, W. (1978). Cause or effect? Chicken or egg? *Mitt. Schweiz. Entomol. Ges.* **51,** 261–268.

Baltensweiler, W., Benz, G., Bovey, P., and DeLuicchi, V. (1977). Dynamics of larch budmoth populations. *Annu. Rev. Entomol.* **22,** 79–100.

Berry, R. J. (1979). Genetical factors in animal population dynamics. *Symp. Br. Ecol. Soc.* **20,** 53–80.

Birch, L. C. (1971). The role of environmental heterogeneity and genetical heterogeneity in determining distribution and abundance. *Proc. Adv. Study Inst. Dyn. Numbers Popul., 1970,* pp. 109–128.

Booij, C. J. H., and Guldemond, J. A. (1984). Distribution and ecological differentiation between asexual gynogenetic planthoppers and related sexual species of the genus *Muellerianella* (Homoptera: Delphacidae). *Evolution* (*Lawrence, Kans.*) **38,** 163–175.

Bottrell, D. G., and Adkisson, P. L. (1977). Cotton insect pest management. *Annu. Rev. Entomol.* **22,** 451–481.

Bovey, P., and Maksymov, J. K. (1959). Le probleme des races biologiques chez la Tordeuse grise du Meleze, *Zeiraphera griseana* (Hb.). *Viertelahrsschr. Naturforsch. Ges. Zuerich* **104,** 264–274.

Boyce, A. M. (1935). The codling moth in Persian walnuts. *J. Econ. Entomol.* **28,** 864–873.

Briggs, G. (1933). "Report of the Virgin Islands Agriculture Experiment Station—1932."

Chanter, D. O., and Owen, D. F. (1972). The inheritance and population genetics of sex ratio in the butterfly *Acraea encedon. J. Zool.* **166,** 363–383.

Charlesworth, B., and Giesel, J. T. (1972a). Selection in populations with overlapping generations. II. Relations between gene frequency and demographic variables. *Am. Nat.* **106,** 388–401.

Charlesworth, B., and Giesel, J. T. (1972b). Selection in populations with overlapping generations. IV. Fluctuations in gene frequency with density-dependent selection. *Am. Nat.* **106,** 402–411.

Chitty, D. (1960). Population processes in the vole and their relevance to general theory. *Can. J. Zool.* **38,** 99–113.

Clarke, B. (1972). Density-dependent selection. *Am. Nat.* **106,** 1–13.

Crow, J. F., and Kimura, M. (1970). "An Introduction to Population Genetics Theory." Harper & Row, New York.

Daniels, N. E. (1977). Seasonal populations and migration of the greenbug in the Texas panhandle. *Southwest. Entomol.* **2,** 20–26.

Day, K. R., and Baltensweiler, W. (1972). Change in proportion of larval colour-types of the larchform *Zeiraphera diniana* when reared on two media. *Entomol. Exp. Appl.* **15,** 287–298.

Dempster, E. R. (1955). Maintenance of genetic heterogeneity. *Cold Spring Harbor Symp. Quant. Biol.* **20,** 25–32.

Diehl, S. R., and Bush, G. L. (1984). An evolutionary and applied perspective of insect biotypes. *Annu. Rev. Entomol.* **29,** 471–504.

Drosopoulos, S. (1976). Triploid pseudogamous biotype of the leafhopper *Muellerianella fairmairei. Nature* (*London*) **263,** 499–500.

Essig, E. O. (1931). "A History of Entomology." Macmillan, New York.

Ewing, K. P., and Ivy, E. E. (1943). Some factors influencing bollworm populations and damage. *J. Econ. Entomol.* **36,** 602–606.

Fernald, M. L. (1950). "Gray's Manual of Botany." American Book Co., New York.

Folsom, J. W. (1936). Notes on little known cotton insects. *J. Econ. Entomol.* **29,** 282.

Foster, S. W. (1910). On the feeding habits of the codling moth. *USDA Bur. Entomol. Bull.* **80,** 67–70.

Futuyma, D. J., Leipertz, S. L., and Mitter, C. (1981). Selective factors affecting clonal variation in the fall cankerworm, *Alsophila pometaria* (Lepidoptera: Geometridae). *Heredity* **47,** 161–172.

Gallun, R. L. (1978). Genetics of biotypes B and C of the Hessian fly. *Annu. Entomol. Soc. Am.* **71,** 481–486.

Gould, H. P., and Andrews, F. (1917). Apples: Production estimates and important commercial districts and varieties. *USDA Bull.* No. **485.**

Graves, P. M., and Curtis, C. F. (1982). A cage replacement experiment involving introduction of genes for refractoriness to *Plasmodium voelii nigeriensis* into a population of *Anopheles gambiae* (Diptera: Culicidae). *J. Med. Entomol.* **19,** 127–133.

Guerin, P. M., Baltensweiler, W., Arn, H., and Buser, H. R. (1984). Host race pheromone polymorphism in the larch budmoth. *Experientia* **40,** 892–894.

Handy, R. B. (1896). History and general statistics of the cotton plant. *In* "The Cotton Plant: Its History, Botany, Chemistry, Culture, Enemies, and Uses" (A. C. True, ed.), USDA Bull. No. 33, pp. 17–66. U.S. Govt. Printing Office, Washington, D.C.

Harrison, R. G. (1980). Dispersal polymorphisms in insects. *Annu. Rev. Ecol. Syst.* **11,** 95–118.

Harshman, L. (1982). Studies on the ecology and genetics of *Alsophila pometaria* (Lepidoptera: Geometridae). Ph.D. Dissertation, State University of New York at Stony Brook.

Harvey, T. L., and Hackerott, H. L. (1969). Recognition of greenbug biotype injurious to sorghum. *J. Econ. Entomol.* **62,** 776–779.

Hatchett, J. H., and Gallun, R. L. (1968). Frequency of Hessian fly, *Mageticla destructor,* races in field populations. *Annu. Entomol. Soc. Am.* **61,** 1446–1449.

Hawkes, J. G. (1983). "The Diversity of Crop Plants." Harvard Univ. Press, Cambridge, Massachusetts.

Heuch, I. (1978). Maintenance of butterfly populations with all-female broods under recurrent extinction and recolonization. *J. Theor. Biol.* **75,** 115–122.

Heuch, I., and Chanter, D. O. (1982). The persistence of abnormal sex ratios in the African butterfly *Acraea encedon. Oikos* **38,** 228–233.

Howard, L. O. (1887). The codling moth. *In* "Report of the Commissioner of Agriculture. Report of the Entomologist" (C. V. Riley, ed.), pp. 88–115. U.S. Govt. Printing Office, Washington, D.C.

Iwao, S. (1968). Some effects of grouping in lepidopterous insects. *In* "L'effet de groupe chez les Animaux." CNRS, Paris.

Kennel, J. (1921). Die Palaearktischen Tortriciden. *In* "Zoologica 54," pp. 547–742. E. Schweizerbart'sche Verlagsbuchhandlung, Stuttgart.

Key, K. H. L. (1957). Kentromorphic phases in three species of Phasmatodea. *Aust. J. Zool.* **5,** 247–284.

Key, K. H. L., and Day, M. F. (1954). A temperature controlled physiological color response in the grasshopper *Koscuiscola tristis* Sjost (Orthoptera: Acrididae). *Aust. J. Zool.* **2,** 309–339.

Kiester, A. R., Nagylaki, T., and Shaffer, B. (1981). Population dynamics of species with gynogenetic sibling species. *Theor. Popul. Biol.* **19,** 358–369.

Kimball, C. P. (1965). "The Lepidoptera of Florida." Division of Plant Industry, Florida Department of Agriculture, Gainesville, Florida.

Krebs, C. J. (1977). "Ecology: The Experimental Analysis of Distribution and Abundance." Harper & Row, New York.

Krebs, C. J. (1978). A review of the Chitty hypothesis of population regulation. *Can. J. Zool.* **56,** 2463–2480.

Krebs, C. J., Gaines, M. S., Keller, B. L., Myers, J. H., and Tamarin, R. H. (1973). Population cycles in small rodents. *Science* **179,** 35–41.

Krombein, K. V., Hurd, P. D., Jr., and Smith, D. R. (1979). "Catalog of Hymenoptera of America North of Mexico." Smithsonian Inst. Press, Washington, D.C.

Lanier, G. H., and Oliver, J. H. (1966). "Sex-ratio" condition: Unusual mechanisms in bark bettles. *Science* **153,** 208–209.

Lea, A. (1968). Natural regulation and artificial control of brown locust numbers. *J. Entomol. Soc. South. Afr.* **31,** 97–112.

Levene, H. (1953). Genetic equilibrium when more than one niche is available. *Am. Nat.* **87,** 331–333.

Lewis, W. J., and Brazzell, J. R. (1966). Biological relationships between *Cardiochiles nigriceps* and the *Heliothis* complex. *J. Econ. Entomol.* **59,** 820–823.

Lorimer, N. (1979). Genetic causes of pest population outbreaks and crashes. *In* "Genetics in Relation to Insect Management" (M. A. Hoy and J. J. McKelvery, Jr., eds.), pp. 50–54. Working Papers, Rockefeller Foundation, New York.

Ludwig, W. (1950). Zur theorie der konkurrenz. Die annidation (Einnischung) als fuenfter Evolutionsfaktor. *Neue Ergeb. Probl. Zool., Klatt-Festschr.,* pp. 516–537.

Mally, C. W. (1916). Codling moth in walnuts. *S. Afr. Fruit Grower* **3,** 3.

Matessi, C., and Jayakar, S. D. (1976). Models of density–frequency-dependent selection for the exploitation of resources. I: Intraspecific competition. *In* "Population Genetics and Ecology" (S. Karlin and E. Nevo, eds.), pp. 707–722. Academic Press, New York.

May, R. M., and Anderson, R. M. (1983a). Parasite–host coevolution. *In* "Coevolution" (D. J. Futuyma and M. Slatkin, eds.), pp. 186–206. Sinauer Assoc., Sunderland, Massachusetts.

May, R. M., and Anderson, R. M. (1983b). Epidemiology and genetics in the coevolution of parasites and hosts. *Proc. R. Soc. London, Ser. B* **219,** 281–313.

Mayr, E. (1963). "Animal Species and Evolution." Harvard Univ. Press, Cambridge, Massachusetts.

Metcalf, R. L. (1980). Changing role of insecticides in crop protection. *Annu. Rev. Entomol.* **25,** 219–256.

Mihok, S., Fuller, W. A., Canham, R. P., and McPhee, E. C. (1983). Genetic changes at the transferrin locus in the red-backed vole (*Clethrionomys gapperi*). *Evolution (Lawrence, Kans.)* **37,** 332–340.

Mitter, C., and Futuyma, D. J. (1977). Parthenogenesis in the fall cankerworm, *Alsophila pometaria* (Lepidoptera, Geometridae). *Entomol. Exp. Appl.* **21,** 192–198.

Mitter, C., and Futuyma, D. J. (1983). An evolutionary–genetic view of host-plant utilization by insects. *In* "Variable Plants and Herbivores in Natural and Managed Systems" (R. F. Denno and M. S. McClure, eds.), pp. 427–459. Academic Press, New York.

Mitter, C., Futuyma, D. J., Schneider, J. C., and Hare, J. D. (1979). Genetic variation and host plant relations in a parthenogenic moth. *Evolution (Lawrence, Kans.)* **33,** 777–790.

Moore, B. P., Woodroffe, G. E., and Sanderson, A. R. (1956). Polymorphism and parthenogenesis in a ptinid bettle. *Nature (London)* **177,** 847–848.

Moore, C. A., and Majors, K. R. (1970). Economic considerations. *In* "Sorghum Production and Utilization" (J. S. Wall and W. M. Ross, eds.). Avi Publ. Co., Westport, Connecticut.

Narbel-Hofstetter M. (1963). Cytologie de la pseudogamie chez *Luffia lapidella* Geoze (Lepid. Psychidae). *Chromosoma* **13,** 623–645.

Owen, D. F., and Chanter, D. O. (1969). Population biology of tropical African butterflies. Sex ratio and genetic variation in *Acraea encedon*. *J. Zool.* **157,** 345–374.

Packard, A. S. (1883). The Hessian fly—its ravages, habits, and the means of preventing its increase. *In* "Third Report of the United States Entomological Commission" (C. V. Riley, ed.), pp. 198–248. U.S. Govt. Printing Office, Washington, D.C.

Painter, R. H. (1930). The biological strains of Hessian fly. *J. Econ. Entomol.* **23,** 322–326.

Pfrimmer, T. R., Stadelbacher, E. A., and Laster, M. L. (1981). *Heliothis* spp. seasonal incidence on cotton in the Mississippi delta. *Environ. Entomol.* **10,** 642–644.

Phillips, P. A., and Barnes, M. M. (1975). Host race formation among sympatric apple, walnut, and plum populations of the codling moth, *Laspyresia pomonella*. *Annu. Entomol. Soc. Am.* **68,** 1053–1060.

Quayle, H. J. (1926). The codling moth in walnuts. *Bull.—Calif. Agric. Exp. Stn.* **402.**

Raffa, K. F., and Berryman, A. A. (1983). The role of host plant resistance in the colonization behavior and ecology of bark beetles (Coleoptera: Scolytidae). *Ecol. Monogr.* **53,** 27–49.

Ricks, J. R. (1932). "Report of the Virgin Islands Agriculture Experiment Station—1931."

Riley, C. V. (1885). "Fourth Report of the United States Entomological Commission." U.S. Govt. Printing Office, Washington, D.C.

Roughgarden, J. (1971). Density-dependent natural selection. *Ecology* **52,** 453–468.

Roughgarden, J. (1972). Evolution of niche width. *Am. Nat.* **106,** 683–718.

Roughgarden, J. (1979). "Theory of Population Genetics and Evolutionary Ecology: An Introduction." Macmillan, New York.

Schneider, J. C., and Roush, R. T. (1987). Genetic differences in oviposition preference between two populations of *Heliothis virescens*. *In* "Behavioral Genetics of Arthropods" (M. Huettel, ed.). Plenum, New York (in press).

Smith, D. A. S. (1975). All-female broods in the polymorphic butterfly *Danaus chrysippus* (L.) and their ecological significance. *Heredity* **34,** 363–371.

Smith, R. E. (1912). Walnut culture in California walnut blight. *Bull.—Calif. Agric. Exp. Stn.* **231.**

Snow, J. W., and Brazzell, J. R. (1965). Seasonal host activity of the bollworm and tobacco budworm during 1963 in northeast Mississippi. *Miss. Agric. Exp. Stn., Bull.* **712.**

Snow, J. W., Cantelo, W. W., Baumhover, A. H., Goodenough, J. L., Graham, H. M., and Raulston, J. R. (1974). The tobacco budworm on St. Croix, U.S. Virgin Islands: Host plants, population survey and estimates. *Fla. Entomol.* **57,** 297–301.

Sosa, O., Jr. (1981). Biotypes J and L of the Hessian fly discovered in an Indiana wheat field. *J. Econ. Entomol.* **74,** 180–182.

Stenseth, N. C. (1977). Evolutionary aspects of demographic cycles: The relevance of some models of cycles for microtine fluctuations. *Oikos* **29,** 525–538.

Stenseth, N. C. (1978). Demographic strategies in fluctuating populations of small rodents. *Oecologia* **33,** 149–172.

Stenseth, N. C., Kirkendall, L., and Moran, N. (1985). On the evolution of pseudogamy. *Evolution (Lawrence, Kans.)* **39,** 294–307.

U.S. Department of Agriculture (1970). Culture and use of grain sorghum. *U.S. Dep. Agric., Agric. Handb.* **385.**

Uvarov, B. P. (1928). "Locusts and Grasshoppers: A Handbook for Their Study and Control." Imperial Bureau of Entomology, London.

Varley, G. C., and Gradwell, G. R. (1970). Recent advances in insect population dynamics. *Annu. Rev. Entomol.* **15,** 1–24.

Werren, J. H., Skinner, S. W., and Charnov, E. L. (1981). Paternal inheritance of a daughterless sex ratio factor. *Nature (London)* **293,** 467–468.

Wille, J. E. (1951). Biological control of certain cotton insects and the application of new organic insecticides in Peru. *J. Econ. Entomol.* **44,** 13–18.

Wilson, F. (1967). Insect abundance: Prospect. *Symp. R. Entomol. Soc. London* **4,** 143–155.

Wood, J. R., Starks, E. A., and Starks, K. J. (1972). Effect of temperature and host plant interaction on the biology of three biotypes of the greenbug. *Environ. Entomol.* **1,** 230–234.

Chapter 20

Evolutionary Processes and Insect Outbreaks

NILS CHR. STENSETH

Division of Zoology
Department of Biology
University of Oslo
Blindern, N-0316 Oslo 3, Norway

I. INTRODUCTION

The term "pest" connotes a value judgment: Densities are higher than what is compatible with the human standard of living. Species become pests when they interfere in one way or another with nature or with products of major concern to people. Thus, pest outbreaks are first of all a problem for social scientists and economists (e.g., Southwood and Norton, 1973; Conway, 1976). Pests are living organisms, however, and thus biological knowledge must be part of any study of pest outbreaks; populations respond, often in a complex and dynamic fashion, to outside interferences. Hence, ecological knowledge is of particular importance. How pests respond to their environmental conditions is determined by their ecological strategies (*sensu* Southwood, 1976), which again is a result of evolutionary historical processes. Hence, evolutionary knowledge is also important in studies of pest outbreaks. Without evolutionary knowledge we are unlikely to understand why some species are pests whereas others are not, or why the same species is not experienced as a pest in all environments.

Ecological changes and evolutionary changes within a species generally take place on greatly different time scales. In general, ecological processes such as pest outbreaks are rapid compared with evolutionary processes such as the evolution of life history tactics, feeding patterns, predator avoidance, or pesticide resistance. Nevertheless, they are closely related. The current ecological setting determines the selective pressure that molds the life history traits and feeding patterns of living species, which in turn determine the ecological processes of the species under consideration. Both intra- and interspecific interactions and interactions between the biotic and abiotic components of the ecosystem are important components of evolutionary as well as ecological considerations.

The differences between ecological and evolutionary time scales have some important consequences for modeling in evolutionary ecology. In the development of ecological models for studying pest outbreaks, certain quantities entering the models can be treated as constants. These are the ecological parameters (*sensu* Lawlor and Maynard Smith, 1976), which determine how the ecological variables (e.g., densities) will change over time. These ecological parameters are, however, a result of factors like habitat structure (whether, e.g., it is simple and uniform, or complex and subdivided), ecological features of coexisting species, and the considered species' evolutionary history. Hence, over evolutionary time these ecological parameters become variables; they become evolutionary variables (*sensu* Lawlor and Maynard Smith, 1976; for general discussion, see Stenseth, 1985a).

Since it simplifies analysis, it is often useful to distinguish between a fast ecological scale and a slow evolutionary time scale [see Lawlor and Maynard Smith (1976) and Stenseth and Maynard Smith (1984)] (but not always; see Section IV,B). However useful, this distinction is rather artificial and must be made with caution. Some evolutionary processes occur, for example, at a pace equivalent to many ecological processes (e.g., viruses and insect pests on forest trees). Similarly, there often may be two or more distinct ecological time scales (see Ludwig *et al.*, 1978). Ludwig and co-workers describe the situation as follows:

> Associated with each state variable is a characteristic time interval over which appreciable changes occur. The budworm can increase its density several hundred fold in a few years. Therefore, in a continuous representation of this process, a characteristic time interval for the budworm is of the order of months. Parasites of the budworm may be assigned a similar, or somewhat slower scale. Avian predators may alter their feeding behavior (but not their numbers) rather quickly and may be assigned a fast time scale similar to budworm. The trees cannot put forth foliage at a comparable rate, however; a characteristic time interval for trees to completely replace their foliage is on the order of 7–10 years. Moreover, the life span of the trees themselves is between 100 and 150 years, in the absence of budworm, so that their generation time is measured in decades. We first conclude, therefore, that the minimum number of variables will include budworm as a fast variable and foliage quantity (and perhaps quality) as a slow variable.

In this chapter I discuss, with reference to pest species, various topics either emphasizing the slow evolutionary processes or the interplay between the relatively slow evolutionary processes and the relatively fast ecological processes. Specifically, I ask whether we can understand why some species, under certain conditions, came to be pests. Even though I cannot defend my position, I believe that, if we understood why some species came to be pests, we might be able to design better preventive methods for avoiding pest outbreaks as well as for treating acute pest problems.

Throughout the chapter, I emphasize the "habitat template" (*sensu* Southwood, 1977b) and the resulting population dynamics of potential pest species. I also discuss how the knowledge of a pest species' life history traits and population dynamics can be used in the design of optimal pest control strategies. Furthermore, I discuss coevolutionary processes of importance in the context of pest outbreaks. I finally discuss how studies of one of the greatest mysteries in evolutionary biology (Bell, 1982), the maintenance of sexual reproduction, can be used for designing effective pest control stategies. I present this as a particularly good example of the fact that pure or basic and applied science indeed may fertilize each other. The pure scientist may provide a conceptual basis for the control of a pest species. The applied scientist may, in contrast, provide a particular case study to be used as an example by the pure scientist. Since the avoidance of pest outbreaks is of great importance and concern to humans, there will—and indeed should—always be economic support for such studies.

II. ECOLOGICAL STRATEGIES AND THE HABITAT TEMPLATE: IMPLICATIONS FOR PEST OUTBREAKS

A. Classifying Ecological Strategies

Answers to both ecological (*how*) and evolutionary (*why*) questions (*sensu* Mayr, 1961; Pianka, 1978) are important in basic science (see almost any modern ecology text), but we seldom realize that answers to both the *how* and *why* questions are of almost equal importance in applied sciences such as studies of pest outbreaks. Most commonly, only ecological aspects are treated in pest outbreak studies. However, the evolutionary aspects are often of paramount importance for the study of pest outbreaks. Only they will help us to understand how we can avoid pest outbreaks. Thus, Southwood and his colleagues, among others, have in several important publications discussed the evolutionary aspects of pest outbreaks (Southwood, 1977a,b; Southwood and Norton, 1973; Southwood and Way, 1970; Southwood *et al.*, 1983). For a discussion of these ideas, see below.

In their classic work on island biogeography, MacArthur and Wilson (1967) expressed the idea that any alternative sets of species traits could be referred to as an *r* strategy or a *K* strategy (see also Pianka, 1970, 1972; Parry, 1981). These ecological strategies were assumed to result from different types of natural selection, that is, *r* selection occurring in highly uncertain and variable environments and *K* selection occurring under more stable environmental conditions. MacArthur and Wilson were not the first to classify ecological strategies into distinct categories; several similar (and mostly older) classifications are listed in Table 1. However, the terms used by MacArthur and Wilson were far more "catchy" than the earlier suggestions. Most important for the acceptance of the *r–K* dichotomy, however, was the fact that MacArthur and Wilson proposed explanations of why these strategies had evolved. Not only did they describe *how* the living world was structured, but they also suggested *why* it came to be as it is. They furthermore tied the idea of ecological strategies in with the neo-Darwinian framework in which the individual within the population is a fundamental unit.

Alternative classifications, also based on evolutionary considerations, have since been proposed. Gill (1974) suggested a three-way classification scheme with *r*, *K*, and α strategies, where α strategies are characterized by high competitive ability. Grime (1977, 1978, 1979) proposed classifying ecological strategies into *R*, *C*, and *S* strategies, where *R* strategists (or ruderals) are adapted to cope with disturbances or destruction of the habitat (as in man-made ecosystems), *C* strategists (or competitors) are adapted to live in a competitive environment (as in the tropics), and *S*

TABLE 1

Some Partial or Complete Synonyms of the *r–K* Spectrum Discussed by MacArthur and Wilson (1967)[a]

r Strategists	*K* Strategists	Reference
Fugitive species	—	Hutchinson (1951)
Vegetative	Sensorimotor	Kennedy (1956)
Opportunists	Equilibrium species	MacArthur (1960)
Denizens of temporary habitat	Denizens of permanent habitat	Southwood (1962a,b)
Pioneering species	—	Wynne-Edwards (1962)
Exploitation competitors	Interference competitors	Miller (1969)
l-Parasites	m-Parasites	Mitchell (1975)
Fast species	Slow species	Smith (1975)
Super tramps	High S-species	Diamond (1975)
Small species	Large species	Connell (1975)
Ephemeral species	Sessile species	Gilbert *et al.* (1976)
Therophytes	Phanerophytes	Raunkiaer (1937)

[a] After Southwood (1977a). In addition there exist several three-way classification systems; see text for a discussion of some of these.

strategists (or stress tolerators) are adapted to cope with severe abiotic environmental circumstances (as in the arctics). Finally, Greenslade (1983) proposed to classify species into *r, K,* and *A* strategies, where *A* strategists are adapted to tolerate adverse environmental conditions. I find Greenslade's proposal rather useful and will therefore rely fairly heavily on his scheme in this chapter; others, however, may find other schemes more useful.

There has been much discussion of the utility of a classification system like the *r*- and *K*-strategy dichotomy (Pianka, 1970, 1972, 1978; Stearns, 1976, 1977; Southwood, 1977a; Parry, 1981; Horn and Rubinstein, 1984). Much of this discussion has been based on a misunderstanding of the original concepts and ideas and has as a result been rather confusing (see Pianka, 1978). First, different authors have failed to interpret *r* and *K* selection (and *r* and *K* strategists) as relative phenomena. A species is an *r* strategist only *relative* to some other species. Only if a species is an *r* strategist relative to the majority of species under consideration can we, with some qualifications, say that it generally is an *r* strategist.

Second, the critics of the *r–K* dichotomy miss the point that, depending on the environmental situation under which a population has evolved and is living, a species may behave as an *r* strategist or as a *K* strategist. That is, the strategy of the individuals within a population is population specific, *not* species specific.

Above all, classification systems of this kind make it easier to understand the diversity of life since the dimensions of the problems under study thus become greatly reduced. Indeed, they may help us link ecological arguments more closely with evolutionary arguments. Knowing whether a population is composed of more or less *r*-selected or *K*-selected individuals will often make us better able to choose among the available pest control options. Similarly, if we have the option of changing the population's strategy (which is always determined partly by genetic properties and partly by environmental properties), such a classification system will make it easier to know what to do to avoid insect pest outbreaks. I return to this topic in connection with the discussion of Tables 2 and 4.

B. Habitat Template

The correlates of *r, K,* and *A* strategies are summarized in Table 2 and Fig. 1. Which of these strategies is favored by natural selection depends upon (1) the predictability of the abiotic components of the habitat, (2) the predictability of the biotic components of the habitat, and (3) the habitat's favorability. In short, it depends on the "habitat template" (Southwood, 1977b). However, in addition to evolution *in situ,* the strategies that can exist in a particular habitat depend on ecological factors such as extinction and ability to invade.

Abiotic and biotic predictability refer to the way in which the habitat template changes over time and space. The favorability of the habitat refers to the average living condition. Figure 2 illustrates the distinction between the "predictability" and "favorability" of the habitat.

Extreme *K* selection occurs in regions with high abiotic predictability, low biotic predictability, and high habitat favorability. Biotic interactions

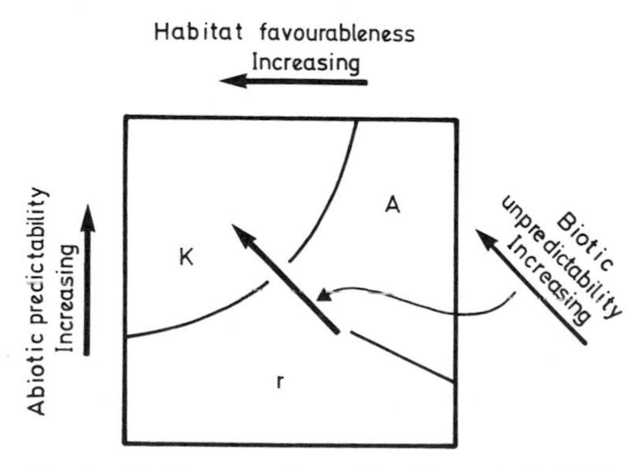

Fig. 1. Habitat template. After Greenslade (1983).

TABLE 2

Correlates of Extreme *r*, *K*, and *A* Selection[a]

Type of selection feature	*r*	*K*	*A*
Properties of the habitat			
Favorableness	Variable	High	Low
Predictability	Low	High	High
Community attributes			
Diversity	Low	High	Low
Interspecific competition	Occasional, sometimes intense, two species	Frequent, continuous, often diffuse	Rare
Investment in defense mechanisms	Low	High	Low
Degree of specialization	Low	High	Low
Population or species attributes			
Capacity for dormancy	Variable	Low	Variable
Selection for migratory ability	High	Intermediate	Low
Geographic distribution	Wide	Restricted	Variable
Selection for parthenogenesis	Variable	Low	High
Length of life	Short	Intermediate	Long
Maturity	Early	Intermediate	Late
Fecundity	High	Intermediate	Low
Population density	Very variable	More constant, near carrying capacity	Variable, below carrying capacity
Rate of increase	High	Intermediate	Low
Density dependence	Weak at low density; strong overcompensating at high density	Moderate, compensating at high density	Weak
Key factors	Adult losses: mortality and migration	Juvenile mortality: variation in fecundity	Mortality at all stages; variation in fecundity and rates of development

[a] After Greenslade (1983).

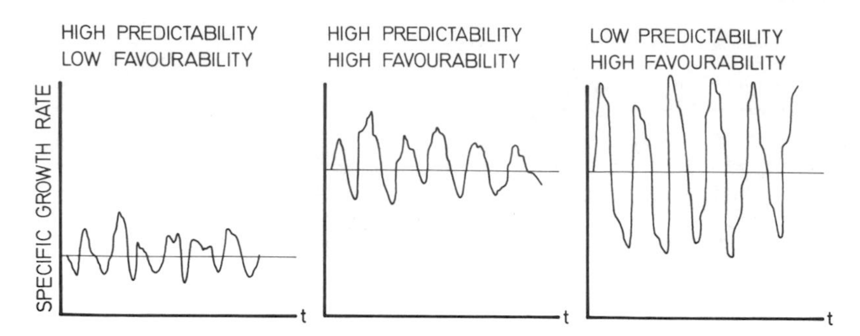

Fig. 2. Interpretation of high and low predictability and high and low favorability.

play a dominating role in the evolution of the *K* strategy. If the abiotic predictability is low, extreme *r* selection will result. Abiotic components of the environment play an important role in the evolution of the *r* strategy. With low habitat favorability but not too low habitat predictability, extreme *A* selection will result. In this case the habitat template is almost like one resulting in *K* selection except that in the case of *A* selection there are weaker biotic interactions and lower habitat favorability. That is, a species that is originally an *A* strategist may be subject to *K* selection if the general favorability of the habitat is improved by, for example, greater food availability or better climatic conditions. The spruce bark beetle, *Ips typographus,* may offer an example of such a transition. This species usually exploits only dead or dying trees (Solbreck, 1986), but due to environmental disturbances caused by current forestry practices it may also be able to exploit living trees (Solbreck, 1986; Berryman *et al.,* 1984). Alternatively, a region may be more prone to invasion by *K* strategists than by *A* strategists if the region's general favorability is improved (see Fig. 1).

Biotic predictability decreases with increasing species diversity. In regions with high species diversity, the coexisting species will experience a continuous change over both ecological and evolutionary time even in the absence of abiotic changes. Any given species will, for instance, have difficulties evolving effective defense strategies against enemies. The biotic predictability will be low.

Abiotic predictability decreases with increasing physical disturbances. Abiotic unpredictability and biotic unpredictability are, of course, qualitatively very different. In the former case, the evolving species are faced with challenges from a habitat template not dominated by biotic processes. In the latter case, however, the evolving species are faced with challenges from a habitat template composed mainly of other evolving species. Commonly, such biotically dominated habitats will be characterized by a Red Queen or arms race type of coevolution in which each of the species has to evolve new adaptations all the time in order to avoid extinction (Van Valen, 1973; Dawkins and Krebs, 1979; Stenseth and Maynard Smith, 1984; Stenseth, 1985a). Intuitively, it seems easier to adapt to an environment dominated by physical factors (unless these are very erratic and have large amplitudes) than to adapt to a biotically dominated environment, since in the latter case the environment changes because of the evolutionary advances made by the coexisting species. After all, we know that the earth's biotic diversity (presumably because of evolution through natural selection) is much greater than the earth's abiotic diversity. These biotic changes, although of an evolutionary nature, may sometimes be rather rapid (e.g., pathogen–host interactions; Pimentel, 1961; May and Anderson, 1983).

C. Population Dynamics

The population dynamics of a species in a given region are influenced by the ecological strategies of the species. That is, the genetic constitution of the organism together with its habitat determine the organism's life history features such as brood size B, survival from egg to reproducing adult s_1, and adult survival, s_2 and how these features depend on the population density. These life history quantities then determine the net specific growth rate R (or fitness) through the expression

$$R = Bs_1 + s_2 \tag{1}$$

(Stenseth, 1984); this quantity may or may not be a function of density (Royama, 1977, 1984). The properties of R are the key to understanding the population dynamics of the species under consideration. That is, let N_t be the population density at time t, and $R = R(N_t)$ the net specific growth rate for the particular ecological strategy in the habitat under consideration. Then

$$N_{t+1} = R \cdot N_t \tag{2}$$

As is well known (Maynard Smith, 1968; Southwood *et al.*, 1974; May *et al.*, 1974; Southwood, 1976), very strong density dependence may in a seasonal environment, particularly when combined with high equilibrium densities, easily generate excessive density fluctuations (Fig. 3). Species with such strong density dependence are those that locally (at least) may become pests. From Table 2 it therefore appears that r strategists are the prime candidates for being pest species.

Some examples of population dynamic patterns that might result from various ecological strategies are illustrated in Figs. 3 and 4. The curves in

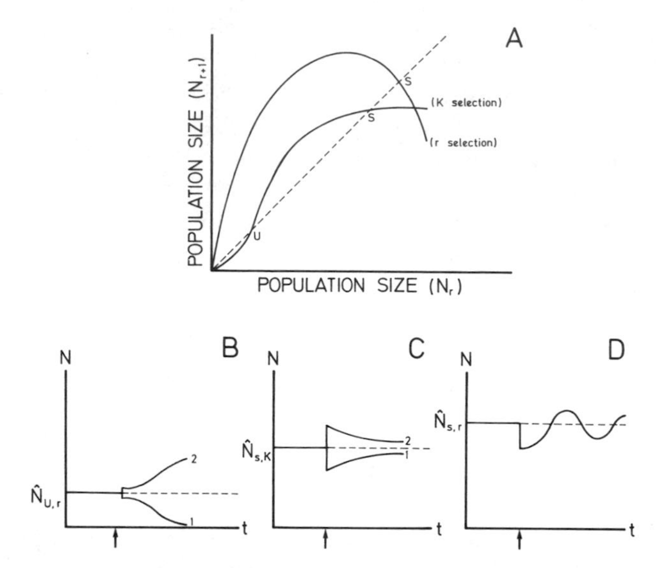

Fig. 3. A, Population growth curves for r and K strategists showing stable (S) equilibrium points and an unstable point, the extinction point U. From Southwood *et al.* (1974). B–D, Time trajectories of the population dynamics after a small distortion from the ecological equilibrium has occurred.

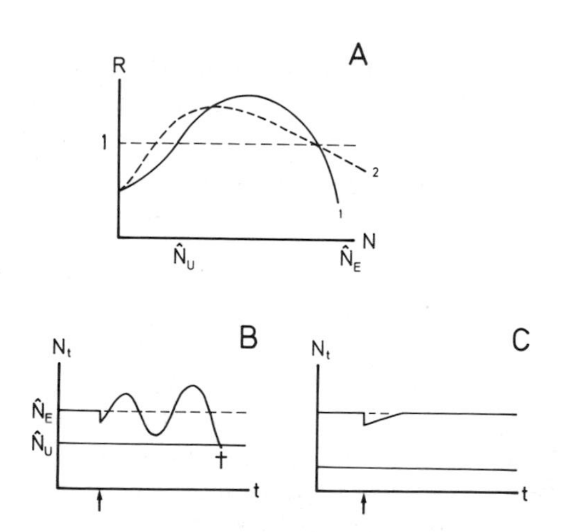

Fig. 4. Net specific growth rate R as a function of density: In graph A, 1 corresponds to an undisturbed natural situation, 2 to a disturbed situation. Graphs B and C depict the corresponding local stability properties around the higher equilibrium.

Fig. 3A are found by plotting successive points with coordinates (N_t, N_{t+1}) for $t = 1, 2, \ldots$. For $N_t = \hat{N}$, found by solving $R(N)_t = 1$, an ecological equilibrium exists. Then, N_{t+1} and N_t will be on the line with a 45° slope in Fig. 2A. If dR/dN evaluated in $N_t = \hat{N}$, is positive, the equilibrium is *unstable*. This corresponds to U in Fig. 3A (see also Fig. 3B). If dR/dN, again evaluated in $N_t = \hat{N}$, is negative, the equilibrium is potentially stable. However, dR/dN cannot be too large and still give rise to a stable equilibrium. That is, density dependence cannot be too strong. If density dependence is very strong, density oscillations with increasing amplitude will generally result. With weaker density dependence, there may be an equilibrium with oscillations having decreasing amplitudes.

As explained above, R may change (with repect to both magnitude and degree of density dependence) through evolution by natural selection and/or through habitat modification. A hypothetical example is depicted in Fig. 4. Let us consider some forest insect attracted to its food and breeding material by chemical stimuli and pheromones. Assume case 1 to be the "natural" or undisturbed situation (i.e., Fig. 4B). As can be seen, this situation is characterized by sufficiently large density fluctuations to cause frequent and rapid (local) extinction of the population. Then if, for example, artificial pheromones are applied "inappropriately," we may in effect redistribute the population more evenly and hence possibly produce an R curve as depicted for case 2 of Fig. 4A. As can be seen, this manipulation may easily stabilize the population at a high equilibrium density (Fig. 4C). Such a situation would, of course, be undesirable if the species under consideration were a potential pest species.

It follows that it is important to have a solid understanding of the habitat template and the evolution of the ecological strategies before we start manipulating the ecological system. Specifically, we need to know how the R function has been molded by evolution and the habitat template (i.e., ecology).

III. PEST CONTROL OF SPECIES WITH VARIOUS ECOLOGICAL STRATEGIES

A. Theory

Elsewhere I have discussed the application of the r–K framework to pest control studies (Stenseth, 1981). Here I review this approach as well as extend it to the r–K–A framework suggested by Greenslade (1983).

The situation I visualize is depicted in Fig. 5A,B. A patchy environment is assumed (see Southwood, 1977a; Hansson, 1977a,b). Within each

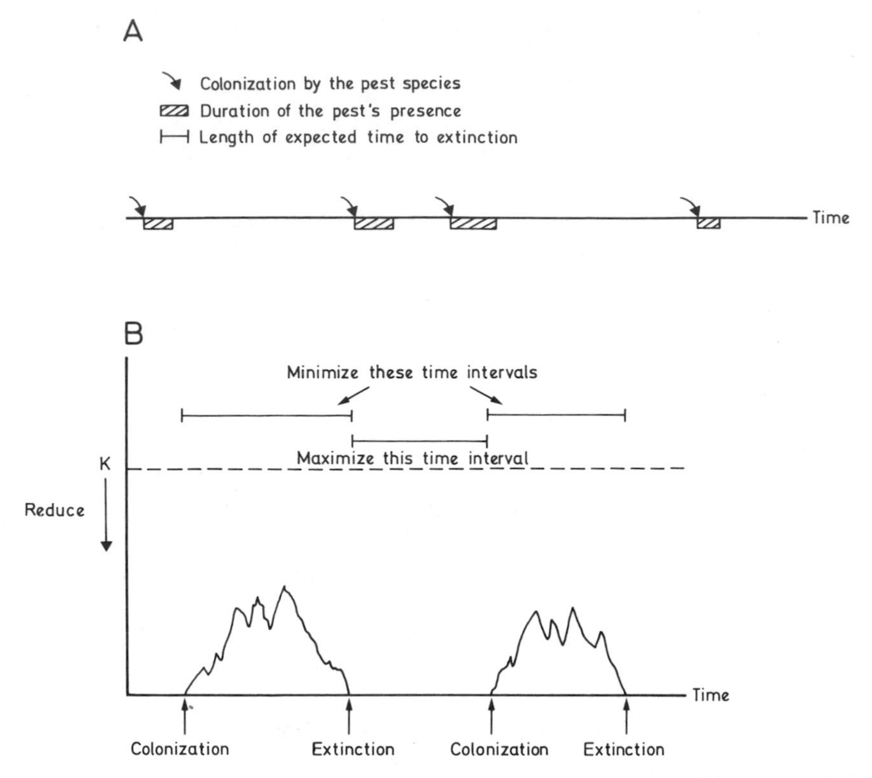

Fig. 5. Situation hypothesized in this chapter. A, Pest outbreak conditions can only be appreciated as such during periods in which individuals are present. Note that situations in which the potential pest species usually occur at low densities but from time to time "explode" may also be considered within the framework discussed here (see text). B, Our objectives for each patch are to (1) minimize periods between "colonization" and "extinction," (2) maximize periods between "extinction" and "colonization," and (3) reduce K. It can be shown that (3) follows automatically from (1) and (2). After Stenseth (1981).

patch we aim at having the potential pest species present for as short a time as possible and absent for as long as possible. In patches suitable for the pest species being considered, control efforts should be applied long before the population has reached densities above some economic damage level, that is, outbreak densities (see Fig. 5B). If control is first applied when high outbreak densities have been reached, it is often too late. The type of control measures I have in mind should have "instantaneous" effects (possibly lasting for some time). Furthermore, they should have no time lags and should act in a density-independent manner.

In a heterogeneous habitat, an optimal control strategy should aim at allocating the limited amount of economic resources in order to minimize

(1) the effective immigration rate (see Levins, 1969; Southwood, 1975, 1977a,b; Stenseth, 1977) and (2) the expected time to extinction of the local population (Stenseth, 1977, 1981). Theoretical results indicate, however, that (1) and (2) may result in conflicting recommendations (Levins, 1969; Stenseth, 1977).

MacArthur and Wilson (1967) concluded that the chance of a single pair of individuals reaching a population density near a patch's carrying capacity K and taking some time T_k to become extinct is about $(\lambda-\mu)/\lambda$, whereas that of rapid extinction at once is about $1 - (\lambda - \mu)/\lambda = \mu/\lambda$ (where λ and μ are the specific net birth and death rates of species, respectively, and where $s_1 B = e^{-\lambda t}$ and $s_2 = e^{-\mu t}$; t is the length of time between reproductive seasons). A good colonizing species is one maximizing the former probability while minimizing the latter. However, to design an optimal pest control strategy, we should aim at minimizing $(\lambda-\mu)/\lambda$ and maximizing $1 - (\lambda-\mu)/\lambda$. The strategy of the species (or population) is defined by λ and μ. These quantities, then, are the link between the strategies discussed in the previous section and the design of pest control discussed in the present section.

In island biogeography theory, we assume that natural selection molds the species to the evolutionary optimal strategy. In pest control, we attempt to affect a species and its environment such that the criteria for optimal control are met. Specifically, we want

$$\tau = 1 - \mu/\lambda \tag{3}$$

to be minimized by manipulating λ and μ. The τ quantity is analyzed with respect to pest control strategies in the Appendix. This then yields predictions regarding the allocation of economic resources and manpower to measures affecting patches suitable for reproduction (λ) and survival (μ) in suitable patches.

A similar analysis of a model developed by Levins (1969) yields predictions regarding how resources should be allocated (optimally) to measures affecting patch extinction (by affecting μ, as explained earlier) or to measures affecting dispersal.

B. Recommendations

1. The r–K Framework

On the basis of the theory just described, I have made the following recommendations for optimal pest control (Stenseth, 1981, p. 785):

1. If there are methods available for reducing immigration rates into empty patches by almost 100%, most resources should be spent on reducing immigration regardless of the demography of the pest. According to

Levins (1969), this treatment should be applied at times as variable as possible from patch to patch in the habitat complex (see also Stenseth, 1977).

2. If, however, no particular methods for reducing immigration exist, separate predictions for *r*- and *K*-selected species in the same habitat result (Table 3).

a. For *r*-selected species, most resources should be devoted to extinction through death and birth; in most cases, most resources ought to be devoted to reducing reproduction rather than increasing mortality. Following Levins (1969), this treatment should be applied in the habitat complex with as little variation in time as possible.
b. For *K*-selected species, most resources should, in most cases, be devoted to increasing mortality.

2. The r–K–A Framework

If effective methods for reducing immigration into suitable patches exist, then recommendation 1 is also valid with reference to the *r–K–A* framework. Otherwise, what it to be regarded an optimal pest control strategy will depend on the magnitude of the undisturbed (or natural) reproductive, survival, and colonization rates. Since the *A* strategy is, in a way, an extreme *K* strategy, it follows from the arguments given in Stenseth (1981) that immigration should be reduced as much as possible.

The more we know about the "undisturbed" life history of the pest species, the more precisely will we be able to predict the optimal pest control strategy. Demographic categories such as *r*, *K*, and *A* strategies

TABLE 3

Principal Control Strategies Likely to Be Optimal for Different Pest Types[a]

Demographic strategy	Effect of reducing immigration[b]	Effect of increasing extinction rate of local populations[c]	
		Increase mortality	Reduce reproduction
r-Selected pests	Little	Some	Much
K-Selected pests	Much	Some	Little

[a] After Stenseth (1981). Note that many predictable exceptions will occur.

[b] Whenever reduction of immigration is predicted to be the optimal strategy, the control treatment should be applied in the habitat complex with as great spatial variability as possible.

[c] Whenever reduction of survival is predicted to be the optimal strategy, the control treatment should be applied in the habitat complex with as little spatial variability as possible.

help us, however, to suggest the appropriateness of utilizing various pest control measures against various pests. This would certainly simplify the pest manager's job.

C. An Alternative Theory

I emphasize that unambiguous predictions regarding optimal pest control of *r*-, *K*-, and *A*-selected species are difficult to obtain (and exceptions will always be found); thus, there will always be some exceptions (often predictable) to any general statement about optimal pest control.

It should be pointed out that Conway (1976) and Southwood (1977b) advocated the application of pesticides (affecting mortality) for *r*-selected species (see Table 4). My analysis suggests that this is unlikely to be the optimal strategy (see Table 3). Furthermore, according to Conway (1976) and Southwood (1977b), the application of pesticides is predicted to be the most effective strategy for *K*-selected pests. My analysis suggests, however, that pesticides will be a better option for *K*-selected (and *A*-selected) pests than for *r*-selected pests. Since many pest species are *r*-selected (Southwood, 1977b), my analysis may explain why the extensive application of pesticides has proved so inefficient in protecting the world's food supply against destruction by pests (Pimentel, 1978; Pimentel and Goodman, 1978).

I recommend reduction of reimmigration for *K*-selected and *A*-selected

TABLE 4

Principal Control Techniques for Different Pest Strategies[a]

Technique	*r* Pests	Intermediate pests	*K* Pests
Pesticides	Early wide-scale applications based on forecasting	Selective pesticides	Precisely targeted applications based on monitoring
Biological control	—	Introduction of and/or enhancement of natural enemies	—
Cultural control	Timing, cultivation, sanitation, and rotation	—	Changes in agronomic practice, destruction of alternative hosts
Resistance	General, polygenic resistance	—	Specific monogenic resistance
Genetic control	—	—	Sterile mating technique

[a] After Conway (1976).

pests, not for *r*-selected pests, as do Conway and Southwood. However, my conclusion depends on the effectiveness of available measures and on the assumption that the *r*, *K*, and *A* pests occur in the same kind of environment and habitat complex. The earlier workers did not make this assumption; indeed, Southwood (1977a) stressed the different dynamics of the habitat of *r*- and *K*-selected species.

The differences in the predictions derived by Conway (1976) and Southwood (1977b), on the one hand, and by myself (Stenseth, 1981), on the other, are due primarily to different assumptions about the patch dynamics. Obviously, we must understand the dynamics of the habitat template. The analyses reported by Conway, Southwood, and myself indeed suggest the importance of considering the ecological habitat template and not only ecological and sociological factors.

My analysis coincides, however, with Conway's and Southwood's analyses in recommending changes in the habitat structure as an effective preventive pest control strategy. This may, however, be incompatible with economic constraints set by the human exploitation of the region in question.

IV. COEVOLUTIONARY INTERACTIONS

A. General

An extensive literature exists on the topic of coevolutionary interactions (Futuyma and Slatkin, 1983). In this section I first provide a brief and general review of such interactions. Then, in Subsection B, I provide a discussion of such coevolutionary interactions in systems with short-lived pests and long-lived hosts; I refer in particular to bark beetles.

Pimentel (1961) studied the genetic feedback mechanism in host–pathogen interactions (see also Schaffer and Rosenzweig, 1978; Rosenzweig, and Schaffer, 1978). The insights gained from Pimentel's work can be summarized as follows: As host (or prey) density decreases from parasitism (or predation), selection for resistance is high, but selection for predation efficiency is low, given that the prey density is not too low. Thus, the predator becomes less effective and the prey becomes more likely to escape. At a low host population and early in its subsequent increase, the parasite is hard put to find its host. Thus, the selective pressure is reversed: The host becomes susceptible and the parasite becomes efficient. Such interaction may lead to cyclic changes of both ecological and evolutionary types. If both host and parasite are genetically polymorphic and if each host homozygote is resistant to one parasite homozygote and susceptible to the other, a limit cycle could easily result (Haldane, 1949; Jayakar, 1970).

The population genetic models of Haldane (1949) and Jayakar (1970) suffer, however, from the unrealistic assumption that only the dynamics of the gene frequencies, and not the dynamics of the population densities, need be considered. It is not clear what sort of dynamic behavior a model including both gene frequencies and densities as variables would have. However, some insight into this problem was obtained by Schaffer and Rosenzweig (1978), who provided a population dynamic model with an implicit consideration of the population genetics. They demonstrated that cycles could occur if the host is long-lived compared with the parasite.

B. Ecological and Evolutionary Time Scales: Short-Lived Pests and Long-Lived Hosts

1. Ecological Model for Bark Beetle–Tree Interactions

Often a pest species has a very fast turnover rate compared with that of its hosts. Consequently, the ecological time scale of the host may approach the evolutionary time scale of the pest. Hence, a special form of evolutionary–ecological interaction results; the pest may undergo many evolutionary changes while the host experiences no evolutionary change. Hence, only evolution in the pest is considered, assuming, of course, that somatic mutations are never heritable. For plants this may not be a generally valid assumption (Whitham and Slobodchikoff, 1981; Whitham *et al.*, 1984; Gill and Halverson, 1984; Antolin and Strobeck, 1985; Rhoades, 1983, 1985; Slatkin, 1985; Fowler and Lawton, 1985; Jerling, 1985). Unfortunately, I cannot provide any overall discussion of cases for which somatic mutations are heritable. Indeed, before such a discussion can be presented, evolutionary theory must be refined.

Bark beetles of the genera *Dendroctonus* and *Ips* and the forest trees in which they breed are examples of systems with greatly different time scales. Berryman *et al.* (1984) developed an ecological model for such systems. The ecological interactions between the trees (defending themselves by producing resinous material) and the attacking beetles (using pheromones to facilitate "coordinated" attacks) were studied by a system of two differential equations,

$$\begin{aligned} dx/dt &= r_x x[1 - x/(c_x R)] - g(x, y) \\ dy/dt &= r_y y\{1 - y/[c_y g(x, y)]\} \end{aligned} \qquad (4)$$

where t is a dimensionless time variable scaled by the generation span of the beetles, x the biomass density of the forest, y the density of the bark beetles, R the resource level available to the trees at the growing site, r_x

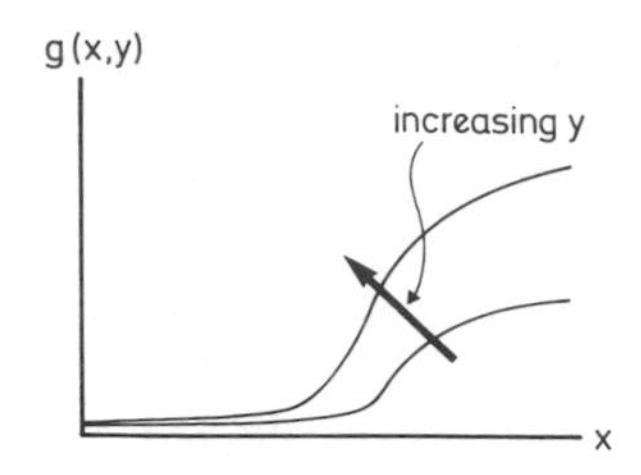

Fig. 6. The $g(x, y)$ function in the bark beetle–forest population dynamic model [Eq. (4)]. Redrawn from Berryman *et al.* (1984).

and r_y are the effective growth rates of the trees and of the beetles, respectively, c_x and c_y are the amount of tree biomass that can be supported by a unit of resource and the density of beetle produced by a unit of host biomass per generation respectively, and $g(x, y)$ represents the loss rate of forest biomass due to beetle attack. The form of the g function is depicted in Fig. 6 (see Berryman *et al.*, 1984, for details). The quantity c_y becomes a coefficient measuring the suitability (in terms of increasing fitness, or the specific net growth rate) of the host material for the bark beetle population.

The ecological dynamics of such systems are depicted in Fig. 7, where the isoclines (i.e., $dx/dt = 0$ and $dy/dt = 0$) of the system are also shown; the point of intersection of the two isoclines represents the ecological equilibrium of the system. In Fig. 7 it can be seen that tree biomass increases only to the left of the tree isocline ($dx/dt = 0$), which intersects the x axis at the tree carrying capacity (c_xR); beetles increase only to the right of the S-shaped beetle isocline ($dy/dt = 0$). The dynamic properties of the system around the equilibrium are illustrated in Fig. 7.

The particular isocline configurations depicted in Fig. 7 occur because

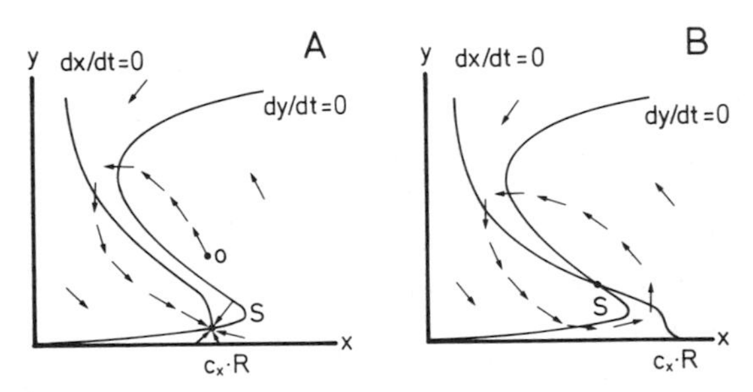

Fig. 7. A plot of the tree and beetle isoclines yields two possible cases for the community equilibrium point (intersection of the isoclines): (A) A single stable node results when $c_xR \approx S$, the x-value for which the S-shaped curve turns. The trajectory starting in 0 shows outbreak and collapse dynamics when the system is displaced from its stable equilibrium by an increase in the beetle population. (B) A limit cycle results when $c_xR > S$. The position of the tree and beetle isoclines can be altered by varying r_x and c_y, respectively. Redrawn from Berryman *et al.* (1984).

the effective growth rate of the tree population, r_x, and the resource utilization constant of the beetles, c_y^{-1}, both take on small values; however, r_y is large relative to r_x because the beetle density is a "fast" variable and the forest biomass is a "slow" variable (notice that r_x^{-1} and r_y^{-1} measure the intrinsic or natural period of the respective populations; see Maynard Smith, 1974). We are back to the issue of analyzing systems in which the dynamics variables have greatly different intrinsic rates of change.

Provided that the trees' carrying capacity c_xR is lower than the density S at which the beetle isocline turns, we usually find a single, globally stable equilibrium (Fig. 7A). The stability of this equilibrium, however, is very sensitive to external disturbances. For example, if the local beetle population is displaced by the immigration of insects, say to point 0 in Fig. 7A, it will first grow rapidly, and as a result, the forest biomass will be dramatically reduced: an insect outbreak has occurred. Notice, however, that the system will return to its stable equilibrium as the beetle population declines from lack of food and the forest will slowly regenerate. There is reason to believe that this picture corresponds to outbreaks like those of *Ips typographus* and *Dendroctonus ponderosa* in Scandinavia and western North America, respectively (e.g., Bakke, 1983; Solbreck, 1986; Berryman *et al.*, 1984).

Another biologically plausible configuration is depicted in Fig. 7B, where $c_xR > S$. In this case the unstable equilibrium is oscillatory; globally, however, it exhibits a stable limit cycle (see May, 1972; Berryman *et al.*, 1984). The position of S is determined by the value of c_y (Berryman *et al.*, 1984); the configuration in Fig. 7B becomes increasingly more likely as c_y increases.

2. Evolutionary Considerations

Which of the two configurations depicted in Fig. 7 is most likely to be the result of coevolution between beetles and trees, or rather of evolution in the beetles? In an abiotically stable or predictable region, evolution in the bark beetle will at high densities (1) change c_y so as to make this quantity as large as possible and/or (2) for any given pair of x and y, change $g(x, y)$ so as to make this quantity as large as possible. This, then, would be a form of K selection (or possibly A selection) because $c_yg(x, y)$ actually represents the carrying capacity for the beetles in the particular environment. Such a strategy would be favored by natural selection because it would make $(dy/dt)/y$ greater under otherwise identical conditions.

In an abiotically unpredictable region, evolution through natural selection in the bark beetle will at low densities increase r_y, possibly leading to a reduced $c_yg(x, y)$. This, then, would be a form of r selection.

The tree can do very little about these evolutionary changes in the beetle population because of their greatly different ecological time scales. In this case there will be no Red Queen coevolutionary race of the kind discussed earlier in this chapter (p. 541). Such a coevolutionary race might, however, take place on a larger time scale. Further analysis is certainly needed before we can be more specific here.

It is rather difficult to draw definite conclusions regarding the outcome of these evolutionary interactions. However, since c_xR remains a rather fixed quantity over the considered time scale, changes in the relative configurations of the isoclines, if any, will probably be from Fig. 7A to Fig. 7B in abiotically predictable habitats (i.e., under K or A selection), and from Fig. 7B to Fig. 7A in abiotically unpredictable habitats (i.e., under r selection) (see Berryman *et al.*, 1984).

That is, the changes occurring in the beetle population under K or A selection seem to destabilize the system. Hence, both human actions (see Berryman *et al.*, 1984) and nature seem to facilitate pest outbreak under such conditions. The resulting ecological changes (i.e., the limit cycle occurring in Fig. 7B) will subsequently make the environment biotically more unpredictable.

The changes occurring in the beetle population under r selection seem, however, to stabilize the system. Hence, human and natural actions seem under these conditions to oppose each other. The ecologically stable system will subsequently make the environment biotically more predictable. Understanding the habitat template is therefore essential for understanding the outcome of coevolutionary interactions and the resulting ecological dynamics. In particular we have found that in abiotically unpredictable regions such as the boreal forest (where, e.g., *Ips typographus* is found), r selection will cause the insect's habitat template to be less prone to outbreak; K or A selection will cause the insect's habitat template to be more prone to outbreak conditions.

These conclusions do not necessarily apply if there are pest-induced resistance reactions in the hosts and/or somatic evolution. Then there may be, in the case of K and A selection, no change over the considered time scale with respect to the phase diagram configuration. This is so because the hosts' reactions will cause both c_y and $g(x, y)$ to be as small as possible. Similar phenotypic changes in the host, however, will in an r-selective habitat further stabilize the pest–forest system.

The overall conclusion is that the habitat template for species like *Ips typographus* in the boreal forest primarily stabilizes its dynamics at a low density not to be considered as a pest condition. Only human activity is likely to destablize this system so as to create pest conditions. Under such circumstances, knowledge about the habitat template and evolutionary processes is of paramount importance if we are to manage our natural resources properly.

Obviously, much further work along these lines is needed. Above all, more data are needed (some have been published by Raffa and Berryman, 1986). Here I have only wanted to draw attention to this form of coevolution. It is important to understand this process in many insect outbreak situations, not only in the *Ips typographus* example discussed above. An understanding of the process might also yield important general insights into coevolution as such. After all, species with greatly different time scales often coexist. Indeed, this is one of the fields where I see a valuable interaction between basic and applied evolutionary ecology.

V. SEXUAL REPRODUCTION AND PEST CONTROL

A. Use of Sex in Pest Control

It is generally agreed among biologists that the problem of sexual reproduction is the "queen problem of biology" (Williams, 1975; Maynard Smith, 1978; Bell, 1982; Shields, 1982, 1983). Even though many models have been proposed to explain the evolution and maintenance of sexual reproduction, none has gained general acceptance. Essentially, we do not yet understand why sexual reproduction is as common as it is. A great deal of basic research is currently being directed toward this topic (for reviews, see Stearns, 1985; Bell, 1985; Bremermann, 1985; Bierzychndek, 1985). I will not discuss here the various theories for maintenance of sex. However, I would claim that this basic research may have important implications for pest control. Essentially, the knowledge emerging from their research might be used to design pest control measures that interfere with regular sexual reproduction. I discuss in the following section how this might be done.

The existence of sexual reproduction has been utilized in pest control earlier. An example is the application of the sterile male technique (Knipling, 1955, 1959, 1964, 1966, 1968; Berryman, 1967; Conway, 1973). Below I will propose another, perhaps much more efficient technique utilizing sexual reproduction as a key factor; this is the application of females that need to be mated but that produce only daughters. Examples are pseudogamous females (e.g., Stenseth *et al.*, 1985) and females with a driving sex chromosome (Oliver, 1971). However, first I will review some observations pertaining to the issue of sexual reproduction and pest control.

B. Pest Control and Sex: Some Observations

The dangers of genetic uniformity and the resulting vulnerability of modern crops to pests have become widely recognized (Day, 1977). Of-

TABLE 5

Ratings of the Biological Control of Weeds Comparing the Degree of Control Achieved with the Predominant Means of Reproduction of the Target Species[a]

Predominant mode of reproduction	Degree of control achieved[b]			
	None	Partial	Substantial	Complete
Asexual	2.3	10	39	48.5
Sexual	13.8	53.5	19	13.8

[a] After Burdon and Marshall (1981).
[b] Percentage of distribution over the classes of success.

ten, asexually reproducing species show far less genetic variability than do sexually reproducing species; this is particularly so if the population started with a few individuals. On this basis we should expect any biological control program to be more successful if the target species reproduces asexually than if it reproduces sexually.

I know of no data on insect pests that could be used for testing this prediction. However, Burdon and Marshall (1981) tested the prediction for weeds. They examined the degree of pest control achieved in 81 different control programs. Their analysis demonstrated a significant correlation between the degree of control achieved and the predominating mode of reproduction of the target plant species; asexually reproducing species were controlled significantly more often than sexually reproducing ones (Table 5). I suggest that a similar relationship also exists for insect pests.

C. A Model and Its Application in Pest Control

An ecological model for a population with both normal and female-producing (or abnormal) females was developed by Stenseth *et al.* (1985; see also Kirkendall and Stenseth, 1987; Stenseth, 1985a,b). Denoting the density of normal and abnormal females by x and y, respectively, this model is given as

$$\begin{aligned} \frac{dx}{dt} &= 0.5b_1ax\frac{px}{px+qy} - m_1x \\ \frac{dy}{dt} &= b_2ax\frac{qy}{px+qy} - m_2y \end{aligned} \qquad (5)$$

where b_1 and b_2 are the expected instantaneous rates of production of adult progeny by mated normal and abnormal females, respectively, m_1

and m_2 are the corresponding adult mortality, a is the number of mated females per male, and p is the relative success of normal females in obtaining sperm defined so that $p/(1-p) = p/q$ is the relative pairing success of normal females.

Analysis of the model demonstrates (Stenseth *et al.*, 1985; Kirkendall and Stenseth, 1987) that the pure sexual population may be invaded to the right of the solid line in Fig. 8. To the right of the dashed line in Fig. 8, the abnormal females will drive the entire population to extinction. Introduction of an abnormal (e.g., pseudogamous) female form into a pest population of only sexually reproducing females would "solve" the pest problem if the initially mixed population fell to the right of the dashed line in Fig. 8. At least the pest problem would be solved until a reinvasion of the sexual form occurred (Fig. 5; Stenseth, 1981, 1985b). Biologically, this means that we would have to find a pseudogamous form of females having a reproductive potential much higher than that of the sexual form and that the two female forms would have to compete for food and other environmental resources.

Abnormal females producing all-female broods will most easily cause the extinction of the whole population (Stenseth *et al.*, 1985; Kirkendall and Stenseth, 1987) if (1) the mating advantages of normal females at low density is negligible, (2) the density dependence in reproductive rates b_1 and b_2 is high, (3) the density dependence in mortalities m_1 and m_2 is low, and/or (4) the density dependence in the male preference p is low. If possible, normal and abnormal females should, in addition, have different dispersal abilities (Stenseth, 1981). Pseudogamous females should be the best dispersers.

Females producing all-female broods do exist. For example, the pest

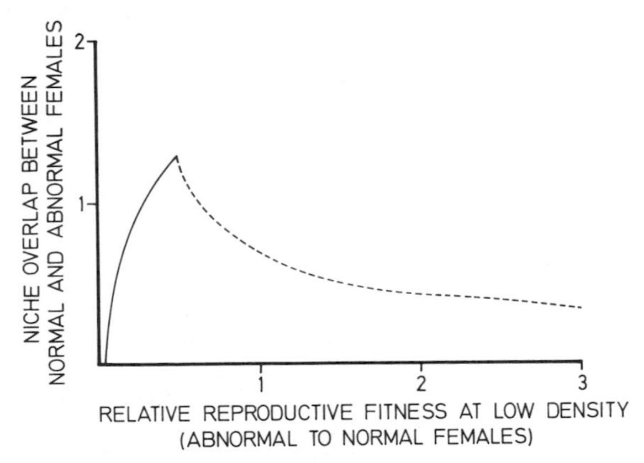

Fig. 8. Stability boundaries of the model given by Eq. (5). Redrawn from Kirkendall and Stenseth (1987).

species *Ips acuminatus* is pseudogamous (Bakke, 1968a,b; Lanier and Kirkendall, 1987; L. R. Kirkendall, personal communication). The African butterfly, *Acraea encedon,* has a female form producing all-female broods due to a driving sex chromosome (Owen and Chanter, 1969; Chanter and Owen, 1972).

An understanding of the underlying cytological mechanism of, for example, pseudogamous reproduction, might give us much more powerful pest control methods than currently available techniques provide. This is so because, by the abnormal female technique, we would be able to cause the population to become completely extinct, whereas this might be much more difficult with, for example, the sterile male technique [which only reduces the quantity a in model (5) and hence only reduces the stable equilibrium to a lower value]. Furthermore, sterile males must be supplied repeatedly, whereas it suffices to introduce pseudogamous females only once and then rely on the intrinsic ecological population dynamics of the species.

These ideas should, of course, be tested under natural conditions. It would seem most efficient to use one of the pseudogamous bark beetle species, for example, *Ips acuminatus,* a pest on Scotch pine (*Pinus sylvestris;* Tragardh, 1939). Field experiments designed to drive the population extinct by introducing pseudogamous females should indeed be attempted.

I now return briefly to the results reported in Section V,B (see also Table 5). Even if the introduced pseudogamous females are not able to drive the entire population to extinction, they will make the population genetically less variable if they constitute a large fraction of the entire population (which they often do; see Bakke, 1968a,b; Kirkendall and Stenseth, 1987). Hence, the population would become easier to regulate by means of traditional biological control utilizing predators.

I believe that this "application" of sex has potential for success in pest control. Obviously much remains to be done (see Chapter 19, this volume). We do not need to understand "why sex" in order to apply pseudogamous females to pest control, but we do need several of the results emerging from basic ecological and evolutionary studies on sexual reproduction before we can apply this pest control technique. Basic and applied ecological work should go hand in hand.

VI. DISREGARDING NATURAL SELECTION

A basic dogma of modern biology is that fitness is maximized by natural selection (Maynard Smith, 1975; Roughgarden, 1979). However, due to coevolutionary pressure from coexisting species, fitness might never be at its maximum possible value (Stenseth, 1985a). Often human actions reduce fitness even further.

In modern breeding, artificial selection is often carried out in order to cultivate some feature of a species that is of economic value. Think about a fruit tree: Generally we would, through artificial selection, try to maximize the fruit crop. This would, in general, reduce some other features of the tree's life history. Following from the principle of allocation (Cody, 1966), increasing reproductive output (fruit crop) might often decrease survival (and the tree's resistance to pest organisms). This will specifically be the case if we, through artificial selection, reduce the genetic variability of the tree population. Following the reasoning in Section IV,B, we might easily make the species far more susceptible to insect outbreaks.

We should always avoid changing natural interactions and relationships without knowing what ecological consequences such changes might have. Obviously, basic knowledge from evolutionary ecology is of essential value to plant breeders so as to avoid disasters due, for example, to pest outbreaks (see May, 1985).

VII. PERSPECTIVES AND RECOMMENDATIONS

In this chapter I have given several examples of the importance of a basic knowledge of evolutionary ecology in solving pest problems and avoiding new and more serious problems. Evolutionary considerations will always be important in pest control studies. It is essential to ask how the evolution of the species will change in response to the application of some pest control measure. By ignoring such knowledge, we might easily create a new problem when trying to solve another.

In addition it helps to understand how the real world operates and is structured and why it evolved to its present state. If, for instance, we had no idea about how the mode of pseudogamous reproduction evolved and is maintained, it would be impossible even to think about how this mode of reproduction could be utilized in pest control.

The thrust of this chapter is that there is mutual benefit between basic work in the field of evolutionary ecology and applied work in insect pest control. For obvious reasons, it is difficult to say in advance which of the ideas of theoretical biology might prove useful. My convinction, however, is that experience in theoretical ecology may provide new and unorthodox approaches to the solution of practical problems. In contrast, work on the practical problem may provide important challenges to theorists and hence stimulate work on new approaches to the many problems that face humankind. It is indeed my hope that theorists and empiricists as well as applied and nonapplied ecologists will work together more often on the same problems. Then we will make much faster progress in both fields.

A closer integration of evolutionary ecology and pest control studies is of great importance. In general it is very difficult to know how to control a system unless one understands how it currently operates and how it came to operate the way it does. People's problems can be solved only by people themselves, and such solutions require the understanding of all aspects of the problem.

ACKNOWLEDGMENTS

My studies on pest species have been generously funded by the Swedish Science Foundations (FRN and SJFR). Support from the Norwegian Science Foundation (NAVF) has also been of great help. I thank the following colleagues and friends for reviewing an earlier version of this chapter: Pedro Barbosa, Alan Berryman, Soren Bondrup-Nielsen, John Greenslade, Jack Schultz, and Sir Richard Southwood. Tony Brown gave valuable input to my discussion of sex and pest control. Listening to a fierce discussion between Alan Berryman and John Borden was instrumental in the development of the ideas in Section VI. I thank them both for their catalytic effect.

REFERENCES

Antolin, M. F., and Strobeck, L. (1985). The population genetics of somatic mutations in plants. *Am. Nat.* **126,** 52–62.

Bakke, A. (1968a). Ecological studies on bark beetles, *Pinus sylvestris* (L.) in Norway with particular reference to the influence of temperature. *Medd. Nor. Skogsforsoeksves.* **21,** 441–602.

Bakke, A. (1968b). Field and laboratory studies on sex ratio in *Ips acuminatus* (Coleoptera) in Norway. *Can. Entomol.* **100,** 640–648.

Bakke, A. (1983). Host tree and bark beetle interaction during a mass outbreak of *Ips typographus* in Norway. *Z. Angew. Entomol.* **96,** 118–125.

Bell, G. (1982). "The Masterpiece of Nature: The Evolution and Genetics of Sexuality." Croom Helm, London.

Bell, G. (1985). Two theories of sex and variation. *Experientia* **41,** 1235–1245.

Berryman, A. A. (1967). Mathematical description of the sterile male principle. *Can. Entomol.* **99,** 858–865.

Berryman, A. A., Stenseth, N. C., and Wollkind, D. J. (1984). Metastability of forest ecosystems infested by bark beetles. *Res. Popul. Ecol.* **26,** 13–29.

Bierzychudek, P. (1985). Patterns in plant parthenogenesis. *Experientia* **41,** 1255–1265.

Bremermann, H. J. (1985). The adaptive significance of sexuality. *Experientia* **41,** 1245–1255.

Burdon, J. J., and Marshall, D. R. (1981). Biological control and the reproductive mode of weeds. *J. Appl. Ecol.* **18,** 649–658.

Chanter, D. O., and Owen, D. F. (1972). The inheritance and population genetics of sex ratios in the butterfly *Acraea encedon*. *J. Zool.* **166,** 363–383.

Cody, M. L. (1966). A general theory of clutch size. *Evolution (Lawrence, Kans.)* **20,** 174–184.

Connell, F. H. (1975). Some mechanisms producing structure in natural communities: A model and evidence from field experiments. *In* "Ecology and Evolution of Communities" (M. L. Cody and F. M. Diamond, eds.), pp. 460–490. Harvard Univ. Press, Cambridge, Massachusetts.

Conway, G. R. (1973). Experience in insect pest modelling: A review of models, uses and future. *In* "Insects: Studies in Population Management" (P. W. Geier, L. R. Clark, D. J. Anderson, and H. A. Nix, eds.), pp. 103–130. Ecol. Soc. Aust., Canberra-City.

Conway, G. R. (1976). Man versus pests. *In* "Theoretical Ecology: Principles and Applications" (R. M. May, ed.), pp. 257–281. Saunders, Philadelphia, Pennsylvania.

Dawkins, R., and Krebs, J. R. (1979). Arms race between and within species. *Proc. R. Soc. London, Ser. B* **205,** 489–511.

Day, P. R., ed. (1977). The genetic basis of epidemics in agriculture. *Ann. N. Y. Acad. Sci.* **287,** 1.

Diamond, J. M. (1975). Assembly of species communities. *In* "Ecology and Evolution of Communities" (M. L. Cody and J. M. Diamonds, eds.), pp. 342–444. Harvard Univ. Press, Cambridge, Massachusetts.

Fowler, S. V., and Lawton, J. N. (1985). Rapidly induced defenses and talking trees: The devil's advocate position. *Am. Nat.* **126,** 181–195.

Futuyma, D. J., and Slatkin, M., eds. (1983). "Coevolution." Sinauer Associates, Sunderland, Massachusetts.

Gilbert, N., Gutierrez, A. P., Frazer, B. D., and Jones, R. E. (1976). "Ecological Relationships." Freeman, San Francisco, California.

Gill, D. E. (1974). Intrinsic rate of increase, saturation density, and competitive ability. II. The evolution of competitive ability. *Am. Nat.* **108,** 103–116.

Gill, D. E., and Halverson, T. G. (1984). Fitness variation among branches within trees. *In* "Evolutionary Ecology" (B. Shorrocks, ed.), p. 105. Blackwell, Oxford.

Greenslade, P. J. M. (1983). Adversity selection and the habitat templet. *Am. Nat.* **122,** 352–365.

Grime, J. P. (1977). Evidence for the existence of three primary strategies in plants and its relevance to ecological and evolutionary theory. *Am. Nat.* **111,** 1169–1194.

Grime, J. P. (1978). Interpretation of small-scale patterns in the distribution of plant species in space and time. *In* "A Synthesis of Demographic and Experimental Approaches to the Functioning of Plants" (J. W. Woldendorp, ed.), pp. 101–124. Cent. Agric. Publ. Doc., Wageningen.

Grime, J. P. (1979). "Plant Strategies and Vegetation Processes." Wiley, New York.

Haldane, J. B. S. (1949). Disease and evolution. *Ric. Sci., Suppl.* **19,** 68–76.

Hansson, L. (1977a). Landscape ecology and stability of populations. *Landscape Plan.* **4,** 85–93.

Hansson, L. (1977b). Spatial dynamics of field voles, *Microtus agrestis,* in heterogeneous landscapes. *Oikos* **29,** 539–544.

Horn, H. S., and Rubinstein, D. I. (1984). Behavioural adaptations and life history. *In* "Behavioural Ecology: An Evolutionary Approach" (J. R. Krebs and N. B. Davies, eds.), pp. 279–298. Blackwell, Oxford.

Hutchinson, G. E. (1951). Copepodology for ornithologists. *Ecology* **32,** 571–577.

Jayakar, S. D. (1970). A mathematical model for interaction of gene frequencies in a parasite and its host. *Theor. Popul. Biol.* **1,** 140–164.

Jerling, L. (1985). Are plants and animals alike? A note on evolutionary plant population ecology. *Oikos* **45,** 150–153.

Kennedy, J. S. (1956). Phase transformation in locusts biology. *Biol. Rev. Cambridge Philos. Soc.* **31,** 349–370.
Kirkendall, L. R., and Stenseth, N. C. (1987). Ecological and evolutionary stability of pseudogamy: Effects of partial niche overlap between sexual and asexual females. *Am. Nat.* (in press).
Knipling, E. F. (1955). Possibilities of insect control or readication through the use of sexually sterile mass. *J. Econ. Entomol.* **48,** 459–462.
Knipling, E. F. (1959). Sterile male method of population control. *Science* **130,** 902–904.
Knipling, E. F. (1964). The potential role of the sterility method for insect population control with special reference to combining the method with conventional methods. *U.S., Agric. Res. Serv., ARS* **ARS-33-98.**
Knipling, E. F. (1966). Further consideration of the theoretical role of predation in sterile insect release programs. *Bull. Entomol. Soc. Am.* **12,** 361–364.
Knipling, E. F. (1968). Population models to appraise the limitations and potentialities of *Trichogramma* in managing host insect populations. *U.S., Dep. Agric., Tech. Bull.* **1387.**
Lanier, G. N., and Kirkendall, L. R. (1987). Karyology of pseudogamous *Ips* bark beetles. *Hereditas* (in press).
Lawlor, L. R., and Maynard Smith, J. (1976). The coevolution and stability of competing species. *Am. Nat.* **110,** 79–99.
Levins, R. (1969). Some demographic and genetic consequences of environmental heterogeneity for biological control. *Bull. Entomol. Soc. Am.* **119,** 237–240.
Ludwig, D., Jones, D. D., and Holling, L. S. (1978). Qualitative analysis of insect outbreak systems: The spruce budworm and forest. *J. Anim. Ecol.* **47,** 315–332.
MacArthur, R. H. (1960). On the relative abundance of species. *Am. Nat.* **94,** 25–36.
MacArthur, R. H., and Wilson, E. O. (1967). "The Theory of Island Biogeography." Princeton Univ. Press, Princeton, New Jersey.
May, R. M. (1972). Limit cycles in predator–prey communities. *Science* **177,** 900–902.
May, R. M. (1985). Evolution of pesticide resistance. *Nature (London)* **315,** 12–13.
May, R. M., and Anderson, R. M. (1983). Parasite–host coevolution. *In* "Coevolution" (D. J. Futuyma and M. Slatkin, eds.), pp. 186–206. Sinauer Assoc., Sunderland, Massachusetts.
May, R. M., Conway, G. R., Hassell, M. P., and Southwood, T. R. E. (1974). Time delays, density dependence, and single species oscillations. *J. Anim. Ecol.* **43,** 747–770.
Maynard Smith, J. (1968). "Mathematical Ideas in Biology." Cambridge Univ. Press, Cambridge, Massachusetts.
Maynard Smith, J. (1974). "Models in Ecology." Cambridge Univ. Press, Cambridge, Massachusetts.
Maynard Smith, J. (1975). "The Theory of Evolution." Penguin, London.
Maynard Smith, J. (1978). "The Evolution of Sex." Cambridge Univ. Press, Cambridge, Massachusetts.
Mayr, E. (1961). Cause and effect in biology. *Science* **134,** 1501–1506.
Miller, R. S. (1969). Competition and species diversity. *Brookhaven Symp. Biol.* **22,** 63–70.
Mitchell, R. (1975). Models for parasite populations. *In* "Evolutionary Strategies of Parasitic Insects and Mites" (P. W. Price, ed.), pp. 49–69. Plenum, New York.
Oliver, J. H. (1971). Introduction to the symposium on parthenogenesis. *Am. Zool.* **11,** 241–243.
Owen, D. F., and Chanter, D. O. (1969). Population biology of tropical African butterflies. Sex ratio and genetic variation in *Acraea encedon*. *J. Zool.* **157,** 345–374.
Parry, G. D. (1981). The meanings of *r*- and *K*-selection. *Oecologia* **48,** 260–264.

Pianka, E. R. (1970). On *r*- and *K*-selection. *Am. Nat.* **104,** 592–597.
Pianka, E. R. (1972). *r*- and *K*-selection or *b* and *d* selection? *Am. Nat.* **106,** 581–588.
Pianka, E. R. (1978). "Evolutionary Ecology." Harper & Row, New York.
Pimentel, D. (1961). Animal population regulation by the genetic feedback mechanism. *Am. Nat.* **95,** 65–79.
Pimentel, D. (1978). Socioeconomic and legal aspects of pest control. *In* "Pest Control Strategies" (E. H. Smith and D. Pimentel, eds.), pp. 55–72. Academic Press, New York.
Pimentel, D., and Goodman, N. (1978). Ecological basis for the management of insect populations. *Oikos* **30,** 422–437.
Raffa, K. F., and Berryman, A. A. (1986). Interacting selective pressures in conifer–bark beetle systems: A basis for reciprocal adaptation? *Ecology* (in press).
Raunkiaer, C. (1937). "Plant Life Forms." Oxford Univ. Press, London and New York.
Rhoades, D. F. (1983). Responses of alder and willow to attack by tent caterpillars and webworms: Evidence for pheromonal sensitivity of willows. *In* "Plant Resistance to Insects" (P. A. Hedin, ed.), pp. 55–68. Am. Chem. Soc., Washington, D.C.
Rhoades, D. F. (1985). Offensive–defensive interactions between herbivores and plants: Their relevance in herbivore population dynamics and ecological theory. *Am. Nat.* **125,** 205–238.
Rosenzweig, M. L., and Schaffer, W. M. (1978). Homage to the red queen. II. Coevolutionary resone to enrichment of exploitation ecosystems. *Theor. Popul. Biol.* **14,** 158–163.
Roughgarden, J. (1979). "Theory of Population Genetics and Evolutionary Ecology: An Introduction." Macmillan, New York.
Royama, T. (1977). Population persistence and density dependence. *Ecol. Monogr.* **47,** 1–35.
Royama, T. (1984). Population dynamics of the spruce budworm *Choristoneura fumiferana*. *Ecol. Monogr.* **54,** 429–462.
Schaffer, W. M., and Rosenzweig, M. L. (1978). Homage to the red queen. I. Coevolution of predators and their victims. *Theor. Popul. Biol.* **14,** 135–157.
Shields, W. M. (1982). "Philopatry, Inbreeding, and the Evolution of Sex." State University of New York, Albany.
Shields, W. M. (1983). Optimal inbreeding and the evolution of philopatry. *In* "The Ecology of Animal Movement" (J. R. Swingland and P. J. Greenwood, eds.), pp. 132–159. Oxford Univ. Press, London and New York.
Slatkin, M. (1985). Somatic mutations as an evolutionary force. *In* "Evolution: Essays in Honour of John Maynard Smith" (P. J. Greenwood, P. H. Harvey, and M. Slatkin, eds.), pp. 19–30. Cambridge Univ. Press, London and New York.
Smith, R. E. (1975). Ecosystems and evolution. *Bull. Ecol. Soc. Am.* **56,** 2–6.
Solbreck, C. (1986). Insect migration strategies and population dynamics. *In* "Migration: Mechanisms and Adaptive Significance" (M. A. Rankin, ed.) (in press).
Southwood, T. R. E. (1962a). Migration of terrestrial arthropods in relation to habitat. *Biol. Rev. Cambridge Philos. Soc.* **37,** 171–214.
Southwood, T. R. E. (1962b). Migration—an evolutionary necessity for denizens of temporary habitats. *Proc. Int. Congr. Entomol., 11th, 1960,* Vol. 3, pp. 55–58.
Southwood, T. R. E. (1975). The dynamics of insect populations. *In* "Insects, Science and Society" (D. Pimentel, ed.), pp. 151–199. Academic Press, New York.
Southwood, T. R. E. (1976). Bionomic strategies and population parameters. *In* "Theoretical Ecology: Principles and Applications" (R. M. May, ed.), pp. 26–48. Saunders, Philadelphia, Pennsylvania.

Southwood, T. R. E. (1977a). The relevance of population dynamic theory to pest status. *In* "The Origin of Pest, Parasite, Disease and Weed Problems" (J. M. Cherrett and G. R. Sagar, eds.), pp. 35–54. Blackwell, Oxford.

Southwood, T. R. E. (1977b). Habitat, the templet for ecological strategies? *J. Anim. Ecol.* **46,** 337–365.

Southwood, T. R. E., and Norton, G. A. (1973). Economic aspects of pest management strategies and decisions. *In* "Insects: Studies in Pest Management" (P. W. Geier, C. R. Clark, D. F. Anderson, and H. A. Nix, eds.), pp. 164–184. Ecol. Soc. Aust., Canberra-City.

Southwood, T. R. E., and Way, M. J. (1970). Ecological background to pest management. *In* "Concepts of Pest Management" (R. C. Rabb and F. E. Guthrie, eds.), pp. 6–28. North Carolina State University, Raleigh.

Southwood, T. R. E., May, R. M., Hassell, M. P., and Conway, G. R. (1974). Ecological strategies and population parameters. *Am. Nat.* **108,** 791–804.

Southwood, T. R. E., Brown, U. V., and Read, P. M. (1983). Continuity of vegetation in space and time: A comparison of insects' habitat templet in different successional stages. *Res. Popul. Ecol., Suppl.* **3,** 61–74.

Stearns, S. C. (1976). Life-history tactics: A review of the ideas. *Q. Rev. Biol.* **51,** 3–47.

Stearns, S. C. (1977). The evolution of life history tactics: A critique of the theory and a review of the data. *Annu. Rev. Ecol. Syst.* **8,** 145–171.

Stearns, S. C. (1985). The evolution of sex and the role of sex in evolution. *Experientia* **41,** 1231–1235.

Stenseth, N. C. (1977). On the importance of spatio-temporal dynamics of rodents: Toward a theoretical foundation of rodent control. *Oikos* **29,** 545–552.

Stenseth, N. C. (1981). How to control pest species: Application of models from the theory of island biogeography in formulating pest control strategies. *J. Appl. Ecol.* **18,** 773–794.

Stenseth, N. C. (1984). Optimal reproductive success in animals with parental care. *Oikos* **43,** 251–253.

Stenseth, N. C. (1985a). Darwinian evolution in ecosystems: The red queen view. *In* "Evolution: Essays in Honor of John Maynard Smith" (P. J. Greenwood, P. H. Harvey, and M. Slatkin, eds.), pp. 55–72. Cambridge Univ. Press, London and New York.

Stenseth, N. C. (1985b). A new hypothesis for explaining the maintenance of all-female broods in the African butterfly *Acraea encedon*. *Hereditas* **103,** 205–209.

Stenseth, N. C., and Maynard Smith, J. (1984). Coevolution in ecosystems: Red queen or stasis? *Evolution (Lawrence, Kans.)* **38,** 870–880.

Stenseth, N. C., Kirkendall, L. R., and Moran, N. (1985). On the evolution of pseudogamy. *Evolution (Lawrence, Kans.)* **39,** 294–307.

Tragardh, I. (1939). "Sveriges skogsinsekter" (in Swedish). Hugo Gebers Forlag, Stockholm.

Van Valen, L. M. (1973). A new evolutionary law. *Evol. Theory* **1,** 1–30.

Whitham, T. G., and Slobodchikoff, L. N. (1981). Evolution by individuals, plant–herbivore interactions, and mosaics of genetic variability: The adaptive significance of somatic mutations in plants. *Oecologia* **49,** 287–292.

Whitham, T. A., Williams, A. G., and Robinson, A. M. (1984). The variation principle: Individual plants as temporal and spatial mosaics of resistance to rapidly evolving pests. *In* "A New Ecology: Novel Approaches to Interactive Systems" (P. W. Price, C. N. Slobodchikoff, and W. S. Gaud, eds.), pp. 15–52. Wiley, New York.

Williams, G. C. (1975). "Sex and Evolution." Princeton Univ. Press, Princeton, New Jersey.

Wynne-Edwards, V. C. (1962). "Animal Dispersion in Relation to Social Behaviour." Oliver & Boyd, Edinburgh and London.

APPENDIX: MINIMIZING τ

The quantity to be minimized in pest control, when one particular patch is being considered, is defined by Eq. (3) in the main text, that is,

$$\tau = 1 - \mu/\lambda \tag{A1}$$

Now, let μ_0 be the mortality due to natural factors in a given environment and μ_c the mortality due to control treatment. Then the total realized mortality rate μ can be written

$$\mu = \mu_0 + \mu_c \tag{A2}$$

(see Stenseth, 1981, p. 776). Furthermore, let $1 - q$ represent the proportional reduction of the effective reproductive output rate λ_0 occurring under natural conditions. The realized birth rate λ under pest control is

$$\lambda = \lambda_0 q \tag{A3}$$

(see Stenseth, 1981, p. 776). The quantitites μ_c and q are called the "control strategy." These quantities cannot, in general, be altered independently of each other; if we spend most of our resources (or money) on decreasing reproduction, there will be little left to increase mortality. Hence, in general there will be some constraint function h defined as

$$\mu_c = h(q) \tag{A4}$$

where $dh/dq > 0$ (see Stenseth, 1981). Substituting models (A2), (A3), and (A4) into model (A1) [or model (3) in the main text] and differentiating to find the minimum of τ with respect to q, it can be shown that the optimal pest control strategy, q^* and $\mu_c^* = h(q^*)$, is found by the line through $(0, -\mu_0)$ having the steepest possible slope but touching the h function (Stenseth, 1981).

Subject Index

Taxonomic Index

D

X

Z